Disclaimer and Co

Legal & Disclaimer

The information contained in this book is not designed to replace or take the place of any form of medicine or professional medical advice. The information in this book has been provided for educational and entertainment purposes only.

The information contained in this book has been compiled from sources deemed reliable, and it is accurate to the best of the Author's knowledge; however, the Author cannot guarantee its accuracy and validity and cannot be held liable for any errors or omissions. Changes are periodically made to this book. You must consult your doctor or get professional medical advice before using any of the suggested remedies, techniques, or information in this book.

Upon using the information contained in this book, you agree to hold harmless the Author from and against any damages, costs, and expenses, including any legal fees potentially resulting from the application of any of the information provided by this guide. This disclaimer applies to any damages or injury caused by the use and application, whether directly or indirectly, of any advice or information presented, whether for breach of

Cover design by Siim Land

Table of Contents

Introduction:

The Stress of Life

"Tough times are an opportunity to learn, grow, become stronger, and prepare for even tougher times yet to come."

Siim Land

Since the dawn of time, living organisms have been exposed to hazardous conditions that stretch their ability to survive. There's always something they're actively adapting to and overcoming. The initial life forms were just microbes and marine animals. After the formation and oxygenation of the atmosphere occurred, the earliest multicellular organisms could evolve. Over the last 500 million years, land vertebrates and plants have spread across the globe. Despite that, there have been several massive extinction events that have wiped out the dominant species of the planet.

In the modern world, humans are faced with different kinds of stressors, such as electromagnetic frequencies, climate change, environmental toxins, air pollution, pandemics, and the ever-imposing threat of nuclear warfare. Total destruction can be hiding behind the next corner. How on Earth has life managed to survive at all amid this chaos?

Charles Darwin said in *On the Origin of the Species* that it is not the strongest or the most intelligent of the species that survives. It is the one that is most adaptable to change. This tends to be the case in the example of civilizations that rise and fall due to their ability to adapt or lack thereof. It's especially relevant in the modern world where mankind is facing exponential change and uncertainty.

Change is something our brains are hardwired to dislike. We'd much rather stay in our comfort zone where it's certain and avoid uncertainty because it can jeopardize our survival. At the same time, we're also prone to be exploratory because it's exciting and can fuel growth. Whether you think it's good or bad is irrelevant to the fact that the floodgates of novelty are open and it's only a matter of time when the world as we know it will be completely different. What you as a person can and must do is prepare yourself for the ever-present stress of life and uncertainty.

This book is not going to give you a response plan to global pandemics or a nuclear apocalypse. Instead, it will teach you the principles of stress adaptation and how to leverage them to become healthier, stronger, less worrisome, more resilient, and confident in your life. Stressful situations and conditions are something we all must deal with in different forms. The world we live in now is not the same as it was in the past and thus our toolkits will also have to change. It's a shame how the average person lacks the ability or knowledge about how to adapt themselves to this changing environment. We need more resilient humans who aren't sick, obese, frail, weak, or on dozens of meds. By leveraging the information here, you can experience both general as well as specific adaptations to the various kinds of stressors you may come across in life.

Here's How the Book Is Structured:

- Chapter One talks about the general stress adaptation phenomena. We will go through the most important past experiments and discoveries of stress as a phenomenon. It introduces key concepts that will be used throughout the book.
- Chapter Two explains how stress and stress adaptation affect aging. Longest living organisms tend to have one thing in common – they live in environmentally unfavorable conditions. Living longer is just a side-effect of being resilient against stress.
- Chapter Three focuses on nutritional stress responses and balancing it with enough nourishment. It's about cycling between periods of deprivation and abundance to cause a beneficial metabolic effect.
- Chapter Four is centered around how different plant compounds can benefit our health by causing a mild stress reaction inside the body. Humans have co-evolved with these various substances and have developed beneficial responses when consuming them in the right amounts.
- Chapter Five teaches you the fundamentals of the immune system and how to strengthen it. With better immunity, you become less susceptible to all kinds of diseases and viruses that we may come across unknowingly.
- Chapter Six may make you shiver because it talks about heat and cold adaptation. This sort of environmental stress adaptation has numerous health benefits.

- Chapter Seven delves into psychology and philosophy. I'm going to explain why stoicism is the most stress-proof mental operating system and how to use it to overcome any obstacle in life.
- Chapters Eight to Nine are about the importance of exercise and muscle growth for both general resilience as well as longevity. There are several implications you must keep in mind, such as what kind of exercises are the best, how it relates to nutrition and recovery.
- Chapter Ten is something we can't fully avoid because it's beginning to surround us everywhere. Specifically, electromagnetic frequencies and radiation. Fortunately, I think we can even adapt to that with the right habits.
- Chapters Eleven to Twelve center around the body's biochemistry and cellular functioning. In order to perform optimally in the macro, you have to be on point at the micro level as well.
- Chapter Thirteen and Fourteen will help you to recover from all this stress by teaching how to optimize sleep and general recovery. I'm outlining many ways to sleep better, sleep faster, and manage stress like a boss.
- Chapter Fifteen includes a list of different supplements and compounds that we might need for optimal health as well as stress adaptation. Nothing replaces quality food, but supplementation can help to fix some of the loopholes.
- Chapter Sixteen is where you'll learn to create your own routine for stress adaptation and management. It outlines

the difference between goals and systems as well as gives a weekly blueprint.

Each chapter is focused on either describing a particular type of stress or method of adaptation. At the end of every section is a short summary of what was discussed as well as additional guidelines for implementation, called the Stress Adaptation Protocol. I've also compiled a source for additional resources and research like articles, podcasts, and videos. You can check it out at https://www.siimland.com/sbsbook. There you can find the product or supplement recommendations I personally use to complement the benefits outlined here.

Ever since my childhood, I've been fascinated with how humans have evolved and adapted to their environment. After spending some time in the military, I got into understanding stress resiliency and how it affects the body. This led me to get a bachelor's degree in anthropology and delving into the world of biohacking my own physiology. My name is Siim Land, and I'm an author, speaker, content creator, and biohacker. I've written several books about the ketogenic diet and intermittent fasting. In my previous book, Metabolic Autophagy, I talked about a specific form of stress adaptation, namely fasting and calorie restriction. This one will include many other types that have an equally beneficial effect.

Chapter One: Of General Stress Adaptation

"Every stress leaves an indelible scar,
and the organism pays for its survival
after a stressful situation by becoming a little older."

Hans Selye

Stress is often described as a negative thing – everyone is just so stressed out, triggered, and anxious that it begins to spread to other people. However, as a phenomenon, stress is indifferent and only a response to certain physiological or psychological stimuli. It's just a kind of imbalance that feels as if something's out of place.

By definition, stress is a disruption to the body's homeostasis or inner balance. This equilibrium was first called the '*milieu interieur'* by Claude Bernard in 1878[1] and '*wisdom of the body'* in 1923[2]. The term 'homeostasis' itself was picked up in 1939 by behaviorist Walter Cannon[3]. It's what all living organisms strive to maintain to stay alive and well.

Here are some examples of homeostasis:

- Stable body temperature at ~37°C regulated by shivering or sweating
- Stable blood pH level at 7.4 regulated by breathing and chemical buffers
- Stable blood sugar regulated by the hormones insulin and glucagon

- Stable electrolyte balance regulated by the sodium-potassium pump
- Stable metabolic rate regulated by your total daily energy expenditure
- The ratio between NAD+ and NADH regulated by energy homeostasis and methylation

Most of these processes are happening automatically and subconsciously. You're never really fully aware of your body temperature or electrolyte balance. Particular inner feelings and physical sensations may give you some insight but you begin to sense an actual difference if there's some sort of a serious disruption to this homeostatic equilibrium. You start to feel stress after it has taken you out of balance to a certain extent.

Whenever your body's homeostasis gets disrupted by a stressor or stimuli from the environment, the limbic system gets activated[4]. It's comprised of the hypothalamus and other regions in the brain that function like your emotional control system.

Here's how your body's response to stress looks like.

- The hypothalamus receives a stress signal from the nervous system and brain stem.
- The hypothalamus sends corticotropin-releasing hormone (CRH) and vasopressin to the anterior pituitary gland.
- The pituitary gland releases adrenocorticotropic hormone (ACTH) via the vagus nerve to the adrenal glands, which are located on top of the kidneys.
- The adrenal glands synthesize corticosteroids from cholesterol and release it into the blood. Cortisol is the main stress hormone that raises adrenaline and norepinephrine to increase alertness.
- Cortisol and other glucocorticoids help to mobilize stored glycogen and fat for energy. The hypothalamus responds to the level of cortisol in the blood and self-regulates the overall stress response by either suppressing CRH production or ramping it up, depending on the situation.

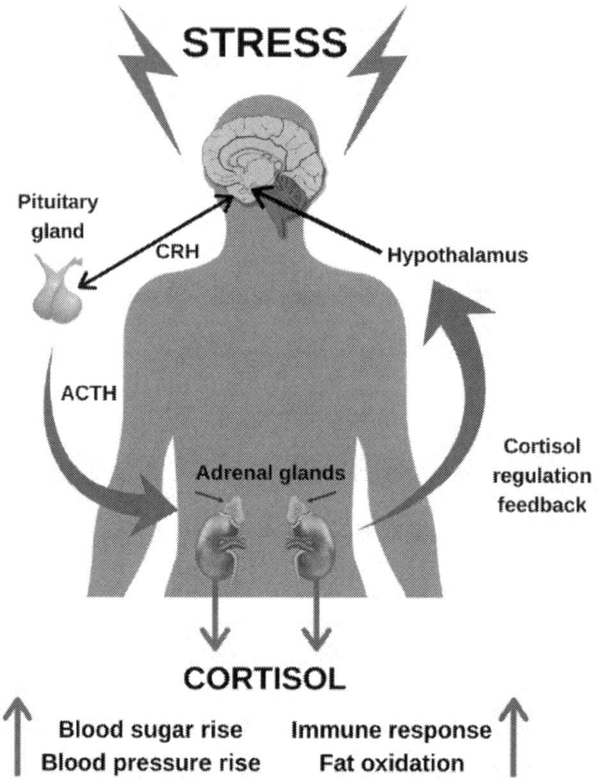

This connection is called the Hypothalamic-Pituitary-Adrenal Axis (HPA), which is a major neuroendocrine system that controls stress reactions and regulates many other bodily processes like digestion, circadian rhythms, immunity, mood, and sexual function[5]. Its role is so fundamental that analogous systems can be found in invertebrates and single-celled organisms.

Corticosteroids are a class of steroid hormones that are synthesized in the adrenal cortex from cholesterol. The most common ones are aldosterone, testosterone, cortisol, cortisone, pregnenolone, progesterone, DHEA, and others. They contribute to the stress response by regulating inflammation, energy metabolism, catabolism, and behavior. With enough steroid hormones, the body has sufficient resources to respond to stressors. Without them, damage or illness may ensue. Glucocorticoids are corticosteroids with anti-inflammatory effects, which is why they're used to treat high fever and infections.

The nervous system consists of several parts and sub-systems that all affect your body's physiological as well as psychological processes. It's the point of communication between you and the outside world that helps to differentiate you from me and safety from danger.

- **Central Nervous System (CNS)** is the central part of the nervous system that consists of the brain and spinal cord. It integrates all the information you receive and coordinates with the rest of the body.
- **Peripheral Nervous System (PNS)** is everything outside of the brain and spinal cord. It consists of nerves and ganglia that connect the limbs, organs, and muscles with the spinal cord and brain.
- **Somatic Nervous System (SNS)** is a part of the PNS that's associated with voluntary control of skeletal muscle and bodily movements. It consists of afferent nerves or sensory nerves and efferent nerves or motor nerves. Afferent nerves relay sensations from the body to the CNS while efferent nerves send out commands from the CNS to the body to stimulate muscle contraction. The 'A-' refers to the prefix 'AD-' (to, toward) and 'E-' to 'EX-' (out of).
- **Autonomic Nervous System (ANS)** is a division of the PNS that supplies smooth muscles and glands, thus

influencing the function of internal organs. It regulates unconscious activity like heart rate, digestion, respiration, pupillary dilation, stress response, urination, and sexual arousal. The hypothalamus functions as an intermediary between the ANS and the limbic system.

The Autonomic Nervous System has three branches: the sympathetic nervous system, parasympathetic nervous system, and enteric nervous system[6]. In most cases, only the sympathetic and parasympathetic are used because they govern the fight or flight response directly.

- **Sympathetic Nervous System** is considered the 'fight or flight system' that regulates stress hormones, energy usage, muscle contraction, alertness, and mobilization of resources. It can be thought of as the 'quick response system'. The adrenal glands release catecholamines like adrenaline and norepinephrine, thus giving you the boost needed to run away from danger. Walter Cannon was the first to describe this phenomenon in 1915[7].
- **Parasympathetic Nervous System** is considered the 'rest and digest system' or 'feed and breed' that regulates relaxation, digestion, recovery, and healing. It can be thought of as the 'slowly activated dampening system'. The vagus nerve connects the CNS with the brain and supplies parasympathetic fibers to all the organs.
- **Enteric Nervous System** is often called the second brain because it governs the gastrointestinal tract. It's capable of

operating independently of the upper brain and spinal cord but does require input from the ANS[8]. Neurons in the gut control secretion of enzymes and neurotransmitters like dopamine and serotonin.

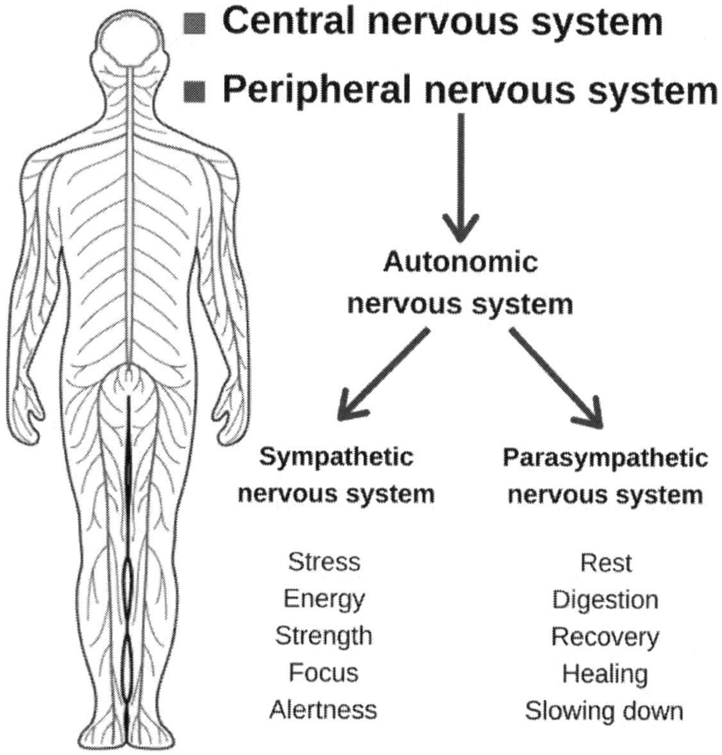

Most functions of the ANS work automatically and involuntarily such as core temperature regulation, nerve firing, heartbeat, respiration, etc. However, you can control it with almost everything you do, starting with changing how you breathe, physical exercise, emotional states, thinking, and food.

When humans were living in the savannah, they didn't have much time to decide if that falling leaf or sound of breaking branches was friend or foe. They couldn't take any chances and thus it was better to be safe than sorry. The ones who stood around being frozen were probably eaten and thus removed from the gene pool. That's why, although the 'fight or flight response' may manifest itself as 'fight-flight-or-freeze', in some people, the biggest effect comes from either fighting danger or fleeing it. Freezing is maladaptive in most situations unless you need to play dead.

Activating the immune response is energetically costly, which is why the body tries to constantly do cost to benefit analysis based on the stressor.[9] Immunity drops when under important stressors that could kill you.

In life-threatening situations like running from predators or starvation, the necessity of immunity isn't as high as just surviving. That's why intense exercise will lead to a short-term drop in immune system function but it recovers afterward[10]. Imposing immune challenges to bumblebees during starvation also accelerates their death because having strong immunity at that point is maladaptive.[11] In a modern society where these threats are less prevalent, social stressors and isolation can cause similar declines in immune parameters[12].

Therefore, the body has to assess the costs and benefits of immunity. The benefits of increasing immune responses are protection against pathogens but the costs include autoimmunity or excessive inflammation. There's also the opportunity cost of using that energy on something else like resource gathering or survival.[13]

For example, creating a fever that's even just 2° degrees above homeostasis would require up to 250 extra calories because of the heat.[14] Other essential organs like the brain and heart require about

350 and 150 calories per day respectively.[15] Producing immune cells and proteins that fight infections also need energy.[16] Sick animals and humans provide that increased demand for energy by exhibiting sickness behavior and reduce their activity. This promotes fatigue, reduces appetite, and makes you less sociable.[17]

Chronic stress increases the risk of developing many diseases such as insulin resistance, obesity, Alzheimer's, and cardiovascular disease[18]. It weakens resistance to infections and can even make the immune system attack itself through an autoimmune response[19].

Stress can change your behavior and psychology by causing anxiety, depression, delusions, sadness, anger, social isolation, panic attacks, and headaches[20]. The hippocampus can also atrophy, which decreases the body's ability to respond appropriately to stressors[21]. In some people, it can also create toxic habits like smoking, alcohol, eating, physical activity, relationships, and drug abuse.

Work-related exhaustion has been shown to damage the DNA of cells and shorter telomeres, which indicate accelerated biological aging[22]. Subjects with the highest stress had the shortest telomeres. Loneliness and social isolation can also impair DNA repair and shorten telomeres[23]. Long-term unemployment in Finnish men has been shown to shorten telomeres and accelerate aging[24]. On the flip side, healthy lifestyle and stress management have helped to lengthen telomeres[25].

Hans Selye and the General Adaptation Syndrome (GAS)

The grandfather of stress science is a Hungarian-Canadian endocrinologist Hans Selye (1907-1982). His experiments demonstrated the existence of biological stress and shaped our understanding of it[26]. He studied the effects of various physical stressors like starvation, exercise, and extreme temperatures on mice.

Selye writes in *Nature* (1936):

> *Experiments on rats show that if the organism is severely damaged by acute non-specific nocuous agents such as exposure to cold, surgical injury, production of spinal shock (transcision of the cord), excessive muscular exercise, or intoxications with sublethal doses of diverse drugs (adrenaline, atropine, morphine, formaldehyde, etc.), a typical syndrome appears, the symptoms of which are independent of the nature of the damaging agent or the pharmacological type of the drug employed, and represent rather a response to damage as such.*[27]

Initially, Selye defined stress as a non-specific neuroendocrine response of the body. However, he later realized that, in addition to the endocrine system, virtually every other system in the body is affected[28]. Regardless of the type of stressor, the animals he was experimenting with experienced similar results, such as ulcers, hypertension, and thymus deterioration. Accordingly, the term stress refers not only to the stressor but to the organisms' overall state as it responds and adapts to the conditions of the environment in which it finds itself in. One of the physicians working with Selye summarized his view of stress to be "*...in addition to being itself, also the cause of itself, and the result of itself*"[29].

Hans Selye introduced the concept of *The General Adaptation Syndrome (GAS),* later renamed to just 'stress response'. It consists of three phases: the alarm reaction, the stage of resistance, and the stage of exhaustion.

1. **Alarm Stage** – This is the mobilization stage that promotes sympathetic nervous system activity and alerts the body of a threat. It's divided into two sub-categories: the shock and the antishock phase[30].
    a. Shock Phase – The body goes into a shock to endure the stressor such as hypoglycemia, hypothermia, hypochloremia, etc. Resistance to the stress drops temporarily below normal and you may experience circulatory shock.
    b. Antishock Phase – Once the stressor has been identified, the body responds by going into an alarmed state. The sympathetic nervous system turns on the fight-flight-or-freeze response by producing catecholamines like adrenaline. This provides short-term energy, arousal, increased muscle strength, the elevation of blood sugar, and blood pressure.
2. **Resistance Stage** – During this stage, the body keeps producing glucocorticoids that can increase the

concentration of fuel substrates in the blood. The body tries to respond to the stressor but after a while, its energy will be depleted.

3. **Exhaustion Stage** – The final stage can be either exhaustion or recovery, depending on how the body managed to deal with the stressor.
    a. Recovery ensues if your body has successfully managed to overcome the stressor or eliminated the source of the stress. The elevated fuel substrates begin to drop down and are being used to repair any damage.
    b. Exhaustion results after all the body's resources have been depleted and it's unable to maintain normal functioning. If this stage is extended, long-term damage can occur, leading to burnout, depression, and chronic fatigue. It also puts you at a higher risk for stress-related diseases by weakening the immune system.

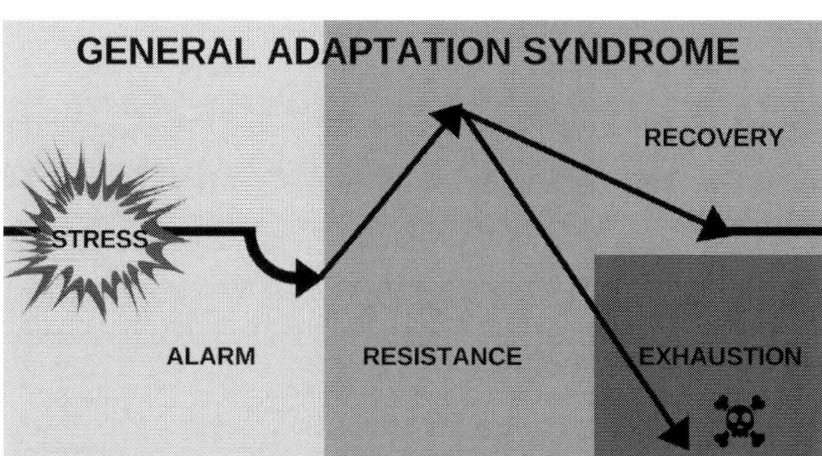

Try to imagine these 3 stages playing out amongst the people you know. Some are already exhausted and burnt out whereas others are still holding on but they're almost shut down and frozen (in resistance). They're showing signs of coping mechanisms like depression, social isolation, or withdrawal. Of course, we all know someone who's just constantly in an alarmed state and making a big deal out of even the smallest of things.

The stress response can happen locally to specific tissues or systems of the body, dubbed Local Adaptation Syndrome[31]. For example, cutting wounds, breaking bones, headaches, gut problems, etc. can inflame only a certain region that's being affected. Despite that, a general low-grade stress response is still increased. It's even possible to trigger a major stress response without experiencing a real physical threat through simply anticipating a threat, getting nervous, or memory recall. The physiological response is equally as damaging.

Adaptation Energy

How well the organism is able to resist a stressor depends on the type of stress, intensity, duration, and the body's resources. The initial response is always strong and virulent but resistance will eventually be diminished. If you're constantly being depleted from the energy needed to endure, then even the smallest of problems can throw you off. What's more, Selye found that during adaptation to a certain stimulus the resistance to other stimuli decreases[32], which means experiencing stress at work reduces your tolerance to stress at home and vice versa. It's sad to think about how many marriages or business projects are destroyed because of this generalized response.

Selye hypothesized that resistance to stress is finite and expandable. He called it 'adaptation energy' and responsiveness to any stimulus is acquired at a cost[33]. Unfortunately, it's difficult to measure this resource because it's not a specific type of currency. Gorban et al (2016) consider adaptation energy *"...an internal coordinate on the 'dominant path' in the model of adaptation*"[34]. Almost like an inner determinant that directs the body towards a certain response, depending on the access to current supply. Whether or not the organism recovers or finds itself exhausted is predicted by the dynamical models of transformation.

Here are Selye's axioms of adaptation energy (AE) as presented by Schkade and Schultz (2003)[35]:

1. Adaptation energy is a finite supply, presented at birth. You have only a certain amount for your entire life.
2. As a protective mechanism, there is some upper limit to how much AE an individual can use at any given moment. It can be focused on a single activity or divided amongst multiple challenges.
3. There is a threshold of AE activation that's required to create a response. You can't respond effectively when depleted. If the stimulus is insufficient it's not worthwhile to react.

4. AE is active at two levels of perception: the primary level where the response occurs at high awareness; and a secondary level at which the response creation is being processed at a sub-conscious level. The former has a high energy requirement whereas the latter is smaller.

Selye's axioms 1-3 are illustrated in the figure below. According to his view, AE is finite and limited to the overall supply obtained at birth. Subsequent stressors exhaust this resource, leading to accelerated death.

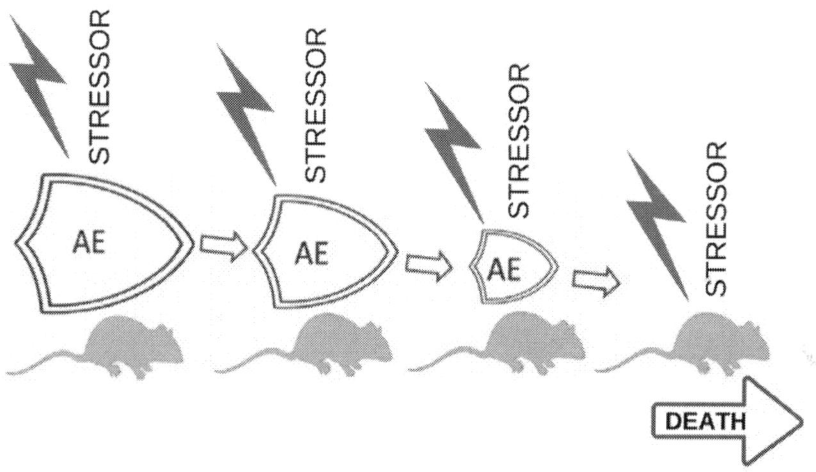

Source: Gorban et al. Journal of Theoretical Biology 405 (2016) 127–139.

Goldstone (1952) proposed that adaptation energy can be restored (to a certain extent) and stored as a reserve[36]. He thought it's possible that, if Selye had spent the AE of his experimental animals at a lesser rate, they could've coped with the stressor indefinitely.

Here are Goldstone's modified axioms of AE, illustrated in the figure below:

1. AE can be re-created, though this energy is generated slower in old age.
2. AE can also be stored as an adaptation capital, though the storage capacity has a fixed limit.
3. If an individual spends their AE faster than it gets created, they will have to draw on their capital reserve.
4. When this storage is exhausted, they die.

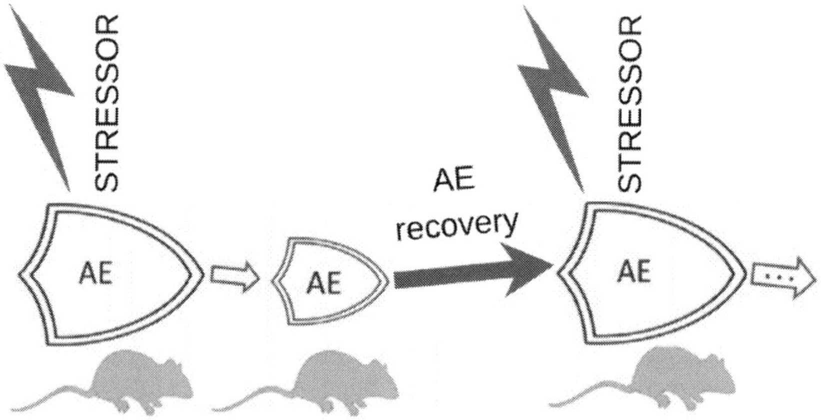

Source: Gorban et al. Journal of Theoretical Biology 405 (2016) 127–139.

In my opinion, Goldstone's model is a more accurate representation of reality and species adaptation. With enough time and right nourishment, you can recover your adaptive energy, thus being able to endure the stressor again. If there weren't a way to replenish AE, then I think humans wouldn't have survived this long.

Both Selye and Goldstone emphasized that there are different levels of supply for adaptation energy. Some have a higher energy expenditure whereas others with lower. Generally, anything from the organisms' level of fitness, state of the nervous system, amount of resources, psychological condition, and previous exposure to a particular stressor determine how much AE gets depleted during an encounter and how fast (if at all) they recover.

Research shows that individual characteristics such as age, personality traits, level of neuroticism[37], childhood experiences, past trauma[38], and well-being before the onset of a stressor also determine the final effect of the stress on an individual[39]. Social connectivity, socioeconomic status, general health, self-esteem, cultural views, and religious beliefs all contribute to a person's coping resources[40].

Things that drain adaptation energy are all kinds of everyday stressors like work, relationships, traffic, sleep deprivation, emotional trauma, as well as sub-lethal environmental toxins like air pollution, heavy metals, and electromagnetic frequencies. AE can be restored with quality nutrition (but not becoming obese), sleeping more, taking a time out, and all kinds of relaxation techniques. Stress management is about improving coping strategies and knowing how to minimize the amount of unnecessary stress.

This is a key point of stress adaptation: you need time for recovery and restoration. Too many people get stuck in chronic fight or flight and burnout just because of not taking a time out. The body seeks homeostasis and deviating from it for too long will result in dis-ease, which eventually leads to disease. Occasional fluctuations and stressors are fine and actually stimulating but in excess, they're still harmful. That describes the entire phenomenon of hormesis, which we'll be talking about shortly.

Good Stress, Bad Stress

In 1975, Selye published a model for dividing stress into eustress and distress[41]. Eustress is something that enhances fitness and functionality. It's derived from the Greek word 'good' as in euphoria[42] and is ultimately beneficial as long as there's enough time for recovery. Persistent or chronic stress is called distress and it comes from Latin, meaning 'dissonance'. Distress can lead to anxiety, depression, fatigue, disease, and even death. The difference is determined by the contrast between an experience (real or imagined), personal expectations, and resources to cope with the stress[43].

In 1993, McEwen and Stellar coined the term *'allostatic load'* to describe the accumulation of repeated stress and its consequences[44]. It's the 'wear and tear on the body' that accumulates from chronic exposure to certain stressors[45]. There are two types of allostatic load:

- **Type 1 Allostatic Load** – Energy demand exceeds supply, activating the emergency survival mode such as starvation, hibernation, and detrimental disease[46]. It's basically the body shutting itself down temporarily to decrease allostatic load and regain adaptive energy. Illness can be both the result as well as the cause of allostatic overload.

- **Type 2 Allostatic Load** – Energy consumption is sufficient or even excessive but is accompanied by social conflict and dysfunction. This happens in human societies and some animals living in captivity. Type 2 affects a person's behavior instead of triggering the survival mode or escape response[47]. It can be counteracted with learning, therapy, and changing the social setting.

Allostasis involves the regulation of homeostasis and predictive stabilization of internal processes by the brain[48]. It was first defined by Sterling and Eyer in 1988 to explain arousal pathology[49]. 'Allo' means 'variable' in Greek so allostasis refers to stability through variability. In a way, it completes the concept of inner equilibrium as the goal is not to preserve a constant internal milieu but much rather to continuously adapt and adjust to the particular milieu the organism finds itself in currently and the one it will be in the future[50]. The advantages of such anticipation are multifold – fewer errors, a higher chance of survival, resource accumulation, better management of adaptation energy, and predictive capacity based on memory.

Here are the 6 main principles of allostasis, proposed by Sterling (2004)[51]:

1. Living organisms are hard-wired to be efficient with their resources
2. Being efficient requires reciprocal trade-offs

3. Efficiency also necessitates the ability to predict future needs
4. Such prediction demands each sensor to adapt to expectations
5. Each effector has to adapt to the expected demands
6. Predictive regulation depends on behavior whilst neural mechanisms also adapt

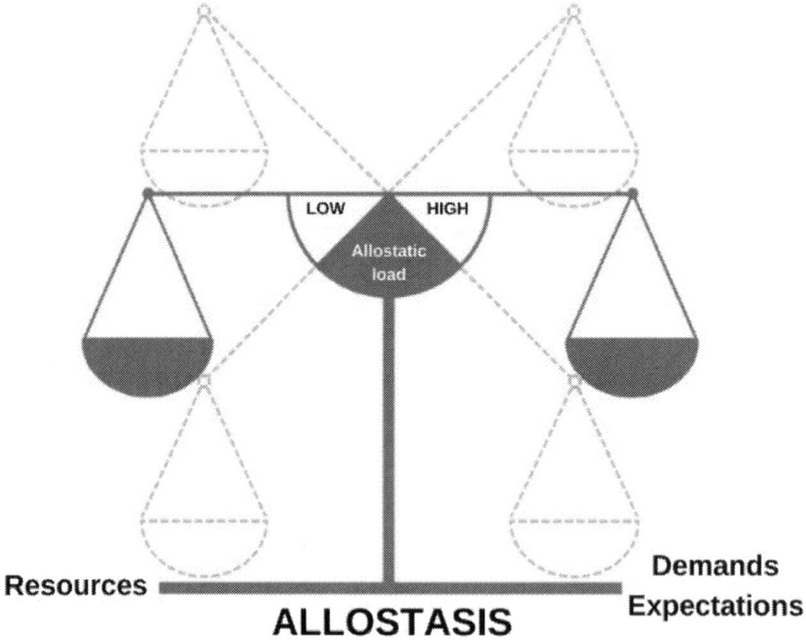

One of the main parts of this regulation is the reduction of uncertainty and unpredictability because humans don't really like surprises. Especially if you're living in nature. Allostasis helps to reduce uncertainty through predictive modeling but it requires a lot of energy. With chronic stress and high levels of uncertainty, allostatic load increases, which reduces the brain's ability to cope with and reduce uncertainty in the future[52]. Essentially, bad memories about uncertainty and lack of competence breed more stress and additional uncertainty, which overwhelms allostatic load. On the flip side, allostatic load can be minimized by seeing how one is able to endure and adapt to uncertain situations, thus increasing one's confidence to do so again in the future. It's the 'rich get richer' and 'poor get poorer' quagmire of your brain.

Allostatic load can be measured by looking at the cumulative strain on different organs and biomarkers related to cardiovascular, neuroendocrine, immune, and metabolic functioning[53]. In the endocrine system, elevated CRH is associated with an activation of the HPA axis[54]. Disrupted patterns of cortisol with low levels in the morning and high spikes in the afternoon or evening indicate a circadian rhythm mismatch that itself causes more stress. Naturally, you're supposed to be wired up after waking up to start the day and drowsy before bed, but chronic stress can invert this pattern. Shortening of dendrites in neurons, as well as increased inflammation overall, indicate overstimulation.

Allostatic load differs by age, sex, and social status of the individual. Low socioeconomic status can increase allostatic load and stress[55]. Protective factors include parental care, higher education, community support, better lifestyle habits, and a sense of purpose in one's life[56].

Past adversity and early life experiences will also dramatically determine how your immune system reacts to stress. There's an association between early life adversity and higher plasma IL-6 in response to acute stress in healthy adults[57]. Early life stress in children is also linked to a lower basal level of cytokines that control immunity[58]. In addition to poverty and abuse, it's also predicted by bullying and peer pressure[59].

Exposure to a stressor "conditions" the system to respond in some way. It's going to go through the steps of GAS and experience either recovery or exhaustion based on the stimulus and reserve of adaptive energy. If the allostatic load is overburdened, under-adaptation, and potentially death will occur. On the other hand, if AE does not get depleted and is restored then the organism may even super-compensate for the stress by adapting to it in the future. That's how allostasis works – predictive adaptation to the ever-changing conditions of the environment based on history.

In the 1970s, a low dose of carbon tetrachloride was found to be protective against much larger subsequent exposures of the same agent[60]. This was called 'autoprotection'. The same phenomenon, albeit named 'adaptive response', occurs in bacteria when a small amount of mutagen prior reduces the response to larger amounts afterward[61]. Situations where being exposed to one chemical reduces the response to other chemicals is called 'heteroprotection'.

The term 'preconditioning' describes how initial low-level stress can provide protection against future more intense stressors. It was coined by Murry et al (1986) who showed that brief episodes of hypoxic training would greatly reduce the magnitude of subsequent myocardial infarctions in dogs[62]. This has been replicated and generalized in other animal models, including the nervous system and various organs. Birds exposed to no previous acute stress or numerous stressful events show higher levels of glucocorticoid-induced oxidative stress (GiOS) in response to acute stress, while birds with moderate prior exposure do not[63]. They've basically built up more adaptive energy thanks to having prior experience with moderate stress and will thus experience less stress-related damage as well.

In 2011, McEwen et al considered that usage of stress as a concept is used inappropriately because it can range from the mildest of challenges to very severe conditions. They proposed that: „*...the term 'stress' should be restricted to conditions where an environmental demand exceeds the natural regulatory capacity of an organism, in particular situations that include unpredictability and uncontrollability.*[64]" That fits the framework of hormesis as well because anything that doesn't fully deplete the organism's adaptive energy isn't actual stress but more like conditioning. It becomes stress in its harmful sense after the body's resources have been depleted and the stimulus turns maladaptive.

Hormesis: What Doesn't Kill You Makes You Stronger

Hormesis was first described by a German pharmacologist named Hugo Schulz in the year 1888. He discovered that a small dose of lethal poison didn't kill off the yeast he was experimenting with but actually made them grow. The term 'hormesis' itself was coined and first used in a scientific paper by Southam and Ehrlich in 1943[65]. It's derived from the Greek word '*hórmēsis*', meaning 'rapid motion, eagerness, to excite' or 'to set in motion.'

Basically, hormesis is a biphasic response to a toxin or a stressor. (1) The initial contact causes injury to the body. (2) The following reaction leads to adaptation or resistance, leaving the body in a better condition than it was before. It's stimulatory in low amounts and inhibitory in large doses. Friedrich Nietzsche said: *"That which does not kill me makes me stronger."* He was a philosopher, not a biologist, but his quote aligns with the principles of hormesis perfectly.

Hormesis is usually U-shaped or with an inverted U-shaped dose-response, depending on the final outcome. If the endpoint is increased longevity or stress resistance, then it would look like an inverted U-shape curve. In the case of disease, death, or underadaptation, the curve would be either U- or J-shaped. For example, the response at low doses might be the opposite to that of higher doses[66]. In a lot of cases, exposure to a stressor in small amounts can be beneficial whereas in high amounts will be detrimental. It's also adaptive and preconditioning, making the body more resilient against subsequent and much larger doses of the same or similar agent.

In 2007, Calabrese et al proposed that a universal stress terminology should include the operational term '*hormesis*', which would be preceded by the type of stimulating agent and whether any prior conditioning was present[67]. For example, if exposure to ischemia by the surgical occlusion of an artery protects against subsequent more damaging ischemic attacks, then it should be called 'physiological conditioning hormesis'. A chemical toxicant reducing the toxicity of a larger radiation or chemical exposure would be called 'chemical conditioning hormesis'. When hormesis or stress adaptation happens without the conditioning element, the 'conditioning' part would be simply removed (e.g., radiation hormesis, physiological hormesis). If the adapting dose is given after being exposed to the larger toxin then it describes 'post-exposure conditioning hormesis'.

In toxicology literature, one percent (195 out of 20,285) of the published articles contain 668 dose-response relationships that meet the entry criteria of hormesis. Subsequent application of evaluative criteria has revealed that 245 (37% of 668) dose-response relationships from 86 articles (0.4% of 20,285) satisfy requirements for evidence of hormesis[68]. The concept that low dose exposure is different from higher doses is widely accepted, but that it could be actually beneficial is met with skepticism by some.

Here's an overview of the most used dose-response models in stress science:

- **Linear Threshold Model** – something becomes increasingly more toxic after a certain threshold i.e. drinking alcohol or drugs.
- **Linear No-Threshold Model (LNT)** – any dose above zero is hazardous. I.e. carcinogens or radiation according to some views.
- **Hypersensitivity Model** – greater risk at lower doses and increased resilience at higher intakes. I.e. someone not

used to drinking alcohol or coffee getting a worse response from smaller doses because their body is not used to it. With more frequent consumption the body adapts and experiences less of an effect.

- **Threshold Model** – threshold dose below which there is no risk and above which risk increases i.e. radiation exposure or UV light.
- **Hormesis Model** – a biphasic response to a stressor with low doses stimulating adaptation but high doses causing weakening i.e exercise, fasting, heat exposure.

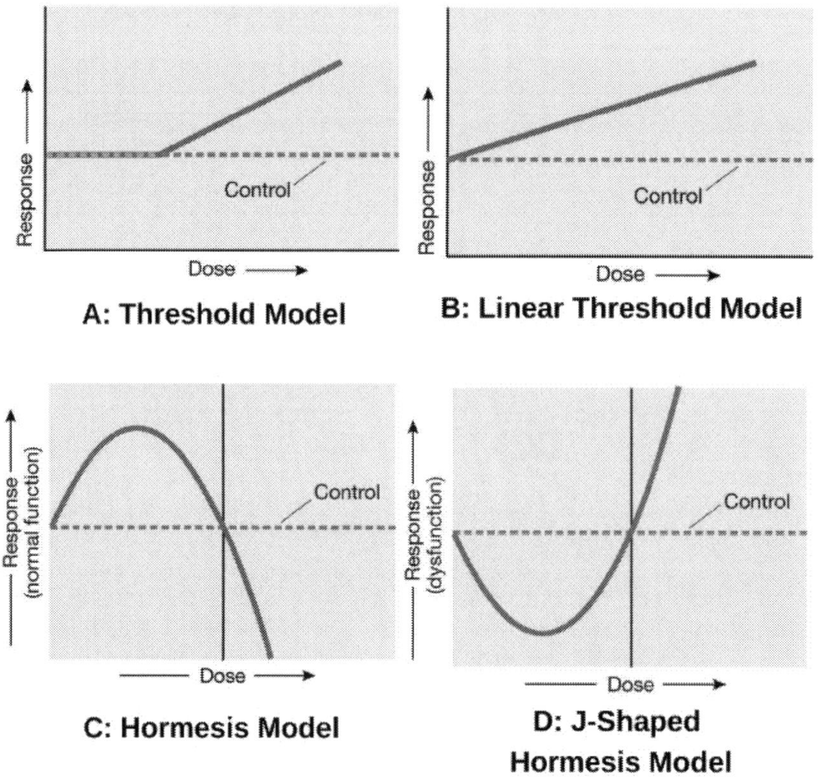

Source: Calabrese (2004) EMBO Reports

Hormetic effects show an overcompensation in response to disruptions in homeostasis, leading to an adaptive response and autoprotection. Hormesis has more evidence than the LNT and standard threshold model[69]. It's at least practical and adaptive in the sense that you can precondition your body against potential threats.

Stress adaptation and hormetic conditioning requires good risk assessment. Although our bodies have the ability to deal and even benefit from low dose toxins, the effect is not universal to all chemicals. Likewise, the beneficial hormetic dose also ranges between stimuli. We should always try to get the optimal dose and focus on the things that will give us the most benefit with the least downside. It's called the 80/20 rule or Pareto Principle.

In 1896, an Italian economist Vilfredo Pareto showed that, in most cases, about 80% of the effects come from 20% of the causes[70]. 80% of the wealth belongs to 20% of the people. 80% of car accidents happen to 20% of the drivers. 80% of the food is consumed and stored as fat in 20% of the world's population, etc. Mainly it's used in economics, but this 80/20 rule, or the law of the vital few, is evident in other areas of life as well.

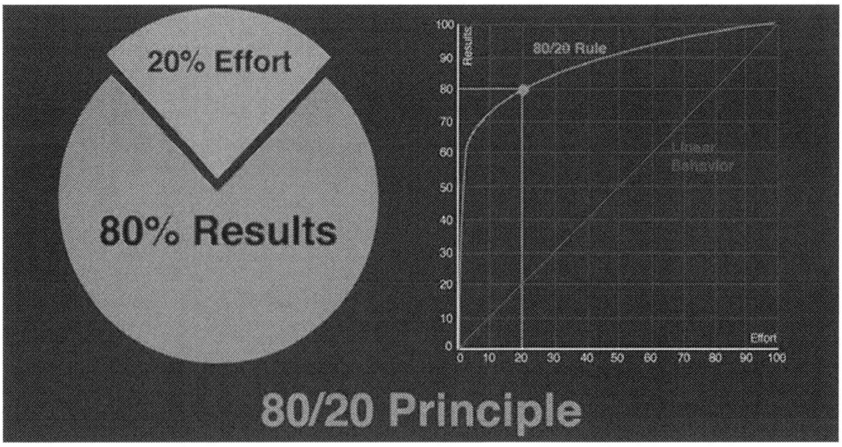

In health and fitness, 80% of your results will come from 20% of the fundamentals, such as eating a clean diet and exercising. More often than not, people focus on the other 80% of the potential things they could do like random supplements which will give them only 20% of the desired outcome. The same applies to hormesis and stress adaptation as well – there's a certain amount of conditioning that's beneficial after which it can become harmful, etc.

Hormesis doesn't mean you should start exposing yourself to random stressors and toxins that don't have a benefit. Instead, the purpose of this book is to share the idea that some things can be taken in hormetic doses and used to increase one's resilience. What those agents and activities are and whether or not they're worth it will be discussed in the following chapters.

Antifragility and Hormesis

Hormesis is an example of antifragility, which is a concept that comes from Nassim Nicholas Taleb's book Antifragile: Things That Gain From Disorder. Taleb is an economist, philosopher, and risk analyst who has come up with several definitions for describing how complex systems such as societies, individuals, and economies can operate. Most people would think that resilience or robustness is the opposite of fragility but this is not the case. The opposite to being fragile is antifragility – the ability to benefit from stress and get better.

Taleb opens his book as such:

> *Some things benefit from shocks; they thrive and grow when exposed to volatility, randomness, disorder, and stressors and love adventure, risk, and uncertainty. Yet, in spite of the ubiquity of the phenomenon, there is no word for the exact opposite of fragile. Let us call it antifragile. Antifragility is beyond resilience or robustness. The resilient resists shocks and stays the same; the antifragile gets better.*[71]

Taleb uses three heuristics (tools/models) to describe his point. The Triad consists of the following:

- **Fragile things break into pieces when put under stress** – they can't tolerate exertion or anything that puts them out of balance.

- **Resilient things resist the stress but stay the same** – they just absorb the shock but remain indifferent – nothing significant happens.
- **Antifragile things adapt and get better** – they don't lose or break under pressure – they don't just endure and absorb it – they adapt to the stress and benefit from it.

I'm going to be using the concept of antifragility in several instances of this book to describe stress adaptation and hormesis. In a way, it's the end goal – to make antifragile humans.

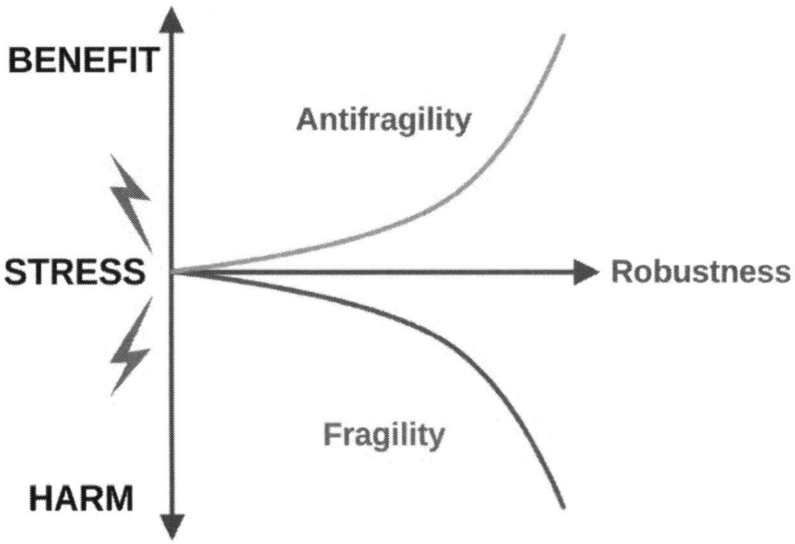

Types of Stress Responses

Environmental stressors cause a wide range of cellular stress responses that change the status of cells both generally throughout the body as well as locally at the site of injury. The purpose is to minimize the damage and adapt to similar conditions in the future[72].

The main trigger for stress responses is the detection of denatured proteins. They lose their healthy structure and indicate the presence of an injury. Injecting unstressed cells under normal conditions with denatured proteins triggers a stress response. The degradation of damaged proteins is governed by the ubiquitin-proteasome system (UPS). It involves two steps: tagging the protein with a polyubiquitin chain and degrading it by the proteasome[73].

Stress responses are mediated by stress proteins. They're divided into two categories: those that are activated only under stressful conditions and the ones that are involved in both stress responses as well as normal cellular functioning. These mechanisms can be found across phyla and are identical in the simplest prokaryotic cells as well as the most complex eukaryotic ones. Different kinds of stressors have their own signature stress proteins, such as heat shock proteins, cold shock proteins, etc.

There are many different types of stress responses with specialized characteristics:

- **Heat-Shock Response** – Exposure to heat triggers a class of proteins called heat-shock proteins. They protect the cell against damage by repairing misfolded proteins and prevent them from becoming denatured[74]. This is a very important role because elevated temperatures on their own can cause the denaturing of proteins.

- **Cold-Shock Response -** Cold-Shock Proteins (CSPs) are multifunctional RNA/DNA binding proteins, characterized by cold-shock domains (CSDs)[75]. They are one of the most evolutionarily conserved proteins found in virtually all organisms[76,77]. During cold shock, cell membrane and enzyme activity decrease, and you reduce the efficiency in mRNA translation and transcription[78].
- **Toxin Stress Response** – Different toxic compounds can trigger cellular stress by denaturing healthy proteins. This includes all the air pollution, heavy metals, chemicals, and toxins we're surrounded by in the modern environment. They also create free radicals, misfolded proteins, and spread inflammation.
- **Unfolded Protein Stress Response** –An increase in misfolded proteins within a specific organelle triggers unfolded protein response (UPR)[79]. This process helps to restore normal functioning of the organelle, halts protein translation, and stimulates the degradation of misfolded proteins[80]. In the endoplasmic reticulum (ER) it's dubbed UPR^{ER} and in the mitochondria UPR^{mt}.
- **Autophagic Stress Response** – Autophagy is the main process that mediates the degradation of cellular components, including entire organelles and protein aggregates. There many specific types of autophagy like lipophagy, aggrephagy, glycophagy, xenophagy, etc. but the three main ones are macroautophagy, microautophagy, and chaperone-mediated autophagy[81]. Some basal autophagy is happening almost all the time but these

processes get magnified under conditions of stress i.e. starvation, fasting, exercise, heat, hypoxia.

- **DNA Damage Response** – DNA damage occurs in response to UV radiation, nuclear radiation, reactive oxygen species (ROS), environmental toxins, and oxidative stress[82]. There are several repair pathways that target a specific DNA lesion. Base-excision repair (BER) remove lesions that affect only a single DNA strand[83]. Nucleotide-excision repair (NER) targets helix-distorting lesions caused by UV exposure and carcinogens[84]. Double-strand breaks (DSBs) can increase the risk of cancer and they're repaired by non-homologous end joining (NHEJ)[85].
- **Oxidative Stress Response** – Oxidative stress results from an imbalanced redox status in the cell – more pro-oxidants than anti-oxidants. It's involved in many pathological processes and disease[86]. The most common free radicals are ROS, superoxide anion (O_2), hydrogen peroxide (H_2O_2), singlet oxygen, hydroxyl radical ($OH^\bullet$), peroxy radical, as well as the second messenger nitric oxide ($NO^\bullet$) which can react with O_2 to form peroxynitrite ($ONOO^-$). They are neutralized by superoxide dismutase (SOD) and glutathione (GSH). ROS can be created from internal and external sources. Too much is bad but not enough is also unwanted.
- **Acute-Phase Response or Inflammatory Stress Response** – Stress increases inflammatory mediators and cytokines that initiate a systemic response to the stressor. This sets off the acute-phase response (APR) or the

inflammatory stress response and the release of acute-phase proteins (APPs), which are essential mediators of inflammation[87]. The most notable ones are interleukins IL1 and IL6, C-reactive protein, TNFα, serum amyloid A and others. Cortisol can also regulate the expression of APPs through IL6[88]. APR is characterized by fever, swelling, inflammation, and increased circulation of white blood cells.

- **Nutritional Stress Response** – Calories and food are a high priority for all living organisms without which life would not exist. Deprivation of certain nutrients as well as fasting cause nutritional stress, characterized by the activation of nutrient pathways that promote stress adaptation and survival. They are AMPK, autophagy, ketosis, Nrf2, and many others. Chronic calorie restriction and starvation are not sustainable nor enjoyable. That's why the purpose of this book is to biohack these processes to your greatest benefit with the least side-effects.
- **Autoimmune Stress Response** – Stress is a trigger for autoimmune disease and it perpetuates autoimmunity[89]. Exposure to stress-related disorders is associated with an increased risk of multiple autoimmune diseases such as lupus and arthritis[90]. Adrenaline and cortisol activate immune cells, making them attack host cells[91,92]. This can lead to allergies, Hashimoto's disease, and other forms of autoimmunity.
- **Psychological Stress Response** – Cognitive, social, or psychological strain of any kind is a stressor that can be as

pathological as physical stress. The brain can't tell the difference between imagination and reality. Imagining a threat lights up similar regions in the brain as experiencing it in real-time does[93]. This is a quintessential point to remember because it means you can become incredibly stressed out just by thinking about things that frighten or infuriate you.

This book will cover all of these categories to a certain extent. More often than not they occur interchangeably and in conjunction with each other. Likewise, hormetic adaptation to these responses can also be created through similar means. For example, the autophagic stress response is also a nutritional stress response if you use fasting as its catalyst. This process will then also promote adaptation to DNA damage, inflammation, and unfolded protein stress response.

Stress responses don't work in isolation but are part of a wider stress network. The pathway that links extracellular stimuli with intracellular processes is the kinase of TOR (target of rapamycin). TOR regulates cell growth, autophagy, proliferation, DNA repair, and translation of proteins[94]. mTORC1 links protein quantity and quality control by sensing the availability of chaperones[95]. A moderate reduction in chaperone availability enhances mTORC1 signaling, whereas complete depletion of chaperones suppresses it. That's how hormesis and growth work – in moderation it sprouts growth and adaptation but in excess nullifies it.

Stress Adaptation and Cyclic AMP

The primary mechanism by which organisms detect their environment has been the heptahelical G protein-coupled receptor (GPCR)[96], also known as 7 transmembrane receptors because they pass through the cell membrane seven times. It detects molecules outside the cell and activates internal cellular responses[97]. The GPCR system has adapted to perceive photons, odorants, lipids, carbohydrates, peptides, nucleic acid, hormones, and neurotransmitters. It's thought that these receptors have also adapted to exogenous compounds that yield a hormetic dose-response[98]. This ability or detective predictability would've yielded a greater evolutionary advantage to exploit these beneficial agents in harsh environments. And in a way this is what many animals in nature already do – they seek out specific plants/berries that help them adapt to their surroundings and climate better.

There are two main signaling pathways involving the GPCR receptors: the cyclic AMP (cAMP) pathway and the phosphatidylinositol (PIP2) pathway[99].

- **Cyclic AMP** is a key regulator of intracellular functions and the metabolism. It regulates the metabolism of glucose, glycogen, and fatty acids[100]. CAMP is generated when an intracellular messenger, such as a neurotransmitter, hormone, chemokine, lipid mediator, or drug, binds to a GPCR that is coupled to a stimulatory G protein α subunit (Gαs)[101].
- **Phosphatidylinositol** consists of different lipids, classed as phosphatidyltriglycerides. They're typically a component

of eukaryotic cell membranes that play an important role in lipid signaling, cell functioning, and membrane trafficking.

GPCRs become desensitized when exposed to the same ligands and messengers all the time. This will lead to certain functions ceasing to work until desensitization is overcome and re-habituated. That's why doing things over and over again in a similar fashion can cause stagnation or plateaus in progress. The way you overcome that is by either changing the dose or taking a break. That's how hormesis also works – you need time for recovery in order to let the signals of adaptation sink in.

Like said beforehand, chronic stress predisposes you to many diseases that shorten lifespan and vitality. It contributes to hypertension and many other conditions related to it such as cardiovascular disease[102]. Stress also contributes to aging directly by promoting many age-related diseases[103]. The key is getting it in the right dose, at the right time, with the right means.

Your Body's Defence System

Here are the body's defense systems that modulate the immune system and increase resilience against disease. For optimal health and stress adaptation, you want to upregulate these factors:

- **Glutathione (GSH)** is the body's main antioxidant produced in the liver. It protects against free radicals and helps to eliminate lipid peroxides as well as toxins. GSH is more powerful and practical than other antioxidants because the body self-regulates it in conjunction with the immune system[104,105].

- **Nrf2 or Nuclear Factor Erythroid 2-Related Factor 2** as it's called is a transcription factor that binds to DNA and expresses particular genes. It is one of the main regulators of antioxidants like glutathione, NADPH, thioredoxin, and cell protection[106].

- **Autophagy** or self-eating is the major cleaning system of the body. It modulates the immune system, eliminates pathogens[107], conducts mitophagy or mitochondrial autophagy, lipid oxidation, DNA repair, and lowers inflammation. Physiological stress or infections trigger the salvage and recycling of damaged particles. More on this in chapter two.

- **Uric Acid** is the most concentrated antioxidant in the human blood that helps to mitigate oxidative stress, especially under hypoxia[108]. In low amounts it's beneficial but in excess can cause gout and fibromyalgia[109]. You obtain it from purine-rich foods like meat, fruit, fish, and grains.

- **NAD+ or *Nicotinamide adenine dinucleotide*** is the body's main co-enzyme that partakes in virtually all cellular processes and energy production. Decreasing NAD+ is linked to aging, disease, and weaker immune system functioning[110]. It's required for carrying out every other defensive response as well as recovery.

- **NADPH or n*icotinamide adenine dinucleotide phosphate* (NADP+)** is a cofactor for anabolic processes such as cellular growth, and nucleic acid synthesis. The extra phosphate group gets added during the salvage pathway of NAD+. NADPH is the reduced form of NADP+. It protects against excessive reactive oxygen species (ROS) and enables the regeneration of glutathione[111]. More on this in chapter ten.

- **Strong gut lining.** Intestinal permeability or leaky gut is associated with autoimmune diseases and the development of several inflammatory diseases[112,113]. Increased low-grade inflammation makes one more prone to infections[114]. Bone broth, tendons, and ligaments have collagen and glycine that promote tissue rejuvenation[115]. Butyrate is also essential as the main source of energy for cells in the large intestine[116]. You can get butyrate mainly from the fermentation of fiber like beans, vegetables, and legumes but also ghee and butter. Avoiding or at least limiting the consumption of gluten and lectins can protect the gut lining[117].

- **Diversity of the gut microbiota** is linked to stronger immunity[118] because microbes have an important role in modulating our body's defense systems. They also help the host adapt to the microbial and pathogenic environment they're in. Almost by proxy of being able to deal with the flora and fauna you may contact.

The goal of self-optimization and -quantification is to get feedback from the state of your nervous system to know what physiological state you're in and then implement certain routines, activities, ingredients, or supplements to influence that in a desired way. Sometimes you want to be more amped up and focused, which would entail going into the sympathetic mode, but at other times you need to be parasympathetic to allow your body to recover and not be damaged by chronic stress.

It's very sad and dreadful to be constantly stuck in fight or flight because it'll not only degrade your health, but also impedes with productivity, relationships, your goals, and general well-being. That's why stress management and proper recovery are so vital for peak performance and a fully optimized life.

General Stress Adaptation Hormesis Protocol

- Your body is built to encounter and endure stress through hormesis. With zero stimuli, resiliency decreases, resulting in fragility. Antifragility stems from occasional stressing of the system so that adaptation could occur. Excessive and chronic stress, however, will drain adaptive energy, leading to exhaustion and metabolic winter.
- There are multiple types of stress that can have both generalized as well as localized responses. Some of them can increase adaptation universally whereas others function only within their domain. The goal isn't maximum exertion or exhaustion but finding the minimal effective dose based on our current situation and condition.

- Some stressors are beneficial whereas others are harmful with no hormetic benefit. As smart biohackers, we should engage in activities that promote hormesis and improve our vitality. The 80/20 rule is fundamental to this as well.

Chapter Two:
Of Mitohormesis and Autophagy

"At 50, everyone has the face they deserve."
— George Orwell

The primary function of this book is to give you a manual for stress adaptation and not leave yourself vulnerable to the uncertainties of life. However, all the strategies and ideas discussed here also have a positive effect on your longevity and lifespan. Hormesis is one of the key phenomena that seems to make organisms live longer. Up to a certain point, of course.

In his book Lifespan, Dr. David Sinclair refers to it as the information theory of aging – that there are certain programs running in your body that read the epigenome, thus regulating all metabolic processes. If the genome picks up dysfunctional or aged data, it's going to cause aging. Epigenetics stands for epi (above) genetics – the process by which the environment or the stimuli the cell gets exposed to affects the fate of the cell and rest of the body.

When I interviewed Dr. Sinclair on my podcast, he said there are two types of information in our body that keep us alive:

(1) DNA or genome – digital information similar to music files on a DVD or the software

(2) Epigenome – machines and structures within the cell that hold the DNA in certain loops and packages that tell the cells what they should be and how they should stay

It's thought that your health and longevity in old age are only 20% genetically determined and 80% is epigenetic, which can be modified by how you live your life. Sinclair and many other researchers are starting to realize that loss of epigenetic information – how the cells read DNA - is just as important or possibly even more important for aging than the genome itself.

Longevity Pathways in Humans

In humans, there are several longevity pathways recognized to control the aging process and its constituent mechanisms. I'm going to outline the ones that are currently most recognized to regulate longevity and lifespan. Then I'll go through some of them in closer detail, showing how they relate to hormesis and stress adaptation.

- **The Growth Hormone/Insulin and Insulin-Like Growth Factor-1 Signaling Pathway**, which regulates cell replication, nutrient partitioning, and storage. Reduced insulin signaling has been found to increase the lifespan of fruit flies, nematodes, and rodents[119]. Knocking out the IGF-1 receptor in mice makes them live 26-33% longer[120]. Additionally, mutated suppression of an insulin receptor substrate (IRS)-like signaling protein called CHICO increases the lifespan of fruit flies by 48%[121]. This seems to be dependent on FOXO proteins, which are transcription factors of longevity.

- **FOXO Transcription Factors,** which includes proteins and transcription factors responsible for energy homeostasis. In mammals, FOXO proteins regulate stress resistance, cellular turnover, apoptosis, glucose-lipid metabolism, and inflammation[122,123]. The FOXO pathway is an evolutionarily viable mechanism for adapting to low levels of insulin and energy deprivation.
- **Sirtuin Proteins and NAD+,** which mediate metabolic functions and DNA repair. SIRT6 overexpression has been found to lengthen the lifespan of male mice by as much as 15,8%[124]. SIRT6 deficiencies in mice accelerate their aging[125]. NAD is critical for converting food into energy, repairing DNA damage, strengthening the immune system, burning fat, and regulating the body's circadian clock[126]. Declining NAD is linked to age-related metabolic dysfunction and is a key contributor to aging[127]. More on this in chapter eleven.
- **Hormesis and General Stress Adaptation** mediated by FOXO proteins and mitochondrial functioning. This phenomenon makes the organism more resilient against environmental stressors. Activation of the stress-response JNK pathway in fruit flies increases their lifespan by up to 80%[128]! Hormesis through heat stress increases the lifespan of flies and worms[129]. It's one of the central components of this book.
- **Yamanaka Factors** – Reprogramming factors (called Oct3/4, Sox2, Klf4, c-Myc) that regulate cell rejuvenation

and reprogramming[130]. They can be used for induced pluripotent stem cells (iPSC), thus influencing the epigenetic clock[131]. In 2011, Lapasset et al reversed centenarian cells aged over 100 back to the age of 20[132]. Yamanaka factors essentially prevent epigenetic alterations that happen during aging and tell the epigenome to be young again. They're primarily turned on in mice with genetic engineering. However, things like niacin, fasting[133], raising NAD+[134], turmeric[135], and spirulina have also been shown to promote stem cell proliferation.[136]

- **The mTOR/AMPK Pathway,** which governs homeostasis between anabolism and catabolism. Basically, is the body going to grow or eat itself? Overexpression of mTOR and its dysfunction is often related to various cancers and genetic disorders[137]. On the flip side, AMPK activation through fasting or calorie restriction extends lifespan by increasing autophagy and sirtuin activity[138,139].

All of these factors slow down aging by sending a signal to the body, telling there's no time to sit on your laurels – you're running away from a lion for god's sake! Minor stress through hormesis makes the body read its epigenetic information in a way that indicates it has to live longer for the time being in order to ensure it passes on the genome. However, if this subliminal allostatic overload becomes chronic or excessive, then it will eventually decrease health and vitality.

Dauer larvae are morphologically specialized roundworms that adapt to harsh conditions such as starvation, temperature stress, and oxygen deprivation. Normally, they would exit the period of conservation according to the influx of energy from their environment, but this cycle can be prolonged in laboratories by regulating the worms' living temperature or access to calories. Dauer larvae can survive up to 8 times longer under laboratory conditions where their energy homeostasis is controlled[140]. That's a huge difference based on a simple elongation of a certain life stage and environmental input.

It's known that reducing total caloric consumption has beneficial effects on aging and lifespan in virtually all animals[141]. Feeding fewer calories to roundworms, flies, mice, as well as monkeys extends their lifespan up to 20-30%[142]. The larvae of skin beetles can also go back into an earlier stage of maturity if they're starved. This cycle can be repeated many times[143].

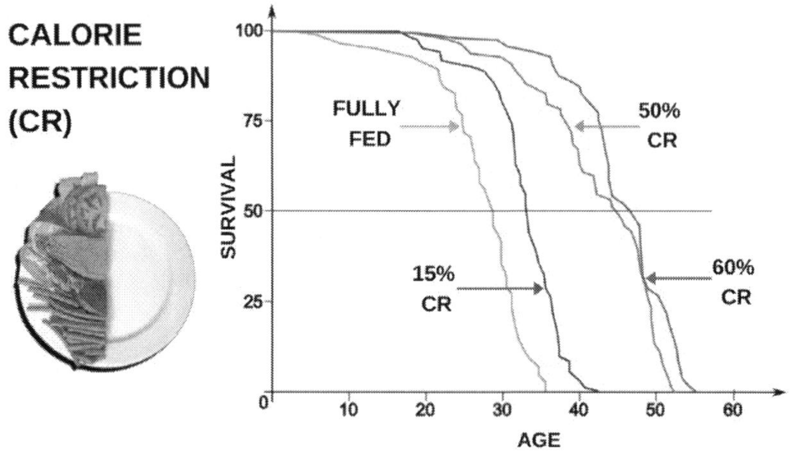

One human study on 3 weeks of alternate day fasting discovered an increase in SIRT1, which is associated with longevity[144]. Why this happens is still unclear, but it's suggested that calorie restriction induces cellular respiration, which increases NAD+ and reduces NADH levels. NADH inhibits Sir2 and SIRT1 whereas NAD+ raises them. SIRT1 has been shown to also activate PGC-1α, which triggers the growth of new mitochondria[145]. SIRT3, SIRT4, and SIRT5 improve mitochondrial function as well[146].

The Mitochondrial Theory of Aging

In 1956, Denham Harman was the first to propose the Free Radical Theory of Aging (FRTA)[147] and furthered the idea in 1970 to describe mitochondrial production of reactive oxygen species (ROS)[148]. This theory states that organisms die because of the accumulation of free radical damage to the cells over time.

Oxidation is a chemical process that creates free radicals and reactive oxygen species (ROS) such as hydrogen peroxide, hypochlorous acid, and free radicals like hydroxyl radical, and superoxide anion. They damage cells by oxidizing DNA, proteins, and causing lipid peroxidation[149]. Antioxidants like superoxide dismutase, vitamin C, E, and A help to neutralize oxidation by donating an extra electron to the oxidized free radical or cell. They keep the body in balance.

Oxidative stress is considered to be one of the main causes of modern diseases such as Alzheimer's, Parkinson's, arthritis, diabetes, and cardiovascular disease[150,151,152]. It's both the cause as well as the result of imbalanced oxidation. Basically, too much damage and not enough healing. Oxidized LDL cholesterol triggers the process of atherogenesis, which results in atherosclerosis and cardiovascular disease[153]. Oxidative damage also promotes cancer and tumor formation.

Mitochondria are the cells' energy manufacturers that generate adenosine triphosphate (ATP). It's the energetic currency needed for life. This process occurs by reacting hydrocarbons from calories or sunlight with oxygen. At the same time, electrons get out of the electron transport chain and react with water to create ROS. In excess, it can promote oxidative stress.

Mitochondrial Free Radical Theory of Aging was introduced in 1980, which implicates the mitochondria as the main targets of ROS damage[154]. Energy generation by the mitochondria damages mitochondrial macromolecules, including mitochondrial DNA (mtDNA), which promotes aging[155]. After a certain threshold, this produces too many reactive oxygen species (ROS), which cause cell death and degradation. It's almost like Hans Selye's model of adaptive energy that once depleted leads to the death of the organism.

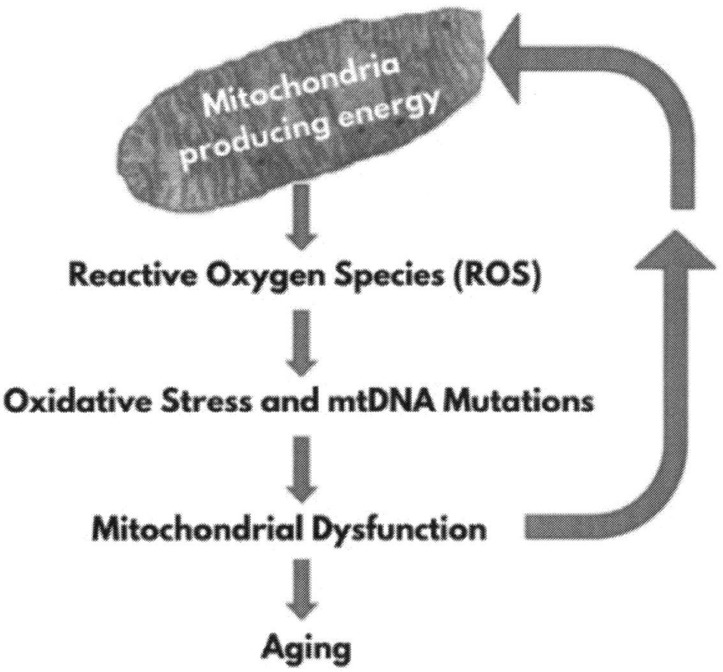

The ability to cope with oxidative stress and other stressors is compromised in aging thus making you more vulnerable to free radicals as you get older. Mutant mtDNA increases with age, especially in tissues with higher energy demands like the heart, brain, liver, kidneys, etc. Aging is associated with elevated heat-shock protein genes, in the absence of other external stressors, which suggests that the process of aging itself generates internal stress responses, inflammation, and denatured proteins[156]. These notions support the theory of mitochondrial aging[157]. However, it's been now shown that mutations in mtDNA can result in premature aging without increasing ROS production by mutating the polymerase Pol-γ that's responsible for mitochondrial DNA synthesis[158].

In some species like yeast and fruit flies, reducing oxidative stress does indeed extend lifespan[159]. However, blocking the antioxidant system in mice doesn't shorten their lifespan in most cases[160]. Likewise, in roundworms, inhibiting the natural antioxidant superoxide dismutase has been shown to actually increase lifespan[161]. What gives?

Taking a lot of antioxidants and lowering oxidative stress with supplements have failed to be effective in fighting diseases and may promote the chances of getting sick. Treatment with high doses of antioxidants like beta-carotene, vitamin A, and vitamin E have been shown to increase mortality[162]. Studies find consuming more fruit and vegetables doesn't seem to have a significant effect on reducing cancer risk[163]. Free radicals and ROS are involved in most human diseases and cancers but their degree of influence is still uncertain. You probably don't want to eliminate all of this oxidative stress because some is needed to maintain cellular homeostasis. Increasing your body's own endogenous antioxidant levels may be a better option for disease prevention[164].

Oxidative stress and free radicals can increase life expectancy in nematodes by inducing a bi-phasic response to the stress. This phenomenon is called *mitohormesis* or mitochondrial hormesis. Sublethal mitochondrial stress with a minute increase in ROS may cause a lot of the beneficial effects found in caloric restriction, intermittent fasting, exercise, and dietary phytonutrients[165].

- If you experience no stress and zero exposure to free radicals, then your body is by default weaker because of having no pre-conditioning from the past.
- If you experience too much stress and excessive accumulation of ROS, then you promote disease and sickness because of depleting adaptive energy.
- If you experience just the right dose of stress, then you'll be able to deal with it, recover from the shock, and thus augment your cells against future stressors. This is hormesis.
- If you block all mitochondrial stress and eliminate free radicals, then your body won't have the time nor the means to promote mitohormesis. That's why antioxidants and zero stress all the time won't have a positive effect.

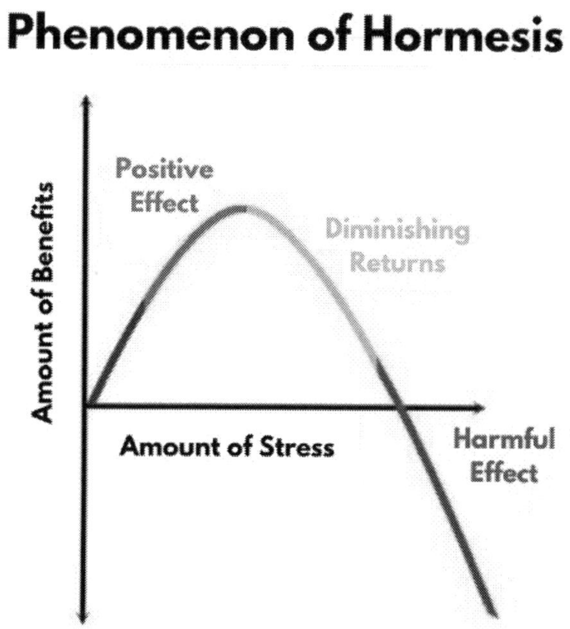

Although the mitochondrial theory of aging has its shortcomings, it's still pretty darn important to be kept in mind. The mitochondria are one of the most important organelles in your body as they govern everything related to energy metabolism and cellular homeostasis. Dysfunctional mitochondria will not only speed up aging but also make you feel more tired, exhausted, lethargic, weak, and experience atrophy.

In 2017, a study published in the journal Cell Metabolism showed that aging and age-related diseases are associated with a decrease in the cells' ability to process energy efficiently[166]. Scientists used nematode worms who live for only two weeks to carry out an experiment on their mitochondria. They found that restricting the worms' calories and manipulating AMPK promoted longevity by maintaining mitochondrial networks and increasing fatty acid oxidation. This happened in communication with other organelles called peroxisomes that regulate fat metabolism (See figure below). Essentially, more fatty acid oxidation from their energy stores led up to living longer because they were put under caloric restriction.

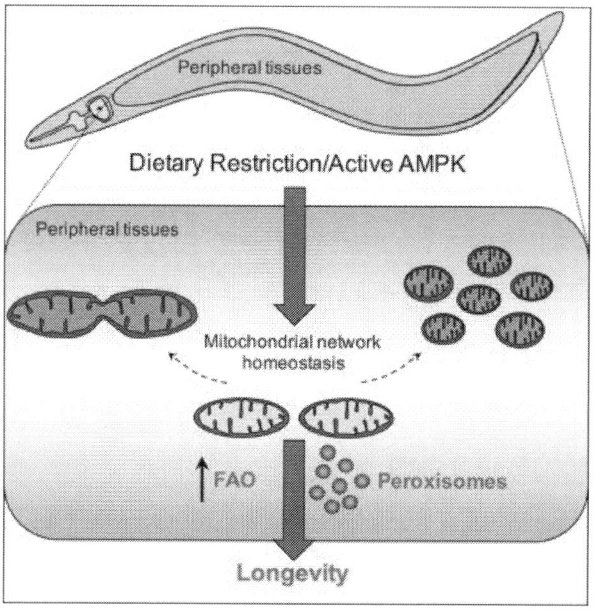

Source: Lamb et al (2013) Nature Reviews Molecular Cell Biology, Vol 14(12), p 759-774.

The researchers proposed that fission and fusion amongst the mitochondria's network and fatty acid oxidation are required for the longevity benefits. For intermittent fasting-mediated lifespan increase, you need the dynamic remodeling of mitochondrial networks, which happen in response to various physiological and pathological stimuli.

The life cycles of mitochondria are characterized by fission and fusion events.

- Fusion states happen when several mitochondria mix and organize themselves into a network. They basically merge into a single much larger mitochondrion.

- Fission states happen when the fused mitochondria get split into 2 out of which the one with a higher membrane potential will return to the fission-fusion-cycle and the one with a more depolarized membrane will stay solitary until its membrane potential recovers. If its membrane potential remains depolarized, it'll lose its ability to fuse and eventually will be eliminated by mitophagy or mitochondrial autophagy.

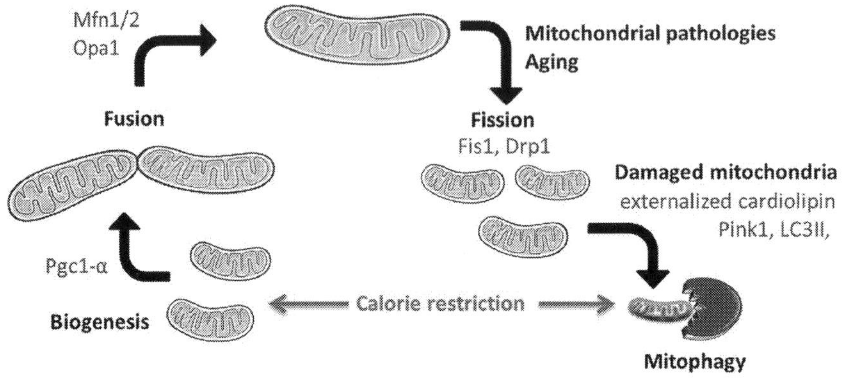

Changes in nutrient and energy availability can make the mitochondria stay in either one of these states for longer.

- Post-Fusion State is called Elongation, which is characteristic to states of energy efficiency, such as starvation, acute stress, caloric restriction, and biological aging (senescence).
- Post-Fission State is called Fragmentation, which shortens the mitochondria and keeps them separate. This is typical to bioenergetic inefficiency caused by high energy supply and extended exposure to excess nutrients.

When your body faces a shortage of energy whether through caloric restriction, fasting, starvation, or anything the like, then you're going to promote the fusion of mitochondria. This lowers your energetic demands because the organelles in your cells are better connected. It will also make you recycle old worn-out cell components and convert them back into energy through the process of mitophagy. The mitochondrial fission-fusion cycles are dependent on autophagy modulating pathways such as AMPK and mTOR[167].

- **mTOR or mammalian target of rapamycin** is responsible for cell growth, protein synthesis, and anabolism. It will make the body build new tissue.
- **AMPK or AMP-activated protein** kinase is a fuel sensor that's involved in balancing energy-deprived states and stimulating fat burning.
- **Autophagy** is the process of self-eating and cellular turnover in which the body recycles its old worn-out components and debris back into energy.

mTOR inhibits autophagy because it makes your body grow, which requires expending energy and upregulating the metabolism, whereas AMPK supports autophagy due to the energy-deprived state. One of the subunits of AMPK, AAK-2, is required for DAF-2 mutations to promote longevity in C. Elegans[168]. The mechanism by which this happens is unknown but overexpression of AAK-2 increases lifespan in worms by 13%.

Nutrient starvation allows unneeded proteins to be broken down and recycled into amino acids that are essential for survival. That keeps the organism alive longer because of increased mitochondrial efficiency. Mice who lack the insulin receptor in adipose tissue live longer because of increased leanness[169]. mTOR interacts with the insulin pathway to regulate the lifespan and development of C. Elegans and fruit flies[170].

However, the goal of stress adaptation as prescribed in this book is not to mimic the chronically restrictive conditions of laboratory animals. Who would want to be cold, starved, and exhausted all the time? That's no fun and not worth it. What's more, things like calorie restriction and fasting work because of activating the body's survival circuit that promotes longevity and hormesis. For instance, if you block the autophagy genes of genetically modified mice, then they won't live longer even under caloric restriction, whereas normal mice who have autophagy activated do[171].

Therefore, the life-extension benefits of caloric restriction and fasting are mostly induced by autophagy and increased sirtuin activity that promote cellular turnover and recycling of old cells. That's an important point because it means you can side-step some of the negative side-effects of prolonged caloric restriction by knowing what you're doing and elevating autophagy with other means. The same applies to other pathways like Foxo factors and AMPK. If there's a better way of turning them on without unnecessary collateral damage, we should focus on that.

Intermittent Fasting and Calorie Restriction

Intermittent fasting (IF) mimics the longevity effects of calorie restriction (CR) without needing to reduce calories that substantially. A study done on mice showed that longer daily fasting improves their health and longevity independent of diet composition or calories[172]. The mice ate once a day vs the 13-hour eating window of calorically restricted mice and the ad libitum mice. One of the researchers de Cabo said:

> *Increasing daily fasting times, without a reduction of calories and regardless of the type of diet consumed, resulted in overall improvements in health and survival in male mice. Perhaps this extended daily fasting period enables repair and maintenance mechanisms that would be absent in a continuous exposure to food.*[173]

For calorie restriction to work you need to reduce your total daily caloric intake by about 30%, but with time-restricted eating you can by-pass that because during the fasting state you experience a much greater degree of energy deprivation. Think of CR as trying to break a wall with your foot – it's gonna hurt and will take a lot of time. IF, on the other hand, is like a sledgehammer that just busts the thing open.

- **In 1946, a study on rats found that fasting one day out of three increased lifespans in males by 20% and in females by 15%**[174]. While they didn't experience any retardation of growth, the death of tumors increased in proportion to the amount of fasting. Other studies on rodents have noted reduced inflammation and fewer age-related health issues[175].

- **Fasting has been shown to increase the lifespan of bacteria and yeast by ten-fold[176].** Yeasts have a very short lifespan of just a few days and weeks, but a 10x boost is still phenomenal.
- **Fasting activates more autophagy than calorie restriction.** Studies in mice have shown that fasting for 48 hours compared to 24 hours produces more autophagosomes[177]. This applies to sirtuins as well. Higher energy stress activates more defense systems. Up to a certain point, of course. Studies on time-restricted eating in humans have shown that eating within 8 hours or less shows a higher expression of autophagy genes, sirtuins, and better insulin sensitivity compared to eating over the course of 12 hours[178]. This will also have a NAD-boosting effect. You do experience some increase in basal autophagy with just calorie restriction but it would get reduce whenever you eat.
- **Heat Shock Proteins** – Fasting increases heat shock proteins[179], which are stress adaptation molecules that strengthen the immune system, fight infections, reduce inflammation, and trigger autophagy. In rats, fasting for 48 hours increases heat shock proteins by about three-fold. They dropped back to normal 6-24 hours after refeeding.[180] Calorie restriction raises these molecules less acutely but

over a slightly longer period of time. Life-long calorie restriction in rats induces heat-shock proteins by 150% compared to age-matched ad libitum fed animals.[181] The challenge is to convince free-living humans into eating 30% fewer calories for the rest of their life, which is why intermittent fasting sounds more appealing.

- **One human study on 3 weeks of alternate day fasting discovered an increase in SIRT1**, which is associated with longevity. This increases NAD+ and reduces NADH levels. In rats, alternate-day fasting improved resistance to neurotoxins that are associated with epilepsy and Huntington's disease, whereas normally fed ones succumbed to the toxins[182]. The ADF animals also lived longer.
- **Fasting puts you into ketosis**. Ketone bodies work like signaling molecules that have anti-inflammatory effects by reducing oxidative stress and inhibiting histone deacetylases (HDACs)[183,184]. Beta-hydroxybutyrate blocks NLRP3 inflammasome-mediated inflammatory disease[185]. When in ketosis or any other form of energy stress like exercise or fasting, you activate AMPK. With low ATP, AMPK signals the creation of NAD through NAMPT, which is one of the rate-limiting enzymes in the NAD salvage pathway[186]. AMPK also enables the activation of autophagy by suppressing mTOR, which is an anabolic pathway of growth that inhibits autophagy. When fasting pretty much guarantees AMPK and increased autophagy, then calorie restriction does not. You could very easily go

into ketosis with a low carb form of calorie restriction but not if you're eating in a way that keeps your insulin and blood sugar elevated all the time.

- **Suppression of mTOR and IGF-1** – Two anabolic signaling pathways mTOR and IGF-1 promote muscle growth, cell proliferation, and immunity. Unfortunately, they may also increase the likelihood of cancers and accelerated aging[187]. Eating low carb keto does lower mTOR and IGF-1 but a complete suppression would require fasting and avoiding all calories.
- **Fasting triggers stem cell regeneration and resets the immune system**, which replaces old cells with new ones. According to an MIT study, a 24-hour fast can reverse age-related loss of stem cell function[188]. This requires the activation of transcription factors called PPARs that switch the body from burning glucose to fatty acids. Glucose restriction also promotes the proliferation of stem cells by raising NAD.[189]
- **Intermittent fasting promotes neurogenesis and growth of new brain cells** by increasing BDNF[190]. Autophagy and ketosis also have neuroprotective effects that prevent neurodegeneration. Short-term calorie restriction enhances hippocampal neurogenesis and improved remote contextual fear memory in mice.[191] Basically, if animals were exposed to a stressful situation during hunger they remembered it better. The researchers speculated that this would increase

the likelihood of survival and refeeding successfully in the future.

- **Less time spent digesting food,** which preserves more of your digestive enzymes and gut health. The body has a certain amount of deposited enzymes that are used to digest food. If you're eating all the time then you'll deplete that reserve faster compared to having some periods of intermittent fasting during which the body can recover. Consequently, you can spend more time doing other things you enjoy.

A 2019 review in The New England Journal of Medicine said that studies in animals and humans have shown that many of the health benefits of intermittent fasting are not simply the result of reduced free-radical production or weight loss[192,193,194]. Instead, IF activates defense systems against oxidative stress and those that repair damaged molecules such as autophagy and sirtuins.

Fasting works as an intermittent stressor. You can't fast forever and you're going to have to eat at some point. During the fasted state you'll experience the activation of autophagy, sirtuins, AMPK, and ketosis much more profoundly than while restricting calories because you're not eating anything. At the same time, you can still feed yourself enough calories to maintain proper hormonal functioning, muscle mass, and nutrient status without becoming frail. Chronic calorie restriction will inevitably lead to a loss of healthy tissue. Of course, you can over-do fasting but the down-side is much smaller than with indefinite reduced caloric intake. Furthermore, with intermittent fasting, you can still build muscle and have more energy despite skipping some meals whereas that's much harder to pull off with chronic calorie restriction. The most important part is that because you fast intermittently, your body won't adapt to it the same way it does to calorie restriction. It's an intermittent stressor that keeps you adapting.

If someone were to tell you that in order to live longer you have to be eating about a third less food for the rest of your life then most people would not want to hear about it. It appears to be true in other species and some evidence hints it might be with humans as well but it's not the only way. Like I've said already, fasting turns on the same pathways as calorie restriction does and even more effectively. The autophagy you get from a 3 day fast is nowhere near as potent as eating a bit of less food for months upon end. Furthermore, you can even be in calorie restriction without activating autophagy by eating too frequently and consuming too many carbohydrates. Insulin inhibits autophagy by raising mTOR and suppressing AMPK even when you're restricting calories.[195] You can also deplete your NAD levels in calorie restriction because of being inflamed or having chronically high blood sugar. Diabetes and high blood sugar offset the redox balance of NADH/NAD+ by overproducing NADH and thus lowering NAD+. Such imbalances if maintained for too long can lead to oxidative damage and contribute to the development of diabetes[196].

If your goal is longevity, then focusing on just calorie restriction instead of intermittent fasting is not a wise choice. You'd have to commit to a lifelong reduction of calories whereas with fasting you do it intermittently yet more effectively.

Fasting VS Starvation

Fasting does not equal starvation because your body is in a distinct metabolic state. Being fasted and fed is quite binary – even small amounts of food will shift you into a fed condition. Fasting isn't entirely the same as caloric restriction either because you can be consuming fewer calories but still not enter into a fasted state.

During World War II, they conducted a study called the Minnesota Starvation Experiment[197] on a group of lean men who reduced their calories by 45% for 6 months[198]. Their diet consisted of primarily carbohydrates, comprising 77% of total calories, and had very little protein to mimic starvation conditions. They ate potatoes, cabbage, macaroni, whole wheat bread while still maintaining their active lifestyle. After the experiment, the men showed a 21% reduction in strength, decline in energy, and vitality. One of them started having dreams about cannibalism. Yikes! This is critical. Very little protein and low intake of essential nutrients promote low satiety as well as catabolism of muscle. Prioritizing nutrient density and adequate protein makes a reduced calorie intake more tolerable.

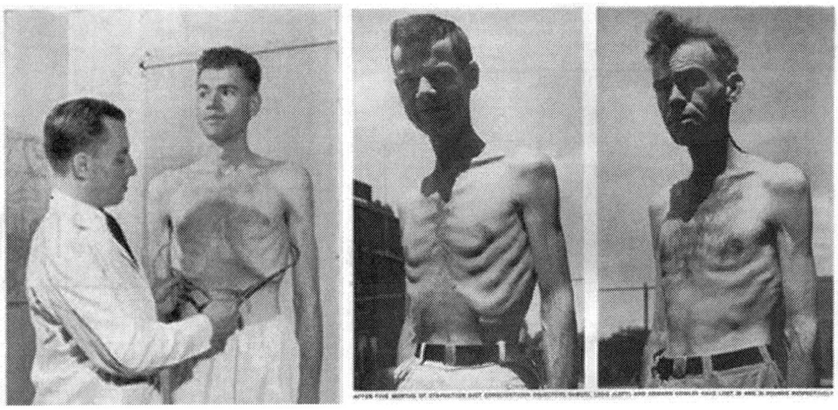

The men in the Minnesota experiment were put under severe caloric restriction that resulted in starvation-like symptoms. Even though they were very malnourished, they weren't fasting because of still eating a significant amount of carbohydrates that kicked them out of ketosis. Their bodies started using the glucose they got from cannibalizing their muscle tissue and organs.

Daily caloric restriction decreases metabolism, so it's easy to presume that this would be magnified as food intake drops to zero. However, this is wrong. Once your food intake stops completely (you start to fast), the body shifts into using stored fat for fuel (ketosis). The hormonal adaptations of fasting will not occur by only lowering your caloric intake. In the case of being fasted, your physiology is under completely different conditions, which is unachievable by regular eating. Malnourishment happens when there is not enough nutrition to be found *i.e.* you go on a weight loss diet and restrict calories. While fasting, the organism is rarely fully deprived of essential nutrients, unless you lose all your body fat. These fuel sources are mobilized from internal resources.

That is why intermittent fasting is a lot better than caloric restriction. If you are feeding yourself, but in inadequate amounts, then your body will most definitely perceive it as a major threat to survival. This will decrease your metabolic rate and creates a new set-point at which you're able to lose weight but it's lower than previously. You'll be causing more damage than good. If you do it the wrong way, you will end up like someone from the concentration camps.

Types of Intermittent Fasting

Fasting has been practiced by virtually all societies and groups of people across history. It's said to have medicinal, cognitive as well as spiritual benefits but in modern society, it's used primarily as a fat loss tool.

Intermittent fasting is a way of eating where you confine your eating in a certain time frame and fast the rest of the day. It's not a specific diet but more of like a habit of food consumption. The idea is to skip meals and have only 1 or 2 of them per day.

People use many different terms and definitions for fasting. There's intermittent fasting, alternate-day fasting, extended fasting, time-restricted eating, prolonged fasting, fasting-mimicking, or just skipping meals. Some of them apply for longer periods than others and it depends on how the person thinks of it. Here are the differences.

- **Intermittent fasting (IF)** is more like the general pattern of eating and not eating, which can apply to daily time-restricted eating, one meal a day, 16/8, alternate-day fasting as well as extended fasting of 3-5 days. Fasting one day of the week as well as doing it every day both are intermittent fasting. The difference is only in their degree. Intermittent fasting is just describing any form of confining your food intake in a short period of time, accompanied by longer times spent fasting.

- **16/8 IF** - You fast for 16 hours and eat over the course of 8. Most people just prefer to skip breakfast and eat lunch and dinner. In my opinion, this should be the minimum fasting length for everyone. You can also do 14/10, 18/6, or 20/4. The idea is to just reduce the amount of time spent in a fed state.
- **One Meal a Day (OMAD)** – It's as simple as it sounds like – eat once a day. Usually, you fast for about 22 hours and eat within a 1-2-hour timeframe. This is great for weight loss because you'll be quite full and thus can effortlessly stay at a caloric deficit. If you do it with proper keto-adaptation, then you'll also preserve more muscle.

- **Time-Restricted Eating (TRE)** refers to eating within a certain time frame in the 24 hour period. It doesn't apply to anything beyond 24 hours. In truth, anyone doing one meal a day, 16/8, or the Warrior Diet is actually doing TRF but it can be still categorized as a sub-group for IF. The difference between them is just a matter of degree.

- **Prolonged or extended fasting (EF)** applies to anything that exceeds 24 hours and usually lasts from 48 hours to 3-5 days and beyond. It can be thought of as a form of intermittent fasting because you're doing it intermittently but at the same time, it's different from the regular time-restricted eating.
- **Alternate day fasting (ADF)** is a form of fasting that's most commonly used in research. You eat normally for one day, the next day you eat either fast completely or eat like 500 calories, and repeat this cycle. It can be thought of as a form of intermittent fasting that's spread out across weeks but the physiological effects are slightly different from time-restricted eating and extended fasting. Unfortunately, most researchers and journalists don't differentiate the differences between these methods.

In my opinion, anyone doing 16/8 fasting, one meal a day or whatever variation of it is actually practicing time-restricted eating, which is just confining your daily eating window within certain hours. It's not as potent of a stimulus for fasting but you still experience some of the benefits. The main one being increased autophagy.

Autophagy and Cellular Turnover

Autophagy is an important intracellular process that eliminates dysfunctional cell components, pathogens, inflammation, senescent cells, and general junk. Inadequate autophagy is said to be a major determinant of aging. Reduced cellular recycling and accumulation of waste inside the cells are common to all aged cells[199].

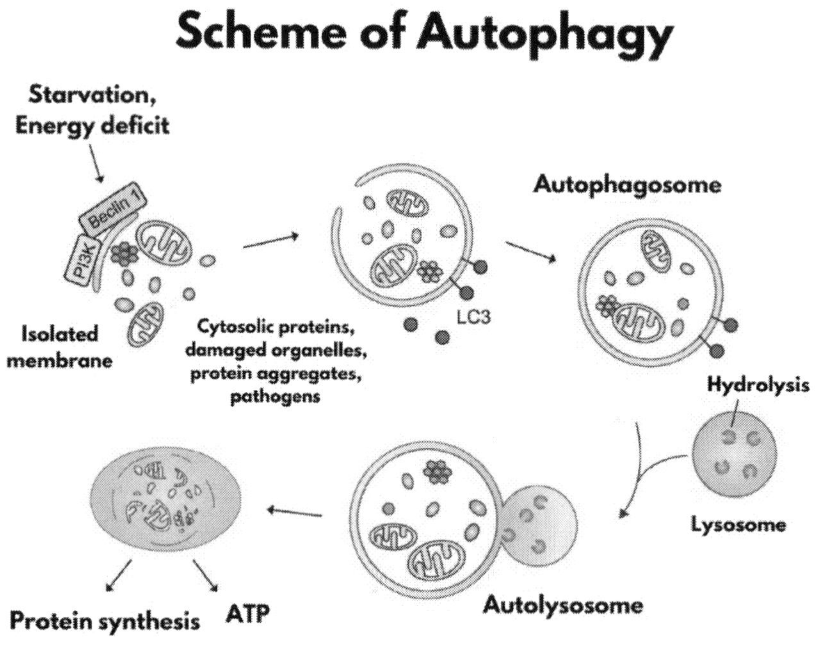

Autophagy has a protective role against many metabolic and age-related diseases such as insulin resistance[200], heart disease[201,202], atherosclerosis[203], inflammation[204], Crohn's[205], bacterial infections[206], neurodegeneration[207], gut health[208], fatty liver[209], and aging in general. Autophagy is the central component to lifeextension seen in caloric restriction[210]. Mice and yeast that are defective in autophagy don't live longer despite consuming fewer calories whereas sufficient autophagy allows these effects to take place[211].

Both fasting and calorie restriction are potent ways of activating autophagy[212]. The difference between them is a matter of degree. Higher energy stress and nutrient deprivation ramp up autophagy almost in a dose-specific manner. Exercise and heat saunas are another top method for this[213,214].

Dark Side of Autophagy

However, although good in many aspects, autophagy also has quite a few negative side-effects. They're not guaranteed nor always detrimental but still possibly regulated by autophagy. That's why we have to understand and interpret both sides of the story.

- **Removal of pathogens by autophagy is called xenophagy[215], which has many immune strengthening benefits.** Bacteria like *Streptococcus pyogenes*[216], or pathogens such as *M. tuberculosis*[217], *Salmonella*[218], and *Listeria monocytogenes*[219] can be eliminated by autophagy. This is good for dealing with infections and bacterial invasions. Autophagy can protect host cells against toxic products generated by pathogens, such as *Vibrio cholerae* cytolysin[220], *Bacillus anthracis* lethal toxin[221], and *Helicobacter pylori* vacuolating toxin[222]. However, some bacteria like *Brucella* use autophagy to replicate themselves[223]. Viruses that escape or block autophagy include herpesvirus[224], HIV-1[225], Human cytomegalovirus[226], and Coxsackievirus B3, B4[227]. In these instances, autophagy is not beneficial but maladaptive.

- **The essential autophagy gene ATG6/BECN1 encoding the Beclin1 protein has been found to suppress tumors in cancer**. However, it's not been found to be that big of a tumor-suppressor as previously thought and sometimes it can even promote cancer due to the self-replicative process[228]. Self-eating can enhance tumor cell fitness against environmental stressors[229], which makes them more resilient against starvation and chemotherapy. It may be that autophagy is better for cancer prevention rather than treatment.
- **It's not clear whether autophagy prevents or promotes apoptosis or programmed cell death[230]**. The outcome turns out to depend on the stimulus and cell type[231]. Blocking autophagy enhances the pro-apoptotic effect of bufalin on human gastric cancer cells, which is a Chinese medical toxin used for tumor suppression[232], through endoplasmic reticulum stress[233]. In this example, less autophagy led to more cancer cell death because the cancer cells were weaker whereas with autophagy they became stronger.
- **Hypoxia-induced autophagy promotes tumor cell survival and adaptation to antiangiogenic treatment[234].** Again, it's not autophagy directly sustaining cancer or tumors. It's the cancerous cells adapting to hypoxia or

states of low oxygen and stress, thus using autophagy to survive. Hypoxia-Inducible Factor (HIF) is a major factor in cell survival to hypoxia that induces autophagy[235]. HIF gets activated when oxygen tension decreases[236], which then increases autophagy activation. Hypoxia increases blood vessel growth that supports angiogenesis[237]. This allows your cells to become more oxygenated because they build a better network of blood vessels. Unfortunately, it can also promote tumor expansion, metastasis, and drug resistance by supplying cancer cells with more oxygen. Recent findings also show that sustaining cardiac function upon hypoxia/reoxygenation (H/R) injury relies on autophagy inhibition thanks to Akt/mTOR signaling[238]. Furthermore, mTOR has a cardioprotective role after (H/R) injury by inhibiting autopahgy[239]. In this example, blocking autophagy helped with functioning of the heart.

Essentially, when malignant tumor cells are put under nutritional stress through calorie restriction or fasting, autophagy may prevent them from dying by inhibiting apoptosis[240]. So, it's not all black and white with autophagy – some viruses and pathogens are eliminated by it whereas others hijack its mechanisms and replicate themselves. It's a double-edged sword that can work both ways.

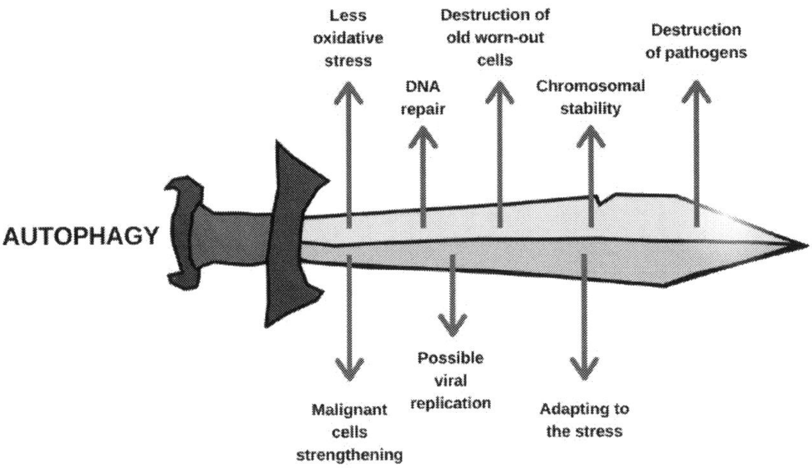

What's more, autophagy has differences in various tissues[241], such as the brain, liver, muscle, and fat. Sometimes it's good, sometimes it's not. You want the beneficial autophagy in the liver and brain to clear out plaque but you don't want to self-eat your lean tissue and muscles. That's what we'll be aiming for in this book – optimal autophagy in the right places in the right amounts.

Thyroid Functioning and Metabolic Rate

It's known that chronic stress and constantly elevated cortisol will shorten telomeres and accelerate aging. Greater stress responses to acute stressors are associated with shorter telomeres as well as higher overnight urinary cortisol and flatter daytime cortisol levels.[242] Healthy lifestyle behavior can alleviate this effect.[243]

Stress accelerates immunological aging and psychological stress can affect the body similar to getting older.[244] The elderly are less able to eliminate cortisol in response to stress. As you get older, your immunity tends to decrease because of losing bodily functionality and metabolic flexibility. This phenomenon is called immunosenescence.[245]

Being constantly stressed out and deprived of energy keeps the body in stagnation and frozen resistance. It's just so pre-occupied with trying to survive and resist the stress that it shuts down all other growth processes, including cell growth, rebuilding, creativity, and openness. This survival mode is what enables organisms to endure and come out of stressful situations alive but it's not the best place to be in for the rest of your life. It should be intermittent.

Energy homeostasis and metabolic rate are regulated by the thyroid located in your throat. Thyroid cells absorb iodine found in some foods and combine it with the amino acid tyrosine, which is used to create thyroid hormones: thyroxine (T4) and triiodothyronine (T3). They are then released into the bloodstream to affect your body temperature, daily caloric needs, heart rate, and metabolic rate.

T3 and T4 are regulated by the hypothalamus and pituitary gland inside the brain. When thyroid hormones decrease, the pituitary gland releases thyroid-stimulating hormone (TSH) to signal the thyroid gland to produce more T3 and T4. With high levels of T3 and T4 in the blood, the pituitary gland releases less TSH so the thyroid gland could slow down. Most of the thyroid hormones comprise of T4 – nearly 80%. T3 is the active form of T4 and it's thought to be more potent in regulating energy metabolism. It exerts its effects through the thyroid hormone receptor (TR). There are also T0, T1, and T2, which act as hormone precursors and byproducts. They all maintain the optimal balance of fuel substrates and oxidation.

The medical condition that describes low thyroid functioning is called hypothyroidism. Hypothyroidism is caused by too low amounts of thyroid hormones in the blood and it usually happens in people who have had Hashimoto's disease, thyroiditis, or have had their thyroid removed. Symptoms include fatigue, weight gain, cognitive decline, intolerance to cold, low body temperature, joint pain, inflammation, constipation, hair loss, dry skin, puffy face, water retention, high cholesterol, low resting heart rate, and depression[246,247,248,249]. That sounds horrible and it may become a huge limiting factor to how good you feel on a daily basis.

Hypothyroidism is diagnosed by looking at the amount of TSH and thyroid hormones in the blood. The pituitary gland is trying to make your thyroid gland produce more thyroid hormones. If your TSH is elevated and T4 or T3 are low, then it indicates hypothyroidism. With both TSH and T4 being in the normal range, the thyroid is functioning normally. Hashimoto's thyroiditis is diagnosed with an elevated TSH between 5-10 mIU/L and normal T4[250]. One quick and easy test you can do at home is to measure your body temperature. If it's below normal despite being in a warm room and you have cold hands or feet, then it shows your body is using less energy to generate heat. Maintaining higher thermal output is less important than surviving.

Low thyroid functioning is quite predominant in modern society. In fact, the average body temperature of people in the US has been gradually declining by 0.03°C per decade since the 1860s[251]. It's nowadays 36.6°C, not 37°C. Lower body temperatures are associated with decreased metabolic rates and obesity[252]. Part of it has to do with nutrient deficiencies and environmental toxins but a generally stressed-out lifestyle is probably the biggest factor. That's why the protocols described in this book aim to give you the optimal hormetic dose for various beneficial eustressors without necessarily depriving the body for too long. We shouldn't want to live in constant deprivation and metabolic winter because it hinders self-actualization.

Stress decreases thyroid functioning by creating adrenal insufficiency[253]. Inflammatory cytokines that are released during a stressful response like IL-6, IL-1 beta, and TNF-alpha decrease the conversion of T4 into T3[254]. In healthy subjects, increasing levels of IL-6 lowers serum T3[255]. Injecting these cytokines into healthy people reduces serum T3 and TSH levels[256]. So, being chronically stressed out, whether that be because of prolonged calorie restriction, too much fasting or working too much, suppresses metabolic rate and predisposes you to a host of metabolic dysfunction. Although a slightly suppressed metabolic rate could promote longevity by reducing the oxidative damage on the mitochondria, it can potentially accelerate aging through other means. For example, if your thyroid is working sub-optimally, you're at a higher chance of gaining weight and getting some sort of nutritional disease just because the body's set point for weight management is lower. That's why there's a fine balance between having a too fast or too slow metabolism.

To prevent low thyroid from fasting or any form of dieting, you want to make sure you get enough calories during your eating window. A slow metabolism isn't the result of time-restricted feeding *per se* but more by not giving the body adequate nutrients. Iodine, tyrosine, protein, and carbohydrates can promote thyroid functioning and ramp up the metabolism again. If you already suffer from low thyroid and experience the symptoms of hypothyroidism, then it would be a good idea to stop longer fasting for a while to allow your body to recover. Instead of going for another 2-3 day fast on a broken metabolism, you would want to stick to just daily time-restricted eating and eat within 10-12 hours. Anywhere between 14-16-hour fasts are still fine and they're not inherently harmful as long as you don't combine it with a bunch of fasted exercise and stimulants.

To cope with stress, you need energy from both calories as well as carbon dioxide (CO2). The rate-limiting factor is usually tissue oxygenation as a lot of people have low thyroid and poor breathing patterns. In the early 1900s, the Danish physician Christian Bohr found that carbon dioxide helps to separate oxygen from hemoglobin in the blood, thus allowing tissues, cells, and organs to absorb oxygen better[257]. It's called the Bohr Effect. Hyperventilation and breathing too rapidly can reduce CO2 in the blood, which promotes anxiety, insomnia, panic attacks, and stress[258]. That is why you see people on airplanes or while under a panic attack breathe into a paper bag – they're trying to breathe in their own CO2 as to calm down and not lose CO2 through excessive exhalation.

Carbon dioxide in the blood will join with water to create carbonic acid. This lowers the blood pH and the nervous system responds by increasing your breathing rate, called “respiratory compensation”. During exhalation, CO2 gets exhaled and pH normalizes. With higher respiration, the metabolic rate stays elevated. You’re basically breathing out calories and that’s directly linked with the amount of CO2 production and exhalation. It maintains a higher metabolic rate and thyroid functioning. Hypothyroidism results in lower carbon dioxide production[259] as you start to breathe less often. To a certain extent that can be beneficial because breathing too fast will also produce more free radicals and damage to the mitochondria but not in excess.

Thyroid hormones are important for producing CO2 and preventing the accumulation of lactic acid. Lactic acidosis is a medical condition caused by decreased tissue oxygenation (no Bohr Effect)[260]. It can cause inflammation, oxidative stress, and growth of malignant cells also known as The Warburg Effect[261]. In the 1920s, Otto Warburg postulated that cancer happens because of metabolic changes towards aerobic glycolysis – burning sugar at rest. You’re supposed to be burning fat when not under high stress or exercise but during lactic acidosis you stay in anaerobic metabolism and are producing more lactate from the use of glucose. T3 maintains proper cellular respiration and ensures the complete breakdown of carbohydrates that creates CO2 instead of lactate.

Naked mole rats or sand puppies are burrowing rodents that live in East Africa. Average mice and rats live about 24-30 months whereas naked mole rats live for decades and up to 30 years.

They're also resistant to cancer[262] and maintain healthy vascular function in later life[263]. One of the reasons they live so long has to do with how they inhabit underground tunnels and are adapted to limited oxygen. They're breathing a lot of their own CO2, which makes their blood hemoglobin to have a high affinity for oxygen[264]. Naked mole rats can even survive up to 18 minutes of zero oxygen with no damage[265]. Living in higher concentrations of CO2 has also led to adaptations that have made them insensitive to pain from acid and capsaicin[266]. This may decrease stress-induced aging from pain. Naked mole rats have a metabolic rate of about 70% of a mouse, which makes it use less oxygen and calories[267]. This can work in an environment where food is scarce but in the modern world it can become maladaptive and harmful to have a low thyroid and slow metabolism.

The thyroid converts cholesterol into steroid hormones like testosterone, vitamin D, DHEA, and progesterone. These hormones have many anabolic properties and beneficial effects on the body, such as improved muscle growth, faster metabolism, stronger bones, fertility, etc. Stress and excess cortisol lower these hormones and decrease thyroid functioning.

Cholesterol is the main building block of all steroid hormones. It helps to make pregnenolone, which then gets converted into either progesterone or DHEA. DHEA is the precursor to other sex hormones. Progesterone promotes other steroid hormones and regulates the stress response. When under high stress, cortisol will steal most of the pregnenolone from DHEA and progesterone. Both pregnenolone and progesterone are endogenous steroid hormones that act as precursors to the other steroid hormones[268].

They're also neurosteroids with neuroprotective properties against stress[269].

You need pregnenolone to make stress hormones and thus cortisol takes up resources from other hormones. It's called the 'pregnenolone steal'. So, when you have high cortisol progesterone as well as the other sex hormones get put on the back burner. The body can't optimize fertility and sexual function in the presence of stress. This will eventually lead to hormonal imbalances and metabolic issues.

It's a very common scenario wherein people with low thyroid see elevated cholesterol and low testosterone because there aren't enough thyroid hormones to complete the conversion and cholesterol stays elevated in the blood. Studies even find that high TSH, indicating downregulated thyroid hormones, can directly raise cholesterol[270]. That's not an ideal situation because you may experience a higher risk of atherosclerosis and other ailments. Hyperthyroidism or overactive thyroid does the opposite and can drop cholesterol too low, creating hormonal imbalances[271].

With a high thyroid function and metabolism, your body's reserve for adaptive energy is also higher thanks to an increase in these steroid hormones. They essentially signal the organism that everything is fine and there's nothing serious to worry about. In low amounts, they signal that adaptive energy is reaching its limit and you're reaching allostatic overload. With an increase in chronic stress, these hormones and overall thyroid functioning decrease, thus reducing the person's stress resilience and adaptive energy. We now know what happened to the mice whose AE was constantly being drained without allowing it to recover.

How to Avoid Metabolic Adaptation

Your metabolic rate is the sum total of your daily energy expenditure (TDEE), basal metabolic rate (BMR), and non-resting energy expenditure (NREE), which is divided into exercise activity thermogenesis (EAT), non-exercise activity thermogenesis (NEAT); and the thermic effect of food (TEF)[272].

Metabolic adaptation is often used to describe changes in the metabolism during dieting or caloric restriction. Eating a certain amount of calories or in a specific way will lead to the body adapting to it by either down-regulating energy expenditure or in other cases speeding things up.

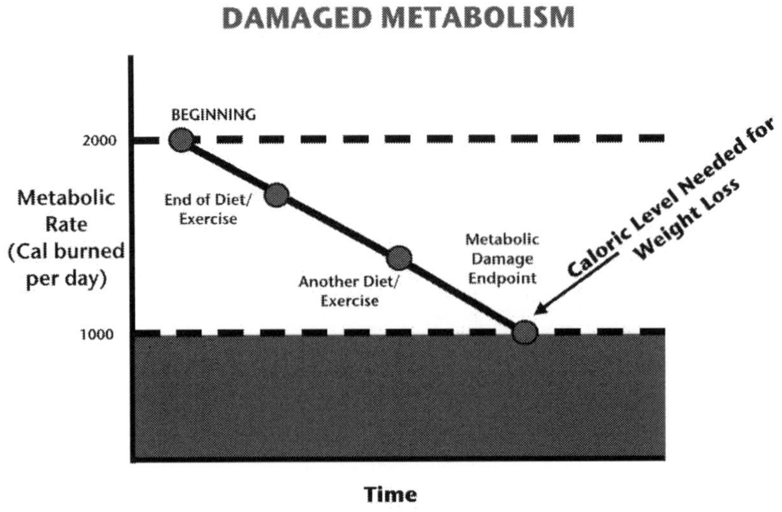

The body adapts to a lower energy intake by increasing mitochondrial efficiency and decreasing metabolic rate[273,274,275]. You'll start to burn fewer calories to fuel the same level of physical activity. This is often accompanied by a lower rate of non-exercise activity thermogenesis (NEAT), which includes spontaneous low-intensity physical movement like fidgeting, moving around, and general motion[276]. Caloric restriction will also increase your desire to eat by promoting hormones, such as ghrelin, and decreasing the satiety hormone leptin[277]. A „slow metabolism" is simply an adaptation to low energy intake, which makes you subconsciously want to move less and eat more.

It's inevitable that to lose weight you're going to have to induce some sort of an energy deficit or metabolic change. Most people have been on some sort of a diet for at least a short period of time. Unfortunately, metabolic adaptations can stick around even after you stop dieting, making it continuously more difficult to maintain the weight loss[278].

The best way to avoid metabolic adaptation, in my opinion, is to practice intermittent fasting because you'll be cycling between periods of nutrient scarcity and excess. Fasting for too long too often will eventually make you more efficient at burning calories i.e. slow metabolism but if you follow it up with strategic refeeds you can avoid this trap. It's also important to not stay in chronic calorie restriction for too long and cycle between periods of refeeding.

Studies show how often you eat doesn't significantly change how many calories you spend on digestion[279]. Eating 6 small meals is not going to stoke your metabolism because eating 2 large meals will have the same thermic effect of food.

The thermic effect of food (TEF), or how much of the food you ate gets burnt off as calories, depends on not the timing but more on the macronutrient ratios of the meal. Protein has the highest thermic effect.

- TEF of protein is 25-30%
- Carbs 6-8%
- Fat 2-3%[280]

If your daily caloric intake is 2500, then you'll still have a TEF of 250 no matter how many meals you have. 2500 calories in total.

- 6 meals of 420 calories each= 42 calories burned per meal (42×6=252 daily TEF).
- 4 meals of 625 calories each= 62 calories burned per meal (62×4=248 daily TEF).
- 2 meals of 1250 calories each= 125 calories burned per meal (125×2=250 daily TEF).

No matter what meal frequency you choose, you'll still be burning the same amount of calories from TEF. To burn more calories during digestion, you would have to eat a higher protein diet and less fat and carbs because protein has the highest TEF and tends to be more satiating. So, how many calories you burn with intermittent fasting also depends on what kind of foods you eat during your eating window.

Here's a chart of the thermic effect of different foods:

Energy Source	TEF	Calories Per Gram	Calories Stored Per 100 Grams
Alcohol	15%	7	85
Exogenous Ketones	3%	4	97
Protein	25-30%	4	70-75
Carbohydrates	7-10%	4	90
Glycogen Spillover	15-20%	4	80
Fat	3%	9	97

Higher protein diets tend to be more inducive of fat loss and improved body composition because they make you waste more calories on digestion as well as fill you up much faster. Increased protein intake also promotes muscle growth, thus raising your metabolic rate. A double whammy victory!

Mitohormesis Longevity Protocol

- Aging is caused by the loss of epigenetic information that would tell the organism to be young and healthy. This loss is caused primarily by oxidative stress, inflammation, poor lifestyle habits, environmental toxins, decreased autophagy, and metabolic diseases. By keeping your body's longevity genes activated you can effectively increase vitality.

- Life-extension in virtually all species is governed by key nutrient-sensing pathways that converge to the phenomenon of mitohormesis and stress adaptation. Organisms that adapt to certain stressors stimulate improved mitochondrial efficiency and reduce free radical damage. This preserves the functioning of mitochondria and lowers oxidative stress.
- Autophagy is the central component of the lifespan extension benefits seen in calorie restriction thanks to recycling damaged cell particles and inflammasomes. This prevents mitochondrial aging, DNA damage, and keeps the cellular machinery working more optimally.
- Intermittent fasting mimics the effects of calorie restriction by increasing autophagy and other longevity pathways like sirtuins and AMPK. It promotes mitochondrial energy efficiency, which lowers oxidative stress and free radical production.
- Hormetic stressors improve mitochondrial energy efficiency. That can lead to a downregulation or suppression of metabolic rate, which can be beneficial for longevity. However, staying in this state chronically or in excess will lower thyroid and hormonal functioning, thus making you more susceptible to gaining weight and other nutritional ailments. That's why by cycling back and forth between stimulating mitohormesis and proper nourishment

you can achieve the optimal balance of growth and survival.

- Prolonged calorie restriction and starvation lowers thyroid functioning and slows down metabolic rate. Chronic low thyroid can predispose you to metabolic diseases, obesity, and stress-induced damage. To prevent that you should engage in hormetic practices in a dose-dependent manner.
- Doing intermittent fasting is a more sustainable and less stressful form of calorie restriction that can mitigate metabolic adaptation. By confining your daily eating window as opposed to reducing total calorie intake you can turn on the longevity pathways without slowing down your metabolism. Fasting after becoming keto-adapted also makes it less stressful and easier. Fasting on a sugar burning metabolism is inherently more catabolic and causes additional stress.
- There are several ways of doing intermittent fasting, but the main idea is to just confine the daily eating window. Instead of spending the entire day in a fed state you can eat your food within 8 hours or less and achieve the minimal effective dose for autophagy and fasting.
- If you show signs of low thyroid or you hit a plateau due to slowing down the metabolism, look at how many calories are you eating and how much exercise are you getting. It might be you need to increase your calorie intake, increase the amount of daily movement, or scale down the intensity of your exercise.

Chapter Three:
Of Keto Adaptation and Metabolic Flexibility

"The three most harmful addictions are heroin, carbohydrates, and a monthly salary."

— Nassim Nicholas Taleb

Nutritional stress is something that humans and all living organisms would experience in nature on a regular basis. Food scarcity, periods of starvation, seasonality, and the high energy expenditure required to fill your stomach are ever-present in an environment where there are no supermarkets or even vending machines. It's actually the norm for our species to experience some sort of shortage or intermittent fasting.

Fortunately, the human body is antifragile, especially our metabolism. We can burn many different fuel sources and adapt to almost any form of nutrient stress. For instance, everyone carries with them thousands and thousands of calories in their adipose tissue, which could keep you alive for weeks if not months. That's the primary function of body fat that enables bears to survive the winter while hibernating and it helped humans to travel long distances with limited food. Unfortunately, modern living rarely requires us to cultivate this ability and thus you see people who can't go any longer than a day without food. They're constantly grazing to keep their blood sugar elevated and suppress even the slightest feeling of hunger. Given that the average person already has slightly too much body fat, the problem isn't lack of energy but instead not being able to access it effectively.

An antifragile metabolism ought to be able to burn many different kinds of fuel, depending on the situation, and make rapid adjustments to changes in food availability. Our body can use many energy sources, such as glucose, which gets stored as glycogen; lactate, which is the by-product of anaerobic metabolism; fatty acids that get converted into ketones, and others. By default, the body and brain burn sugar for energy because it's quickly absorbed and prevents the oxidation of fat, which is supposed to be preserved for food shortages. However, doing only that is extremely fragile and subject to failure whenever your glucose stores run out. With enough metabolic flexibility, you ought to just switch over into burning fat and ketones after sugar gets depleted. There shouldn't be this massive energy crisis that forces you to run to the nearest vending machine to grab a candy bar or carry around several Tupperware worth of food.

Therefore, the fundamental pillar stone that holds up any nutritional antifragility is keto-adaptation and a strong fat-burning engine. It's what helped humans adapt in the past and would be very beneficial in the modern world as well. If your body doesn't know how to efficiently use its own body fat, then you'll always be dependent on food for energy and maintaining lean muscle. When in ketosis, these problems become almost non-present or at least greatly reduced because you're tapping into that reserve of tens of thousands of calories. Furthermore, ketosis is also greatly beneficial for the mitochondria and everything related to mitohormesis as discussed in the previous chapter.

The metabolic state of ketosis is characterized by elevated ketone bodies in the blood. Ketones are by-products of fat oxidation that can be created by either burning body fat or eating low carb high fat. You produce them while fasting or when restricting carbs. Ketone bodies have some pretty unique benefits:

- Reduced inflammation and inhibition of inflammatory markers like NF-kB, TNF-alpha, and COX-2[281]
- Neuroprotection against seizures, Alzheimer's and Parkinson's[282]
- Protection against oxidative stress[283]
- Blood sugar and energy stabilization[284]
- Improved mitochondrial function[285]
- Activation of the Nrf2 antioxidant system and glutathione[286, 287]
- Suppression of appetite and cravings[288]
- Decreased growth and proliferation of cancer cells[289]
- Inhibition of HDACs, which are enzymes that are associated with cancer, aging, and oxidative stress[290]
- Elevation of sirtuins and NAD, which are associated with longevity[291]

Ketone bodies are more than a fuel source – they're also signaling molecules that affect your physiology on an epigenetic level. They turn on different pathways and genes that have a beneficial effect on the body regardless if they're coming from fasting or eating keto. On a metabolic level, they're pretty much the same.

Humans evolved to be in ketosis virtually most times of the year – they would frequently fast, miss out on meals, experience famine, and got to eat carbs only during certain times of the year. Even babies are born in ketosis and they get ketones from their mother's breast milk[292]. Imagine how many infants would've died to starvation in the past if they missed a meal. As crazy as it might sound, people didn't have fresh fruit or granola with them year-round. Being able to burn one's body fat is an essential part of survival, especially for foragers and their offspring.

Ketogenic diets were initially created to treat epilepsy in children[293]. Doctors noticed that their patients who fasted didn't experience any seizures and their symptoms improved. This happens because ketones start to replace glucose in the brain and provide sufficient energy to all cells of the body. The problem is you can't fast indefinitely and need to eat at some point. In the 1920s, a physician at the Mayo Clinic named Dr. Russel Wilder proposed that a very low carb, low protein, high fat diet called the ketogenic diet can produce similar results to fasting[294]. He demonstrated its effectiveness on many of his patients. Nowadays it's used by many epileptic people to control their seizures.

Running out of glycogen without being keto-adapted causes stress to the body, which can result in muscle catabolism, low thyroid, and all the other symptoms related to allostatic overload. Keto-adaptation turns your body antifragile in terms of needing energy from food – you become a self-sustaining powerplant. This also raises the threshold at which you begin to experience the negative side-effects of energy shortage because your mitochondria are working more energy efficiently. It promotes mild mitohormesis and even improved mitophagy to a certain extent.

Ketosis promotes brain macroautophagy by activating Sirt1[295]. Ketone bodies also stimulate chaperone-mediated autophagy (CMA), which targets only specific amino acids and substrates[296]. Beta-hydroxybutyrate and other ketones tend to be high during fasting and starvation but they're also elevated when eating the ketogenic diet. A ketogenic diet can decrease neuronal injury via autophagy and mitochondrial pathways in seizures[297]. It mimics many aspects of fasting.

Autophagy is essential for the synthesis of ketone bodies[298]. Autophagy deficient mice have decreased production of ketones by the liver. Gluconeogenesis isn't affected by this, which makes autophagy an important part of shifting into ketosis and becoming keto-adapted. Without adequate autophagy, you'll stay more in the sugar burning state.

Autophagy doesn't require ketosis to be activated as you can fast for up to 3 days and still not be in ketosis, depending on how keto-adapted you are. However, usually, being in ketosis already meets a lot of the prerequisites of autophagy like low insulin, low blood glucose, and lower mTOR. You just have to base it on how long you've been fasting for. If you're not consuming too many carbs or protein daily, then you can expect to go into autophagy faster than someone who has to burn through those calories first. In a way, the mild increase in autophagy on a ketogenic diet can be great for getting just the right amount and not causing the collateral damage as discussed in the previous chapter which might happen if you're in too deep of autophagy.

You do get autophagy on keto but just have to make sure you're not eating too frequently, you're practicing some form of time-restricted eating, you're not eating too much protein, and you stay physically active. However, those same principles apply to other diets like a vegan diet, carnivore, paleo, or whatever else. The difference is in only how long you have to fast to see the same effects.

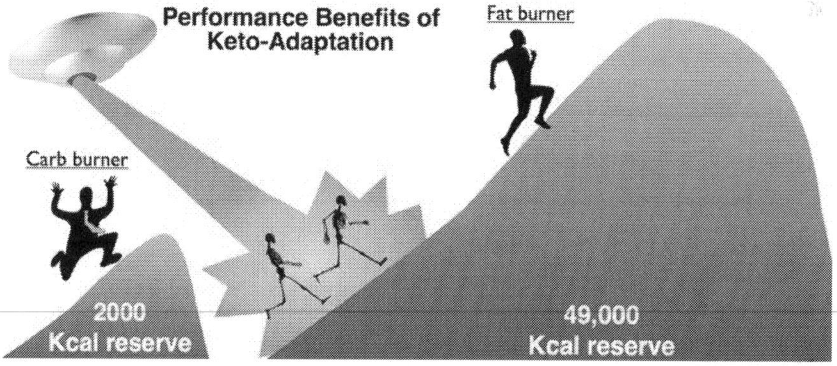

The figure above depicts the notion of burning carbs and fat quite well. There's only a limited amount of glycogen our bodies can store compared to the vast fat reserve. To access the fatty acid tank, we need to prime keto-adaptation and condition the body to run on ketones in most situations. The chasm in between can be somewhat difficult to cross, which is why some may fail completely.

Ketosis and Keto-Adaptation

Ketosis and keto-adaptation aren't mutually inclusive, and they have their differences. Here's how I define the two.

- **Being in ketosis is the actual metabolic state with the appropriate levels of blood sugar and ketone bodies**. It's said that ketosis begins at 0.5 mmol-s of blood ketones but having 0.3 mmol-s already is quite good. The average person will rarely reach even 0.1 mmol-s because they're neck-deep in sugar burning mode all the time. You can be in mild ketosis already after fasting for 24-hours but it doesn't necessarily mean you're successfully using fat and ketones for fuel.
- **The keto-adaptation process makes your body adapt to utilizing fat and ketones as a primary source of energy.** It means you don't have to rely on glucose and can thrive on consuming dietary fat or by burning your stored body fat. You can tap into your fat reserves very fast whenever liver glycogen gets depleted and won't experience negative symptoms from staying low carb.

On the standard ketogenic diet, your macronutrient ratios would be somewhere along the lines of 5-10% carbs, 15-25% protein, and 70-80% fat. In medical practice or for disease prevention, the ketogenic diet has to be kept quite strict because the purpose is to be in deep therapeutic ketosis with low blood sugar and high ketones between 1.5-3.0 mmol-s. However, people who want to simply reap the benefits of a low carb keto way of eating don't have to be that restrictive and they can safely get what they want by focusing on becoming keto-adapted.

How much fat you're able to burn and how much protein you'll compensate with depends on your level of keto-adaptation.

- Eating the high-carb-low-fat-high-protein diet is making your body quite dependent on glucose and frequent eating. The same applies to a high carb, high fructose diet. You have to eat very often to not go catabolic.
- Eating slightly lower carb, like a paleo approach where 30-50% of your calories come from carbs leaves some room for burning fat but it's still making you burn some glucose because you're eating more carbs. Shifting over to keto would take less time because the body is already used to a limited intake of carbohydrates.
- Eating a strict low carb high fat ketogenic diet is the furthest you can promote keto-adaptation with diet. It'll keep you in a state of nutritional ketosis wherein the body is geared towards using ketones as a primary fuel source.
- Eating a modified ketogenic diet with higher protein or occasional carbs maintains sufficient keto-adaptation but can also promote additional metabolic flexibility or the ability to use both fuel sources in different situations. This is the most anti-fragile and adaptable way of doing it.

Keto-adaptation results from nutritional ketosis but it's not needed to maintain it. To become keto-adapted, you have to go through a period of being in ketosis where your liver's enzymes and metabolic processes change so you could have the ability to burn fat for fuel at a higher rate. However, it's not necessary to be in ketosis all the time to maintain keto-adaptation. You can briefly dip in and out of ketosis for a day or two without fully losing your fat-burning abilities.

When ketosis is quite binary – you're either in it or you're not – then keto-adaptation is more of like a matter of degree – a wide range of efficiency. There isn't a specific point where you can draw a line and go 'now I'm completely switched over to a fat-burning engine'. Therefore, the pursuit of getting into ketosis with a low carb ketogenic diet is going to facilitate keto-adaptation, which will enable you to shift your metabolic machinery more towards producing energy from fatty acids.

The goal of the ketogenic diet for most people is not necessarily to be in ketosis but to become keto-adapted. However, in order to make it happen, you still need to go through the keto-adaptation process wherein you get into ketosis and eat a low carb ketogenic diet. Otherwise, you'll never cross the chasm between burning glucose and ketones but will stay in the periphery all the time.

Here's how the keto-adaptation process looks like:

- **Carb Withdrawal** – you go on a ketogenic diet and remove all carbohydrates from your diet. On keto, you eat leafy green vegetables, low carb tubers, fatty meat, fish, eggs, and some other fats.
- **Keto Flu Period** – you may experience some fatigue and exhaustion because the brain doesn't know how to use ketones for energy that efficiently yet. This may last from a few days up to several weeks, depending on your sensitivity.
- **Getting Used to Ketones** – you begin to feel better and more energized from eating low carb high-fat foods. The process can be accelerated by implementing intermittent fasting and making sure you're not starving yourself. This may last from 2 weeks up to several months and the longer you do it the better it gets.
- **Fat Burning Mode** – your exercise performance will improve or at least you'll regain the vigor you might have initially lost during keto flu. Here you can begin to see increased time to exhaustion, faster recovery from workouts, less fatigue during the day, mental clarity, and reduced hunger.
- **Keto Adaptation** – you can run very efficiently on dietary fat as well as your own body fat without needing carbohydrates to perform or feel energized. Thanks to burning ketones, you don't get that hungry, and whenever you do it's temporary.

- **Metabolic Flexibility** – you can also use carbohydrates for fuel and you're not going to get brain fog from being kicked out of ketosis. This is the ultimate goal of keto-adaptation – to not be dependent on ketones nor carbs and to use both in various situations. More on this shortly.

The process of becoming keto-adapted requires about 2-4 weeks or even up to 3-6 months. How long it's going to take depends on how easily your body begins to accept ketones and fatty acids as a fuel source. Metabolically healthy and insulin sensitive people may need only a week or two whereas someone who's diabetic or obese will need more time.

Some of the side-effects you might experience include losing water weight because of low levels of insulin, increased thirst, a slightly metallic and fruity keto breath, mild fatigue, and lack of appetite. Fortunately, these things can be quickly overcome and even avoided. They're not permanent and will pass away shortly. Some of the good signs of keto-adaptation include no hunger whatsoever, mental clarity, high levels of energy all the time, increased endurance, reduced inflammation, stable blood sugar, and no muscle catabolism.

To know whether or not you're in ketosis, you can measure your blood ketones using Keto-Mojo. Optimal measurements are between 0,5 and 3,0 mMol-s[299](See figure below). The same can be done with a glucometer. If your fasting blood glucose is under 80 mg/dl and you're not feeling hypoglycemic then you're probably in ketosis. Ketoacidosis occurs over 10 mMol-s, which is quite hard to reach.

BLOOD GLUCOSE CHART

Mg/DL	Fasting	After Eating	2-3 hours After Eating
Normal	80-100	170-200	120-140
Impaired Glucose	101-125	190-230	140-160
Diabetic	126+	220-300	200 plus

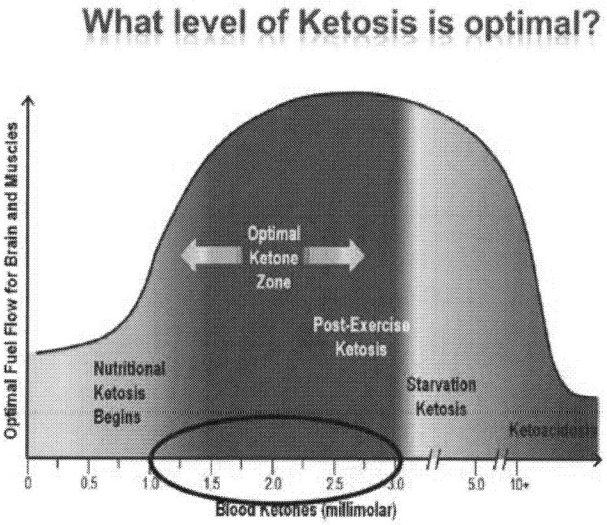

Page 91: The Art and Science of Low Carbohydrate Performance
Jeff S. Volek and Stephen D. Phinney

Low blood glucose in the context of a non-ketogenic diet generally indicates hypoglycemia, wherein you may feel tired, lethargic, and your brain can't get access to energy. However, it's not a stress response when in ketosis. If your blood glucose is lower than 60 mg/dl or 3.0 mmol/L and you're feeling energized, then it means your brain is getting an alternative fuel from beta-hydroxybutyrate and you're not going to pass out. That's why people on the ketogenic diet have much lower blood glucose than is considered normal. In my opinion, lower blood glucose is also better for your health and longevity. To a certain extent of course.

When glucose goes down, then a metabolically healthy person should see an elevation of ketones as the body shifts into a fat-burning mode. If you're not keto-adapted, then you're going to crash and feel exhausted. Tapping into ketosis increases that buffer-zone and protects the body against low blood sugar or lack of calories from food. It can preserve adaptive energy and raise the threshold at which the brain goes into a stress response during fasting or calorie restriction.

Physiological Insulin Resistance

One of the main causes of cardiovascular disease and diabetes in society is insulin resistance or hyperinsulinemia. In most cases, hyperinsulinemia is both a result and the driver of insulin resistance[300]. The two essentially describe the same thing.

- Hyperinsulinemia is a condition where there's excess circulating insulin in relation to the amount of blood glucose. It's associated with hypertension, diabetes, obesity, and metabolic syndrome[301].
- Insulin resistance is a condition where the cells don't respond normally to the elevation of insulin and are resistant to picking up glucose, thus causing high blood sugar. The beta cells in the pancreas continue producing

more insulin but to no avail. This keeps both blood sugar and insulin levels elevated for a longer time.

Symptoms of insulin resistance or glucose intolerance include uncontrollable hunger, increased thirst, high blood sugar, hypertension, brain fog, lethargy, lightheadedness, easy weight gain around the stomach, stubborn belly fat, elevated triglyceride and cholesterol levels. It's pretty much synonymous with hyperinsulinemia.

Although chronic insulin resistance is quite harmful, it also has a beneficial adaptive mechanism. Under harsh metabolic conditions, insulin resistance helps to give the brain glucose that would otherwise be taken up by muscles[302]. It's also thought to be a normal physiological response to a sustained caloric surplus and overeating to protect the body against the accumulation of lipids[303]. Animals who eat a lot more calories than needed develop rapid insulin resistance and get obese very fast[304], especially if fed a lot of grains and carbs.

However, from the perspective of longevity, it's not a good idea to have continuously elevated levels of blood sugar or insulin. First of all, it promotes fat gain and raises triglycerides, but secondly, it also keeps the body in a continuous state of dysfunctional mTOR signaling. Elevated blood sugar and fasting insulin throughout the day keep the body in a constantly anabolic state, which accelerates aging. You want to be more insulin sensitive so you'd clear your bloodstream from any excess glucose as fast as possible and get back into ketosis.

Fast food that combines a lot of salt, sugar, and fat with sweetened drinks is a perfect recipe for insulin resistance and over-eating because of the high caloric content and palatability[305]. Trans fats and oxidized vegetable oils are also responsible for this[306]. Elevated levels of fatty acids and triglycerides are associated with insulin resistance as well[307]. However, the prolonged elevation of insulin and triglycerides is most commonly caused by the combination of carbs and fats that keep the blood sugar jacked up for longer. People who eat more fat tend to also eat a lot of carbs. Someone eating fat on a low carb diet will not spike insulin as much and they've excluded one of the offending macronutrients.

A 2018 study claimed that short-term feeding on a ketogenic diet causes greater severe hepatic insulin resistance than an obesogenic high fat diet[308]. However, the study was done on mice who were fed pure glucose in a fasted state. During the fasted state, ketogenic mice were metabolically healthy whereas the high-fat junk food mice had higher levels of insulin. The glucose challenge showed that both ketogenic mice and high-fat junk food mice were glucose intolerant. Glucose uptake in muscles and other tissues remained the same. Insulin was suppressed because of the higher rates of fat oxidation that blunt insulin's ability to lower blood glucose. The researchers concluded that this physiological insulin resistance observed in the ketogenic mice is probably not linked to any damage and is reversible. These mice weren't diabetic, they just were underadapted to utilizing glucose because their body was burning fat. If the same mice were to take another glucose tolerance test later, they would show a much better ability to utilize that glucose as their body has already gotten kicked out of ketosis. An 8-week study on ketogenic mice showed that their glucose control decreased when consuming a high carbohydrate meal but those effects were quick to reverse after returning to a regular diet[309]. The same phenomenon occurs in humans.

A 2-week ketogenic diet on obese diabetic individuals resulted in a 30% reduced calorie intake, weight loss of about 1.65 kg-s, a 75% improved insulin sensitivity, and a decrease in some vital biomarkers like triglycerides and cholesterol by 10-35%[310]. Part of the effects may have come from caloric restriction as the subjects were consuming about 1000 fewer calories than normally. Another 2009 study compared 3 groups of isocaloric diets over 8 weeks. The percentage of fat loss wasn't significant between them. However, they did differ in terms of fasting insulin and postprandial blood glucose response[311]. Researchers concluded that isocaloric very low carb diets are effective in improving triacylglycerols, HDL cholesterol, and insulin resistance.

If you're not eating carbs for a while then of course your body downregulates its insulin production because there's no necessity for it. It's only a short-term adaptation induced by increased fat oxidation that favors using fat for fuel instead of carbs. You could call it insulin resistance but it's not characterized by any of the pathological symptoms like elevated fasting insulin, high blood sugar, or excess triglycerides. The more accurate term would be 'decreased glucose demand' or 'glucose sparing' because that's what's essentially happening.

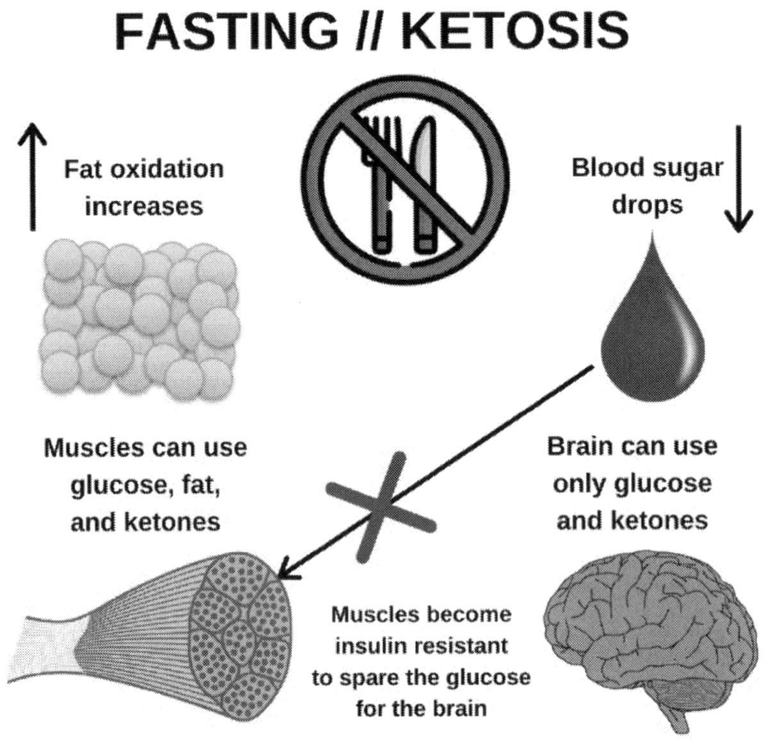
FASTING // KETOSIS
Fat oxidation increases
Blood sugar drops
Muscles can use glucose, fat, and ketones
Brain can use only glucose and ketones
Muscles become insulin resistant to spare the glucose for the brain

It's been shown that after prolonged exposure, fatty acids will eventually inhibit glucose-induced insulin secretion[312]. Fatty acids and ketones have a glucose-sparing effect in the absence of glucose, which is an essential survival mechanism for the brain during starvation[313]. This will also preserve pyruvate and lactate both of which are precursors to gluconeogenesis[314]. Thus, the demand for glucose production also decreases in a state of ketosis or while fasting. It makes sense because the body shifts into using ketones instead and taps into the vast storage of adipose tissue.

After eating a high-fat meal, plasma concentrations of fatty acids increase, rerouting the unoxidized glucose to glycogen re-synthesis. This also explains why during starvation or diabetes people's glycogen stores are still full – their body spares glucose and stores it as glycogen while burning primarily fatty acids and ketones.

Physiological insulin resistance can also be seen during exercise where the use of fatty acids increases. The glucose that is not oxidized gets rerouted to glycogen, which may explain the rapid re-synthesis of muscle glycogen post-exercise[315]. Pascoe et al (1993) showed that you can refill your glycogen stores after fasted exercise even without consuming any calories and continuing to fast[316].

Causes of Insulin Resistance:

During states of high-fat oxidation (starvation, fasting, keto, exercise), you're burning fat, which suppresses insulin production. Introducing carbohydrates as a supplement may show short-term insulin resistance but it's just the body's way of using that glucose for certain tissues like the brain and preserving fat utilization in muscles.

Insulin resistance in which your body starts to uncontrollably produce endogenous glucose is the real concern. This may happen in diabetics or with alcohol poisoning. The most common reason this happens is because of eating processed food that's high in fats and carbs.

Here are the causes of pathological insulin resistance that will lead to disease:

- **Eating fats and carbs together ala the Standard American Diet**. This was explained by Randle et al (1963). The Randle Cycle also called glucose fatty acid cycle is a metabolic process where glucose and fatty acids compete for oxidation[317]. It's thought that the Randle cycle can explain the reason for type 2 diabetes and insulin resistance[318,319]. The hypothesis is that combining fats and carbs promotes obesity and metabolic disorders because of confusing the body's priorities for burning these fuel sources. That's why the go-to obesogenic diet for lab animals is fats+carbs and it seems to work on humans as well who eat primarily junk food. I wouldn't even combine healthy carbs like potatoes with healthy fats like avocadoes or steak. Instead, to prevent metabolic dysregulation and improper fuel partitioning you'd want to eat either low carb higher fat or low fat higher carb. Things like calorie restriction and fasting can ameliorate this phenomenon to a certain extent but it's quite pathological in an energy surplus.
- **Excess visceral fat and obesity.** Abdominal visceral fat is strongly correlated with insulin resistance and type 2 diabetes[320]. It's both the cause and effect. A 2015 study from Yale found that the creation of new fat cells was governed by a key nutrient-sensing pathway called phosphoinositide 3-kinase PI3-kinase/AKT-2[321], which is a part of the mTOR/insulin/IGF-1 pathway. Insulin promotes the storage of nutrients and the growth of new cells through mTOR and AKT-2[322]. It can grow both muscle and fat

cells. Visceral fat decreases glucose tolerance and spreads inflammatory cytokines throughout the body. Chronic inflammation also promotes insulin resistance[323]. Giving insulin resistant rats anti-inflammatory omega-3 supplements alleviates their pathology[324].

- **Not enough exercise.** Physical activity promotes insulin sensitivity and increases glucose uptake by cells. It's one of the biggest determinants of how insulin sensitive you are[325]. The main cause is thought to be muscle contractions causing the glucose receptor GLUT4 to translocate to the membrane. GLUT4 can improve the uptake of glucose through a distinct mechanism than that of insulin[326]. So, even if you do experience physiological insulin resistance while fasting or when on a keto diet, you can mitigate that with exercise-mediated GLUT4. This will lower blood glucose and insulin, preventing hyperinsulinemia.

- **High amounts of cortisol.** Chronic stress and cortisol are known to raise blood sugar, blood pressure, and insulin, which over the long term will lead to insulin resistance[327]. Cortisol impairs the uptake of glucose by reducing the translocation of glucose transporters such as GLUT4[328]. It also maintains that stubborn visceral fat around the belly. Mindfulness-based stress-reducing activities like meditation and yoga have been shown to improve insulin resistance[329].

- **Sleep deprivation and insomnia.** Not enough sleep inhibits glucose tolerance, raises blood sugar and cortisol, and promotes insulin resistance. Even just a single night of insufficient sleep makes you borderline pre-diabetic in the short term[330]. After a bad night's sleep, your glucose tolerance for the next day is going to be drastically lower. If you didn't sleep enough, then you should say no to all cheat foods or donuts because your body's ability to metabolize them is severely hindered.

- **Trans fats and vegetable oils.** Things like margarine and canola oil promote oxidative stress, inflammation, and insulin resistance[331]. They're highly inflammatory, oxidized, and with no hormetic benefit. In fact, they make your future attempts to practice hormesis more harmful because while the body is burning these fats for fuel it experiences increased oxidative stress in the short-term. More on this in chapter eleven.
- **Smoking and drinking alcohol.** Smoking also induces insulin resistance and causes atherosclerosis[332]. Getting drunk and intoxicated promotes inflammation and oxidative stress the body has to spend extra energy to fix. This drains your adaptive energy for dealing with other stressors and also inhibits future adaptation.
- **High fructose consumption.** Excess fructose can be very easily converted into triglycerides that will cause a similar situation of high fats and high carbs in your bloodstream. Fructose-sweetened beverages especially are linked with insulin resistance[333]. Natural fruit is fine in moderation because of its other co-enzymes and fiber but eating it

beyond your liver's capacity can still be harmful. I wouldn't recommend eating more than a few servings of fruit per day, preferably from lower sugar sources.

Inhibiting glucose utilization by fatty acids resembles physiological insulin resistance but it's a temporary adaptation to the influx of fatty acids and lack of glucose. This often happens in a fasted state or when eating the ketogenic diet. Insulin resistance caused by fasting or keto is a temporary adaptation to preserving glucose for certain vital organs like the brain by blocking the uptake of glucose in muscles that would otherwise steal the glucose from the brain.

The general guideline is that the longer you do the ketogenic diet the easier it gets and the better you'll start performing. However, if your goal is metabolic flexibility, then you would want to deliberately kick yourself out of ketosis every once in a while to break insulin resistance and promote metabolic flexibility.

Metabolic Flexibility and Keto Adaptation

Using carbohydrates strategically will not only improve your performance but overall health as well. Here are a few reasons why you should occasionally get out of ketosis.

- **Low Thyroid Functioning** – If you can't meet the body's demands for glucose, it'll decrease thyroid functioning. When in ketosis, the requirement for glucose is reduced greatly but even then your glucose demand may increase cyclically. For example, high-intensity exercise, pregnancy, childbirth, or higher stress will make your body need more glucose. Testosterone and other steroid hormones will also drop because of that, thus decreasing the body's ability to tolerate stress. At that point, glucose can work as an antioxidant that will lower stress and reverse low thyroid. Carbohydrates are the main fuel source for T3 and leptin that both raise metabolic rate[334]. Ketogenic diets reduce T3 and raise reverse T3[335]. Your body can indeed create all the glucose it needs by converting fat or protein into glucose through gluconeogenesis but it doesn't deny the fact that straight up sugar or carbs are effective for reversing low thyroid. At least in some situations.

- **Carbs and Insulin Raise Leptin** – Leptin the satiety hormone regulates glucose homeostasis, satiety, and energy expenditure. High leptin speeds up the metabolism and makes it easier to lose weight. Low leptin results from chronic calorie restriction and low thyroid. Raising leptin by eating carbs and spiking insulin can improve insulin sensitivity and weight loss[336]. Restoring leptin has been shown to normalize blood sugar and insulin resistance[337]. Excess stress resulting from chronic ketosis can just keep the body in metabolic hibernation.
- **Some Cells of the Body Need Glucose** – After keto-adaptation, the brain can derive up to 50-70% of its energy demands with ketone bodies[338]. The remainder of energy can be covered by converting fat and protein into glucose

through gluconeogenesis. However, red blood cells and corneal lenses can't use ketones for fuel. This requirement isn't nearly as high as you'd think but it will rise in situations of increased energy demands i.e. stress and exercise.

- **Low Carb Can Reduce Gut Lining Integrity** - The main source of energy for the colon and bacteria are short-chain fatty acids (SCFAs), mainly butyrate[339]. It's anti-inflammatory and improves insulin sensitivity. You create SCFAs from the digestion of fiber but you can also get it from animal fats, albeit in smaller amounts. Ketone bodies like BHB can also feed intestinal cells[340]. However, being too low carb for too long may decrease gut permeability. Eating more different vegetables and fiber will also increase your metabolic flexibility and resiliency against plant compounds, which we will discuss in the next chapter.
- **Some Cancers Can Feed Off Ketones** – Most cancerous cells use glucose, lactate, and glutamine for fuel[341]. However, some tumors can also feed off ketone bodies[342]. And as you remember, the same applies to autophagy in some instances. It's an adaptive mechanism by which malignant cells adapt to the nutritional environment and stress they're put under. That's why an optimal strategy is cyclical and targeted.
- **Carbs Improve Exercise Performance** – At low intensities of exercise (below 65% of VO2 max), the body

can easily burn just fat and ketones. Above that threshold, at higher intensities, the body starts using glycogen. You don't need to eat carbs to replenish muscle glycogen because the body can convert fat into glucose through gluconeogenesis. It's also been shown that keto-adapted athletes can raise this ceiling and still burn ketones even at intensities of 80%[343]. However, if you're a hard-charging CrossFit athlete, a bodybuilder, or strongman, then you'd greatly benefit from eating more carbs. The average person who doesn't have very ambitious training goals can safely workout on keto but even then, introducing some carbohydrates strategically would improve their performance.

- **Carb Refeeds Maintain a Higher Metabolic Rate While Dieting** – One 7 week study put one group of resistance-trained individuals on a continuous 25% energy restriction (CN) and the other on a refeed protocol (RF) with 5 days at a 35% calorie deficit followed by 2 days of higher carb intake at 100%. Their macros and carbohydrates were identical during the week (~120 g carbs, 140 g protein, and 60 g fat). The results showed that the RF group preserved more muscle, dry fat-free mass, and resting metabolic rate compared to CN[344]. This was probably due to changes in leptin, thyroid, and glycogen re-synthesis that translated over to better performance and higher satiety in the RF participants. Perhaps they saw this improvement because their body didn't go into ketosis and was still dependent on carbs.

- Another study by Wilson and Lowery et al (2015) compared cyclic ketogenic dieting (CKD) to normal ketogenic dieting. The calorically restricted subjects by 500 calories a day, and the cyclic subjects had a normal carbohydrate diet on Saturday and Sunday. All participants did high intensity exercise and resistance training. Both groups lost 3 kilograms of body weight but the keto group with no refeeds lost nearly only fat, while the CKD group lost 2 kg-s of lean mass. Although this might indicate that continuous ketosis is better in preserving muscle during a deficit, it doesn't mean that a CKD protocol done at maintenance or a surplus would have the same effect. Carb cycling has been used by bodybuilders to build muscle and burn fat for decades.

- **Carbs Produce More CO2** –The Bohr Effect describes how tissue oxygenation results from CO2 releasing oxygen from hemoglobin and releasing it into the blood. This has an anti-anxiolytic and stress-reducing effect. Carbohydrate oxidation consumes 50% less water and generates 50% more CO2 than fat oxidation.
    - During glycolysis, which is the oxidation of carbohydrates, pyruvate gets created. The decarboxylation of pyruvate creates one CO2 molecule and acetyl CoA. Acetyl CoA enters the citric acid cycle, consumes 3 water molecules, and creates 2 CO2 molecules. During beta-oxidation of

fatty acids in the mitochondria, only one water molecule is consumed and acetyl CoA gets created. No additional CO2 gets produced, thus pyruvate metabolism yields an extra CO2 molecule. Eating more carbs raises CO2, increases breathing rate, speeds up the metabolism, and makes us exhale more of that CO2.

- Vitamin K uses CO2 to enable proteins bind to calcium. This regulates blood clotting, prevents tissue calcification, promotes mineralization of bones, growth, and hormonal health. I think this doesn't mean that a low carb diet hinders optimal health due to the lower CO2 production but it does point out that CO2 has a pretty important role in the body. Fat oxidation does generate 2 CO2 molecules but it's not the 3 molecules created from carbohydrate metabolism. Therefore, if a person is constantly deprived of energy they may just run into a deficiency of CO2. It also means that if you want to shut down the stress response more efficiently and faster, then the easiest and quickest way to do so is to consume some carbohydrates because they're going to raise CO2 and thyroid functioning.
- Ketogenic diets have been shown to lower partial pressure of CO2 in the arterial blood and reduce the period of ventilation in artificially ventilated patients[345]. This puts less stress on the lungs by

reducing carbohydrate metabolism and increased energy efficiency. So, keto is great if you suffer from some respiratory disorder or are put into a state of hypoxia while free-diving or in an airplane. Ketosis decreases the stress induced by hypoxic and poorly oxygenated situations and speeds up recovery.

- **Chronic Ketosis is Fragile** – Being in ketosis all the time causes physiological insulin resistance as we discussed and reduces the ability to use glucose for fuel. That's not a very antifragile place to be because you'll feel awful whenever you get kicked out of ketosis. By cycling in and out of ketosis you maintain higher levels of insulin sensitivity, thyroid functioning, and metabolic flexibility.
- **Restrictive Diets Can Cause Food Intolerances** – If you're eating a very restrictive diet like keto, carnivore, or vegan, then you may develop food intolerances and potential allergies to certain foods. Imagine, you've been keto for 6 months straight with no cheats and then you have some cheesecake at your parents' birthday. You just want to be nice and savor the moment. Unfortunately, you start to feel awful within the next few hours, your heart is pounding, blood sugar is sky-high, and it feels like you're dying. What's going on? Well, because of not eating grains or gluten, your body has lost some of its ability to deal with them, thus it triggers an autoimmune response. If you were to "micro-dose" gluten and other potentially allergenic foods regularly you wouldn't develop this reaction but

would maintain your antifragility instead. Of course, this strategy doesn't work if you have actual autoimmune disorders or serious medical conditions like asthma, diabetes, etc. However, healthy people who don't have actual allergies shouldn't avoid entire food groups just because they think it's healthy. They'll just lose some of their metabolic flexibility and tolerance to "bad foods". It also doesn't mean you should start eating donuts and pizza on a regular basis. Instead, eat just a little bit of bread or cake on special occasions like birthdays or other celebrations to get your hormesis. And yes, you can use it as a get out of jail free card when eating out with your biohacker friends.

- **Too Low Carb Can Reduce Gut Diversity** – Gut health, the microbiome, and microbial diversity are very loaded terms and I think even experts in this field don't really know what they truly mean. There is an association between better health, longevity, and gut diversity[346]. Dysbiosis is a situation of too many "bad bugs" in relation to the good ones and it results from over-eating, insulin resistance, antibiotics, chronic stress, and certain foods[347]. The ketogenic diet has been shown to modulate the gut microbiota in a beneficial way by reducing ROS, weight loss, decrease lactate but it also lowers alpha-diversity and richness in gut bacteria[348]. This may not be an issue if you were to keep eating the same diet for the rest of your life. However, in order to maintain robustness and prevent autoimmunity, it's better to occasionally eat the foods you normally wouldn't eat. Think of it has predictive allostasis modeling and hormetic conditioning. More on this food-related hormesis in the next chapter.

And of course, it's nice to sometimes eat foods that aren't bacon or vegetables. Although I'm the kind of guy who could eat just steak and eggs for the rest of my life, I still agree that some variety is not only healthy but also beneficial for antifragility. But don't worry, getting kicked out of ketosis doesn't mean you'll lose keto-adaptation. You'll still be able to effectively burn fat for fuel. It's just that you'll gain some of the other benefits of metabolic flexibility.

How to Increase Metabolic Flexibility

The foundation for metabolic flexibility is the ketogenic diet because you need to be able to burn fat as a primary fuel source. On a high carb diet without keto-adaptation, you're only capable of burning glucose while not being able to use ketones. But to avoid bonking after your glycogen runs out and to not get the keto flu whenever you eat some carbs, you need to go through a period of keto-adaptation.

Metabolic flexibility is a matter of degree like keto-adaptation – someone with greater muscle glycogen stores and improved insulin sensitivity can absorb more glucose than the one with lower energetic demands. If you do more resistance training, then you'll have a bigger buffer zone for eating carbs but you'll also increase your basal metabolic rate.

Here's how you can improve metabolic flexibility and stay keto-adapted while eating carbs.

- You have to establish nutritional ketosis by doing a low carb ketogenic diet for at least 2-4 weeks.
- After the first period of keto-adaptation, you can start tinkering with some carbohydrates to improve your performance.
- The fact of the matter is that you still want to be eating relatively low carb, especially at times when you're not exercising.
- If you're able to go without food for over 24 hours and not experience hypoglycemia or muscle weakness, then that's a good indicator of keto-adaptation.
- At this point, your physical performance at all intensities is generally the same and you don't need carbs to fuel your training. However, you can still use a few hacks that include strategic carbohydrate consumption.
    - The Targeted Ketogenic Diet (TKD) involves consuming a small dose of about 5-10 grams of carbohydrates during your most intense workouts.
    - The Cyclical Ketogenic Diet (CKD) involves eating keto for 5-6 days, having a day of eating more carbohydrates, and then returning to keto.

- There's also something called Carb Backloading (CBL) where you eat low carb all day, then you go to the gym to have a muscle glycogen depleting workout that makes you more insulin sensitive, and then have dinner with a few extra carbs like a sweet potato, a bit of fruit, or some rice.

The vast majority of people would still do best on a regular ketogenic diet, at least most of the time. Both the CKD and TKD are viable ways of boosting your performance, improving your metabolic flexibility, and upregulating the metabolism.

You can use both the CKD and TKD for promoting any physical endeavor, whether that be ironman triathlon, bodybuilding, Olympic weightlifting, or Crossfit. Of course, this doesn't apply to walking, jogging, yoga, disc golf, or something else that doesn't tax your glycogen stores. Therefore, you want to be using these methods only as tools for becoming stronger, faster, more enduring, or resilient, not as an excuse to simply eat some carbs.

The problem with strict therapeutic ketosis is that it's not necessarily going to ensure keto-adaptation as you can be in ketosis without using those ketones for fuel and you can have very high ketones without being able to perform at your best. More ketosis doesn't equal more keto-adaptation as it has to involve the aspect of mitochondrial density and energy production. It can also neglect some of the performance-enhancing benefits of carbs.

Of course, you'd have to be eating a low carb ketogenic diet the vast majority of time to maintain ketosis and become keto-adapted. However, your goal doesn't have to be ketosis as it's not going to ensure metabolic health or performance. A keto-adaptation diet would include high-quality low carb foods that build up the person's fat-burning engine and then add some occasions of higher carbs as a leverage point for improved metabolic flexibility.

Oxidative Priority and Ketosis

Respiratory quotient (RQ) is a number that estimates your basal metabolic rate. It's calculated from the ratio of CO_2 produced by the body to oxygen consumed. RQ values show which macronutrients are being burned and how they affect energy balance[349]. The RQ of a mixed diet is 0.8 - fats have an RQ of 0.7, alcohol 0.67, protein is 0.6-1.17 and carbs are 1.0[350]. A higher RQ means you're burning more carbs/glycogen whereas a lower RQ shows the body is burning more fat.

Oxidative priority describes which fuel substrates the body prefers to burn at any given moment and how it affects your metabolism in general. To ensure survival, you need appropriate allocation and use of resources. Otherwise, you'll end up missing out on key opportunities and/or will just starve.

Here's a table of the oxidative priority of the body's main energy stores[351]:

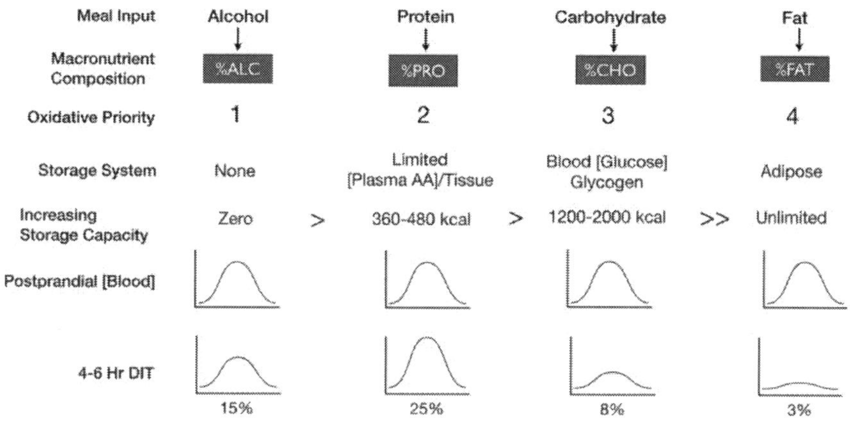
Meal Input
Alcohol
Protein
Carbohydrate
Fat
Macronutrient Composition
%ALC
%PRO
%CHO
%FAT
Oxidative Priority
1
2
3
4
Storage System
None
Limited [Plasma AA]/Tissue
Blood [Glucose] Glycogen
Adipose
Increasing Storage Capacity
Zero
>
360-480 kcal
>
1200-2000 kcal
>>
Unlimited
Postprandial [Blood]
4-6 Hr DIT
15%
25%
8%
3%

Source: Cronise et al (2017) Metab Syndr Relat Disord. 2017 Feb 1; 15(1): 6–17.

As you can see, the #1 oxidative priority is alcohol because it's a toxin the body tries to get rid of ASAP. That's why it's known that ethanol inhibits the utilization of other macronutrients and causes insulin resistance[352]. You're essentially suppressing the oxidation of fats as well as carbs when drinking alcohol until you're not intoxicated anymore.

Ketosis does increase the rate of fat oxidation but it doesn't make you necessarily burn more calories. Calories arc burned with energy production and physical activity. A 2014 study by Alan Aragon and Brad Schoenfeld found that body composition changes in people who trained fasted vs fed were the same as long as they were eating the same amount of calories at a deficit[353]. Although the people who were fasting may have shown higher rates of fat oxidation, their overall fat loss was still governed by how many calories they ate. Instead, being in ketosis changes the hierarchy of oxidative priority, making your body use fat and ketones as a fuel source instead of glucose. As a result, your body experiences higher rates of fat oxidation, thus being able to tap into its fat stores more easily as well. To lose weight you still need an energy deficit but the oxidative priority shifts more towards burning fat.

Eating fat in the presence of other macronutrients with a higher oxidative priority may result in higher fat storage due to decreased fat oxidation. The body is burning other fuel sources i.e. alcohol or carbs first before it starts to burn fat. That's why mixed diets tend to be more obesogenic and energy dense.

In order to raise the oxidative priority of your body fat (lose fat, not muscle), you have to maintain semi-ketosis during weight loss. For that, you should practice some intermittent fasting, time-restricted eating, and control your carb intake. Eating too many carbs kicks you out of ketosis, thus keeping the body burn carbs instead of fat.

If you want to lose more body fat then cutting down on your fat intake is the most reasonable thing to do. It has the smallest thermic effect of food and burns the least calories for digestion. Reducing protein intake isn't a smart idea because you may lose lean muscle tissue, which slows down your metabolism. The only thing to cut down on is your fat intake. You don't have to go zero fat but just don't eat extra fats from oils, butter, or some keto junk food. Instead, just get your fat from whole food protein.

Metabolic Flexibility Protocol

- The foundation of an antifragile metabolism is keto-adaptation because it conditions the body to use its own fat and ketones for fuel. This raises the threshold of how well your body can endure nutritional stressors and has other longevity boosting benefits. Being able to burn only sugar leaves the body more fragile and vulnerable to stress because there's only a limited amount of carbs the body can store at once. Ketosis will have a hormetic effect by improving mitochondrial efficiency, reducing ROS, and making you bonk-proof.

- Long-term keto creates physiological insulin resistance or glucose intolerance, which isn't pathological although it can be a problem. Cycling in and out of ketosis regularly maintains higher insulin sensitivity and metabolic flexibility. It won't lose keto-adaptation as the body still maintains its fat-burning pathways.
- Cyclical ketosis also keeps the thyroid and leptin levels higher. Carbs produce more CO2, which promotes tissue oxygenation and raises metabolism. Increasing CO2 with glucose is also a faster way of lowering stress and getting out of fight or flight. Eating some carbs will prevent food intolerances and gut problems related to chronic carbohydrate restriction.
- To improve metabolic flexibility you can do the cyclical ketogenic diet, targeted ketogenic diet, or carb backloading. How many carbs you need or how often to cycle out of keto depends on your level of insulin sensitivity and physical activity. A diabetic who's trying to fix insulin resistance can't do it as often as a high-level athlete. Generally, it's necessary to keto-adapt for about 2

weeks by eating a low carb ketogenic diet, and then transitioning over to either having carbs once a week or only on your hard training days. Here are the recommended ratios:

   - Standard Ketogenic Diet: 50-70% fat, 25-40% protein, 5-15% carbs
   - Cyclical Ketogenic Diet: low carb keto for 5-6 days and 1 day of low fat high carb
   - Carb Backloading: low carb keto or intermittent fasting throughout the day before training and low fat high carb post-workout in the evening
   - Targeted Ketogenic Diet: low carb keto all the time and 5-15 grams of carbs during training

- Don't combine fats and carbs together because it promotes insulin resistance and metabolic dysfunction. When you're eating more carbs, keep the fats low, and vice versa. Protein intake should still be relatively high on both diets.
- The best time to eat carbs is around working out and preferably post-workout. This would promote the re-synthesis of muscle glycogen stores as opposed to fat storage. You can also get back into ketosis faster if the carbs get shuttled into muscle cells.

Chapter Four:

Of Xenohormesis and Mithraism

„All things are poison and nothing is without poison.
Only the dose permits something not to be poisonous."
Paracelsus

Amongst biohackers and die-hard healthy eaters, it's very easy to see signs of orthorexia and eating disorders. Everyone is just afraid of having foods that contain gluten, grains, soy, or anything else that isn't keto or organic. This can turn into a subconscious source of stress that kind of defeats the entire purpose of their food restrictions. The health outcomes from the slight anxiety are actually more harmful than the supposedly unhealthy ingredient.

Bartrina (2007) describes orthorexia as such:

> *Orthorexia is an obsessive-compulsive process characterized by extreme care for and selection of what is considered to be pure 'healthy' food. This ritual leads to a very restrictive diet and social isolation as a compensation. Orthorexics obsessively avoid foods which may contain artificial colours, flavours, preservant agents, pesticide residues or genetically modified ingredients, unhealthy fats, foods containing too much salt or too much sugar and other components.*[354]

According to Bartrina, groups at the highest risk of orthorexia are women, adolescents, and those who do physique sports like bodybuilding. I would also include some people who are on a very niche diet i.e. vegan, paleo, keto, carnivore, etc. Now I'm the kind of guy who doesn't care whether or not others think my diet is too strict or borderline an eating disorder. However, for the sake of antifragility, I'm aware of the fact that sometimes the most antifragile thing you can do to your superhuman biohacker lifestyle is to break the cycle and eat something completely out of the ordinary. This chapter will explain exactly why.

Mithraism and Antifragility

Around 100 BC, Asia Minor was governed primarily by the kingdom of Pontus. It's an ancient Greek colony and considered to be the original home of the Amazons by Herodotus.

Mithridates VI of Pontus was the king of Pontus and Armenia Minor at about 120-63 BC. Before taking the throne, his father was poisoned. Mithridates thought it was his mother who killed the old king by having small amounts of poison to be added to his food. The young king fled into the wild where he began to self-ingest non-lethal amounts of toxins and mixing them into potions to make himself immune to poisoning[355]. He invented a "universal antidote" against toxins that was called *Antidotum Mithridaticum* or *mithridate* by Medieval writers[356].

Unfortunately, when Mithridates was attempting to escape Roman invaders, he failed to kill himself with poison because he had built an immunity towards it[357]. Talk about bad luck... or, much rather, such antifragility.

Mithraism is the practice of developing resistance to certain poisons by consuming them in small non-lethal amounts. You can find examples of toxicity resistance in the ability to handle more alcohol without passing out, being able to digest lactose, gluten, FODMAPS, and not feeling tired after consuming too much food.

Mithridates VI was so paranoid of being poisoned that he took small doses throughout his life to build up an immunity. When he was finally captured by the Romans, he tried to kill himself with poison but failed because he was immune.

The problem with being very restrictive and tidy with your diet is that you negate the potential benefits of hormesis. Sure you can avoid all gluten and carbs for the rest of your life but it's going to make you vulnerable to the randomness of life. It can also lead to physiological insulin resistance as discussed in the previous chapter. You could plant your feet into the ground and say 'NO' indefinitely but imagine what would it do to your relationship with others and food. A much more antifragile approach is to embrace the chaos every once in a while and not let it stress you out. Your mental wellbeing would be stronger as a result. This doesn't mean you should schedule regular cheat meals and binge. Instead, savor those situations for some special moments or celebrations with family. Yes, it means you can have your birthday cake and eat it too. At least reasonably and in the right setting.

I do agree that you don't want to be eating potentially harmful foods like bread or sugar all the time. Hell, doing it „in moderation" like the rest of the population does is also problematic because you can't quantify how much is too much. You can end up making excuses for bad habits. These foods can also keep you in a state of sub-optimal performance and slump because they do indeed promote inflammation and gastrointestinal problems. If it takes you an entire week to recover from a splurge, then it's not worth it. However, introducing them into your diet every once in a while as hormetic conditioning is a good idea. You don't want to become hypersensitive to these allergens just because of avoiding every potential danger and being too strict. I consider myself a picky eater with high standards for the quality of what I consume but I have to admit that from an antifragility perspective 'micro-dosing' gluten and other „bad foods" is quite a viable strategy, especially in the modern world where we're already going to be coming across these ingredients at an increasing rate. It's not a matter of discipline or self-control either. I've done the die-hard keto diet for 3 years straight in the past with no slip-offs and I'm able to do it again. It's just that I value higher metabolic flexibility and antifragility instead of some sort of an ideology or belief.

Hunter-gatherers would forage a lot of wild plants and other edibles that had some medicinal benefits despite their toxicity. Different culinary techniques like sprouting, soaking, fermenting, and plain simple cooking can also reduce the number of antinutrients in them. People in the modern world have simply become way too soft and domesticated that they can't handle even just a little bit of digestive strain. The perfect example is antibiotic resistance and lack of microbial diversity in the gut. Children who aren't exposed to bacteria from dirt, animals, grains, and other sources are much more likely to develop autoimmune disorders later in life[358]. C-section newborns are more prone to suffer from all chronic diseases like diabetes, obesity, and asthma[359]. Another reason to get your hands dirty on a daily basis.

The most common allergens include dairy, eggs, fish, nuts, seeds, gluten, grains, soy, mustard, celery, peanuts, shellfish, sulfites, strawberries, pollen, and the common suspects. Ideally, you don't want to be intolerant to any of these foods and have a gut that's capable of handling everything but that's not possible for a lot of people. You definitely shouldn't eat raw beans or legumes because they can be dangerous. Likewise, there are a lot of lethal berries and plants. It's not advisable for someone who's already suffering from autoimmune disorders or inflammatory diseases either.

Xenohormesis

The entire concept of hormesis and stress adaptation applies to both environmental conditions as well as nutrition. It's beneficial to be consuming these plants and herbs that have a mild hormetic effect so that your body could get stronger and more resilient. What makes a poison deadly is the dosage.

In 2012, Mark Mattson explained that cells respond to bioenergetic stressors by increasing DNA repair proteins, antioxidant enzymes, and the production of neurotrophic factors (such as BDNF)[360,361,362]. It's also believed that this is the reason eating vegetables, tea or coffee can improve brain health. There's evidence that habitual tea drinking has positive effects on brain efficiency and slows down neurodegeneration.[363] The polyphenols in coffee have also shown to reduce the risk of diabetes[364], Alzheimer's[365], dementia[366], and even liver cancer[367].

Plants produce natural pesticides to fight off predators. They're often bitter-tasting and supposed to make animals sick or lethargic. These chemicals aren't lethal – they're just supposed to motivate the intruders to stop eating the plant and go away. Bittersome leaves especially tell us not to eat too much of these kinds of plants because they taste bad and that's usually a sign of potential toxicity. That's why most people by default don't like bitter vegetables and much rather prefer things that taste sweet. Sweetness indicates high amounts of easily accessible energy that can be stored and burned off quickly whereas wild plants are low in sugar.

Consuming plant-based toxins in low amounts causes hormetic stress to the body and mimics calorie restriction or fasting to a certain degree by stimulating the same pathways, such as sirtuins, autophagy, and AMPK[368]. Thanks to the constant evolutionary arms race between us and the plant kingdom, we've developed counter-adaptations in the form of hormesis and will trigger a small beneficial response when eating these foods. Your body gets stronger in a similar fashion – it's a hormetic exercise.

Xenohormesis is the idea that certain molecules like plant polyphenols, tannins, carotenes, etc. can have a beneficial effect when eaten[369]. Plants produce these chemicals to endure environmental stress and predation. They can act like signaling molecules that indicate either forth-coming famine, the ice age, or just reflect the conditions of the particular environment. Eating the plant makes the body hunker down and prepare for the hardships as well. This may have a stress adaptive, longevity-boosting, and immuno-modulating effect.

Xenohormesis described by Hooper et al (2010):

> *Xenohormesis is a biological principle that explains how environmentally stressed plants produce bioactive compounds that can confer stress resistance and survival benefits to animals that consume them. Animals can piggy-back off products of plants' sophisticated stress response which has evolved as a result of their stationary lifestyle. Factors eliciting the plant stress response can judiciously be employed to maximize yield of health-promoting plant compounds. The xenohormetic plant compounds can, when ingested, improve longevity and fitness by activating the animal's cellular stress response and can be applied in drug discovery, drug production, and nutritional enhancement of diet.* [370]

Plants contain multiple sirtuin-activating compounds that co-ordinate sirtuin-mediated defense[371]. In fact, many of the polyphenols that promote stress adaptation get synthesized during famine, dehydration, or infection. Animals also synthesize these mechanisms during fasting or exercise but they can also turn them on by consuming certain compounds. That’s a great way to mimic conditions of semi-starvation and fasting without having to actually go through such conditions.

Eating foods rich in plant polyphenols and probiotics can activate many antioxidant pathways including Nrf2 and be a powerful tool for disease prevention[372]. However, they don't work as anti-oxidants but more so as pro-oxidants that trigger mild oxidative stress, leading to increased glutathione.

Nrf2 sits in the cytosol of cells where it cannot directly interact with DNA. It's waiting there to be released and held in place by proteins called Keap1 and Cullin 3[373]. Oxidative stress and free radicals whether from exercise or eating plants trigger Keap1's sensors for reactive oxygen species, which releases Nrf2 into the nucleus[374,375]. Once inside the nucleus, Nrf2 combines with some small Maf proteins and binds with the 'Antioxidant Response Element' or ARE[376]. After Nrf2 has been released into the nucleus and bound with ARE it can promote powerful antioxidant pathways such as glutathione and NQO1. This will help the body deal with oxidative stress and toxins.

The goal of xenohormesis isn't to chew on bitter-tasting leaves or eat thorns. Instead, it's about consuming certain compounds and foods that would mimic some of the effects you get from calorie restriction, exercise, and fasting by stimulating the same pathways. It's mimicking some aspects of those activities in the short-term. For example, there's a whole list of ingredients that turn on autophagy, Nrf2, glutathione, or sirtuins.

Compounds That Mimic Calorie Restriction and Autophagy

A lot of the foods that promote Nrf2 also stimulate autophagy like curcumin, sulforaphane, coffee, green tea, polyphenols, and flavonoids. Nrf2 and Autophagy work together to fight inflammatory compounds that float around your cells and eliminate them[377]. Both of them get activated under oxidative stress and nutritional deprivation[378]. One of the autophagy proteins p62 intersects autophagy and Nrf2 through Keap1[379]. Here are the compounds that promote this process:

- **Green tea is rich in polyphenols** such as catechins, quercetin, and myricetin, which mimic the effects of exercise and caloric restriction on the body. One of its compounds EGCG activates Nrf2[380], autophagy[381], and PGC-1α[382]. Green tea is one of the most widely consumed beverages in the world's longevity hotspots. It's probably the healthiest drink after water.
- **Sulforaphane found in cruciferous vegetables,** especially in broccoli seeds and sprouts activates the Nrf2 pathway and has many other health benefits. Sulforaphane induces autophagy through ERK activation in neuronal cells[383]. Cabbage, Brussels sprouts, kale, collard greens, radishes, turnips, watercress are all great.
- **Grape seed extract activates PGC-1α,** which inhibits oxidative stress and mitochondrial dysfunction[384]. In vivo,

grape seed proanthocyanins induce autophagy and inhibit tumor growth[385].

- **Medicinal mushrooms** like Reishi and Chaga induce autophagy and can fight cancer[386]. They also have many adaptogenic or stress modulating properties.
- **One of ginger's compounds 6-Shogaol** induces autophagy by inhibiting the AKT/mTOR pathway in human cell lung cancer[387]. Ginger phenylpropanoids and quercetin increase the Nrf2-ARE pathway in human fibroblasts[388].
- **Curcumin induces autophagy** by activating AMPK[389]. It also has many antioxidant and anticarcinogenic effects[390]. In rats, curcumin induces Nrf2 translocation and prevents glomerular hypertension, hyperfiltration, oxidative stress, and decreased antioxidant activity in response to renal injury[391].
- **Polyphenols regulate autophagy**[392]. They can be found in dark vegetables, berries, and leafy greens. Various herbs and spices show potential in neuroprotection against cognitive decline by modulating the Nrf2-ARE pathway[393]. Coffee induces autophagy by promoting ketosis, regulating blood sugar, and thanks to its high polyphenol count[394].
- **Resveratrol inhibits breast cancer stem-like cells and induces autophagy[395].** Resveratrol is found in dark grapes, berries, cherries, and red wine.

- **Excess zinc can also support the stimulation of autophagy[396].** Foods rich in zinc are oysters, liver, red meat, eggs, salmon, pumpkin seeds. Too much, however, causes nausea and other problems.

- **Fermented foods** like pickles, sauerkraut, olives, vegetables contain alkyl catechols which the bacteria convert into Nrf2[397]. They also contain precursors for NAD+.

- **Beans have phytonutrients** and other compounds that promote autophagy and can reduce oxidative stress[398]. Legumes and beans are also a staple in the longevity hotspots across the globe.

- **Salidroside (SAL)** is an active component of *Rhodiola Rosea* with many antioxidant properties but it can also stimulate mitochondrial biogenesis[399].

- **Luteolin is a flavonoid that induces Nrf2**[400]. It's found in radishes, peppers, greens, celery, red lettuce, and artichoke.

- **Docosahexaenoic acid (DHA)** is partially associated with ROS production, which activates the PI3K/Akt/Nrf2 signaling[401]. It also upregulates SIRT1[402]. Foods high in omega3s and DHA are mackerel, salmon, oysters, chia seeds, flaxseeds, and walnuts.

- **Hydroxytyrosol (HT) in extra-virgin olive oil** is considered one of the most important polyphenolic compounds responsible for the health benefits of the Mediterranean diet for lowering incidence of cardiovascular disease. It can also promote mitochondrial function by stimulating mitochondrial biogenesis[403]. Polyphenols in olive oil like oleuropein and oleocanthal trigger autophagy and suppress mTOR[404,405].

- **Quercetin is the most widely consumed flavonoid in the human diet[406]** with many antioxidant benefits and it activates the SIRT1. It's found in elderberries, red onions, hot peppers, cranberries, kale, and tomatoes[407].
- **The metabolic state of ketosis whether through fasting or nutrition** increases PGC-1α, AMPK, and SIRT1, which causes mitochondrial biogenesis[408]. Ketone bodies are modulated by PGC-1α and they have anti-inflammatory effects on the body. Exercising in ketosis has a greater hormetic effect.
- **Carotenoids** such as lutein and zeaxanthin found in foods like carrots activate the Nrf2 antioxidant system[409].
- **Butyrate** is a short-chain fatty acid that has anti-inflammatory effects on the colon and can activate Nrf2[410]. It's created by digesting fiber in the gut but also found in animal fats like butter and tallow in smaller amounts.

The amounts of these foods are also relevant. Eating too many berries or dark grapes for their resveratrol will still eventually inhibit autophagy if you consume too many carbs. Likewise, a smaller serving of salmon or oysters won't necessarily break autophagy completely either if you still stay at a large caloric deficit with minimal protein intake.

Plant Hormetics – Friend or Foe?

However, as with almost anything in nature, there are trade-offs to these plant hormetics. Some might even consider them dangerous, damaging, and too potent, if you will, whereas others think they're the best thing since sliced bread.

There are many classes of compounds plants produce to protect themselves: polyphenols, phenolics, alkaloids, tannins, terpenoids, flavonoids, isoflavonoids, and many others. They're even considered pesticides and 99.99% of the pesticides in the American diet by weight are produced by the plants themselves[411].

Excessive consumption of phytonutrients can cause hormonal imbalances, liver damage, digestive problems, thyroid malfunctioning, DNA damage, and just sub-optimal wellbeing[412]. That's why it's not a good idea to overconsume them, especially if you're the kind of person who likes to take things to the extreme and you get way too enthusiastic about the promising effects of plant hormetics.

Animals know this and they go out of their way to carefully choose which kind of plants they consume and what's their relative toxicity. For example, while grazing, ruminants engage in cautious sampling, diet rotation, mixing their plant intake to dilate toxins, upregulating detoxification pathways, or just simply spitting out the ones that don't sit well[413]. Nevertheless, herbivores still consume plants because of their availability, palatability, nutritional qualities, adaptability, and addictive qualities. That's what their bodies and digestive tracts have evolved to do and it's their sole means of survival. Fortunately, animals do develop a tolerance to specific compounds the more they consume them, which reduces their relative toxicity and improves digestibility[414]. For example, deer are more tolerant of locoweed than antelope, and elk are more tolerant of ponderosa pine than bison[415]. However, they resort back to the more toxic plants during times of food scarcity and prefer to eat the ones with less toxicity.

Human societies also know this as they engage in various cooking, sprouting, fermentation, and other preparing practices that ought to reduce the plants' toxicity and improve digestion. Many vegetables, beans, legumes, and lentils are indigestible when eaten raw and can cause serious gut damage. Some of them like cassava, bitter apricot seeds, and bamboo shoots contain toxic cyanide, which is not compatible with human digestion. Fortunately, there are many ways to prepare plants in a way that makes them relatively harmless and non-toxic. Modern cultivation practices also reduce the toxicity of crops as they're far safer than wild plants. In the future, we might be able to engineer foods that have only the beneficial compounds with no negative ones.

In 2009, an older diabetic woman consumed over a kilogram of bok choy a day in an attempt to treat her condition but unfortunately she developed hypothyroidism and went into myxedema coma[416]. Chinese cabbage and other cruciferous vegetables contain glucosinolates, which are known to inhibit iodine uptake and impair thyroid functioning[417]. The problem was her large consumption of bok choy raw. When eaten raw, brassica vegetables release the enzyme myrosinase, which accelerates the processing of glucosinolates whereas cooking largely deactivates this[418]. That's why you ought to know what's the relative toxicity of different plants, how much is safe to consume, and how to prepare them before eating. A lot of it is individual and context-dependent as tolerance to various compounds varies even within the same genus[419]. The phrases 'what makes a poison deadly is the dosage' and 'one man's food is another man's poison' are 100% accurate.

Here's a list of the most common plant phytonutrients and chemicals that may have a hormetic effect. I'm going to wage both their claimed benefits as well as documented side-effects so you could know whether or not to consume them:

- **Glucosinolates are compounds found in many pungent plants like mustard, cruciferous vegetables, and horseradish.** They work as defense against pests and have a characteristic bitter flavor[420].
    - Eating too many of these vegetables inhibits iodine absorption, lowers thyroid functioning, and may cause goiter[421]. Cooking lowers the amount of goitrogenic compounds and it's not a good idea to be eating these plants solely raw.
    - Some studies find that feeding 500 grams of raw broccoli a day to pigs increases DNA damage in colon cells[422]. However, previous ones are showing a decrease in DNA damage, whereas broccoli cooked in a microwave does not[423,424].
    - The anti-carcinogenic effects of cruciferous vegetables are often attributed to their glucosinolate content. In experimental models, they've been shown to protect against colorectal cancer but epidemiological studies have shown inconsistent results[425].
- **Sulforaphane (SFN) is a sulfur-containing compound with many antioxidant effects[426].** It's derived from glucoraphanin and found in cruciferous vegetables like broccoli, cauliflower, cabbage, and Brussels sprouts[427]. When certain plants get damaged from either chewing or cutting, they produce myrosinase that transforms

glucoraphanin into sulforaphane. After this process, SFN starts to degrade quite rapidly[428]. Cooking at temperatures below 284˚F (140˚C) also produces sulforaphane[429] but a complete mash destroys it.

- Research has shown sulforaphane to be beneficial in managing type-2 diabetes[430], improve blood pressure[431], promote liver detoxification[432], lower LDL cholesterol[433], boost the immune system[434], inhibit bacterial and fungal infections[435], lower inflammation[436], support liver functioning[437], increase BDNF that promotes growth of brain cells[438], and reduces risk of cancer[439].
- Sulforaphane inhibits HDAC enzymes, which increases the activity of Nrf2[440], leading to the activation of glutathione, thioredoxin, and NQO-1. The same HDAC inhibition leads to the activation of autophagy as well[441]. Sulforaphane promotes energetic stress, autophagy, and cell death[442]. SFN causes mild oxidative stress, lipid peroxidation, and generates ROS[443], which are considered to be a part of its chemo-protective properties by up-regulating glutathione and autophagy[444].
- Excessive sulforaphane can cause some genome instability and over-activation of white blood cells[445]. In mice, sulforaphane increases susceptibility to seizures and other symptoms of toxicity[446]. However, that was done with a dose at 200 mg/kg, which is pretty hard to reach by eating just

vegetables or broccoli. You'd have to take a sulforaphane supplement, which I would say is not necessary or advisable anyway.

  - Uncooked Brussels sprout juice in rats causes DNA damage and apoptosis or programmed cell death[447]. However, this effect was confined to animals injected with dimethylhydrazine (DMH) 48 hours earlier. Their levels of apoptosis and DNA damage were already high. I don't know anyone who ever drinks raw Brussels sprout juice, do you?
  - There is a lot of research about the effects of sulforaphane in vivo but the evidence in humans tends to be lacking[448]. Taken the potential side-effects, cruciferous vegetables aren't necessary to be consumed all the time by everyone although I would say they can be useful whenever you need to upregulate glutathione. They're still a great low carb alternative to many other foods and are quite filling.

- **Curcumin is the main active compound in turmeric.** It's been shown to have anti-inflammatory, antioxidant, antimicrobial, antiseptic, antimutagenic, hepatoprotective, and immunostimulating properties[449]. The efficacy of curcumin is comparable to corticosteroids and other anti-inflammatory drugs but without the side-effects[450]. It's also similar to drugs that improve endothelial function and reduce the risk of heart disease[451].

- Curcumin scavenges ROS and boosts the body's antioxidant enzymes[452]. However, at higher concentrations, it can increase ROS[453,454]. In studies, a high dose is considered as high as 1 gram because that would be higher than most people could obtain from consuming the spice in of itself[455].

- Pre-treatment with curcumin can also protect against toxins like mercury[456], arsenic, cadmium, lead[457], and 2,3,7,8-tetrachlorodibenzo-*p*-dioxin (TCDD)[458]. It also restores memory impairments caused by sleep deprivation and can protect the brain against poor sleep[459]. That's why I use it on days I didn't get enough shut-eye.
- Curcumin protects the brain against neurodegeneration by increasing BDNF and reversing the effects of chronic stress[460]. It also promotes neuronal autophagy in chemotherapy[461] and prevents hippocampal damage[462]. Curcumin-induced autophagy contributes to the decreased survival of oral cancer cells[463] and it has chemopreventative effects[464].
- The dark side of curcumin is that it can cause chromosome aberrations[465], alterations in redox balance[466], and DNA damage[467,468]. In mice, it's reported to promote lung cancer[469] and an increased incidence of carcinomas in the small intestine[470]. However, those findings were found on daily doses

of 2,600 mg/kg for up to 15 months! That's about ½ teaspoon per kilogram of bodyweight. I'll let you do the math on how many teaspoons you would have to take to reach any harm. No other biologically significant differences or toxicity were observed.

- In humans, taking up to 8 grams of curcumin a day doesn't show short-term toxicity but doses between 0.9-3.6 g/day for 1-4 months cause nausea, diarrhea, and an increase in serum alkaline phosphatase and lactate dehydrogenase[471]. Epidemiological studies suggest that the lower incidence of colorectal cancer in India might be due to their curcumin-rich diet. It's estimated that people there consume about 0.15 grams of curcumin a day[472]. The amount of curcumin in turmeric powder isn't that high either – around 3.14% by weight[473]. From one teaspoon of turmeric powder, you can expect about 0.15 grams of curcumin. In curry powder, it's even less because curry contains other ingredients like coriander, cumin, and chili. So, consuming regular turmeric or curry as seasoning is probably harmless unless you megadose it with supplements. Megadoses, like you get from supplements, haven't proven to be safe and they can cause negative side-effects. Even then the threshold of toxicity depends on your antioxidant and ROS status.

- In rats fed a poor iron diet, curcumin causes anemia because it's a potent iron ghelator[474]. That's a problem if you're not getting enough iron from your diet but beneficial if you're getting too much. Seasoning your high iron meals with some curry or turmeric will lower the iron overload. Excessive iron is a risk factor for heart disease[475].
- Curcumin isn't very bioavailable and is poorly absorbed. Consuming it with black pepper, which contains piperine, enhances the absorption of curcumin by 2000%[476]. It's also fat-soluble so adding fat improves absorption. I like to use curry in broths and as seasoning on fried foods to lower the potential oxidative stress.

- **Flavonoids and isoflavonoids are a class of plant polyphenols and fungus metabolites.** They fulfill many functions in plants, mainly floral pigmentation, UV filtration, and nitrogen fixation. Flavonoids also work as chemical messengers and symbiotic cell regulators between the roots and host plants[477]. They, such as quercetin and flavan-3-ols, are the most common polyphenolic compounds in the human diet[478]. You can find them in foods like onions, blueberries, parsley, citrus fruit, black tea, green tea, cocoa, and buckwheat[479]. Compared to other plant compounds they're with very low toxicity.
    - Flavonoids can be anti-inflammatory by inhibiting ROS and reactive nitrogen species (RNS)[480]. They also suppress the pro-inflammatory activity of

enzymes involved in free radical production[481]. However, flavonoids have negligible antioxidant effects due to poor bioavailability and they increase antioxidant capacity through the production of uric acid in response to flavonoid depolymerization and excretion[482]. So, they work along the same lines as sulforaphane by turning on the body's defense mechanisms, not due to the inherent benefits of the compound itself.

- Cell studies show flavonoids to have anti-inflammatory[483], anti-allergenic[484], anti-microbial[485], antifungal[486], antiviral[487], anti-cancer[488], and anti-diarrheal effects[489]. However, these outcomes haven't been researched in living organisms and the FDA nor the European Food Safety Authority has approved any health claims for flavonoids or approved any flavonoids as pharmaceutical drugs[490]. Dietary flavonoid consumption is associated with reduced risk of gastric cancer in women[491], significantly reduced risk of aerodigestive tract cancer, and marginally lower risk of lung cancer in smokers[492]. Unfortunately, these are epidemiological studies that shouldn't be interpreted as causative.
- Flavonoids have been shown to reduce risk factors of cardiovascular disease by lowering blood pressure, improving endothelial function, modifying blood lipids, reduced oxidative stress,

inhibition of coagulation, platelet aggregation, and thrombus formation[493]. They also regulate glucose metabolism and insulin sensitivity, which protects against metabolic syndrome[494]. However, population-based studies have found no strong benefit, which might be due to the low flavonoid intake in most people's habitual diet[495].

  - Plant flavonoids have estrogenic effects[496], which can cause hormonal imbalances and endocrine disruption. The isoflavones in soy are known to be estrogenic and they disrupt the endocrine system[497]. Some of the beneficial effects include reduced risk of breast cancer and improved metabolic profile but the evidence is inconsistent and indirect[498]. In men, consumption of soy is associated with infertility and poor sperm quality[499]. Swapping out meat with soybean like tofu results in lower testosterone and higher sex-hormone-binding globulin (SHBG), which binds to free testosterone, reducing it even further[500].

- **Resveratrol (RSV) is a plant polyphenol, specifically a stilbenoid.** It's concentrated mostly in the skin and seeds of grapes, berries, fruit, and some vegetables[501]. Some think that resveratrol from red wine explains The French Paradox and why people in France didn't have nearly as high rates of heart disease as other countries despite eating a lot of saturated fat and cholesterol[502].

- Resveratrol lowers oxidative stress and inflammation but also activates many antioxidant pathways like glutathione, Nrf2, and sirtuins[503]. By activating SIRT1 and suppressing mTOR, resveratrol activates AMPK, which triggers many catabolic pathways related to caloric restriction and longevity[504,505]. Resveratrol is a calorie restriction mimetic that may reduce age-related diseases[506].
- Resveratrol inhibits the NF-kB pathway, which is one of the most important activators of inflammation in the body[507]. RSV lowers blood sugar and improves glucose metabolism[508]. It also benefits liver health[509]. Resveratrol can protect against seizures, stroke, and neurodegeneration[510].
- Resveratrol increases nitric oxide, which is a molecule that promotes blood flow and oxygen uptake[511]. It also lowers blood pressure and protects the heart[512]. RSV attenuates TMAO-induced atherosclerosis by regulating TMAO synthesis and bile acid metabolism via remodeling of the gut microbiota[513].
- There isn't enough evidence to support the full efficacy of resveratrol[514]. In one study, daily oral RSV supplementation of 1000 mg-s for 16 weeks didn't improve inflammatory markers or glucose homeostasis in middle-aged men with metabolic syndrome[515]. It's also been shown to lower androgen precursors like DHEA, DHEAS, and

androstenedione but circulating testosterone, free-T, and dihydrotestosterone were unaffected[516]. A placebo-controlled, randomized clinical trial on high-dose resveratrol treatment (1,5 grams/day) for 6 months didn't alleviate NAFLD[517].

  - Resveratrol acts as a pro-oxidant in the presence of copper ions and causes DNA damage[518]. In rat tissue, resveratrol has the opposite effects on lipo-peroxidation: it's a pro-oxidant during the day and an antioxidant at night[519]. Rats have the opposite diurnal rhythm to humans.

- **Salicylates are plant compounds used in medication, such as aspirin, toothpaste, and food preservatives.** You can get salicylates from foods like raisins, dried fruit, spices, vegetables, tea, licorice, chewing gum, olives, etc[520]. They're produced to defend against insects and fungus but they can cause adverse reactions in some people[521]. Salicylate sensitivity is thought to be caused by overproducing leukotrienes linked to inflammatory conditions such as asthma, arthritis, and IBD[522]. Symptoms include sinus infections, diarrhea, abdominal pain, swelling, and nasal problems[523]. A low salicylate diet can treat aspirin-exacerbated respiratory disease (AERD), bronchial asthma, and nasal polyposis[524]. For everyone else, salicylates are benign.
- **Oxalates or oxalic acid is an organic acid found in certain plants**, such as leafy green vegetables, spinach, rhubarb, beetroot, chard, cocoa powder, some fruit,

turmeric, almond flour, nuts, and seeds[525]. You don't need to eliminate all of the oxalate-containing foods from your diet although dietary contribution to oxalate formation is much bigger than previously thought[526].

- When you consume oxalates, you may get an adverse reaction because oxalates bind together with calcium, iron, and other minerals. Too many oxalates prevent calcium from being absorbed into the body[527]. Combining oxalates with eating fiber can also hinder the absorption of minerals[528].
- Some people think that oxalates contribute to autism but studies show there's no link there[529]. However, a „green smoothie cleanse" has been shown to cause acute oxalate neuropathy in a 65-year old woman[530].
- Oxalates get formed as a by-product of regular metabolism to a certain extent already, but consumption of too many oxalates may cause kidney stone formation and digestive issues[531]. About 80% of kidney stones are made of calcium oxalates.
- In 1919, a woman died due to oxalic acid poisoning from eating rhubarb soup with leaves and stalks for supper[532]. The leaves were fried for greens and the stalks boiled. Unfortunately, she started getting stomach pain, bloody vomiting, exhaustion with no signs of fever the next morning. That's why you

definitely shouldn't eat foods super-high in oxalates like rhubarb leaves.

- Another case of death from oxalic acid poisoning happened in 1989 from eating sorrel soup[533]. The victim was a 53-year old man with insulin-dependent diabetes, who smoked, drank alcohol heavily, had metabolic acidosis, and hypocalcemia. So, he was already pretty sick to start with.
- Before starting to restrict your oxalates, you should know if your urine has too high levels of oxalates. In the case of too many oxalates, most urologists prescribe a low-oxalate diet with less than 50 mg per day[534]. Oxalate restriction is recommended for people with kidney stones, high uric acid, low calcium intake, magnesium, and sulfur deficiencies. People who have had gastric bypass surgery also show higher levels of oxalates in their urine[535].
- Eating slightly more calcium-rich foods like cheese, dairy, fish, and broccoli can reduce the oxalate load in your body by making calcium bind together with oxalates in the gut[536]. Cooking, sprouting and soaking vegetables before eating also lowers their oxalate content. If oxalates are causing you digestive problems, then you either overconsumed them or have poor gut health to begin with.
- Certain bacteria like the *Oxalobacter formigenes* can reduce the number of oxalates your

body absorbs by using it for energy and thus reducing the risk of kidney stones[537]. The problem is that antibiotics and general inflammation in the gut have been shown to lower the amount of these anti-oxalate bacterial colonies in the gut[538]. The woman who got neuropathy from an oxalate smoothie had previous history of gastric bypass and prolonged antibiotic therapy. So, there you have it. Foods that can support the growth of this beneficial anti-oxalate *Oxalobacter formigenes* are fermented foods, sauerkraut, raw kefir, and animal fats.

- **Lectins are plant sticky proteins that attach to other food molecules and cause inflammation[539].** This can lead to leaky gut syndrome[540], brain fog, autoimmune disorders, weight gain, arthritis, and other inflammatory conditions[541]. They can even contribute to atherosclerosis due to intestinal permeability and arterial damage. However, there are no studies to prove those claims. You can find lectins in sunflower seeds, cashews, peanuts, beans, legumes, grains, wheat, oats (gluten is a lectin), nightshades, tomatoes, cucumbers, peppers, and potatoes. However, some animal foods also contain lectins, such as pork, anchovies, eggs, lobster, milk, and honey. If you're sensitive to these foods and get bloated or inflamed after consumption, then you may want to consider restricting lectins. Even if you're not it's probably a better idea to avoid eating a diet excessively high in lectins. Lectins can be blocked by simple sugars and oligosaccharides[542], which might be why plant-based eaters don't seem to report a lot of issues with them. Cooking and soaking foods also lower their lectin content. Eating raw beans or potatoes can be lethal in large quantities but even small doses will cause gastrointestinal stress and food poisoning[543].

- **Tannins are polyphenolic compounds that bind to amino acids and alkaloids.** They create the taste of astringency and puckery in your mouth when consuming red wine, teas, or unripened fruit[544]. You also get tannins from berries, herbs, spices, nuts, chocolate, legumes, beer, coffee, and fruit juices[545]. Although some of these foods can have beneficial phytonutrients and hormetic effects, tannins can impair digestive processes by inhibiting digestive enzymes[546]. High amounts of tannin-rich foods like fatabean, cereal, and sorghum can reduce protein and amino acid digestibility in rats by up to 23%[547]!
- **FODMAPs are short-chain carbohydrates that don't get absorbed by the small intestine that well.** It stands for „Fermentable Oligo-, Di-, Mono-saccharides And Polyols" and they can create digestive problems, bloating, autoimmune conditions, irritable bowel syndrome, and leaky gut. FODMAP foods aren't the cause of these issues but a low-FODMAP diet can help to improve the symptoms. FODMAP foods include wheat, rye, barley, onion, garlic, artichokes, cabbage, beans, lentils, tofu, tempeh, and some fruits like apples, apricots, avocados, cherries, plums, and chicory.

The reason I made such a long list has to do with the contradicting information you hear about plant polyphenols and phytonutrients nowadays. You've got a lot of people saying they're amazing and you should want to consume them as much as possible whereas others say they're the devil and will immediately cause irreversible damage to your cells. As with almost everything, the truth is somewhere in the middle. There's plenty of evidence to show the benefits of curcumin, sulforaphane, and resveratrol but you can also find negative side-effects. In most cases, the negative outcomes are derived from megadosing, too frequent use, or individual intolerance.

We don't need vegetables or polyphenols to maintain an optimal antioxidant status or hormetic adaptation. High fruit and vegetable consumption have been associated in epidemiology with a lower incidence of cardiovascular disease due to their antioxidant and anti-inflammatory properties[548]. However, studies have found no difference in immune function and antioxidant status between consuming 800 grams of vegetables a week and 4200 grams[549]. Although eating more plants increases circulating levels of carotenoids and other nutrients in healthy subjects, it does not affect antioxidant status or lymphocyte DNA damage[550]. No change in oxidative stress or DNA damage has been observed between eating an antioxidant-free basal diet and 600 grams of vegetables a day, or a supplement with corresponding amounts of vitamins and minerals, or placebo over the course of 24 days[551]. Antioxidant supplementation has not been shown to prevent cardiovascular disease or cancer[552].

Nevertheless, I think it's a good idea to still consume some of these xenohormetic plant foods as hormetic conditioning and to battle the unnatural amounts of oxidative stress we're exposed to in our current environment. They're not going to be any more potent than exercise or fasting, which I think every person interested in improving their health and fitness should be already doing. However, they can work as calorie restriction mimetics and simulators of nutrient shortage. Most people experience too much oxidative stress and their bodies don't produce enough endogenous antioxidants. That's why these compounds can be useful assistance. Someone who's already exposed to different hormetic stressors like intermittent fasting, exercise, cold, heat saunas, etc. is probably not going to see a significant improvement from supplementing additional sulforaphane or resveratrol.

More often than not the negative side-effects of consuming polyphenols or plant phytonutrients comes from either improper preparation methods (i.e. raw and juiced), overconsumption (eating 2 kilos of broccoli a day) or supplementation (high concentration compounds unobtainable from real food). Even if sulforaphane is a potential treatment for cancer, you probably shouldn't engage in therapeutic cancer treatment every day, right? Of course, anything in excess can be harmful, even healthy food, which is why you should take everything in a hormetic dose. The way you achieve that is by getting your xenohormetic compounds from whole food sources and natural seasoning as opposed to macro dosing supplements.

Another indirect mechanism by which plant compounds and vegetables can promote longevity-hormesis is through mimicking calorie restriction. Plant-based meals higher in fiber and lower in fat tend to decrease daily calorie intakes[553]. Low energy density diets can reach satiety at an energy intake one half of a high energy density diet (1570 vs 3000 calories)[554]. That's a massive difference in terms of the final outcome on longevity and body composition.

Even healthy whole foods can make you obese and sick if you overconsume them. The particular nutrient does matter but the quantities are equally if not more important. Over-eating calories is bluntly put not healthy in the long-term and would also negatively affect your lifespan. That's why incorporating some plant hormetics and low-energy foods enable you to reach higher satiety and fullness with fewer calories. At least for some periods of the year. In so doing you also stimulate the longevity pathways that you can't do while eating animal foods.

Xenohormesis Protocol

- Eat a variety of different xenohormetic plant compounds in small amounts. Don't make them the staple of your diet but use it as conditioning. How much is too much depends on your tolerance, situation, and goals. Generally, eating more polyphenolic vegetables and dark pigment berries is relatively benign for most people. It's pretty hard to over-eat them to the point of danger unless you literally eat 2-3 kilos. Here are some points to keep in mind concerning the foods we discussed.

- **Sulforaphane** – Taking an SFN supplement may cause negative side-effects and excessive damage. You should get your SFN from primarily cruciferous vegetables where the amounts are much smaller. Don't eat vegetables raw and always have them either steamed or slightly cooked. Massive raw salads will be harmful to the thyroid. Daily consumption is also not advisable and you should cycle them between days you're working out and not.
- **Oxalates** – Don't drink green smoothies, don't drink nut milks or other milk alternatives because they're highly concentrated in oxalates. At least not in massive quantities. Cook or soak your vegetables before consumption. If you experience stomach pain or kidney stones, you may want to go on an oxalate-free diet for a certain period of time. However, if you don't, then just be mindful of not over-eating oxalate-rich foods like beets, spinach, almonds, cacao, cassava, etc. Don't eat them every day. Adding a bit of calcium like cheese or dairy, citrus like lemon juice or oranges, and fermented foods to your meals will also help to break down oxalates.
- **Curcumin** – Taking a curcumin or turmeric supplement is probably not recommended. At least not daily. You should use it only whenever you're under higher inflammation or oxidative stress, such

as after sleep deprivation, jet lag, or toxicity. On an everyday basis, you can use 1-2 teaspoons of turmeric or curry powder as seasoning on food. Remember that it chelates iron and other metals. If your iron levels are high that's great but not if you're anemic.

- **Flavonoids** – Foods high in flavonoids like onions, peppers, vegetables, garlic, tea, and buckwheat are benign, as long as you're eating reasonably. Just pay attention to your symptoms and how you react to these foods. Then adjust your intake accordingly but I think most people don't have to restrict their intake. In fact, I think they're more beneficial because of stimulating longevity pathways.
- **Herbs and Spices** – Various herbs and spices are amazing for cooking and they impose additional health benefits. For instance, rosemary protects against high-temperature oxidation and glycation. Peppers like cayenne and black pepper will also raise your metabolic rate but can cause stomach problems when consumed in excess. Herbs like thyme, parsley, oregano, dill, etc. are also concentrated with nutrients but you should use them moderately for seasoning because they also contain some oxalates. Then again, you'd have to eat unreasonable amounts for them to become a problem.

- **Dark Pigments** – Blackberries, bilberries, purple cabbage, etc. are high in polyphenols and resveratrol. They're great to make salads out of or as dessert. Keep in mind that these beneficial plant compounds are more present in foods that have been "stressed out" during cultivation by the elements. That's why the GMO engineered red grapes at the supermarket have little to no resveratrol. They're grown to be much higher in sugar and smaller in nutrients. Instead, try to get preferably wild berries or organic produce that tastes more bitter and savory. The same applies to other GMO vegetables that don't cause the same hormetic increase in glutathione because they haven't been stressed out as if grown organically.
- **Lectins** – Moderate consumption of lectins shouldn't be a problem for most people unless they have some genetic predisposition or allergy. With that being said, you shouldn't make lectins the staple of your diet and it's not recommended to eat them in excess. Signs of eating too many lectins are bloating, inflammation, achy joints, pain, chronic fatigue, digestive problems. Cooking and heat reduce the amount of lectins in a given food.
- **Tannins** – The main source of tannins is tea or coffee. They inhibit the absorption of other nutrients, which is why you may not want to have them with meals. At the same time, green tea and

coffee can also chelate iron, which can be used to diminish the iron overload you might get from a particular meal. Too much tea or coffee can also cause digestive problems, dehydration, and addiction.

  - **FODMAPs** – Histamine and FODMAP-rich foods like sauerkraut, kimchi, beef jerky, cheese, and leftovers can cause allergic reactions in quite a lot of people. That's why it's not recommended to eat them every day and in large quantities. They should be thought of as condiments or savory treats. Signs of reacting negatively include bloating, constipation, water retention, brain fog, sleeping problems, inflammation, and irritability. At the same time, these FODMAP foods like sauerkraut or cheese can also break down oxalates. Everything has its pros and cons.

- To know which foods you react the worst to, you have to be doing a lot of self-experimentation and pay attention to how you feel after eating a given meal. Taking a food sensitivity or allergy test is also a smart idea. However, you should still rely primarily on your first-hand experience.

Chapter Five:
Of the Immune System

"Nature is a numbers game. We need all the support we can get as our immune systems and health are under assault from pollution, stress, contaminated food and age-related diseases as our lifespans increase."

Paul Stamets

In addition to the elements, humans in nature would also get exposed to countless bacteria and bugs that can be both beneficial or harmful. To endure the onslaught of different stressors and intruders, you need to have a strong immune system. Otherwise, you'll be wiped out by even the smallest contagions. This chapter is solely dedicated to preventing that and making sure you stay indomitable.

The word "immunity" comes from the Latin word '*immunis'*, meaning exempt. It describes the ability of multicellular organisms to fight infections, disease, pathogens, or other invaders. There are two categories that make up all of the body's defense systems:

1. **Innate Immunity** is the one you're born with. It's comprised of primarily bone marrow cells that recognize and react to foreign substances as well as vaccinations. The innate immune system is mediated by antigen-presenting cells (APC), neutrophils, and the complement defense system.

2. **Adaptive Immunity** is what you obtain over a lifetime of exposure to various pathogens. It's composed of advanced lymphatic cells, T-cells (killer cells) and B-cells (producers of antibodies). Adaptive immunity can be acquired naturally through exposure or artificially (vaccination). These sub-categories can be further divided into active and passive, depending on whether the host developed immunity itself (active) or through transfer or injection (passive).

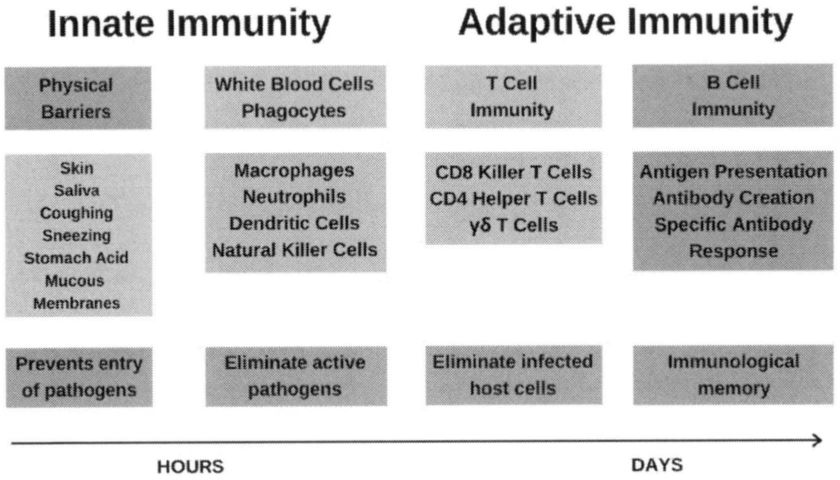

It's known for centuries that mild fevers and infections especially in childhood can be protective and healthy in adulthood. A 2019 study found that children who got influenza early in their life developed a stronger immunity against mutated forms of the same future virus[555]. However, in other cases, you can experience the opposite. For example, measles can wipe out the immune system's memory, leaving the person vulnerable to other diseases[556]. So, it's not always black and white but more so dependent on the particular situation.

Immune cells are constantly trying to meet the challenges of the environment. That's why they have receptors for hormones like cortisol and adrenaline to mobilize an immune response.[557] During stress, immune cells change their sensitivity to these hormones, developing glucocorticoid resistance and other inflammatory diseases[558].

Acute stress mobilizes certain immune cells into the blood and raises pro-inflammatory cytokines.[559] Brief psychological stressors like taking an exam tend to suppress cellular immunity but preserve humoural immunity.[560]

Chronic stress, on the other hand, raises inflammatory markers like CRP and IL-6.[561] Inflammation is necessary in the short-term for eliminating pathogens, initiating healing, and adaptation but if chronically elevated it will promote diseases like atherosclerosis or osteoporosis[562]. Psychological stress is also implicated in rheumatoid arthritis.[563] It can increase the risk of developing an autoimmune disease.[564] Symptoms of irritable bowel syndrome are associated with elevated cortisol.[565]

Chronic stress also activates dormant viruses that will begin to undermine the immune system, thus leaving you more vulnerable to additional infections. It's thought human immunosuppression could even be contagious because of how certain viruses like cytomegalovirus interact with the host's immune cells[566].

Strategies for a Stronger Immune System

You can increase resilience against viruses and infections with healthy regular lifestyle interventions. Here are some strategies for strengthening the immune system.

- **Regular exercise** stimulates the body's defense mechanisms and strengthens immunity by activating Nrf2[567]. It also promotes lymph and blood circulation. Exercise improves arterial function, which protects against the development of atherosclerosis and acts as an anti-inflammatory.[568] Exercise improves gut microbiome diversity, where most of the immune system is located at.[569] Regular exercise enhances immunosurveillance lowers basal inflammation, which improves protection against pathogens and other ailments.[570]
    - With age, the immune system goes through immunosenescence, which refers to a deterioration in immunity, and exercise can alleviate that.[571] It boosts the function of several immune cells.[572] Exercise prior to influenza vaccination increases the effectiveness of the vaccine by fueling anti-body response and lowering prevalence of the infection[573].
    - Exercise bouts less than 60 minutes increase the circulation of anti-inflammatory cytokines, immunoglobulins, neutrophils, and others that all have an important role in fueling immunity.[574] Even just 30 minutes of walking has shown to cause a

rise in immune parameters.[575] Moderate exercise elevates IL-6, which has anti-inflammatory effects, improves blood sugar management, and lipid metabolism[576,577]. However, heavy physical exertion has been associated with transient immunodeficiency, elevated inflammation, and increased risk of upper respiratory tract infections.[578, 579] Prolonged intense output suppresses immunoglobulin A (IgA), natural killer (NK) cells, T and B cells. Although exercise increases inflammatory biomarkers transiently, physically fit people have lower levels at rest.

- During exercise, natural killer cell activity increases but it drops to a minimum 2 hours later and returns to pre-exercise levels in 24 hours[580]. The intensity of exertion determines the degree of this release. A 60-minute bike ride has also been shown to decrease T cells but they return after recovery.[581] Carbohydrates and plant polyphenols have been shown to counter-act this decline in immune cell metabolism.[582] Protein powder enriched with green tea and blueberries has been shown to reduce the risk of exercise-induced virus infections.[583]
- Exercising regularly has been shown to reduce the prevalence of upper respiratory infections[584]. Light and moderate exercise during sickness can even be beneficial in fighting the disease[585]. While exercising, the body also releases endogenous

pyrogen, which is a protein that mediates fever and trace metal metabolism during an infection.[586] However, overtraining will make you more vulnerable to getting sick[587]. If you have a fever and you feel sick, do not exercise but other than that keep yourself physically active all the time. This should include both cardiovascular and resistance training, which we'll talk about in chapter eight.

- **Regular sauna.** Sweating eliminates toxins, heavy metals, and infections by improving lymphatic drainage and blood circulation[588]. It's one of the few ways to detox that actually works. Both the traditional and infrared sauna kill pathogens thanks to the heat. Regular sauna bathing will strengthen the immune system[589]. Going to the sauna more than 2 times per week has been shown to reduce the risk of acute and chronic respiratory conditions, including pneumonia[590]. In mice, short-term heat shock provides beneficial anti-HPAIV H5N1 ("the bird flu") properties, which can offer an alternative strategy for non-drug prevention for influenza infection[591].
    - Rising body temperature strengthens the immune system by mobilizing white blood cells, producing interferon, and modulating lymphocytes, which all help with antiviral activity. Hotness and fever also induce heat shock proteins that inhibit viral replication[592,593,594], and reduce pro-inflammatory cytokines.[595] Therefore, mild elevations in temperature or a small fever can be beneficial for

fighting infections or preventing them from being established.

- **Cold exposure.** Moderate exposure to cold doesn't increase the susceptibility to infections but instead increases resilience against them[596,597]. Winter swimming studies show it lowers uric acid and increases glutathione[598]. Just make sure the exposure time to cold or wind is not too long. If you don't feel uncomfortable or frail from the cold, then it can be useful. Preferably protect your neck, lungs, and stomach. Overdoing it or getting caught in a draft may weaken immunity because of overstimulation. Chapter six talks about heat and cold adaptation.
- **Intermittent fasting** is an effective strategy to build resilience against pathogens. Time-restricted eating can upregulate glutathione and autophagy to protect against sickness[599]. The metabolism related to fasting also protects against bacterial inflammation but not all viruses[600]. It can be useful for prevention but it is not effective if you are already infected with a virus. Refer to chapter two for a list of viruses and infections that have been shown to benefit from autophagy and which ones do not.
    - One study in particular done by one of the leading researchers of fasting Valter Longo *et al* showed that you can reset your immune system by fasting[601]. Mice and chemotherapy patients who didn't eat for several days saw a significant reduction in white blood cell count. This then turned on signaling pathways for hematopoietic stem cells (HSC),

which are responsible for the generation of blood cells and the immune system. It means during a prolonged fast like 72 hours, the immune system drops slightly but it rebounds afterward. That's why you may not want to expose yourself to potential pathogens or highly infectious venues while fasting for longer. Daily time-restricted eating is fine and immuno-strengthening.

 - 48-120 hour fasts reduce pro-growth signaling and enhance cellular resistance to toxins[602]. It inhibits mechanisms of cell replication and directs resources towards hunkering down instead. This decreases the potential of spreading infections and promotes starving them out.

- **Nutritional ketosis** activates the Nrf2 pathway that lowers inflammation and oxidative stress[603]. Being exposed to stress, jet lag, infections, and pollution is safer in a state of ketosis whether that be with fasting or eating a ketogenic diet because the body is in a heightened state of self-defense[604].
  - Beta-hydroxybutyrate, one of the main ketone bodies, is a powerful anti-inflammatory that blocks inflammatory pathways like NF-kB, COX2, and NLRP3 inflammasome[605]. Over-reactive inflammasomes can damage the body during an infection[606]. Ketogenic diets also show potential for reducing thermal and neuropathic pain by lowering inflammation[607]. Burning fat and ketones causes

less oxidative stress as a by-product of its metabolism.

- AMPK is the main fuel sensor that regulates energy balance and the allocation of resources under stress. AMPK has a major role in controlling viral infections and their replication independent of its role as an energy regulator[608]. Activation of AMPK restricts replication of coxsackievirus B3 by inhibiting lipid accumulation[609]. Inhibiting AMPK leads to hepatitis C replication and lipid accumulation[610]. In other studies, AMPK promotes hepatitis B virus replication by increasing autophagy[611]. Eating low carb, fasting and exercise activate AMPK in response to burning through glycogen. As glycogen gets depleted further, the body shifts into ketosis and starts producing ketones. Some microorganisms can become more virulent and replicate faster in high glucose environments[612].
- Ketogenic diets lower blood sugar and improve insulin resistance. Excess glucose reduces the ability of neutrophils (a type of white blood cell) to ingest and kill bacteria[613]. In a 1973 study, subjects consuming 100 grams of carbs from different sources after an overnight fast reduced the effectiveness of neutrophils by about 40 % for 5 hours[614]. This applied to sucrose, fructose, and

honey as well as glucose. The least impactful was starch.

- In a 2019 study, mice fed a ketogenic diet were better able to combat the influenza virus than those who were fed carbohydrates[615]. One of the mechanisms was an increased expansion of γδ T cells in the lungs that improved barrier functions and antiviral resistance. Blocking NLRP3 also prevents inflammasomes from damaging healthy cells during the infection.
- A ketogenic diet could lower your immune system if you're nutrient deficient, insulin resistant, overweight, super-inflamed, or have some sort of a gut dysbiosis. You could also weaken your immunity in the short-term by losing weight too fast. Malnourished people have a higher risk of infections and sickness because they lack the necessary resources for optimal immunity[616]. Countries with both high poverty as well as affluence suffer from key deficiencies[617]. That's why you ought to cover your essential nutrients for the immune system, such as vitamin D, zinc, B3, vitamin C, and selenium.
- A low carb ketogenic diet isn't nutrient deficient because one of the most nutrient-dense foods on the planet are keto-proof. Organ meats, egg yolks, vegetables, fish, and herbs are packed with all the nutrients you need. That's why a bacon and burger

style keto would not suffice if you want to cover all of your essential nutrients.

- **Regular sunlight.** The most bioavailable source of vitamin D is the sun[618]. You should get daily sunlight exposure as often as you can but avoid getting burnt. The best time for sunlight is in the morning to help in balancing the circadian rhythm, which also influences the function of the immune system[619]. The immune system has its circadian clocks and when disturbed, the immune system also is disrupted[620].
- **Reduce cortisol.** Chronic stress is one of the major contributors to an imbalanced immune system and predisposition to diseases[621,622]. Patients with viral infections show elevated cortisol[623]. As a reminder, the body perceives psychological stressors as physical and thus even anxiety, negative thoughts, anger, and afflictive emotions can also weaken your immunity.
    - Optimism is associated with a stronger immune system when there's some stress but it's not too demanding.[624] However, when circumstances are too difficult or uncontrollable, optimism is associated with weaker immunity. This might be because coping with harder stressors imposes higher energy demands on optimists when maintaining optimism itself is already energetically costly.[625] During brief, acute stress, optimists have better immune function than pessimists but this relationship gets reversed under more difficult stressors. The next chapter talks about augmenting

your mental operating system with stoicism, which is a more functional way of thinking under higher stress than just pure optimism.

- **Adequate sleep.** Melatonin, the sleep hormone, is also a powerful antioxidant that modulates autophagy and deep cell repair during sleep[626,627]. The body repairs itself primarily in deep sleep and sufficient sleep is extremely important for immune system health.
    - Sleep improves T cell functioning, which are killer cells that eliminate pathogens and viruses. During sleep, T cells are able to stick to infected particles more easily and then remove them. Stress and wakefulness inhibit this process, which is why stressed-out people are more susceptible to infections like the common cold[628].
    - There's a bidirectional relationship between sleep and immunity against infections[629]. In one study, people who slept less than 5 hours were nearly 3 times more likely to get sick compared to those who slept 8 hours[630]. What's more, low sleep efficiency increased the likelihood of getting the cold by 5 and a half times.
    - Sleep deprivation reduces natural killer cells. A single night of sleep deprivation reduces NK efficiency by 75%[631]. This weakens the body's ability to respond to invaders. In fact, it's found that poor sleep reduces the effectiveness of vaccination by negatively affecting antibody function[632]. On the

flip side, sleep enhances antibody synthesis and responsiveness to vaccines[633].

- Sleeping is important for the adaptive immune system, which works like an immunological memory[634]. Basically, sufficient sleep helps the body remember how to respond effectively to infectious agents and how to fight them. Memory consolidation occurs primarily in REM and slow-wave sleep, which also affects immunity[635].
- Melatonin acts also on both the innate and specific responses of the immune system via combined mechanisms that mainly involve the modulation of cytokines and the production of oxidative stress[636]. Additionally, infection-fighting antibodies and cells are reduced whenever you don't get enough sleep[637]. Supplementing with melatonin at bedtime may be useful in the early stages of infection to facilitate proper immune system response by increasing T-helper cell production, particularly of CD4+ cells. Production of IL-2, IL-6, and IL-12 are also stimulated by melatonin[638].

All of these strategies are based on optimizing lifestyle. If you follow these practices regularly, your chances of catching a cold or getting sick from any virus will be drastically reduced. I don't recall the last time I was seriously ill. Maybe I'll have a runny nose a few times a year but nothing that would put me in bed for the entire day. I'm not doing anything special besides being consistent with the things discussed in this book.

Nutrients Needed for Stronger Immunity

It's irrefutable that food has a major impact on your overall health, including immunity. Getting the right nutrition is important for governing the body's defense systems as well as dealing with intruders on a more visceral basis. Malnourished people are more vulnerable to infections and sickness because their immune system lacks the resources to function properly[639]. This makes third world countries more susceptible to infections. However, nutrient-deficient diets and poor food choices can also weaken the immune system[640].

Fixing deficiencies in vitamin D, zinc, iron, and vitamin A have shown to help prevent as well as treat pneumonia, especially in children[641]. That's why eating a nutrient-dense diet and proper supplementation can be beneficial.

Here are the most important nutrients and compounds that support the immune system:

- **Vitamin D** is central to the body's immune system. Low levels of vitamin D are associated with increased risk of infections[642] as well as autoimmune diseases[643]. In one study, amongst 19 000 subjects, people with lower vitamin D levels were more likely to suffer from upper respiratory tract infections[644]. A 2012 study found that giving young children 1200 IU of vitamin D a day reduced the risk of influenza[645]. Several systematic reviews of daily vitamin D supplementation have shown it to be protective against respiratory tract infections[646,647]. Taking a few thousand IUs during sickness may speed up recovery. Vitamin D toxicity is quite rare and happens only if supplement extremely high doses (>10 000 IU) over a long period[648].

- Research between 1988 and 1994 amongst 19000 participants showed people with vitamin D below 30 ng/mL were more likely to self-report upper respiratory tract infections compared to those with normal levels, even after controlling for other variables.[649] A 2017 large meta-analysis showed that vitamin D supplementation reduces the odds of developing a respiratory infection in people with 25-hydroxyvitamin D below 25 ng/mL[650]. However, a more recent study on older subjects didn't find the same effect.[651]
- Indonesian research has found that 98.9% of COVID19 patients with a vitamin D deficiency (< 20 ng/ml) died, 88% with insufficient vitamin D (20-30 ng/ml) died as well but only 4% of people with sufficient vitamin D (<30 ng/ml) died.[652] In the Philippines, every increased standard deviation of vitamin D was associated with an 8 times more likelihood of having mild rather than severe COVID19 outcome and a 20 times higher chance to have mild instead of critical results. [653] These regions also have more sunlight, which might

explain why the pandemic didn't break out as much as it did in places like New York or Northern Italy. Nearly 70% of Americans have vitamin D insufficiency.[654] Some researchers think that the higher mortality in Spain and Italy compared to North Europe might stem from a vitamin D deficiency.[655]

- **Vitamin K2** and D work synergistically and co-dependently. Vitamin D regulates calcium in the blood and K directs it into the right place such as the bones. High vitamin D with low vitamin K may increase blood vessel calcification and heart disease because of keeping the calcium at the wrong place. Vitamin D toxicity can cause hypercalcemia and blood vessel calcification, which causes heart disease[656]. Low levels of vitamin K are also linked to heart disease[657]. Supplementing vitamin K can reduce the risk of heart disease and calcification[658].
    - RDA for vitamin K1 is 90 mcg, which can be easily obtained from leafy green vegetables and cruciferous.
    - RDA for vitamin K2 is about 120 mcg but there isn't a real specified recommended intake. You should aim for about 150-200 mcg a day from animal foods like organ meats, fermented foods like sauerkraut and natto, egg yolks, dark poultry, and fermented cheese.

- **Selenium** is an essential mineral that is a cofactor in glutathione production[659]. It's also important for hormonal balance, antioxidant defense redox signaling and redox homeostasis (balancing oxidative stress in the body). Viral infections increase the production of reactive oxygen species (ROS)[660]. Excess production of ROS induces oxidative stress by overwhelming the antioxidant defense system. This, in turn, can enhance viral replication and raise the need for selenium. Selenium deficiency has been linked to the pathogenicity of several viruses[661] and survival from pneumonia due to influenza[662].
- **Vitamin B3** or niacin supplementation increases NAD+ biosynthesis[663]. Pharmacological doses of niacin may help the immune system fight against severe infections like HIV and tuberculosis, but the research is still preliminary[664].
- **Vitamin C** is an antioxidant that animals produce in response to stress[665]. Humans have lost that ability during our evolution and have to obtain it from diet. Vitamin C helps to recycle oxidized glutathione back into active glutathione[666]. People who take vitamin C regularly won't prevent the cold completely but they can expect shorter colds (by 8 % in adults and 14 % in children) with slightly less severe symptoms. Athletes who take vitamin C regularly are half as likely to catch a cold as athletes who do not[667].
    - A 1999 study done on 463 students showed that mega-dosing vitamin C may help treat flu

symptoms right after their appearance with hourly doses of 1000 mg of Vitamin C for the first 6 hours and then 3 times daily thereafter. Overall, reported flu and cold symptoms in the test group decreased by 85% compared to the control group after administrating a megadose of Vitamin C[668]. Nobel laureate Linus Pauling has said that as soon as you feel the symptoms of sniffles, a cold, or the flu, taking oral doses of thousands of milligrams of vitamin C is helpful[669].

- In stressed mice, mega-dosing vitamin C helped to prevent influenza (H1N1) induced pneumonia[670].
- A recent 2020 meta-analysis published in the Journal of Intensive Care showed that 1–6 grams of intravenous vitamin C per day shortened the ventilation time of patients needing intensive care on average by 25 %[671].

- **Ubiquinone (Coenzyme Q10)** acts as a contributor to the electron transport chain. It's a fat-soluble compound that helps to generate ATP (adenosine triphosphate) and hence energy. It has been used for decades as a dietary supplement. Low cellular ubiquinone levels may be a predisposing factor for various illnesses due to insufficient aerobic energy production in the cells[672]. With low energy production, the body will not be able to fight the intruders. CoQ10 reduces oxidative stress and preserves macrophages in the immune system[673].

- **Zinc.** In humans, zinc is required for the function of more than 300 enzymes and over 1000 transcription factors (proteins that regulate the function of genes). It acts in enzymatic reactions as a catalyst to accelerate their actions[674].

    - Zinc also plays an important role as a structural agent of proteins and cell membranes preventing oxidative stress[675]. Zinc is important for hormone production and immunity. It's also known for fighting against infections. Low zinc status can cause gastrointestinal problems and increase the risk of pneumonia[676]. However, high zinc supplementation can lead to toxicity and stomach pain[677].
    - Zinc acetate and zinc gluconate lozenges have been shown to inhibit cold viruses from latching onto the cells and shorten flu duration. Lozenges are beneficial only in the early stages of infection. The optimal dosc according to studies is 75–90 mg/day divided into multiple doses taken 2–3 hours apart. Best results are achieved when starting within 24 hours of first symptoms[678]. According to studies in children, regular use of zinc can prevent the flu[679]. It's been shown to inhibit the replication of viruses like SARS and arterivirus[680]. Don't exceed 100 mg of zinc per day for any longer than two weeks. Avoid nasal sprays as they might cause a lingering loss of smell perception.

- **Nitric Oxide (NO)** is an important signaling molecule that's been shown to fight against some viruses and bacterial infections[681]. Supplementation with NO boosters (such as L-arginine and L-citrulline as well as foods rich in nitrates like beets and leafy greens) may inhibit viral replication of SARS coronavirus and others[682]. However, excessive arginine can activate an underlying herpes infection when you're not getting enough lysine[683].

Foods That Strengthen the Immune System

Here are foods and specific supplements that can fortify the immune system:

- **Collagen** is an important building block for connective tissue as well as in various immune system functions. For example, collectins (collagen-containing C-type lectins) found in liver, lungs, placenta, and kidney, have been shown to mediate innate host defense against influenza virus infections and prevent secondary infections[684]. Collectins are a vital part of the innate immune system in the lungs[685]. They generally mediate pathogen clearance via complement activation and by aggregating cells[686].
- **Licorice Root.** Influenza is characterized by an acute lung inflammatory response (hypercytokinemia) and high oxidative stress that contributes to the virus-induced lung damage and morbidity. *Isoliquiritigenin* (ILG) found in licorice has been shown to be a potent inhibitor of

influenza virus replication in human bronchial epithelial cells and an inhibitor of inflammatory cytokines. Administration of ILG reduces the morbidity of mice infected with the H1N1 virus[687]. *Glycyrrhizin*, an active compound of licorice root, has also been used to inhibit the replication of SARS-associated coronavirus[688].

- **Lactoferrin.** Many studies have demonstrated the antiviral activity of lactoferrin against viral pathogens that cause common infections[689,690]. Lactoferrin consumption may protect the host from viral contagion by inhibiting the attachment of a virus to the cells, replication of the virus in the cells, and enhancement of systemic immune functions[691]. Lactoferrin-derived peptides are being researched as potent therapeutic inhibitors of influenza virus infections[692]. Additionally, hydrolyzed whey protein, which contains many bioactive peptides, has been shown to induce macrophage activity and activate anti-inflammatory functions[693].

- **L-Glutamine** is the most abundant amino acid in the bloodstream that accounts for 30-35% of the nitrogen content of amino acids in the blood. The body is able to synthesize glutamine itself, but glutamine is also needed in the diet, especially when under chronic stress, high physical activity, and many different medical conditions[694]. Glutamine is used by various immune system cells and is required to support optimal lymphocyte proliferation and production of cytokines by lymphocytes and macrophages[695].

- Glutamine, among other things, has been found to help people with food hypersensitivity by reducing the resulting inflammation of the gut surface[696].
- Glutamine can help repair leaky gut and hence also improve the immune system[697]. Getting enough glutamine from diet or by using a supplement helps protect intestinal epithelial cell tight junctions, which prevents intestinal permeability[698].

- **Elderberries and Dark Berries.** Dark pigmented berries have polyphenols and antioxidants that strengthen the immune system by modulating the gut microbiota[699]. Based on a 60 people study, taking 15 ml of elderberry extract 4 times per day 48 hours after the onset of influenza virus A and B can relieve the symptoms on average 4 days earlier than without[700]. Elderberries have also been shown to reduce symptoms of flu on other trials, but the evidence is weakened by small sample sizes[701].
- **Consume Fruits and Vegetables.** Regular consumption of fruits and vegetables may be beneficial for the immune system function. A higher intake of fruit and vegetables has been shown to lead a reduction in pro-inflammatory mediators and an enhanced immune cell profile[702]. For example, one 2012 study found that increased fruit and vegetable intake improved antibody response to a vaccine that protects against *Streptococcus* pneumonia in older people[703].

- **Probiotic Foods.** Bacteria like *Lactobacilli* and *Bifidobacteria* have been shown to improve gut health and immunity[704]. You can get them from fermented foods such as sauerkraut, kimchi, kefir, and fermented dairy[705]. Lack of fermented foods in the diet has been shown to cause a fall in innate immune response[706].
    - Humans have a built-in security system called SIgA (secretory immunoglobulin A), which is present in mucosal membranes that line for example the nose and upper respiratory tract as well as the gut[707]. IgA can prevent cold and flu viruses from entering the system. Based on experimental studies, probiotics may provide antiviral effects directly through probiotic-virus interaction or by stimulating the immune system[708].
    - Older people especially can benefit from long-term use of an oral blend of probiotics including *Lactobacillus plantarum, Lactobacillus rhamnosus,* and *Bifidobacterium lactis,* which enhance secretory immunity and increase IgA antibodies[709]. In another study, a probiotic strain *Bacillus subtilis* was shown to stimulate IgA in the elderly to reduce the frequency of respiratory infections by 45%[710]. *Lactobacillus plantarum* has been found to enhance human mucosal and systemic immunity and also prevent NSAID-induced (such as ibuprofen) reduction in T regulatory cells[711].

- **Sulfur-Rich Foods.** One way to increase glutathione levels is by eating sulfur-rich foods such as eggs, beef, and dark leafy greens[712]. Cruciferous vegetables, like broccoli and cauliflower, are also beneficial sources of sulfur[713].
- **Pelargonium sidoides.** The plant *African geranium* contains tannins that can help bacteria from attaching to the lining of the throat and lungs. It can reduce coughing and treat acute bronchitis according to several studies. It has also shown to reduce the duration and severity of colds, but the evidence is still preliminary[714].
- **Patchouli Oil.** An alcohol extract of *Pogostemon cablin* or Korean mint (*Huo Xiang* in Traditional Chinese Medicine) has been shown to inhibit viral replication of various influenza viruses and reduce plaque formation by 75% in vitro[715]. One mouse study has confirmed these particular anti-influenza effects in vivo. Whether these effects carry over to humans is not known. You can use it as an essential oil with antiviral and antibacterial properties on airborne pathogens[716].
- **Essential Oils.** Eucalyptus, cinnamon, clove, pimento, thyme, and rosemary oils have been shown to have antibacterial and antiviral properties on airborne pathogens[717].
- **Olive Leaf Extract (OLE).** Olive leaf extract (not to be confused with olive oil), contains polyphenols, notably *oleuropein,* and *hydroxytyrosol*, that have antiviral, antibacterial, anti-inflammatory and antioxidant properties that may reduce rates of upper respiratory infection[718].

Based on a 2019 randomized controlled trial on young athletes, OLE decreases the duration of upper respiratory infection in high school athletes[719].

- **Allium and Garlic.** Allium vegetables like onions, leeks, and shallots increase glutathione levels[720]. Garlic is a natural antibiotic and antimicrobial agent that kills viruses directly and strengthens immunity in multiple studies[721,722,723]. To activate its benefits, you have to crush garlic and consume cold because heating destroys its beneficial compounds like allicin. Alternatively, use 9000 mg-s of high-allicin garlic supplement daily. Aged garlic is also effective and has slightly different immunomodulatory effects[724].
    - Black garlic is created by fermenting fresh garlic. It contains many more antioxidant compounds such as polyphenols, flavonoids, tetrahydro-β-carboline derivatives, and organosulfur compounds, including S-allyl-cysteine and S-allyl-mercaptocysteine, compared to fresh garlic. Fermentation has been shown to enhance the bioactivity of black garlic. A growing body of evidence demonstrates the therapeutic effects of black garlic including antioxidant and immunomodulatory effects[725]: black garlic extract supplement impedes serum TNFα, interleukin-6 (IL6), and interleukin-1 β (IL1β) production and prevents mice from LPS-induced death[726].

- **Oregano and Other Herbs and Spices.** Oregano and especially oregano essential oil is an effective antifungal and antibacterial substance[727]. Other herbs with similar properties include thyme, rosemary, clove, lemon balm, and cat's claw[728]. Spices like cayenne pepper, chili pepper (containing capsaicin), and black pepper can also kill pathogens directly[729,730].
- **Teas.** Green tea, black tea, and herbal teas have medicinal properties, such as polyphenols, that boost antioxidant defense systems and fight infections[731,732,733].
- **Raw Honey and Bee Pollen.** Honey has antimicrobial peptides (including bee defensin-1, defensin-2, hemenopectin, and apidaecin) and medicinal properties (such as flavonoids, polyphenols, vitamins, and minerals) that strengthen the immune system[734]. It can also inhibit the growth of pathogens such as E. coli and salmonella[735]. Using raw honey as a sweetener is a great alternative to sugar and syrups. Bee pollen is also a powerful modulator of immune system function[736]. Honey is an effective treatment for cough caused by an upper respiratory tract infection[737]. Be wary of giving honey to infants as it can cause botulism[738].
- **Myrrh** is a resin or sap-like substance extracted from several small, thorny tree species of the genus *Commiphora*. It's used to treat hay fever and wounds and has antibacterial as well as antimicrobial properties[739].

Myrrh essential oil has been shown to fight infections when combined with frankincense[740].

Foods That Weaken the Immune System

Not everything is good for you. Here are the foods you should avoid or limit during times of higher contagion:

- **Cigarette Smoking.** Smoking undermines the immune system and increases the risk of respiratory infections, pneumonia, and the risk of death from these diseases[741]. Smokers also incur a 2- to 4-fold increased risk of invasive pneumococcal disease. Influenza risk is several times higher and is much more severe in smokers than nonsmokers[742].
  - Recent studies found that 2019-nCov and SARS-nCov share the same receptor, ACE2. There is a significantly higher ACE2 gene expression in people who smoke cigarettes than in those who do not smoke. This indicates that smokers may be more susceptible to 2019-nCov[743].
- **Excessive Alcohol.** Consumption of alcohol impairs the immune system and increases the vulnerability to lung infections[744]. In folk medicine, prescribing small doses of strong spirits like vodka and herbal tinctures is used to kill pathogens locally. It may be beneficial in small therapeutic doses causing a hormetic response in the body[745]. However, having multiple drinks is probably too much.

Sugary alcohol, beer, cider, cocktails, and rum are also higher in sugar and carbs, which will not improve hormesis.

- **Inflammatory Oils and Rancid Fats.** Canola oil, margarine, sunflower oil, and seed oils, in general, are highly inflammatory and damage cell membranes when rancid[746]. Most cooking oils in restaurants and fast food places use vegetable oils that oxidize easily. Even healthy fats like olive oil or roasted nuts can become rancid[747]. You should use minimal heat when processing fats or meats to avoid the production of trans-fats, oxidized fats, and various carcinogenic compounds such as AGEs (advanced glycation end-products)[748].
- **Gluten and Grains.** Even if you're not gluten intolerant, excessive consumption of grains can damage the gut lining and cause chronic inflammation[749]. Refined carbs like pastries and white bread are worse than traditional *sourdough* bread because the latter has bacteria that essentially pre-digest the gluten and increase the nutritional content[750,751]. Chronic avoidance of gluten isn't a smart idea because your body may lose its ability to deal with it as we discussed in the previous chapters. Reducing the consumption of grains drastically will improve gut health and lower inflammation, especially if you are sensitive to gluten and grains in general[752]. During pandemics or higher risk of contagion, it's smarter to avoid these foods until the situation gets better.

- **High Sugar Consumption.** Excess glucose reduces the ability of neutrophils (a type of white blood cell) to ingest and kill bacteria[753]. In a 1973 study, subjects consuming 100 grams of carbs from different sources after an overnight fast reduced the effectiveness of neutrophils by about 40 % for 5 hours[754]. This applied to sucrose, fructose, and honey as well as glucose. The least impactful was starch. Excessive amounts of carbohydrates can also promote low-level inflammation, increase the risk for type 2 diabetes, and make one more prone to infections, for example by increasing nuclear factor-κB activation[755,756,757]. It's also smarter to not eat a lot of carbs or sugar during pandemics because of how it lowers the body's responsiveness to infections.
- **Poultry.** Chicken, turkey, and poultry, in general, have quite an unfavorable fatty acid profile. They are predominantly high in omega-6 fats, especially if the animals have been fed corn or grains. It's not an issue if poultry is your only source of omega-6 but if the diet is already high in omega-6 then it can make things worse. Too high omega-6 to omega-3 ratio has been linked to increased inflammation[758]. Factory-farmed birds are also more prone to infections and viruses due to living in confinement[759].
- **Processed Meat.** Bacon, sausages, dumplings, canned meat, and other processed meats have fairly often sugar, grains, and preservatives added to them which can be pro-inflammatory. Nitrates present in processed meats can

cause harmful compounds such as nitrosamines to form in the gut in the absence of vitamin C[760,761].

- **Toxic Seafood.** Most seafood is high in mercury and other pollutants. Environmental toxins such as dioxins and PCBs are concentrated in fish fat. Toxins become concentrated in long-lived and large predatory fish. Therefore, avoid large fish like tuna, shark, pike, halibut, and trout because they accumulate more heavy metals due to their size and eating habits. Smaller fish like salmon, pollock, krill, sardines, and oysters are lower in heavy metals[762]. Farmed fish can be fed antibiotics as well as grains and other inflammatory foods that produce an unfavorable fatty-acid profile[763].
- **Heavy Metals Such as Cadmium.** Environmental pollution in the form of cadmium (Cd) has been shown to disrupt mitochondrial function and potentiate pulmonary inflammation in animal studies. Cadmium elevates inflammatory IL-4 levels and alters metabolites associated with fatty acid metabolism, leading to increased pulmonary inflammation during viral infection[764].

Supplements That Show No Effect

Over-the-counter availability of supplements and drugs for treating flu symptoms are common, but most of them are ineffective:

- **Multivitamins.** In elderly subjects, multivitamin supplementation of B-vitamins, vitamin E, folate, and vitamin C has not been shown to have any meaningful protection against common infections[765,766]. However, it may be useful, if you have deficiencies or an underlying medical condition, such as type 2 diabetes[767]. Therefore, take a blood test before.

- **Vitamin E** is a common antioxidant and fat-soluble vitamin that improves cellular functioning by preventing the oxidation of cell wall protein structures[768]. It's quite difficult to be vitamin E deficient if you're eating some healthy fats and vegetables. Based on an observational study on 72 000 participants, dietary vitamin E was associated with reduced risk of lung cancer, but increased vitamin E supplementation raised the risk[769].
- **Echinacea** has immunostimulating properties, especially in treating upper respiratory infections[770]. However, more recent systematic reviews have found several trials of poor quality and the health benefits lacking greatly in statistical relevance[771].
- **Fish Oil.** Several animal studies demonstrate that although fish oil is anti-inflammatory, it can impair the production of immunoglobulin A[772] and delay recovery from influenza[773]. However, findings in mouse models may not be directly translated into humans since both organisms have differences at the immunological and metabolic levels. Based on a comprehensive scientific review done in 2019, omega-3 fatty acids ALA, DHA, and EPA exert an inhibitory effect on the activation of immune cells from both the innate and the adaptive systems. Still, some

specific immune functions are promoted by dietary omega-3 fatty acids in specific immune cell types (for example phagocytosis by macrophages and neutrophils or T-regulatory cell differentiation), suggesting that omega-3 fatty acids do not act as immune suppressors[774]. In conclusion, the jury is still out on the use of fish oil in infection prevention.

- **Coughing Medications.** Coughing is a protective mechanism that clears the airways of mucus and pathogens. None of the common over-the-counter drugs such as codeine[775], dextromethorphan (DXM) or antihistamines[776] are effective against the flu or coughing[777].
- **Inhaling Menthol.** Menthol may feel refreshing, but it will not help on nasal patency or cough prevention[778]. The same applies to menthol gums or candies that are said to improve throat infection or the cold.
- **Paracetamol and NSAIDs.** The use of painkillers for flu is common but they are not effective in either shortening the flu or treating the symptoms. Acetaminophen or paracetamol has been shown to increase the duration of colds as it reduces the natural antibody response[779]. In cases of pneumonia, NSAIDs are harmful to the patient and impair neutrophil intrinsic functions, their recruitment to the inflammatory site, and the resolution of inflammatory processes after acute pulmonary bacterial challenge. This means slower recovery from the disease itself[780]. Even more, a 2014 study found that fever suppression with

NSAIDs or other pain killers increases the expected number of influenza cases and deaths in the US[781].

Antifragile Immune System Protocol

- The immune system can develop tolerance to certain infections and viruses by being exposed to them. It's not recommended to deliberately try getting sick but at least you can find some solace in the fact that your body may become stronger against sickness in the future if it goes through some illness every once in a while.
- To keep your immune system strong, you should exercise regularly but not over-train, get enough sleep, take a regular sauna, and expose yourself to sunlight as much as you can. Mild cold exposure can be beneficial but also detrimental in excess.
- Intermittent fasting and nutritional ketosis are also protective against getting sick and they will reduce the time you stay infected. Going into potentially infectious areas or stressful environments in a fasted state is probably safer thanks to increased glutathione and autophagy that would counter-balance any potential viral load. Having ketones present will also reduce overall stress on the body and brain.
- For optimal immune system function, you need vitamin D, zinc, selenium, and CoQ10. Things like vitamin C, E, and B3 can be used to catch infections in their infancy but they are less effective if taken all the time. High-dose

antioxidants can weaken immunity if you're not sick because of reducing the beneficial ROS signaling that's needed for hormesis.

- The best foods for a stronger immune system are sulfur-rich vegetables, eggs, meat, organ meats, and fermented foods. Certain herbs and spices like turmeric and licorice root are also great but not if taken excessively. If you're not sick, stick to the baseline intake of these plant hormetics as discussed in the previous chapter. Increasing their consumption during infections, or sickness may work in accelerating recovery.
- Eating processed foods, fried foods, a lot of sugar, and oxidized fats will not only weaken your immune system but also slows down healing. You should avoid putting extra ROS on your body by smoking or drinking. Limiting the ingredients you react to like gluten, lectins, or FODMAPs during sickness will probably reduce the likelihood of catching something during times of higher contagion as well as alleviate the symptoms.

Chapter Six:
Of Heat and Cold Adaptation

"Because as it turns out, exposing your body to
less-than-comfortable temperatures
is another very effective way to turn on your longevity genes."

— David A. Sinclair

„Okay, I now officially don't feel my toes anymore," said the high-school version of me at the back of an open cargo truck. We had already been driving for about 2 hours at a temperature of -30ºC (-22ºF) and there was about as much time to go. Even worse, the wind was howling from all angles, penetrating the bones of me and my companions.

I was at the back of an army Unimog, transporting a small squad of 8 sharpshooters which I belonged to. There was an entire company of us from different battalions in other trucks, heading to the same location. We were chosen to get our advanced training from actual snipers.

It was one of the coldest feelings I've had to endure in my life – several hours of being completely jackhammered (a condition in which you can't stop shaking) and thinking I was going to lose my toes. One of the guys from another truck almost did when he fell into a small pond during a toilet break. Being the fool he was, he didn't tell this to anyone and continued to ride along as if nothing had happened. When we had reached our destination, his toes were literally blue and he had to be transported to a local hospital. Fortunately, he kept all his extremities.

When we arrived at our then to be home for the next week, none of us felt better. It was the pit of an abandoned strip mine with a stone floor and a wind tunnel. The temperatures were as unforgiving as -30ºC and it got even down to -35ºC at midnight. And yeah, we had to do night patrol for at least an hour until we could change shifts with someone else. Sleeping in a tent with a small stove was the only time we could stop shivering and enjoy a brief moment of warmth.

During the day we trained sharpshooting the details of which I won't get into. Basically, camouflage (aka lying motionless in the snow and trying to crawl towards a target), measuring distances with binoculars, making landscape maps by drawing on a piece of paper, firing long distances, and so much more. The unfortunate part was that you were much more accurate and successful if you didn't have any gloves on. It was much more effective to pull the trigger with your bare fingers, but the consequences were also higher. Even several years afterward my fingers tend to flare up with redness during the winter when they're exposed to the cold, resembling some mild neuropathy. Nevertheless, it was something I had to do in order to be a good sharpshooter.

Time spent in the military sparked my interest in cold exposure and how humans have managed to adapt to various climates. Me and my squad were constantly either drenched in the rain, sleeping underneath a tree, crawling through the snow, making our way through the swamp, crossing waist-deep rivers, or just walking for hours through the forest in the middle of a cold winter night. It definitely conditioned me to handle any kind of temperatures I might come across, but it also taught me unknowingly how the human body responds to stress.

In this chapter, I'm going to talk about the science of cold exposure and what's the hormesis protocol for practicing it yourself. The other side of the coin is heat and humidity, which I'm not a stranger to either. We'll cover the benefits of taking a sauna, what kind of other high-temperature hormetics you can try, and how to adapt to thermal stress.

Cold Exposure and Cold Shock

Living organisms have evolved many mechanisms for dealing with environmental stressors and changing conditions. During cold shock, cell membrane and enzyme activity decrease, and you reduce the efficiency in mRNA translation and transcription[782]. Everything kind of slows down.

Cold-Shock Proteins (CSPs) are multifunctional RNA/DNA binding proteins, characterized by cold-shock domains (CSDs) that fix misfolded proteins and RNA[783]. They're one of the most evolutionarily conserved proteins found in virtually all organisms[784, 785]. In humans, the predominant proteins from CSD are the Y-box protein family. The most known of them is Y-box protein-1 (YB-1). It's a potential target in cancer therapy[786]. Other cold shock proteins in humans are Lin28, calcium-regulated heat-stable protein 1 (CARHSP1), PIPPin, and upstream of N-RAS (UNR)[787].

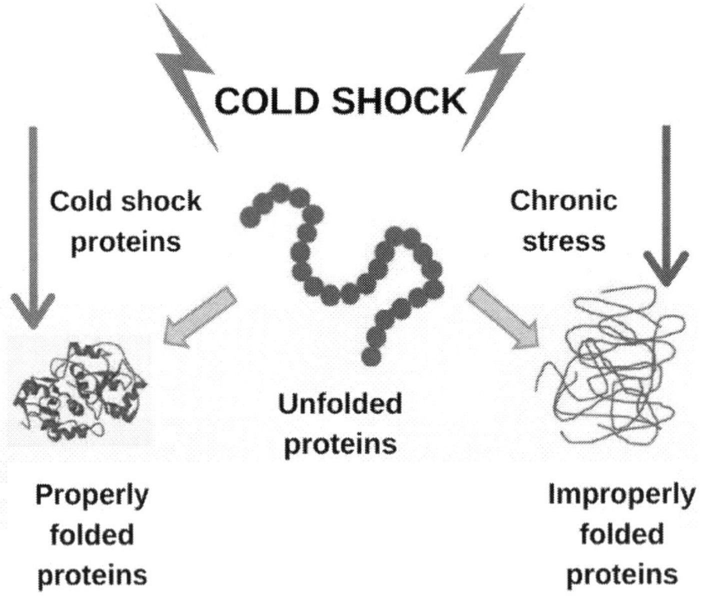

The human body tries to maintain a homeostatic temperature around 36-37°C (98°F)[788]. Deviations disrupt homeostasis and cause adaptive changes to stay in that range. Proper conditioning leads to increased tolerance and stress adaptation through hormesis. You'll begin to tolerate either a higher or lower climate based on the exposure your body has received.

Cold metabolism creates two types of reactions: insulative actions (redirection of blood flow away from extremities) and metabolic actions (increased metabolic rate to produce heat)[789]. If one of them is limited i.e. you're not shivering, then the other will compensate for it by raising metabolism[790]. Insulative reactions include things like goosebumps, seeking warmth, and decreased blood flow[791]. Once insulative reactions are exhausted, the body recruits non-shivering adaptive thermogenesis (NST), which regulates mitochondrial uncoupling and energy production in skeletal muscle[792]. Shivering burns mostly fatty acids but in higher intensities can tap into glycogen[793,794].

Benefits of Cold Exposure

When cells experience stress related to coldness, they activate cold shock proteins, which begin to regulate gene expression. The benefits include:

- **Cold exposure increases norepinephrine, which promotes focus, vigilance, attention, and mood[795].** Taking cryotherapy for 3 weeks has been shown to lower depression and anxiety in people with depressive disorders[796]. This heightened adrenaline is useful during some parts of the day but not chronically, especially not at night.
- **Coldness can decrease inflammation seen in arthritis[797].** Patients with arthritis report significant reductions in pain by just taking a 2-minute cold shower every day for a week[798]. Taking an ice bath or a cold shower is amazing for reducing muscle soreness or any pains. It's great for recovery as well as making you more supple again.
- **Therapeutic cold therapy can help with neurodegenerative disease** by suppressing neuronal apoptosis and inflammation[799]. One of the proteins CARHSP1 binds to and stabilizes tumor necrosis factor (TNF)[800]. Cooling babies to 33°C for three days after birth is used to treat hypoxic-ischemic encephalopathy or poor

oxygenation in the brain[801]. It reduces brain damage and increases the infant's chances of survival.

- **Cold increases adiponectin, which is a protein that helps with blood sugar regulation[802].** Coldness also promotes glucose uptake similar to GLUT4 and can be used to prevent type-2 diabetes[803]. Sitting in a room at 58°F for 2-6 hours a day for 10 days improves insulin sensitivity in T2 diabetics by 43%[804].
- **Cold shock protein YB-1 is important for embryonic development and survival[805].** UNR maintains the pluripotent state of embryonic stem cells[806]. Mild algidity signals the body how to prepare for the environment in which it will find itself in. From an allostatic perspective, it ought to equip one to handle any potential drops in average temperature i.e. the winter, seasonality, or an unexpected ice age.
- **Coldness activates brown adipose tissue, which improves mitochondrial functioning, metabolism, and thermoregulation[807].** It increases energy expenditure and metabolic rate[808]. Cold exposure stimulates lipid metabolism, burns white adipose tissue, and decreases triglycerides[809]. UCP1 or thermogenin gets increased, which promotes stress adaptation, redox balance, and browning of white fat into brown fat[810].
    - A 10-day experiment on obese men showed that sitting in a cold room (58°F) for 6 hours a day increased their metabolic rate by 14%[811]. Another

study put 11 young lean men into a cold room (67˚F) while wearing a cooling vest (62˚F) for about 90 minutes. After just 30 minutes their metabolic rate had increased by 16.7% with fat-burning rising by 72.6%[812]!

- There is a hypothesis called *Uncoupling to Survive,* which suggests that increased mitochondrial uncoupling and thus increased energy expenditure might increase longevity by preventing the formation of reactive oxygen species (ROS)[813]. Expression of uncoupling protein 1 (UCP1) in skeletal muscle mitochondria increases lifespan considerably[814]. UCP1 or Thermogenin is an uncoupling protein found in the mitochondria of brown adipose tissue that's used to generate heat through non-shivering thermogenesis. Fasting, exercise, heat, and cold exposure promote UCP1, which in turn increases heat-shock proteins and lipid metabolism.

- **Sleeping in slightly colder temperatures improves sleep onset, time spent in deep sleep, and overall sleep satisfaction[815].** It can also help with the production of melatonin and regulation of the circadian clock[816]. In my opinion, sleeping in a hot environment is almost impossible and very uncomfortable. The best night's sleep I ever had was during the military when we were forced to crawl into a fetal position to get our shut-eye.

- **Winter swimming lowers uric acid and increases glutathione**, which promotes detoxification and the body's antioxidant system[817]. Immersion in modestly cold water lowers heart rate, cortisol, systolic blood pressure, and inflammation after the fact. It's also been shown to improve general well-being[818] and activate brown fat[819]. However, it also increases oxidative stress and lipid peroxidation in the short-term as a way of adapting to the stressor[820].
- **Cold exposure activates the immune system similar to exercise and can strengthen immunity[821].** In rats, the cold predisposes sickness to infections but there's no link in humans[822]. A daily hot to cold shower for 30 days resulted in a 29% lower self-reported absence from work due to sickness in healthy adults[823]. There was no difference in illness days.

I want to take a closer look at a study about 1-hour immersions into waters of different temperatures (32°C, 20°C and 14°C). At 32°C, rectal temperature and metabolic rate did not change, but heart rate lowered by 15%. Plasma renin activity (PRA), plasma cortisol and aldosterone concentrations were also lowered by 46%, 34%, and 17%, which regulate your blood pressure. Diuresis (excretion of urine) was increased by 107%. 20°C caused a similar decrease in PRA, heart rate but increased metabolic rate by 93%. 14°C raised heart rate and systolic and diastolic blood pressure by 5%, 7%, and 8%, respectively. So did diuresis (by 163%) and aldosterone (23%), while cortisol decreased. Now for the real kicker: immersion in 14°C water for 1 hour increased metabolic rate by 350%, norepinephrine by 530%, and dopamine by 250%[824]! That's pretty damn amazing – a storm of neurotransmitters and excitement – the kind of wokeness I prefer.

Cold Exposure and Longevity

When E. Coli are exposed from 37 Celsius to 10 degrees, they experience a 4-5-hour lag phase after which they resume growth at a reduced rate[825,826]. During the lag phase, the expression of cold shock proteins increases by 2-10-fold[827]. This is similar to the Dauer stage in roundworms or metabolic hibernation that may have anti-aging effects.

It's been found by Miguel et al (1976) that flies who live at 21°C rather than 27°C live twice as long and those who live in 18°C three times longer[828] (See figure below). There's a linear association between survival and time spent at 21°C or lower. Roundworms who live in 5°C lower temperatures have a 75% increase in lifespan[829]. Many studies in insects find a negative relationship between temperature and lifespan[830,831].

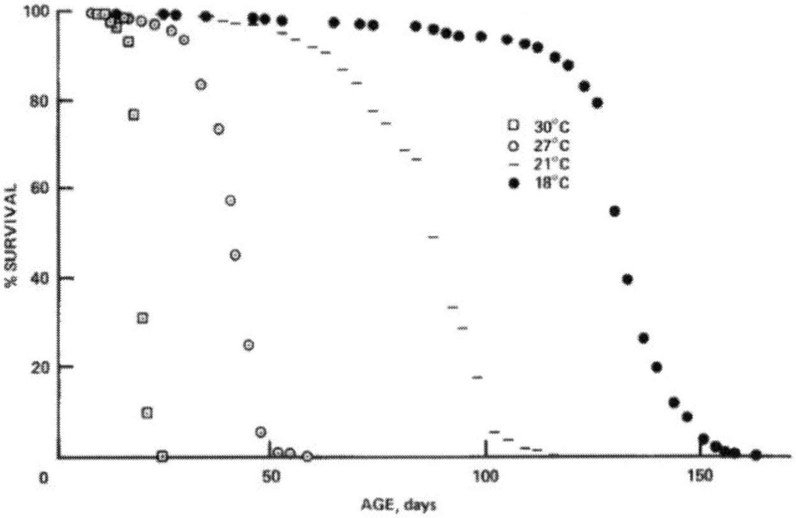

Source: Miguel et al (1976) Mech Ageing Dev. 1976 Sep-Oct;5(5):347-70.

However, this doesn't entail a freezing, infertile, and fragile life as one might think. The same study by Miguel et al (1976) showed that cooler conditions increased vitality as expressed by mating (See figure below). Hotter temperatures like 27°C show a substantially lower rate of copulation than cooler ones. Of course, this is a study done on flies, but it at least refutes the idea that living in slightly harsher conditions immediately disables the possibility to have sex or anything the like.

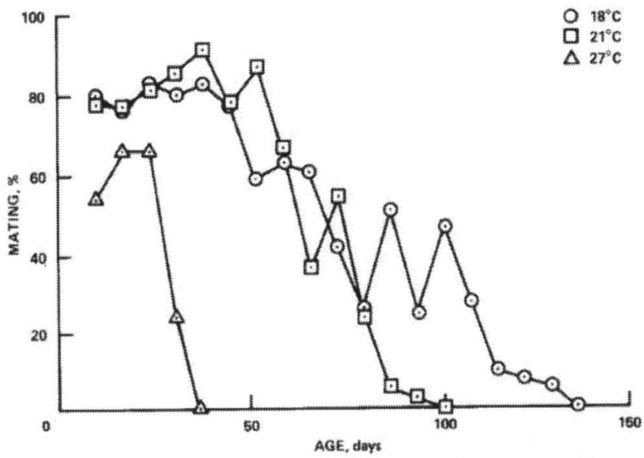

Fig. 8. Effects of exposure to various temperatures on vitality, as expressed by mating. The points indicate the percentage of males which started copulation during a period of 11 min. The results are similar to those shown in the preceding Figure.

Source: Miguel et al (1976) Mech Ageing Dev. 1976 Sep-Oct;5(5):347-70.

Cold shock triggers the assembly of stress granules in mammals, which are implicated in many neurodegenerative diseases[832]. They form when mRNA translation is stalled or blocked[833]. When the breakdown of stress granules is defective, the granules that would undergo autophagy become lysosomes instead. The accumulation of these stress granules can promote disease. Even if you are doing cold therapy, it's important to have adequate autophagy because it will help to eliminate the stress granules that assemble during stress. Perhaps doing cold exposure while fasting with elevated autophagy is better for quick elimination of these stress granules and aggregates. That might also explain why most people may react negatively to any form of temperature stress or hormesis. It's because their basal autophagy is already too low due to inhibiting it with diet and eating too frequently. This, in turn, leads to the accumulation of stress granules that would be removed by autophagy otherwise, thus leading to potential disease or other side-effects.

Both cold and heat stress have similar properties, such as hormesis, thermoregulation, reduced inflammation, and improved cardiovascular functioning. Heat-shock proteins can also induce autophagy[834]. However, exposure to cold does not increase autophagy directly. It can encourage it through other pathways, which we'll talk about later in this chapter.

FOXO proteins are transcript factors that regulate longevity through the insulin and insulin-like growth factor signaling[835]. They are activated by hormetic stressors like fasting[836], calorie restriction[837], exercise[838], heat[839], and cold[840]. FOXO3 induces FOXO1-dependent autophagy activating the AKT1 signaling pathway[841]. FoxO transcription factors promote autophagy in cardiomyocytes[842]. AMPK, another stress adaptation sensor, contributes to autophagosome maturation and lysosomal fusion[843]. This regulates autophagy through mTOR and ULK1[844].

Combining ice baths or cold showers with saunas is an amazing way to boost the effectiveness of the heat. Winter swimming with saunas has been shown to trigger hormetic adaptation as well as increase lysosomal enzymes[845]. Exposure to mild cold and rewarming induces autophagy[846]. Cold shock proteins and heat increase a protein called LC3, which is associated with the inner membrane of autophagosomes and increased autophagy[847].

Dark Side of the Cold

However, too much stress and cold can have really bad side-effects. Lin28, one of the cold-shock proteins in humans, can promote cancer stem cells[848]. YB-1 can also promote pancreatic cancer metastasis[849]. It might have to do with how freezing thermogenesis accumulates stress granules in the absence of autophagy. In that case, the risk of negative side-effects would be much lower in already healthy people with sufficient metabolic flexibility. However, we should still be wary of all the potential consequences even if you're on point.

Here are the other negative side-effects of cold exposure you should be aware of:

- **Coldwater immersion before exercise decreases maximum power output, heart rate, and overall performance in elite cyclists[850].** You want your muscles to be warmed up and supple to prevent injuries. Some studies show cold therapy can reduce muscle soreness after exercise[851], but systematic reviews find no significant difference[852]. Personally, I feel they work amazing for accelerated recovery, but I know it can come at a cost.
- **Post-workout cold water immersion can attenuate the anabolic signaling and long-term adaptations in muscle to strength training[853].** It's similar to antioxidants that block the ROS and inflammation, which you need in some amounts to make the body adapt. That's why I'm never taking cold showers or anything similar after a workout.

- **Cold exposure burns calories and increases metabolic rate, but it can also make you hungrier[854].** It doesn't matter how much fat you're burning if you still compensate for it by over-eating. Furthermore, if you over-do the cold or become stressed out, then you can also lower thyroid functioning and slow down the metabolism.
- **Although the cold doesn't appear to directly suppress immunity, it might do so when combined with high physical exertion or over-exposure.** Intense exercise is known to transiently decrease immune functioning[855]. So, if you deplete your adaptive energy with cold exposure, you may just get sick more easily. Especially if you get exposed to the wind or a draft afterward.
- **Viruses are more stable in colder and drier conditions, which makes them survive for longer[856].** Influenza strands prefer winter climate, which is why we experience them seasonally[857]. Human Rhinoviruses like the common cold replicate more effectively at temperatures lower than 37°C, or 98.6°F[858]. So, if you're constantly freezing or hypothyroid, your body may cool down to below 36°C, which makes it easier for viruses to survive. Another reason a higher metabolic rate and thyroid functioning are beneficial. Cooling yourself down during a fever may lower your body temperature but it also blunts some of the healing processes the increased heat is trying to accomplish.
- **Keeping the testicular area between 31-37°C (88-99°F) improves sperm production and DNA, RNA, and**

protein synthesis[859]. That might explain why the flies engaged in more mating at lower temperatures. However, cold exposure seems to have no effects on testosterone and it may decrease T levels in men[860][861]. In my opinion, it's due to increased cortisol and stress. So, you have to be aware of the other stressors in your life.

- **Chronic cold exposure can cause neuropathy in your fingers and toes**[862]. That's what I'm still battling with every once in a while because of my army days. During the winter and autumn, my hands tend to flare up and turn red whenever they're exposed to the cold. Chilblains, frostbite, and urticaria are just another phenomenon of this.

All these hormetic stressors like fasting, heat, and cold are most effective as disease prevention, not treatment. You want to keep doing them while you're already healthy as it'll protect against illness but chronic exposure is not optimal. What's more, they shouldn't be followed in a constantly linear progression, meaning you're consistently increasing your dose. It will make you more resilient for sure but at the same time will make it harder to keep making progress. That's why there are certain situations and times of the year better suited for training as well as recovery.

How to Do Cold Exposure

The easiest and most obvious place to start would be with cold showers. It would be too difficult to jump straight away to 2-hour ice baths but if you dedicate yourself completely to this practice then someday you might be able to pull it off. As you progress through the stages of the cold exercises, you will begin to understand the body on a deeper level.

- **Phase 1 Cold Shower After a Hot One.** Try to control your breaths and lungs. Instead of gasping, breathe with ease. Regular habituation will improve the entire vascular system. Whenever you have a shower, turn it all the way down for at least 10-30 seconds and start conditioning yourself that way. There's no reason why not to have this small cold shock.
    - **Contrast Showers** are in my opinion the best ones for training as well as hormesis. You're essentially switching between 10-30 seconds of warm water and 10-30 seconds of cold. It stimulates the lymph system much better and feels more like a workout rather than complete freezing. In total, you can alternate these contrasts for 2-5 minutes, depending on how much time you have for taking a shower. And yeah, always finish with cold.
- **Phase 2 Straight to Cold Showers**. Before you even begin, your body temperature can already drop because of your mind anticipating the shock. Breathe naturally and you will be able to steer your mind towards adaptation and

consciously regulating the autonomic nervous system. Pure force doesn't work because the body will begin to fight back. Having just a cold shower without the warm one is supposed to condition you to endure coldness for longer periods as well as teach your body how to start generating its own heat both during and afterward. If you come out of a cold shower your skin is still wet and somewhat chilly. This requires a different kind of mental adaptation. Instead of drying yourself off immediately, let the water evaporate itself or just wipe it off with a light towel. You don't have to do it every day, but I'd try to have one at least once or twice a week, especially if you don't have access to an ice bath or cryo-chamber.

- **Phase 3 Cold Water Immersion.** You need to either find a lake, stream, pond, sea, or fill up your bathtub with colder water. Adding ice cubes or snow in there might be inconvenient and costly if you have to do it frequently. As a shortcut, you can use pre-packaged ice cubes in bags or dedicate a separate freezer for your plunges.
    - Same kind of breathing but add in visualizations. Visualize heat generating within your body just before you enter. With every breath, make this sensation more intense and keep your mind focused on the heat. Stay in the cold water for as long as it feels comfortable. If you feel pain or uneasiness, then it's time to get out. It takes at least a minute for the adaptations to set in.

On a daily basis, I take a brief contrast shower or a warm one for relaxation. Once or twice a week or after the sauna I'll either have an ice bath or just a cold shower. When I was doing polyphasic sleeping for 100 days, I had also been taking a cold shower every morning for 2 years straight. No excuses. I remember waking up at 1 AM, walking to the shower, and shocking myself with some glacial water every day. It was brutal and quite masochistic in hindsight. So, I'm the kind of guy who doesn't recommend purely cold showers because I couldn't tolerate it. I just think a contrast shower is healthier and more effective in stimulating lymph flow and mitigating stress.

To maintain control over the core temperature, you have to influence the body by steering the hypothalamus, the thermostat in the brain. Imagine heat in your lower stomach. With each breath, you're inhaling fire, that fills your body and exhaling out the cold. Go somewhere warm with your thoughts.

There are two parts to resisting cold:

- **The first is physical,** the pure fact of getting used to it. When we're habitually conditioned in a way that promotes lower temperatures then it will gradually have lesser of an effect on us. This can be accomplished exposing ourselves to the weather.

- **The other part is mental**, which derives from our perception of the experience. This is also the result of changing our mindset. By default, cold is uncomfortable and associated with pain. It forces us to spend more energy which the brain is trying to preserve. An evolutionary response which, unfortunately, is working against us most of the time.

When we're shivering, we'll inevitably try to run away from the situation by any means necessary. At first, it will not be enjoyable. Our muscles will tense up, we'll start to shiver, our thoughts will scream: *„Let's get the hell out of here!"* This is only our habitual response. Instead of following our reaction, we should experience the cold for what it truly is. Feel the hair of our skin stand up, how the wind cuts through our bones, what breathing in does to our nose etc. By yielding to the situation and becoming the observer we can notice ourselves in the midst of it all and take control of our urges. It's an enlightening experience and conditions our willpower in other areas of our life.

Once we've taken control of our breath, we can do so with the urge to escape the situation. It's the result of becoming conscious in the present moment. The most important thing is to remain calm in both body and mind. Rather than tensing up like we normally would we need to let loose and yield to the cold. By cramping up we create a habitual pattern of reaction which will always make the experience uncomfortable. If, however, we begin to interpret it as something enjoyable then the necessary change in mindset will eventually happen.

Benefits of Sauna and Heat Shock Proteins

Next up, I want to cover the opposite of the cold, which is heat exposure. Fortunately, it's something I've developed a close relationship with as well. You see, I'm born and raised in Estonia where saunas have been a part of the culture for centuries. I recall my first sauna experiences dating back to kindergarten when me and my brother would sit in the heat. Since that time, I've taken a sauna for at least once a week for over 20 years.

Many cultures have practiced both fasting and some form of heat therapy for over a millennia. Saunas are basically sweat lodges or heated rooms. It's not just ancient broscience. Now we have a lot of studies showing their effects on longevity. Hyperthermic conditioning as it's called has multiple amazing health benefits.

- **Taking a sauna has been shown to improve cardiovascular functioning and lower the risk of heart disease[863].** One of the main killers in modern society are heart disease and strokes. In the US, over 610 000 people die to heart disease every year – that's 1 out of 4 deaths[864]. Heart disease is caused by chronic inflammation, too much stress, poor blood circulation, high cholesterol levels, and not enough exercise.
    - In a Finnish study, people who used the sauna 2-3 times a week had 22% less chance of dying to a sudden cardiac event than those who used it only once a week[865]. Those who went to the sauna 4-7 times a week were 63% less likely to experience cardiac death and 50% less likely to die from

cardiovascular disease compared to those who used it once a week. They were also associated with a 40% reduced risk of all-cause mortality[866].

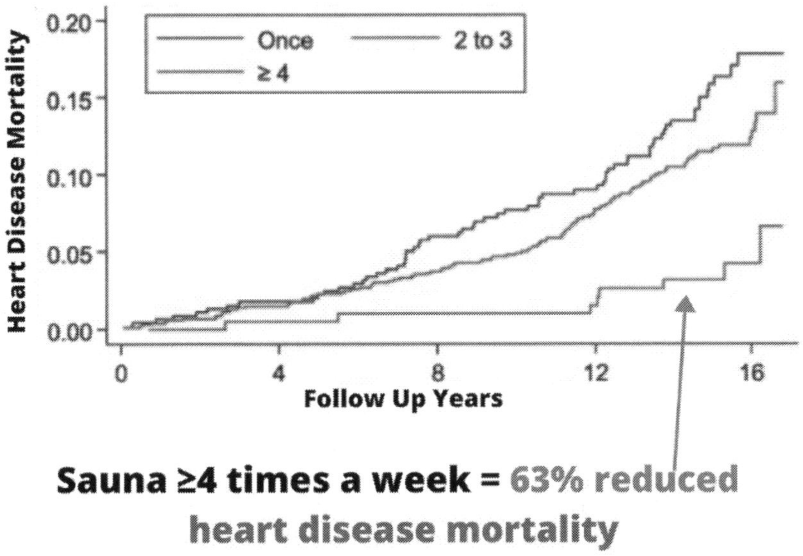

Source: Laukkanen et al (2018) BMC Medicine, 16, 219

- **Saunas improve insulin sensitivity** by increasing the expression of a glucose transporter called GLUT4 that helps to clear the bloodstream from sugar and directs it into muscles. Just 30-minutes of hyperthermic conditioning 3-times a week for twelve weeks has resulted in a 31% reduction in insulin and blood sugar levels[867]. This can be useful for managing glycemic variability and symptoms of diabetes.
- **Better blood circulation and blood flow to skeletal muscle[868],** which can reduce the rate of glycogen depletion and increases the efficiency of oxygen transport to muscles[869].

Hyperthermic conditioning has been shown to reduce muscle glycogen use by 40-50%[870] and lower lactate accumulation during exercise[871]. It promotes physical endurance by increasing the heart's stroke volume.

- 30-minute sauna sessions after working out 2 times per week for 3 weeks have been shown to increase the participants' run until exhaustion by 32% compared to baseline[872]. It can also enhance plasma volume by 7.1% and red blood cell count by 3.5%. This not only improves endurance but can also help with muscle growth and resistance training.
- Heat adaptation can reduce the amount of protein degradation during exercise, which helps to establish a positive *net* protein synthesis for the day, resulting in muscle hypertrophy. In rats, 30-minutes of hyperthermic conditioning at 41°C (105.8°F) increases heat shock proteins in muscles, which correlated with 30% more muscle regrowth than the control group during 7 days after they had been immobilized[873]. Basically, taking an easy sauna helped the rats to regrow their damaged muscles faster.

- **Saunas are amazing for speeding up recovery from exercise and any kind of stress**. It helps you lower inflammation created from working out, shuts down muscle soreness[874], and also relaxes you completely so your body could switch from the sympathetic mode into the parasympathetic rest and digest mode. This doesn't seem to impair recovery the same way antioxidants or cold exposure does.

- **Heat stress also releases massive amounts of growth hormone, which will inhibit protein breakdown further[875, 876].** Growth hormone stays elevated for several hours after the sauna and it has incredible anti-catabolic effects that prevent muscle breakdown and can promote additional fat burning. If you're going for extended long fasts, then it might be a good idea to take a short sauna in the fasted state to preserve more lean tissue.
    - Two 20-minute sauna sessions at 80°C (176°F) separated by a 30-minute cooling period can raise growth hormone two-fold over baseline[877].
    - Two 15-minute sauna sessions at 100°C (212°F) dry heat separated by a 30-minute cooling period increase growth hormone five-fold.
    - Two 1-hour sauna sessions a day at 80°C (176°F) dry heat for 7 days was shown to boost growth hormone by 16-fold on the third day[878].
- **Strengthens the immune system and increases white blood cell count.** It also flushes the lymph system from toxins and pathogens. As a result, you'll get sick less often and have clearer skin. Sweating, in particular, helps to eliminate bioaccumulated toxins, heavy metals, and other infectious particles[879].
    - Sauna therapy has been used to combat influenza since at least the 1950s.[880] One study found that

taking the sauna 2-3 or more than 4 times a week resulted in a 27% and 41% reduction in respiratory disease compared to those who did it once a week[881]. They also saw a 33% and a 47% reduced risk of pneumonia[882]. The main preventive strategy against typhus in Finland during World War II was a regular sauna[883]. Consistent sauna practice for a few months has been shown to cut episodes of the common cold in half compared to no heat exposure[884].

- The WHO has stated that heat at 56°C kills the SARS coronavirus at around 10000 units per 15 min, which is a very significantly rapid reduction[885]. At room temperature, the virus can remain stable for 1-2 days and 21 days at 4°C and -80°C. Although the temperature isn't going to get that hot inside your lungs where the virus lingers, it can still speed up its destruction.
- During the 2003 SARS pandemic, a Hong Kong doctor noted that higher body temperatures may kill the coronavirus[886]. They saw that patients who had a temperature above 37°C saw a milder immune response and recovered faster compared to those who were below 36°C. A higher fever can kill the virus, which is why it's recommended to not use fever lowering medication – it's the body's response to dealing with the infection.

- **Hyperthermic conditioning lowers the risk of dementia and Alzheimer's disease[887].** It increases endorphins and beneficial brain neurotrophic factors. Being in a sauna feels like easy cardio that's great for the brain.
- **Saunas make you tired in a good way.** It doesn't make you more energized because more often than not you'd like to take a nap afterward. My sleep scores based on my OURA ring are already pretty high but I notice a consistently high deep sleep from using the sauna.

In the past, saunas were used for cleaning and self-washing. It was the cleanest place in the household – children were given birth in saunas and the dead were washed in saunas before burial as well. Even some business and political deals are made in the sauna where everyone is deemed to be equal. Especially when you're sitting there naked…

Effects of Heat Shock Proteins

When you're exposed to high heat whether that be in a sauna or while exercising, you're generating mild oxidative stress. The body responds by turning on its defense mechanisms, many of which we've talked about already.

Heat shock proteins (HSPs) are a family of proteins that get produced in response to heat stress and they help to adapt to the stress through hormesis. They get released under environmental conditions of inflammation, heat stress, starvation, hypoxia, or even water deprivation[888].

- HSPs prevent the accumulation of free radicals and cellular damage[889]
- HSPs repair damaged, misfolded proteins similar to autophagy[890]
- HSPs promote cellular antioxidant capacity with glutathione[891]
- HSPs are involved in macroautophagy and cellular turnover[892]
- Hsp20 phosphorylation correlates with smooth muscle relaxation and has a significant role in cardiac myocyte function and skeletal muscle insulin response[893894]
- HSP mediated heat exposure has been shown to increase the lifespan of flies and worms up to 15%[895,896,897]

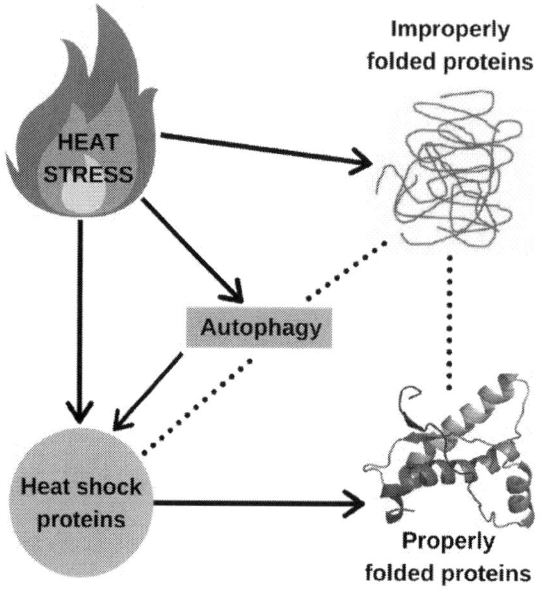

Hormetic heat stress and HSF-1 induce autophagy to improve survival and proteostasis in C. elegans[898]. Worms who were deficient in autophagy failed to benefit from the heat-shock whereas the ones with intact autophagy did. This may indicate a lot of the benefits of heat-shock proteins are mediated by autophagy.

Furthermore, you can see greater benefits from heat stress with elevated autophagy (while fasting) as in the case of cold exposure. Mitochondrial autophagy protects against heat shock-induced apoptosis, which makes it even more important for stress adaptation[899]. If you just sit in the sauna while blocking autophagy you may be causing more cellular damage than necessary. That's why exposure to heat may be less stressful with elevated autophagy similar to the cold.

Dark Side of the Heat

As with all of the practices discussed in this book, the sauna and excessive heat can have some negative side-effects. We just have to be aware of them and know where our tolerance lies.

- **Staying in the sauna for too long can cause dehydration, arrhythmia, heat exhaustion, electrolyte imbalances, hypertension, and even stroke**. If you have cardiovascular issues, then keep the temperatures mild and don't combine it with a cold plunge. There are reports of people getting heart attacks from this sort of drastic alterations[900].

- You can also lose a lot of electrolytes and minerals through sweating, especially if you don't hydrate properly or aren't getting enough salts. This will add an additional burden on the cardiovascular system and may manifest as chronic fatigue. I'd take at least ½ teaspoons of salt mixed in water before a sauna session and drink plenty of water afterward.

- **In Finland, one out of four burns is sauna-related and at least one sauna burn a day requires hospitalization[901].** Those crazy fins with their saunas… These cases aren't usually serious but one 64-year old man in Germany fell face-first onto his sauna stove and died from the injuries[902]. It's especially important to ensure you don't fall asleep or pass out in the heat because you might not wake up again.
- **In 2010, a Russian man named Vladimir Ladyzhensky died at the Sauna World Championships[903].** He was in the finals with the Finnish five-time champion Timo Kaukonen and they both passed out after six minutes of 110°C (230°F). Kaukonen survived with severe burn injuries but Ladyzhensky had to be dragged out and he almost immediately went into cramps and convulsions. After this tragedy, the event would no longer be organized. It comes to show how seriously people in Russia and Finland take the sauna.

- **Too much heat, especially around the testicles, can reduce sperm count[904].** If you're concerned, you may want to occasionally splash some cold water onto the testicular region while sitting in the sauna. Taking a sauna during early pregnancy may also cause embryonic or fetal abnormalities[905].

Saunas aren't going to make you burn fat but you may lose some water weight. The heat and increased heart rate does burn a little bit of calories but not any more than regular walking. So, you shouldn't think of it as a weight-loss tool. Unfortunately, there are reports of people with bulimia combining the sauna with laxatives, and diuretics, which made them experience serious dehydration[906]. That can be dangerous and cause body dysmorphia.

How to Take a Sauna

Hyperthermia kicks in once body temperature rises above what's normal. In general, it starts at higher than 37.5°C (99.5°F)[907]. The average temperature for the human body is around 36-37°C (98°F)[908]. From a clinical perspective, fevers are considered significant after they reach 38°C (100.5°F). Hyperthermic conditioning is just a way to simulate the effects of high heat and elevated core feverishness. The minimal effective dose for thermal hormesis appears to be about 38°C (100.4°F).

Based on the research I've just outlined, the optimal dose for taking a sauna is about 15-30 minute sessions at about 70°C to

100°C (156-212°F) about 2-4 times per week. Even once every 7 days will be better than nothing. More isn't going to be necessarily better and you won't gain increasingly more health benefits, especially when it comes to the temperatures. You will only become more resilient against the heat and can physically tolerate it better. If you get too used to the warmth, then you'll have to induce higher stress to gain the effects of hormesis. That's why I don't think it's a good idea to have a sauna every day. Having it a few days apart is probably wiser.

Me, I like to have about 30-minute sauna sessions on my rest days to promote recovery from workouts. If I've just done resistance training, then I might hop in for 10-15 minutes but at that point, I don't necessarily want to cause additional heat stress on the muscles because it may interfere with recovery by shutting down the beneficial adaptation of working out. Although research shows the heat can accelerate recovery and additional hypertrophy, I choose to be conservative with it post-workout and more aggressive on rest days. In total, I may get about 2-3 sauna sessions per week. Having a sauna once a week is better than nothing and you can still gain the benefits. Hell, even once a month would be still worth it. In that case, you may want to have a 15-20 minute session followed by a short cooling off and then another 15-20 minute session.

Children between the ages of 2-15 show similar cardiovascular and hormonal benefits as adults[909,910]. However, the ability to maintain stroke volume may be impaired in younger children (age 2-5) who have the highest heart rate[911]. Generally, it's advised to let children spend only about 5-10 minutes in the sauna and have them sit on the lower benches where the temperature isn't that high. That's how I and my brother started.

You shouldn't drink alcohol before or during the sauna because it can promote stroke and death[912]. The heat itself already increases your heart rate and the intoxication from alcohol will raise your blood pressure, which can be bad for the heart. I know it's pretty common to have sauna-beer and -booze here in Northern European countries but it's pretty irresponsible. Taking a sauna when sick or with a fever isn't a good idea either because it will impose too much stress on the body. The fever is already a natural heat shock protein response that tries to heal but over-doing it is probably harmful.

It's important to keep yourself hydrated before and after the sauna as well to prevent dehydration, fainting, hypoglycemia, or loss of electrolytes. That's why have some water nearby to drink in between sessions. However, don't drink the water from sauna buckets. An Algerian university analyzed the water from ten Turkish baths and found that 50% of them had fecal contamination[913]. Yikes!

Traditional Sauna VS Infrared Sauna

Lastly, let's talk a bit more about the differences between infrared saunas (IR) and regular sauna. The effects of a hot bath are completely different so we'll stick to the saunas.

- Both improve blood circulation and cardiovascular health. The higher temperatures at a traditional sauna do put more stress on you, which can be great for exercise performance and heat tolerance. However, if you have hypertension or blood pressure issues, then you have to be careful with not overdoing it because you can easily pass out.

- Both promote relaxation, reduce inflammation, and soreness. They're best taken after a workout but any other time of the day works. Generally, I like to have it in the afternoon. When it comes to the infrared sauna, then I think you can safely do it every day because it targets a different system and isn't as hot. It's an amazing tool for relaxation and daily detoxification.
- Both can make you sleep better and speed up recovery. Heat exposure too close to bedtime may disrupt your sleep by keeping your heart rate elevated. Too much heat can keep your body in a stressed-out state.
- Traditional sauna can be heated up much higher up to 100°C whereas IR stays at 60°C. However, at that temperature you're not getting increased health benefits, you're just making your body endure the heat much more, which has primarily an exercise performance effect but not deeper detoxification or anything the like.
- Infrared saunas penetrate deeper into the skin and joints where they'll trigger mitochondrial density and collagen synthesis. This is better for the skin and cell clean-up.
- You have to heat the traditional sauna for at least an hour. If you're using firewood or even an electrical stove, you generally have to wait at least 45-60 minutes until the

temperature in the room reaches the right zone. Otherwise, you're just sitting in a cold room with no clothes on. The benefits of IR sauna come from the infrared wavelengths that don't require a specific temperature to work. Generally, it takes about 30-45 minutes for the IR sauna to reach 60°C but you could even just sit in there right after turning it on and still get the collagen benefits from the wavelengths. It's just that most people prefer to sweat a little bit as well, which is why they'll wait until the temperature is around 50-60°C.

However, most infrared lamps emit a small amount of electromagnetic radiation, which doesn't make sitting in a cubicle that attractive. The best low-EMF brands are Clearlight and Saunaspace. They can be fit in small compartmental spaces much better than traditional saunas, but nothing beats a good wooden-heated sauna.

When it comes to steam rooms and hot baths, then they can reduce inflammation, induce relaxation, flush out some toxins, and release heat-shock proteins. However, their effect is probably modest compared to an actual sauna. Nevertheless, inhaling steam is probably better for the respiratory tract and general breathing.

The dark side of steam rooms and public saunas is that there's a lot more microbes, fungus, and pathogens. That's why you should wear flip flops and use soap to clean yourself up afterward. In fact, one man reported fever and chills after regular sauna use because he poured water over the sauna stove from a bucket that contained mold[914]. Fortunately, the contraction of sexually transmitted diseases via sauna bench surfaces is deemed to be highly unlikely in countries of high-level hygiene[915].

In conclusion, I think everyone would benefit from some form of hyperthermic conditioning or sauna regularly. You don't have to do it every day but at minimum 2 times a week would be great. Nowadays, I'm still using the infrared sauna almost every day and have the traditional sauna maybe 2 times a week.

How to Sauna Without a Sauna

But what if you don't have access to a sauna? Are you forever doomed to premature aging, cardiovascular disease, and high toxicity? Definitely not. There are many other strategies to turn on the same pathways as the sauna does. The idea is to increase heat shock proteins and cause hyperthermia through other means. Here are the easiest examples:

- **Higher Body Temperature** – As was shown by the SARS study, a higher temperature can accelerate the killing of viruses. Heat shock proteins will be recruited in response to rising heat stress. Having a fever isn't the same as taking a sauna but a higher body temperature would increase the basal HSP response. You can turn on the heating up to 23°C and wear warmer clothes. A faster metabolic rate will also raise body temperature and energy expenditure.
- **Physical Exercise** – The body turns on HSPs during exercise and they're needed to gain the benefits[916]. Both

cardio and resistance training will do. Cardio ends up raising your body temperature and promotes sweating while submaximal resistance training stimulates HSPs in the muscle and can improve insulin resistance[917]. Eccentric contractions such as the negative portion of the lift or running have been shown to increase more HSPs[918]. You should do something physical for at least 30 minutes a day and get a sweat on. If you don't have access to a gym you can do calisthenics, blood flow restriction training, or jumping jacks in one place.

- **Isometric Contractions** – Contracting the muscle near your maximum effort releases HSPs and raises body temperature. That's why doing yoga or calisthenics is a simple way of getting similar benefits from less time. You can hold planks, front levers, deep lunges, horse stance, planches, etc. If you take it to near failure and end up holding a position for about 30-60 seconds, then you'll start to sweat and generate heat quite profusely.
- **Longer Fasting** - Fasting increases heat shock proteins in response to the stress[919]. It also improves lipid profile and lowers triglycerides. Although your metabolic rate may drop slightly during the fast, you'll experience a higher rebound effect after breaking the fast.
- **Water Deprivation** – Dehydration causes heat stress and upregulates autophagy by inhibiting mTOR signaling[920]. I'm not recommending you do dry fasting if you lack the experience but it can still turn on HSPs without exercise or

taking a sauna. It might be a shortcut for getting the benefits of the heat and exercise in situations where you don't have access to those things.

- **Sun Exposure** – UV light stimulates HSPs in a dose-dependent manner[921]. Higher intensities and LUXes release more because the body tries to protect itself and repair the damage[922]. Regular walking outside and getting exposed to the sun will not lead to an over-expression or damage but sunbathing might. You should try to spend at least an hour or two outside where the light environment is brighter to synchronize the circadian rhythms but using blue light face lamps can also be useful for that.
- **Spicy Food and Spices** – Different herbs and spices like cayenne pepper, curry, peppers, ginger, ginseng, and garlic can raise your metabolic rate and body temperature[923]. That's an indirect way of boosting some HSPs as well. Using spices as seasoning on food will definitely make you hot and even sweat. You can also mix them into hot water or just take by themselves.
- **Sauna and Hot Baths** – Exposure to higher temperatures than normal will upregulate HSPs. I think it will do so in a linear fashion, meaning higher temperatures will lead to more HSP activation but more isn't always better. Even a warm shower can increase HSPs slightly but your body will eventually adapt to that heat and it becomes less effective. That's why an actual sauna or a sweat lodge will be still more powerful.

Anything that involves the heat stress response system will be accompanied by an upregulation of HSPs. They start repairing the damage and also upregulate the body's resilience for the future.

In general, I think you should want to raise your body temperature several times during the day to keep the metabolism going. Working out for at least 30 minutes daily is the first thing you can do but some other form of heat whether from drinking tea, eating spicy food, taking the stairs, or doing jumping jacks for a few minutes.

Hot and Cold Stress Protocol:

Here's the hormesis protocol for hot and cold stress adaptation:

- The benefits of cold exposure start at about 6°C below comfort, which is around 16°C (60ºF). It can increase energy expenditure by 4-6%[924]. Most cryotherapy facilities and ice baths tend to drop lower than that. As you get more advanced this number may decrease slightly but you don't need to keep pushing your adaptation to get the minimum effective dose. Lower temperatures will just increase stress adaptation, fat burning, and cold tolerance but it may have negative side-effects.
- Signs of too much cold exposure include not feeling the benefits anymore, low thyroid, stubborn fat loss, problems sleeping, anxiety, constant shivering, getting sick easily.
- Taking a contrast shower every day is a quick and easy way to get daily cold therapy. Just alternate between hot and cold for a few minutes but always finish with cold. Going for an ice bath or polar plunge once a week should be enough, but you can condition yourself to do it every day as well.
- Going to the sauna in a fasted state promotes deeper cellular detox and autophagy. Autophagy also helps to mitigate the potential damage from heat-shock proteins by promoting repair. It also prevents the accumulation of stress granules that accumulate when you're exposed to the cold. Therefore, doing hot and cold therapy while fasting is the way to go. Just don't get dehydrated or push yourself to

the point of faint. Rewarming the body after cold exposure leads to additional autophagy.

- Try to get to the sauna at least once or twice a week. Doing it every day is probably too much but you can alternate between the infrared and traditional sauna to avoid any negative adaptations.
- Taking niacin or beta-alanine before a sauna creates a flushing effect that can increase sweating, blood flow, and lymph stimulation. Just make sure you're hydrated enough because otherwise there isn't enough water to form sweat. Drinking water afterward is also important.
- Don't drink alcohol or take any other drugs before or during hot-cold therapy. You'll put yourself in danger and at a higher risk of passing out.
- Thermal adaptations aren't permanent and they will soon diminish with disuse. That's why some form of hot and cold therapy should be a regular part of your weekly hormesis if you want to maintain this kind of temperature tolerance.

Chapter Seven: Of Hedonic Adaptation and Stoicism

"Never let the future disturb you. You will meet it, if you have to, with the same weapons of reason which today arm you against the present."

— Marcus Aurelius, Meditations

Stress adaptation requires as much psychological conditioning as it does physical. If your mind is working against you, then it doesn't matter how many other tricks you have because you'll literally undermine your efforts. That's why it's important to spend some time talking about mindset and optimizing our mental allostasis. In this chapter, I'm going to explore the philosophy of stoicism as a tool for becoming empowered by stress and adversity as opposed to being wiped out.

Hedonic Adaptation

Will Durant, a famous 20th-century historian said: *"A nation is born Stoic and dies Epicurean"*. Stoicism is a branch of Hellenistic philosophy that emphasizes personal ethics, logic, and virtuous living whereas Epicureanism places pleasure as the greatest good. This quote can be seen to be true throughout history. Great civilizations of the past like Ancient Greece, the Roman Empire, Mesopotamia, and the French monarchy of Louis

XIV all fell after reaching the peak of their glamour and grandiosity. As these nations became wealthier, their people turned softer and thus more vulnerable to foreign invaders or upheavals. In other cases, they were destroyed by their internal power struggle, sparked by ambitious megalomaniacs. Essentially, they became the victims of their own hedonic downfall. To an extent, you can start seeing this with the United States as well.

Hedonic adaptation or the hedonic treadmill describes how humans are able to maintain a stable level of happiness despite the ups and downs of life. No matter the fortunate or disastrous events that may happen, we're all going to maintain the same homeostatic level of happiness we were at previously. Here are a few examples:

- You win the lottery and you're winning – a new house, nice cars, fancy clothes, traveling the world and all that fun stuff. Then, after a few months, you adapt to this new lifestyle and it becomes normal. You're back at baseline.
- You break up with someone and feel heartbroken. In a few weeks, you've gotten over it and are back in the game.
- You lost your job and had to cut down on your expenses. At first, it's uncomfortable and displeasing but soon you'll get used to it. Instead of eating $200 dinners every night, you're satisfied with eating less glamorously.
- The economy goes into recession. Money becomes tighter and you don't have that much disposable income for luxury goods. However, you can still feed yourself and will thus stay satisfied.

- You start smoking only one cigar a day but over the course of a year you've built up to an entire pack to get the same effect. Now you're burning through multiple cigarettes before breakfast, but you're still not pleased.

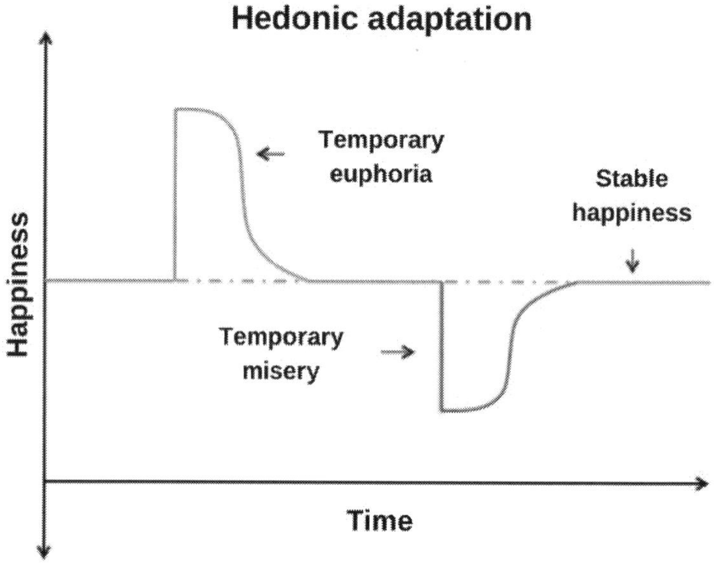

Basically, hedonic adaptation determines your happiness setpoint, which becomes your default state. In the field of positive psychology, it's said to be determined by your genetic heritage and how you've reacted to life's events[925]. However, your genes don't have to determine your destiny if you know what to do. For example, my DNA analysis reveals I'm predisposed to melancholy and anxiety if deficient in serotonin, thus I focus on optimizing my levels. Although I would say I am melancholic by nature, I've gone through periods in my life where I've been different and I'm always changing as a person in some aspects.

As a conscious human being who seeks to optimize their health and overall life, you should want to pay close attention to where your hedonic setpoint is. This way you can recognize whether or not your lifestyle is becoming too lavish i.e. fragile or you're borderline depressed i.e. too robust. The goal should be to have an antifragile homeostasis that gains from both the ups and downs of random events that are going to happen eventually. Soon enough the good times will be replaced by the bad ones and vice versa. We just have to know which part of the cycle we're at and adjust our setpoint accordingly.

Enter Stoicism

Stoicism was founded in Athens by Zeno in the early 3rd century BC and became more popular later in the Roman Empire. It's a philosophy based on personal ethics, logic, and views on the natural world. Accordingly, happiness or eudaimonia is found in accepting the present moment as it is, not being controlled by fear or desire, working in alignment with nature, and by treating others fairly. This is one of the best antifragile mental operating systems to have for reasons I'll explain shortly.

The Urban Dictionary defines the word *'stoic'* as such:

> *Stoic (n)* - Someone who does not give a sh#t about the stupid things in this world that most people care so much about. Stoics do have emotions, but only for the things in this world that matter. They are the most real people alive. [926]

If you were to put anyone else into that situation like a modern-day keyboard hero or social media justice warriors, then they would've freaked out, started crying, made an inflammatory Facebook post, perhaps even sued the kid, and initiated a campaign to prevent children from going out in public without their parents. Of course, it's not nice to call names at others and anyone who does so should be taught a lesson of some kind. However, it shouldn't matter what someone thinks or whichever word they use to describe you. If you feel like it does, then your sense of self-worth is being held hostage by other people and it's bound to break into a million pieces like all things fragile eventually will.

For the stoics, destructive emotions and feelings result from errors in judgment and not living in accord with nature. For a good life, one had to understand the natural order of things because everything was rooted in it. In stoicism, virtue is considered the only good, and external things such as wealth or pleasure are not bad in of themselves. They have value as material for virtue to act upon and it matters how a person behaves not what they say.[927] Being rich or wealthy doesn't make you a bad person. What you do with it does and that's based upon your character.

Stoic thinking enables you to understand that *"sh*t happens"* and you can't do much about it. You're inevitably going to miss busses, lose your wallet, face challenges, experience stress, have to work hard, and get your heart broken many times. It's a way to pre-emptively come to terms with the fact that adversity is bound to happen and you can actually benefit from this. Neither the ups nor downs of life will affect you negatively because you understand they're only temporary and are making you a stronger person. It's a psychological immune system that wards of toxic people, inflammatory situations, negative emotions, apathy, and hedonic cravings. Writers like Seneca and Epictetus considered the sage (the one who knows virtue and acts according to it) immune to misfortune because virtue itself is sufficient for happiness.

Nassim Taleb is also a big believer in the antifragility of stoicism. He has said: *"A stoic is someone who transforms fear into prudence, pain into transformation, mistakes into initiation, and desire into undertaking."* It's the antifragile mental operating system that turns misfortune into prosperity.

In a way, stoicism is just a form of stress adaptation to the challenges of life – a shield of indifference and antifragility. It's not about becoming numb or emotionless like a piece of cold stone. It can manifest as such but in reality, you're more flexible and embrace conditions that are less than ideal. The difference is that you're not overtaken or controlled by your emotions. Being emotional just for the sake of it isn't virtuous because it's not guaranteed to solve the problem. Although that's what society would like you to believe. Prudence, reason, and making the right decisions amid chaos and disorder is what gets things done and maintains order. That requires an antifragile mindset and flexible stoicism.

As was said before, optimism is associated with a stronger immune system when the stress is not too demanding and just enough[928]. However, in more difficult, uncertain, uncontrollable, and harsh situations optimism is associated with weaker immunity.

This is because staying optimistic in the face of already higher energy demands and stressors will require additional resources. [929] Optimists thrive during acute stressors because they have sufficient buffer against the burden but when things get more challenging this capacity gets ever depleted. In fact, from my own experiences with other people, I'd say being overly positive and optimistic when all the world around you is melting down can be maladaptive. The reason is that with optimism you're kind of expecting a happy ending and you're almost certain everything will be just fine. However, that's not how the dice of life roll. The best thing to do would be to direct your mental powers towards solving the situation by embracing stoicism. It'll not only ground you to reality, keeping the fact that you could come crumbling down any moment at the back of your head but also reduces the perception of stress and pain to a certain extent.

You can become stoic because of life events, your conditioning, education, choosing, or by nature. Some people like myself naturally gravitate towards this way of thinking because of genetics as well as past experiences. I can clearly recall having the 'work hard and do the right thing' kind of mentality since my childhood and it got instilled deeper during the military. Back then, tricking your brain into not feeling sad about all the physically hard and cold things we had to do was the only way to find at least some solace. Others that couldn't create this sort of inner bliss either became depressed, sad, or very angry, none of which are productive. After starting to read the great stoic philosophers I quickly realized what I had been doing and understood the power of this way.

Marcus Aurelius who is considered the last of the Five Good Roman Emperors is also one of the most important stoic philosophers. His reign was characterized by constant military conflict with the Sarmatian rebels in the east and Germanic tribes who he fought on the borders of Rome. During that time, the empire was also suffering from the Antonine Plague that killed on estimate 5 million people or 1/3rd of the population. Marcus Aurelius was considered a philosopher-king and philanthropist because of his blameless character and temperance[930]. One of his quotes says: *"Waste no more time arguing about what a good man should be. Be one."* In his personal diary collected and titled *Meditations,* he writes:

> *Say to yourself in the early morning: I shall meet today ungrateful, violent, treacherous, envious, uncharitable men. All of the ignorance of real good and ill...*
>
> *I can neither be harmed by any of them, for no man will involve me in wrong, nor can I be angry with my kinsman or hate him; for we have come into the world to work together...*[931]

He was the world's most powerful man at that time and could've done anything he wished yet he chose to live a virtuous life characterized by temperance, justice, and deep inquiry. His teachings are studied even nowadays by politicians, entrepreneurs, scholars, and anyone else who wants to live a meaningful life. Meditations is one of those books I find myself going back to at least once a year and it's eye-opening every single time.

Voluntary Hardship as Pre-Conditioning

One of the more renowned writers of stoicism Seneca had an exercise where he voluntarily practiced poverty. He said:

> *Set aside a certain number of days, during which you shall be content with the scantiest and cheapest fare, with coarse and rough dress, saying to yourself the while: 'Is this the condition that I feared?'*

Seneca was the richest banker in Rome, yet he deliberately put himself through difficult situations, discomfort, and pain. This not only made him resistant against turmoil but also allowed him to calibrate his hedonic homeostasis as he wished. His happiness wasn't dependent on anything external because he found it within himself. Seneca achieved 3 things:

1. The difficult experience was constantly etched into his mind, which made him not take his fortune for granted.
2. He was capable of surviving in adverse conditions, which made him bulletproof against anything that might have happened.
3. His happiness improved because his state of mind was satisfied with virtue, not material belongings, and his actual affluence was exponentially higher than he was expecting to have.

I'm not telling you to become a beggar or act like one. Instead, the main idea is to just do uncomfortable things in a world where discomfort is becoming a rarity. If we were to always get what we wanted in unlimited amounts, then we're teaching our brain to be satisfied with nothing less. It pushes our hedonic adaptation through the roof.

Stoicism isn't the same as minimalism either in terms of how many possessions you have. The key difference is your attachment to externalities. A minimalist deliberately gets rid of their stuff because they don't want to get distracted or dependent. A stoic doesn't care whether or not they have a lot of luxurious items because they can stay happy with or without them. It's more like carefree-ism where outside phenomena don't affect you. Seneca and Marcus Aurelius were both incredibly wealthy but they're one of the most famous philosophers in history. They were in control of their riches but not controlled by them. Marcus Aurelius auctioned off many of his and his family's treasures to finance the defense against invading forces instead of raising taxes. It was probably a public relations act to show that even the emperor has to make sacrifices during difficult times but few other people in his shoes would've done the same. He was just more concerned with the safety of his country than his personal affluence.

Someone trying to live minimalistically may actually get anxious or distraught if they don't live up to their minimalistic expectations: *"Oh, god! I should get rid of my lamp because I don't need it. While I'm at it, I ought to throw away all my clothes as well because it's not woke enough."* It's a very fragile state and kind of self-defeating because you can always be more of a minimalist until you've got nothing left. Stoicism, on the other hand, is like comfort distancing or intermittent adversity conditioning, in which you disrupt your hedonic setpoint to prevent it from climbing too high. In a way, it's just like fasting. Here are a few examples:

- Extended fasting resets your taste buds and makes healthy food taste amazing. Junk food will actually become too stimulating. Even plain vegetables without seasoning taste amazing after a longer fast.
- Avoiding caffeine for a certain period will lower your tolerance to it. You'll get more energy from less coffee when you do enjoy a cup. Instead of having to drink 5 cups a day you're satisfied with only one or two.
- Not consuming social media and entertainment for a while will give your brain a break from being constantly stimulated and triggered. This will help you to become more mindful and focused in life. It also prevents getting too caught up with the opinions and likes of others.
- Sleeping on the floor or outside every once in a while reminds you how fortunate you really are for having even just a roof over your head. It can also condition you to hold a better posture. This, in turn, teaches you to sleep in unconventional spaces and positions so you can take a nap on the airplane, public transport, or anywhere else. It's a prerequisite for antifragile napping.
- Doing the things you're afraid of but know are good for you makes your mind more stoic and body stress adapted. Plus, you'll also get it over with and won't waste time on procrastination.

If you understand this concept, then you can see that it's a massive life hack for happiness. You'll become happier with less. In order to appreciate this and make it effective, you need to change your perspective.

The Power of Perception

We use our senses to receive stimuli from the environment, which then get turned into perceptions and sent as messages throughout the body. Those catalysts create responses that interpret our position in space and time, hence our reality. They can come in any shape or form - auditory, olfactory, tactile, and visual senses all contribute to this.

What the work of the biologist Bruce Lipton shows, is that our perceptions directly create our reality and determine epigenetics. We're all made up of millions of cells that influence how certain genes get switched on or off. This happens when the cell receives stimuli from the outside world and sends signals throughout the organism to conduct specific processes, i.e. protein synthesis. As a result, we will either begin to build new tissue or break it down. The brain-part of the cell isn't the nucleus but the membrane - the outer layer through which it experiences the surrounding world. It's also the point where perceptions are created.

Certain actions can cause different reactions according to the situation and the choice of interpretation. The discrepancy

depends on not the stimulus but the response. For instance, the body experiences all types of stimuli as the same. It can't tell the difference between the sources of temperature but just realizes that it's subject to certain degrees and acts accordingly. For the senses, there isn't a difference between being in a sauna and walking in the desert. It's hot and burns the skin. That's all that matters. However, the body and mind react differently in both of those situations. Heat regulation happens no matter what, but the hormonal and mental response varies. Being in a sauna is deliberately created and voluntary. The mind will remain calm and enjoys the process. When lost in the desert, the response is completely different. There might be no water around, you get dehydrated, your head gets filled with feelings of despair, hopelessness, etc. As a result, it turns into a negative experience because you've concluded the direness of your future. The same stress response can be triggered by someone not wanting to be in a sauna and *vice versa* or if they think it's harmful.

When it comes to experiencing our reality, the biggest thing that contributes to this is how we choose to perceive it. Every action is followed by an appropriate reaction and every stimulus creates a response. The part where our reality gets created does not lie in the first part, but in the second, where our perceptions have translated the message. What's best about it is that we are in total control of this.

Negative Visualization

There's the misconception that happiness is about constant positive thinking, no discontent, and high vibration energy. Any feeling of sadness or disdain is negative and to be avoided. Otherwise, there's something wrong with you. Indeed, in some respects being happy entails having a level of contentment, well-being, satisfaction, and joy[932]. It's what the stoics called eudaimonia. However, it doesn't have to be this feeling of excitement or motivational hysteria many people mistakenly think. I would say a calm bliss would be more accurate.

The problem is that the universe doesn't care about how much positivity you think you're emanating. There are still some inevitable parts of life that are going to happen. You can't visualize away the fact that you're going to eventually fail, get hurt, lose some of your most valuable possessions, see the death of your loved ones, etc. Change is a constant, almost lawlike rule, that influences how the world functions.

Seneca writes in Moral Letters to Lucilius:

> *Nothing happens to the wise man against his expectation, nor do all things turn out for him as he wished but as he reckoned — and above all he reckoned that something could block his plans.*

If you were to think only positively, then you would expect only the good because that's how you think the world works. Anything less than ideal will be a dreadful decrease in what you're used to.

Once *sh*t hits the fan,* the ground you're standing on will be swept beneath your feet and you'll face plant to the floor. Solid pavement. Now, if you were to keep the fact that everything won't go according to plan in the back of your head, you're conditioning yourself to be more grounded to reality. Trust me, it's going to be a struggle in the future with unexpected colossal disasters threatening the entire existence of our species. However, similar events will happen in our everyday life as well. Just on a smaller scale.

Herein comes into play the mental technique of negative visualization, which I've defined as such:

> *Negative visualization is the use of worst-case scenario, or at least negative, less than ideal, type visualizations and expectations, in which you're envisioning yourself encountering adverse, unexpected and volatile conditions, situations and events, and seeing yourself reacting, behaving and overcoming them, to bulletproof yourself against "sh*t hits the fan" circumstances.*

I'm not trying to shy you away from positive thinking. Instead, I don't want you to neglect the polar opposite, which is beneficial in the grand scheme of things. Here are some examples of negative visualization:

- **Memento Mori** – One of the biggest fears people have in life is the fear of death. This is a stoic meditative exercise that's supposed to remind the person that everything in the world – the body, career, reputation, even family – should not be the primary focus of our minds, nor the source of our happiness because these things can be swept away by death at an instant. Reminding ourselves that one day we're going to die is almost enlightening and will benefit our happiness. It prevents us from getting distracted by things that don't serve us nor contribute to our existence. We attain a new perspective and desire to follow our calling and pursue becoming more conscious. However, you shouldn't let it become a source of anxiety or worry. Instead, just spend a little bit of time thinking about these things and structure your life in a way to avoid regrets.

- **Premortem** – A technique designed by the psychologist Gary Klein. Before beginning a project or an endeavor you envision what could go wrong? Stoics called it *premeditatio malorum* (premeditation of evils). This way you'll always prepare for disruptions and can work those obstacles into your game-plan.

Instead of constantly trying to expect positivity, occasionally envision experiencing negativity. Come to terms with the fact that it's going to happen someday. Adversity may be hiding around the next corner or it might happen in 20 years. It's not a matter of if but when.

Epictetus describes negative visualization as such:

> *When you are going to perform an act, remind yourself what kind of things the act may involve. When going to the swimming pool, reflect on what may happen at the pool:*

some will splash the water, some will push against one another, others will abuse one another, and others will steal. Thusly you have mentally prepared yourself to undertake the act, and you can say to yourself: "I now intend to bathe, and am prepared to maintain my will in a virtuous manner, having warned myself of what may occur."

Do this for every act, so that if any hindrance does emerge, you can think: "I did not prepare myself only to undertake the act, but also for this hindrance that has occurred, and also to handle this hindrance virtuously & keep my will conformed to nature — and this will be impossible if I become vexed."

This is a completely different paradigm through which to look at the world and your relation to others. What Epictetus said in the second paragraph almost hints that if something negative does happen to you or if someone else bids you wrong it's actually your fault to a certain extent because you didn't prepare enough beforehand. Now that's a controversial statement and I can certainly imagine situations where this is not the case. However, it does tell you that the way you respond to adverse situations and other people is solely dependent on your own reaction. Your responses to anything are the result of how you've taught yourself to behave and that's completely in your control.

Here's an exercise for you. Whenever you're preparing for an upcoming event or situation, envision what the worst-case scenario would be. Instead of wanting it to happen, simply see everything going the opposite to your desire.

- Family troubles – a loved one gets sick or needs a lot of support.
- You lose your job – no financial independence anymore.
- Relationship problems – your spouse becomes hard to deal with.
- Organizing an event – some people or other companies jump out on you.
- Taking an exam – the questions are the opposite of what you prepared for.
- Creating a business – first product launch fails.
- Crossing the street – get hit by a speeding bus
- Global pandemics – a novel virus is spreading.
- World War – a military conflict breaks out with your country involved.

The effect does not lie in seeing things turning out bad. Instead, it's about them going worse than you expected, thus having a distorted space between your hopes and what could be. To get any benefits, you would also have to envision yourself behaving in these adverse situations and overcoming them. Use these guidelines to enhance your negative visualization:

- **How are you going to react in response to a specific negative event?** If you were to in some distant future scenario host a public event and the catering jumps out on you the last minute, what are you going to do? Will you quit, or do you already have a back-up plan?
    - What are you going to do when you've lost the doc file of your essay and the deadline is within an hour? Do you have a backup of some sort?
    - How are you going to deal with your spouse or family members who have problems? Are you going to work it out or run away?
    - What are you going to do once you lose your job? Declare individual bankruptcy, become homeless, or take massive action towards starting your own business?
    - What are you going to do to prevent your first product launch from failing? If it does fail, how are you going to address the issue?
    - How are you going to react to a global pandemic, military conflict, alien invasion, or zombie apocalypse? You might think it's a joke but so do all the other millions of people who are going to die from these events. Have you done anything at all to be prepared?
- **Still want things to go well.** Don't become pessimistic. Optimism is still healthier. Simply use negative

visualization to keep your expectations in congruence with reality. Expect to succeed in everything you do, but don't blindfold yourself to the resistance you'll inevitably encounter.

- **Don't feed your subconscious mind the wrong signals**. The potential danger lies in how your thoughts can begin to influence your overall mindset. Don't think negatively because that's what your subconscious mind will begin to see as the real thing. The dark side can begin to poison you if you fail to notice its effects. Still see yourself as an awesome human being (all the while phrasing: *I'm amazing*) who goes out to accomplish extraordinary success but isn't attached to the results. It's about accepting adversity as the omnipresent antagonistic force that can fall upon you any moment but you're preparing for it in advance.

You should also incorporate some positive visualization along the same lines. What are you going to do if something amazing happens like a new business deal, romantic endeavor, or any opportunity? Most people aren't even prepared for that because of not having trained themselves in advance. That's why lottery winners tend to lose their money within a few years because they don't know how to use it whereas someone who's gained the skills through hard work will stay wealthy for the rest of their life thanks to having the financial competence.

Negative visualization isn't about becoming pessimistic. I think it's actually an enhanced version of learned optimism because you'll be wanting to get the best results, but at the same time, are grounded to reality. The choice is yours: you can either be pessimistic, seeing the worse in everything, a positive thinker, only expecting the good, or incorporate the best aspects of both ends of the spectrum, which is stoic negative visualization.

As Seneca said: *"Luck is when preparation meets opportunity."* Do you want to be lucky? Or would you rather be prepared?

The Amygdala Hijacks Back

All this talk about negative events that may or may not happen can be quite exhausting and fear-provoking. That's not the goal here – living in fear and anxiety is not antifragile but can leave you more vulnerable to various stressors. Instead, you should free yourself from the outcome and focus on the things you can control.

Stress has a profound impact on the brain. As I've mentioned before, psychological stress can trigger the same physiological response in the body as any other physical stress. It raises adrenaline, releases cortisol, shuts down digestion, increases alertness, and makes you anxious. You can literally sit quietly in a chair and freak out internally without the blink of an eye. I'm sure you've felt that before whether due to some negative thoughts, rumination, or fear.

In his book *Why Zebras Don't Get Ulcers,* Stanford University biologist Dr. Robert Sapolsky explains that in animals, like zebras, stress is episodic and brief – you run away from a lion every once in a while but spend most of the time grazing[933]. For humans, however, stress is more chronic and low-grade because of the fast lifestyle, work-related responsibilities, relationship problems, traffic jams, EMFs, etc. That's why wild animals are less susceptible to stress-related diseases like ulcers or cardiovascular disease. There's no opportunity to become chronically stressed out because your likelihood of getting eaten and dying are higher.

Glucocorticoids are great for short-term survival and adaptation but chronic exposure to stress can cause serious metabolic, immune, and psychological issues, ultimately leading to disease[934]. Acute, uncontrollable stress increases catecholamines in the pre-frontal cortex, which lowers cognitive abilities and the firing rate of neurons[935]. On the flip side, they also strengthen connections in the amygdala and striatum, which govern primary motor control and fear response. Therefore, you decrease your reasoning faculties and become more reactive out of anxiety. It makes your brain more anxious and paranoid by default.

The amygdala hijack describes a situation where your emotions take over and you freak out. It's an immediate emotional and overwhelming response to a perceived threat imagined or real. Daniel Goleman coined the term in his 1995 book *Emotional Intelligence: Why It Can Matter More Than IQ*[936].

Your sense organs communicate with the thalamus, which is connected to the amygdala. Whenever we receive a stimulus, the amygdala sends out a corresponding signal to the rest of the brain. The neocortex is referred to as the 'human brain' because it governs rational thinking and higher executive functioning. That's why we all have multiple inhibitory mechanisms working all the time, such as not screaming in public, dancing on tables, or punching the police. Our pre-frontal cortex is preventing our primal urges from taking over because we would like to think we're in control. However, if the amygdala perceives the stimulus to be threatening, it's going to trigger the fight, flight, or freeze response. The amygdala then hijacks the neocortex and activates the HPA axis[937]. As a result, you may act irrationally, destructively, and will experience a lot of stress. In hindsight, you'll feel as if something took over and you couldn't recognize yourself. That's where the majority of people live – stuck in a traffic jam while being hijacked by their amygdala.

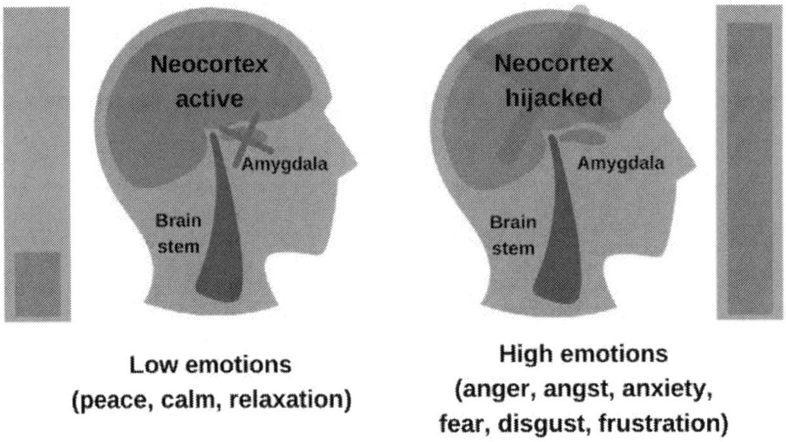

The amygdala hijack is a good thing because it will help you to immediately run away from danger or jump up to a tree whenever you come across a snake. It's just somewhat maladaptive in an environment where the amount of actual dangers is close to zero and you're surrounded by only minor stimuli that aren't that life-threatening.

The problem is that the amygdala reacts to any threat a hundred times faster than the neocortex. It takes about 3 seconds for information to reach the rational parts of the brain whereas the amygdala receives it within a few milliseconds. That's why even the toughest guys can get scared for a moment if you scare them – they can't help it. Most people are also too eager to react out of fear or anger because they're guided by the immediate instructions from the amygdala as opposed to their human brain.

If information about a threat reaches the neocortex and it's deemed to be real, then the entire brain will respond accordingly. Experiences of past events with a similar stimulus will make the amygdala react based on that reference. This knowledge is held in the hippocampus that stores memories. If there's no prior data or experience, then the amygdala will try to act based on the directions received from the neocortex. You'll react according to your conditioning and imagined best course of action.

Limbic system hypersensitivity describes an acquired hypersensitivity to various stimuli in the brain's limbic region where the amygdala is also located. This condition precipitates triggers that cause states related to chronic stress and disease such as oxidative stress, hormonal imbalances, hypertension, and low-grade inflammation[938,939,940]. Usually, it results from exposure to some acute or persistent stressors like chemical exposure, toxicity, trauma, viral infection, radiation, abuse, or other psychological turmoil, which then disrupts the endocrine and nervous system of the body[941]. All of these initiate a downward spiral into imbalances in the person's ability to perceive and cope with stress. They just become increasingly more sensitive and less tolerant of anything that sets them off.

Getting hijacked by the amygdala is a learned behavior that gets instilled the more often you get hijacked. For some people, it's a habitual response to even the most minute situations like getting triggered on social media or being cut in front of in traffic. They're going to freak the *f#k* out really fast because that's how they've wired their brain to respond. Unfortunately, this chronic triggering may lead to stress-related diseases like hypertension and ulcers. Not to mention the damage it does to their relationships or public image.

Psychological resilience is the ability to cope with crises, stressors, and uncertain events emotionally. It's characterized by the ability to protect oneself from the potential negative side-effects of stress[942]. Basically, it allows a person to stay calm and in control to move towards more antifragile outcomes. Someone with high psychological resilience will be able to contain their emotional outbursts and maintain sanity in situations of adversity.

How to Not Get Emotionally Hijacked

Neuroplasticity refers to the brain's ability to restructure its neural connections[943]. Whenever you execute a certain behavior or learn something new, you're literally changing the structure of your brain and are creating new patterns of thinking and behaving. The more often you do something, the more habitual it becomes. This means, if you get scared or distracted easily, then you're strengthening these bonds and creating a vicious circle for repeating that same behavior in the future.

BDNF helps to recover from chronic stress by promoting neuroplasticity and making the brain more maluable[944]. If you're in the survival mode all the time, you become rigid and less flexible. Your exploration of new things is decreased because the brain tries to preserve energy and safety. To enable novelty and creativity, you first have to get out of the stressed-out state and facilitate growth factors for the brain.

Here are some ways of preventing the amygdala hijack:

- **Stoic Negative Visualization** - You already know that sooner or later someone is going to cut you off or make you angry. By meditating on that you can make peace with the fact before it even happens and can prevent your brain from getting hijacked. Use both memento mori and premeditation of evils.
- **Mindfulness Meditation** - Meditation is a technique that conditions your mind to perceive sensory experiences more deeply. Instead of just reacting to everything that happens, you become more proactive. There's an empty space between you and the event – a temporal buffering – during which you can choose to create a much more logical and preferable response. Meditation teaches how to widen that gap. It basically slows down how fast you react to the stimuli received by the amygdala while speeding up the

response of the neocortex. To meditate, you can just sit down in a quiet room and start focusing on your breath or any other bodily sensation. The key is to just become the observer of what's going on inside you and in the surrounding environment.

- **Stress Management** – Active stress management keeps your buckets of stress empty. With a vacant bucket, you can pour in a lot more stress before it starts to overflow. Most people's buckets are constantly full and they never get emptied. Instead, they explode onto other people or themselves like a ticking time bomb.
    - **Sleep** plays a huge role in BDNF production and stress adaptation[945]. If you're sleep deprived, then your emotional bandwidth decreases quite substantially and you're more prone to over-eating, over-reacting, and losing your grip on things.
    - **Exposure to sunlight** also raises BDNF and circadian rhythm alignment is crucial in mood regulation and all metabolic processes[946]. Bright daylight is quite important for keeping the brain working properly.
    - **Acupuncture therapy** improves neurological recovery after traumatic brain injury by activating the BDNF/TrkB pathway[947]. Using a simple acupuncture mattress can help you to relax and improves sleep.

- **Music** can increase BDNF[948] by lifting mood and getting you out of fight or flight. Having fun and enjoyable activities do so as well.
- **Exercise** increases BDNF greatly[949]. However, too much exercise may cause chronic stress so you would have to keep it moderate.
- **Curcumin** can reverse the effects of chronic stress on the HPA axis and BDNF expression[950]. It lowers inflammation and promotes relaxation
- **Cold and heat** thermal regulation require BDNF developmental plasticity[951]. It's also critical to not over-react to these stressors as it may embed a negative memory into your psyche. That's why you have to stay calm when in the ice bath and associate it with something positive.

- **Develop Emotional Intelligence** – Having more emotional intelligence prevents amygdala hijacking by teaching you how to read the emotions of yourself and others. Emotionally intelligent people can put themselves into the shoes of someone else and analyze their feelings. It reveals the actual context of the situation. You can develop this skill through meditation, journaling, self-inquiry, reading books, watching movies, social communication, and creating deep relationships with others.
- **Social Support** – Support from others helps to lower stress and deal with it better. In a study, married women who were able to hold their husband's hand while getting

painful shocks to their ankles showed a reduced pain response in many areas of the brain[952]. Holding a stranger's hand had a slightly smaller effect but it was still helpful. A lot of people, especially men, tend to carry their stress and burden with them without sharing it. This builds up tension and visceral stress that can lead to disease or burnout. I think you shouldn't be that guy who just complains all the time but sharing your troubles with family or a therapist is an effective way of relieving the burden.

- **Human Touch** –Touching someone creates deep connection and a feeling of safety. Research has found that touch is essential for brain development and emotional learning[953]. Even just 20 seconds of hugging someone releases oxytocin the hormone of connection and love. It also reduces blood pressure, heart rate, and stress[954]. You should hug your loved ones at least a few times a day. Showing affection with physical contact also creates more trust and a sense of support.
- **Letting Go** – We all have past trauma of some kind both conscious and subconscious. By letting go of our injuries and spite, we can achieve a higher state of freedom and eudaimonia. Forgiving others and yourself is liberating. It gets rid of the fear and anxiety that's dragging you down. Instead of carrying around this handicap, let it go and move on. In his book *Letting Go: The Pathway to Surrender,* Dr. David Hawkins writes that this process: "*…involves being aware of a feeling, letting it come up, staying with it, and letting it run its course without wanting to make it different*

or do anything about it [955]". First, you have to let the feeling be there with no resistance or judgement. If resistance does arise, ignore it. Any thought is just a rationalization trying to explain feelings and they're more often than not related to basic fear. To be surrendered is to have no strong emotion about a thing and not have remorse.

- **Self-Management** – Have you noticed that people who are all over the place and hyperactive tend to also be more stressed out and anxious? There's also an association between type-A personalities and cardiovascular disease[956]. It's due to chronic stress and hypertension but research also finds that being stressed out impairs efforts to be physically active and healthy[957]. Not being able to manage yourself or your time increases the likelihood of experiencing stress because you're always busy. That's why you should never be in a hurry. Instead, prepare in advance and manage your time better. Create a schedule, don't push off your responsibilities until the last minute, avoid procrastination, take care of yourself, and establish proper routines.

With practice, your ability to respond and emotional control will improve. It's not about becoming emotionless or completely irresponsive. Instead, we just want to cultivate the skill of cutting through the noise and not letting stress get the best of us.

Learned Helplessness

In 1972, Martin Seligman did an experiment on 2 groups of dogs who both got electric shocks at random times.[958] One group could end their torment by pressing a lever, the other didn't and therefore thought of the shock as inescapable or determined. In the second part of the study, the dogs were tested inside a shuttle-box apparatus with an electrifying floor that gave them shocks in a similar fashion. They could've easily escaped the electrocuted room by jumping over a small fence. The dogs who previously got to press the lever quickly learned this and made it through. Those who were conditioned to think there was nothing they could do to stop getting shocked simply laid there and kept getting more shocks. They gave up.

In humans, three groups of participants were put into different rooms. One group was subjected to loud noises, but they were able to turn it off by pressing a button four times. The second group experienced the same scenario, but their button wasn't working and the third group heard no noise at all. Later, when all the subjects were under the same loud noise and given a button to turn it off, those who had no control during the first part of the experiment didn't even try to press the button whereas the others did.[959] Just like the dogs.

Learned helplessness is the result of a real or perceived absence of control over the outcome of a situation[960]. It's usually observed in subjects who have experienced repeated adversity beyond their ability to handle it. The idea is that prior learning and condition can result in drastic behavioral changes and responses to stimuli. If you've learned that there's nothing you can do about the electrified floor, your bad genes, the government, or anything else harmful for that matter, then you're not going to even try to do something about it.

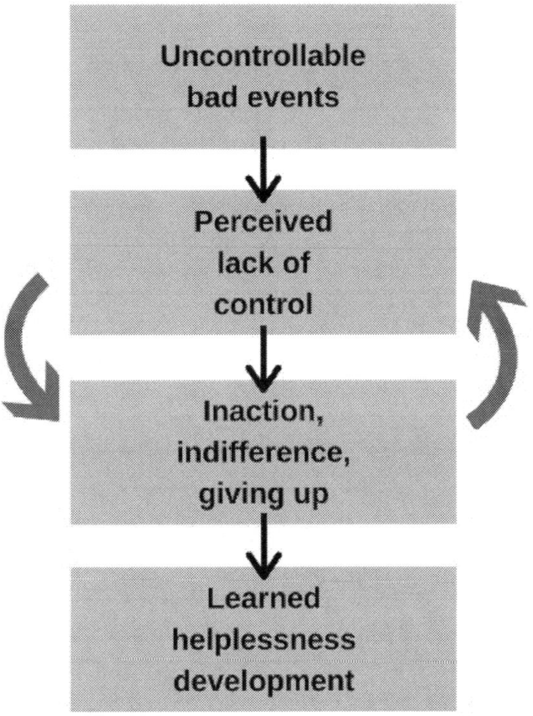

Neuroscience has since discovered that the brain's default state is to assume that you have no control and you learn the presence of helpfulness[961]. The default state is to think you can't do much about your situation and only after learning that events are controllable you shift from passivity into proactivity. With that kept in mind, you can teach yourself to respond to stimuli in a way that won't harm you as much and to get out of the weakened state faster.

People who have learned helplessness show emotional imbalances, passivity, aggressivity, depression, burnout, and problems doing cognitive tasks[962]. They experience poor work satisfaction and relationships[963]. Pessimists also have weaker immune systems and are more susceptible to health problems like heart disease[964].

Here are the causes of learned helplessness:

- Child abuse and neglect [965]
- Poor parenting and punishment
- Emotionally abusive relationships
- Social anxiety and shyness [966]
- Poor performance at school or work [967]
- Aging and losing independence [968]
- Experiencing prejudice and stereotypes [969]
- Dependency on social support
- Chronic stress, especially about finance [970]
- Heightened pain response [971]

Research has found that learned helplessness involves key areas in the amygdala, hippocampus, and hypothalamus[972]. In effect, you derive information from past experiences and emotions to assess how much power you have over a given situation and based on that data you either take action or give up. It's characterized by suppressed neocortical activity and increased 5-HT or serotonin[973]. One of the coping mechanisms for stress and perceived powerlessness is to trick yourself into thinking that it's fine and you ought to relax because there's nothing you can do. Stress releases 5-HT as a natural anti-depressant as a means to put you into hibernation. A small feeling of urgency is healthy because it increases your vigor and makes you want to spring into action. The same applies to cortisol.

One possible treatment for learned helplessness involves stimulating the ventromedial prefrontal cortex and inhibiting the brain stems dorsal raphe nucleus with medication, electrical stimulation, trans-magnetic stimulation (TMS), or therapy[974]. In recent studies, TMS seems to be effective in treating depression already and it might work on learned helplessness as well[975]. Exercise has also been found to prevent learned helplessness in rats[976]. Turns out the amount of exercise is not that important. Exercising even just a little bit will already give you the effect. Unfortunately, as we've found out, stressed out or depressed people are less likely to exercise.

People can be immunized against learned helplessness by increasing their awareness about past events when they were able to control the outcome[977]. You have to remind yourself of the time where you had power because it might be forgotten. Cognitive therapy can also show that one's actions do make a difference and boost self-esteem[978]. Pre-conditioning yourself to handle stress and solve it is a protective adaptive mechanism or confidence booster, which facilitates behavior of learned helpfulness as opposed to helplessness.

Growth Mindset VS Fixed Mindset

In 1975, Carol Dweck showed in her study that failure can also prevent learned helplessness in subjects who all showed maladaptation to failure. Participants were divided into 2 groups: one group experienced only success while the second group repeatedly failed their tasks, but they were taught to take responsibility for it and attribute the failure to a lack of effort. As a result, the success-only group showed no improvement in their responsiveness to failure whereas the group who failed did improve.[979] They also put more emphasis on insufficient motivation versus ability as a determinant of failure.

Dweck is a renowned psychologist who's written the best-selling book *Mindset: The New Psychology of Success*. She distinguishes two types of mindsets people have:

- **The FIXED mindset** is rooted in the belief that your traits and potential are predetermined, thus limited.
    - People with this one see their abilities as set within a certain range of parameters that have been created there whether by the genome, the environment, or their personality. *„I'll never be able to do that; I'm just not that kind of a person; I don't know how could I ever do that? It's just the way I am."*
    - The fixed mindset also applies to traits you'd consider positive, like being good at sports or showing better results in school. Here the difference is that with a fixed mindset you see your abilities in

any vocation as a part of your innate genius, not the hard work you put in. *„I'm just born for this; I have those genes of excellence; I'm better than others because it's who I am."*

- **The GROWTH mindset** is based on the belief that your abilities can be developed and cultivated.
    - People with this one see themselves capable of achieving anything as long as they commit and put in the effort. They respond to setbacks and failures as challenges that lead to individual growth. *„I'm willing to learn and improve myself; If I dedicate wholeheartedly to this, then I'll increase my chances of getting the results I desire."*
    - They don't see their pre-requisite conditions, like upbringing, history, or social role as detrimental. Instead, they're looked upon as steppingstones that made them the person they are today. *„This is difficult, but it will prove to be an invaluable lesson in the future. I'm able to change my bad habits because they're not bound to me."*

The power of mindset and the consequences of believing in the potential of one's abilities have a profound effect on a person's success in life. Dweck found in her years of research that people with a growth mindset showed better results in their work, relationships, personal development, and happiness.

Both of these mindsets are learned ways of thinking – you exhibit one or the other based on how you've grown up, what your parents taught you about believing in oneself, how your teachers encouraged you to behave and what kind of interpretations you made between your actions in the world and the results you've been getting. If you think that you're a person with a fixed mindset, then rest assured you are because you've already made up your mind about it. Having a growth mindset will enable you to not only accomplish the things you desire despite your current condition but will also give you much more agency in life.

As the stoics knew – there are things in our life that we can control and the things we cannot. Epictetus, one of the philosophers, opened his book The Enchiridion with the following paragraph:

> *Some things are in our control and others not. Things in our control are opinion, pursuit, desire, aversion, and, in a word, whatever are our own actions. Things not in our control are body, property, reputation, command, and, in a word, whatever are not our own actions.*

There are also the things we can influence – the things we can't control entirely but we can influence to a certain degree – like the mental states of others, the outcomes of certain events, and how we can feel in the near future. This gives us the 3 circles of control – the things you can't control, the things you can, and the things you can influence (See figure below). Notice how big the circle of the things you can't control is – it's enormous in comparison to the things you can control. However, the core is the strongest and most effective place to focus on.

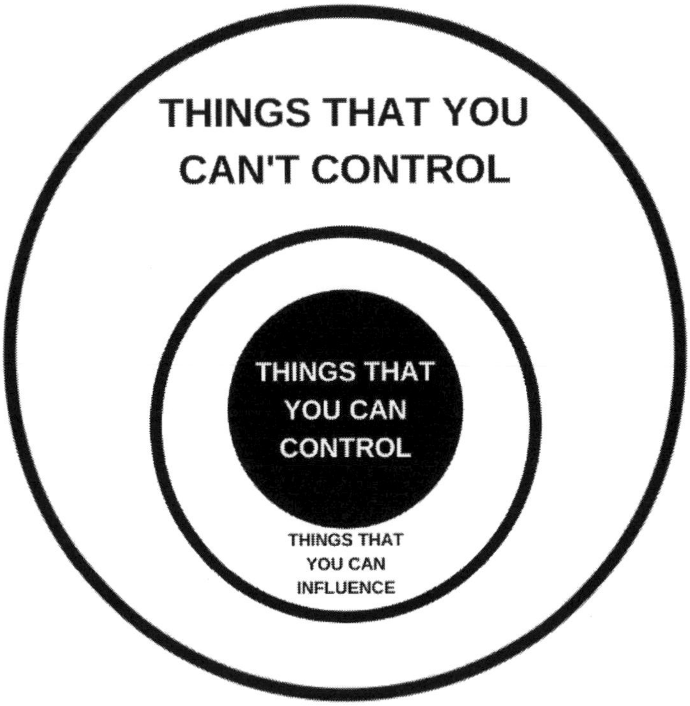

Stoicism, in a way, is founded upon a growth mindset – you base your decisions on the principle that you cannot control the things that happen to you, but you can control (or at least influence) your reactions to those events. You can't control how you feel or how your body responds to certain stimuli i.e. you feel like a pile of cow dung in the morning or what you think about waking up, but you can control how you act despite of it.

A similar mindset was showcased by Viktor Frankl, a Jewish psychiatrist who spent 3 grueling years in Auschwitz during the Second World War. He said in his book Man's Search for Meaning:

Between stimulus and response there is a space. In that space is our power to choose our response. In our response lies our growth and our freedom.

What it means that your willpower is unlimited – you can always make the right decision despite the circumstances you're in. The problem is that we're highly affected by external stimuli, like random events, other people, our mood, how well we slept, does our brain have access to energy, and whether or not we're even aware of the possibility of changing our perception. This kind of thinking is a stoic approach to learned optimism that teaches you to find bliss and positivity even in the face of turmoil and despite the unpredictable events of life.

After the Buddha learned that suffering is a part of life, he realized that to end suffering, he needed to find out what causes it. He came to the conclusion that suffering is caused by desire and ignorance – craving for attachment and being ignorant about the possibility of liberation from it. Ignorance doesn't refer to being uneducated. It's the inability to see the truth of things. This is accepted in Stoicism as well where it's thought people suffer from the things they can't control because of not knowing that they can't control them in the first place. Focusing on the things one can control, striving to become a virtuous character, and living by this truth will result in peace of mind and less stress. A person with a growth mindset focuses on the things they can control – their perceptions and actions.

Hedonic Adaptation and Stoicism Protocol

- Your happiness setpoint is being constantly regulated by your lifestyle, your level of comfort, what you eat, in what amounts, and a general sense of satisfaction. It functions like the homeostasis of any other bodily process that can be changed and conditioned. Too much hedonic adaptation can lead to dissatisfaction, uncontrollable cravings, and irrational desire. Stoic practices like intermittent fasting, practicing poverty, and abstention from pleasure helps to lower that setpoint.
- Unexpected and adverse events are inevitable – they're going to happen eventually even to people who think positively. Negative visualization conditions your mind to come up with solutions to the worst-case scenarios and grounds you in reality. You shouldn't think about bad things happening to you but focus on how you're going to respond.
- To prevent your amygdala from getting hijacked, engage in active stress management, personal inquiry, develop emotional intelligence, and nourish deep relationships with others. Don't trust your assumptions or immediate response because they may be misleading.
- You can choose to believe you are fixed or able to grow. This predicts either learned helplessness or helpfulness. By focusing on the things you can control, which are your own thoughts and actions, you can dictate your luck.

Chapter Eight:

Of Muscle Mass and mTOR Signaling

„I think you should bear in mind that entropy is not on your side."

Elon Musk

All stress manifests itself in your body. It's affected by both physical and psychological factors but ultimately your biological health is the most important thing for overall vitality and well-being. In this chapter, I'm going to talk about strengthening and maintaining your meat-suit so that you could live longer and adapt to stressors better.

I hate to break it to you but your body is slowly wasting away and you're on the verge of complete deterioration. Aging is characterized by a progressive loss of skeletal muscle, a process called sarcopenia. It's identified as a universal characteristic of the aging process in many species, from worms to humans[980].

After the age of 40, lean tissue and strength keep reducing at a rate of 1% per year and 3-8% per decade[981]. Even very athletic people's bodies consist of only about 30-40% muscle, which isn't that much, and the average person has approximately 20-25%. This makes maintaining your musculature an incredibly vital component of healthy aging and longevity.

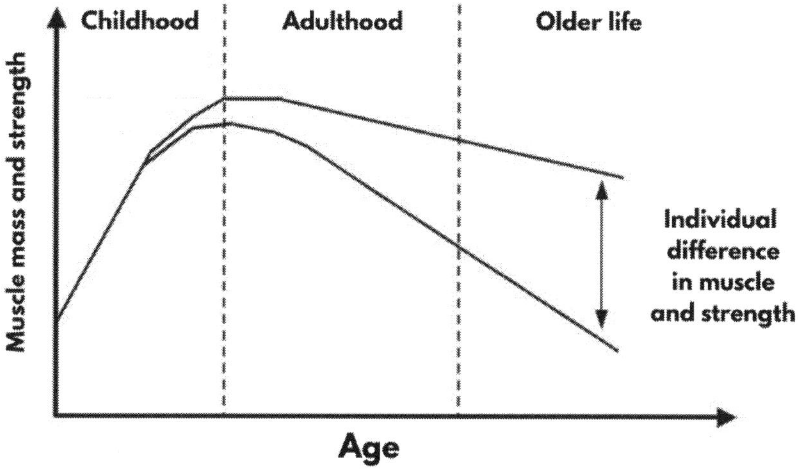

To make things worse, it becomes increasingly more difficult to re-build that lost tissue because the body becomes more resistant towards anabolic growth. Rates of increased sarcopenia are often similar to the declines in growth hormone and testosterone, which are essential for building and maintaining lean muscle. (See figure below).

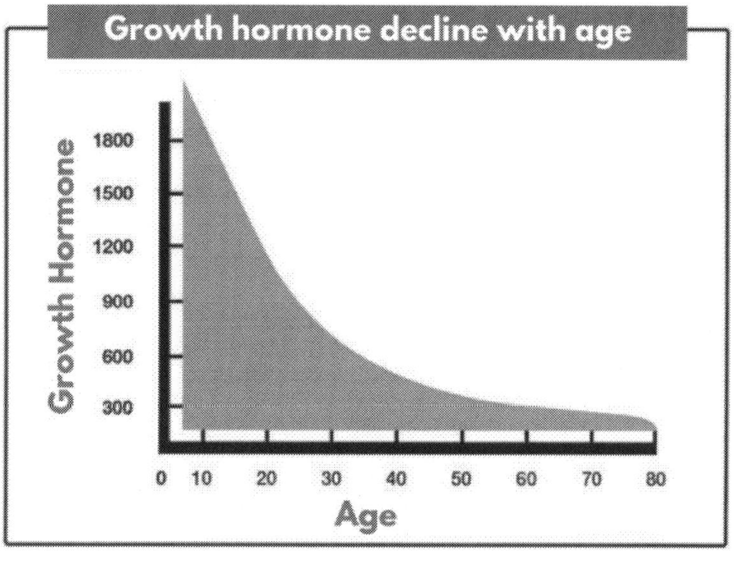

Sarcopenia increases damage from falls, promotes sarcopenic obesity, poor metabolic health, insulin resistance, and all-cause mortality[982]. Higher muscle mass and strength, on the other hand, has the opposite beneficial effect by increasing insulin sensitivity, glucose disposal, body composition, bone strength, cognition, and longevity.

Muscle wasting and decreased fitness are very much caused by disuse and a sedentary lifestyle. Older people start to move less, which exacerbates sarcopenia due to loss of necessity. That's very well illustrated by hospital bed rest and weightlessness in space where people lose a lot of weight. Even astronauts in space have a hard time maintaining their lean tissue albeit they're exercising. Low musculature combined with poor nutrition is a major risk factor for age-related diseases and mortality from all causes[983].

Physical inactivity and aging also decrease muscle protein synthesis (MPS), which promotes further anabolic resistance. Fortunately, it's well demonstrated that acute resistance exercise enhances MPS in both younger and older individuals[984]. Even a single bout of resistance training can increase MPS by 2-3 times, which can be enhanced further with a protein-rich diet[985]. What you don't use you'll lose, especially muscle and strength.

A 2016 paper found that amongst a large cohort of 65 and older, mortality rates were significantly lower in individuals who did regular strength training[986]. This should be very encouraging for the elderly to continue exercising but it ought to also tell younger adults to keep working out as part of their everyday lifestyle.

For anti-aging and heart health, the American Heart Association (AHA) recommends aerobic activities, such as walking, jogging, swimming, or cycling[987]. They also propose some strength training, but the vast majority of people aren't actively engaged in any form of resistance exercise. However, strength training may be much more important for longevity than just aerobics. There's a lot of research suggesting that muscular strength is inversely and independently associated with all-cause mortality[988]. Essentially, more muscle and strength reduces your chances of dying and can increase your lifespan, especially in older people.

A recent meta-analysis found that resistance training is more effective than cardio for fat loss as well as overall health[989]. In addition to that, both high intensity and resistance exercise tend to suppress your appetite a lot more than low-intensity cardio. That's another hidden contributing factor for longevity – you'll subconsciously eat less because of working out smarter.

There's evidence that skeletal muscle acts similarly to an endocrine organ, just like the adipose tissue does. Muscle cells (myocytes) produce certain cytokines called myokines that fight inflammation and maintain healthy physiological functioning[990]. Visceral fat, on the other hand, releases pro-inflammatory cytokines. That's another potential mechanism by which muscle uncoupling protects against chronic diseases. Being fit and muscular also enables the person to maintain a slightly lower resting heart rate and blood sugar, by reducing inflammation levels and burning off unhealthy fat.

Muscle mass, essentially, is like a solid pension fund that is associated with increased longevity, healthy aging, greater healthspan, and improved physical well-being throughout your entire life.

How to Build Lean Muscle Mass

The worst thing about age-related muscle loss is that you'll keep the fat. Sarcopenia decreases the number of functional muscle fibers and proteins in your body while accumulating excessive intra- and extracellular lipids underneath the skin[991]. You'll be the same bodyweight but with less muscle and more fat, making you look skinny-fat.

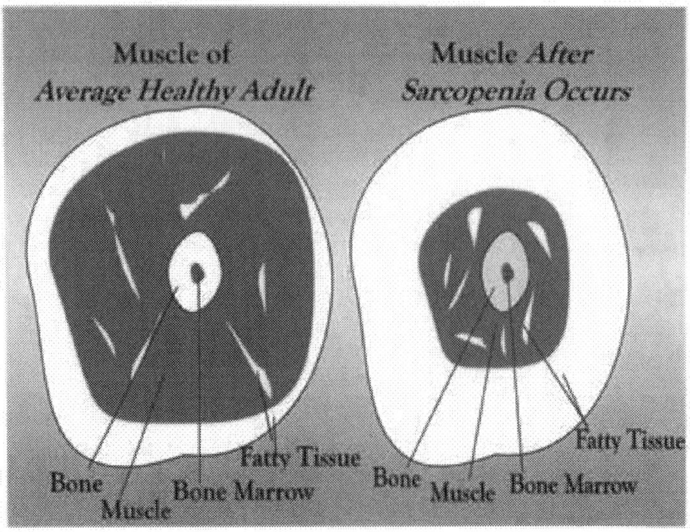

One of the biggest benefits of muscle mass is that it improves your insulin sensitivity[992], which will protect against diabetes, obesity, promotes carbohydrate tolerance, and enables the person to eat more calories without gaining weight. Muscle is like a massive sponge for calories and it comprises the majority of whole-body glucose uptake. Muscle mitohormesis is thought to promote longevity because of repressing insulin signaling and increased insulin sensitivity[993]. That's why training and having more muscle can be preventative for insulin resistance.

To build muscle, the body has to have a reason for doing so. This entails applying mechanical stress that would stimulate muscle protein synthesis and turn on the mTOR pathway. Mild distress on the muscle preserves its mitochondrial function, which will prevent sarcopenia. It's another hormetic adaptation that prolongs lifespan.

mTOR Signaling and Muscle Growth

Mechanistic Target of Rapamycin or Mammalian Target of Rapamycin or mTOR is a protein kinase fuel sensor that monitors the energy status of your cells[994]. It's involved in every aspect of cellular life and existence.

There are two mTOR complexes – mTORC1 and mTORC2. They stimulate cell growth, proliferation, DNA repair, protein synthesis, new blood vessel formation (angiogenesis), muscle building, the immune system, and everything related to anabolism[995].

- **mTORC1 functions as a nutrient sensor that controls protein synthesis[996].** mTORC1 is regulated by insulin, growth factors, amino acids, mechanical stimuli, oxidative stress, oxygen levels, the presence of energy molecules (ATP), phosphatidic acid, and glucose[997]. It's a key factor in skeletal muscle protein synthesis[998].
- **mTORC2 regulates the actin cytoskeleton, which is a network of long chains of proteins in the cytoplasm of eukaryotic cells[999].** It also phosphorylates IGF-1 receptor activity through the activity of the amino acid tyrosine protein kinase[1000].

mTORC1 detects many extra- and intracellular signals and growth factors[1001], whereas mTORC2 is known to be activated only by growth factors[1002]. Growth factors would be things like insulin, mechanical muscle stimulus, while nutrient factors would be things like amino acids and glucose that promote growth factors such as insulin and IGF-1. mTOR is intrinsic to the functioning of insulin, insulin-like growth factors (IGF-1 and IGF-2), and amino acids[1003][1004].

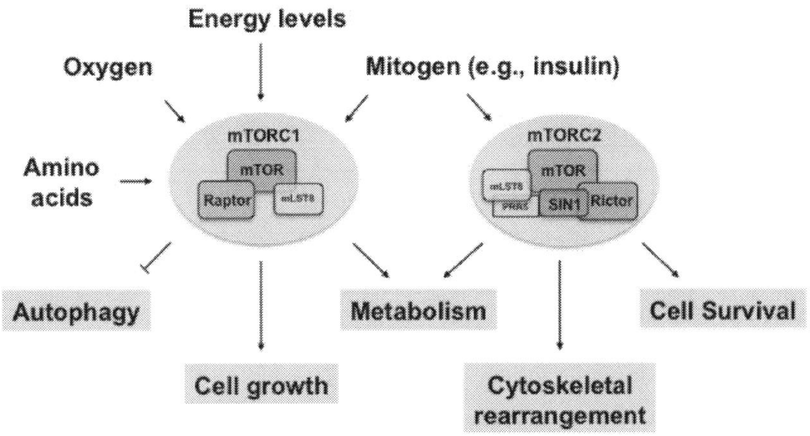

When there are plenty of nutrients around, mTORC1 binds to ULK1, which is an autophagy activating kinase and inhibits the formation of autophagosomes which would initiate autophagy[1005]. When energy gets depleted, mTORC1 becomes inactive and releases itself from the ULK1 complex, thus freeing up the formation of autophagosomes. The daily balance between mTOR and autophagy – anabolism and catabolism – determines muscle protein synthesis or breakdown.

Myostatin is a growth factor that inhibits muscle growth by suppressing hypertrophy[1006]. It's basically your body's attempt to not build too much muscle as a way to preserve energy from an evolutionary perspective. Myostatin reduces Akt/mTORC1/p70S6K activity and protein synthesis, which blocks muscle cell copying[1007]. Therefore, for muscle growth, you need the Akt/mTORC1/p70S6K pathway to be active.

Here are the other mTOR activating nutrients and factors:

- **Amino acids promote mTORC1 activity[1008] without affecting mTORC2 activity[1009].** Leucine specifically activates mTORC1 the most[1010]. Young muscle seems to be sensitive to just 1 gram of leucine to turn on MPS but older muscle requires over 2 grams[1011]. Maximum MPS can be reached by obtaining 2.6-3 grams of leucine per meal. Some evidence also hints that leucine's by-product HMB may have a similar anabolic effect through the signaling pathway of mTORC1[1012]. Although I'd say you'd get the same effect directly from leucine. Animal protein is higher in leucine and IGF-1, which both stimulate mTOR and muscle protein synthesis.

- **Mechanical stimuli from resistance exercise, especially eccentric contractions, increases the levels of mTORC1[1013].** Emphasizing the slower negative proportion of an exercise releases more growth hormone as well compared to faster repetitions[1014]. That's why cold exposure can sometimes also activate mTOR in muscles – you freeze up and contract the muscles.
- **Phosphatidic acid enhances mTOR signaling** and resistance exercise-induced hypertrophy[1015]. It gets regulated by exercise which activates mTORC1[1016]. You can find it the most in cabbage leaves, radish leaves, and herbs[1017] but also in supplements.
- **Ursolic acid stimulates mTORC1 after resistance training in mice[1018].** It stimulates anabolism via PI3K/Akt pathways. Ursolic acid can be taken as a supplement or found in apples, bilberries, rosemary, lavender, thyme, oregano, and other foods.

- **Creatine may potentially promote mTORC1** by increasing IGF-1 activity after exercise but doesn't further potentiate mTORC1 several hours later[1019]. So, the best time to take it is with your post-workout meal. It's one of the most well-studied supplements that works. Plus it's very cheap and effective.
- **Testosterone and androgens can also signal mTOR** and induce muscle hypertrophy[1020,1021]. Testosterone has many anti-catabolic as well as anabolic properties, which is why high cortisol tends to wreak havoc on this hormone. Low thyroid lowers testosterone and inhibits muscle anabolism.

Overexpression of mTOR or its dysfunction is often related to various cancers and genetic disorders[1022]. Inhibiting mTOR also promotes autophagy. High mTOR activity may promote tumor growth because of stopping autophagy from removing cancerous cells[1023]. Patients with Alzheimer's disease also show dysregulated mTOR activity in the brain and connection with beta-amyloid proteins[1024,1025].

mTOR has its benefits for performance as well as some supportive aspects for longevity. It's going to help to build muscle and burn fat but also regulates the immune response.[1026] Constant mTOR inhibition suppresses certain immune cells[1027]. However, it's not optimal to have it elevated all the time for obvious reasons. That's why knowing how to cycle mTOR is a vital thing for your long-term health. If mTOR is suppressed most of the time like during fasting, then that's going to allow autophagy to actually kick in and give you the other lifespan-boosting benefits. It's much more effective and easier to fast and then stimulate mTOR only within a very small time frame.

You can also dictate the location of mTOR activation and whether or not it becomes beneficial or malignant. If you're insulin resistant, obese, or with cancer, then turning on mTOR will be pathological. For someone who is healthy and working out, mTOR stimulation will just lead to increased muscle hypertrophy and strength but we still ought to be mindful of how much it's turned on. More on this balancing later in this chapter.

What Determines Muscle Hypertrophy

Muscle hypertrophy is a phenomenon that increases the size of skeletal muscle by enlarging and expanding its extracellular matrix[1028]. Hyperplasia is slightly different because it increases the number of fibers within a muscle.

Most of exercise-induced hypertrophy results from increased units of striated muscle tissue (sarcomeres) and myofibrils (unit of muscle cells)[1029]. It's mediated by muscle satellite cells that become active when under sufficient mechanical stimulus[1030] as well as the activation of the Akt/mTOR pathway[1031]. After that, they proliferate by fusing with already existing cells or amongst themselves to create new myofibers (muscle cells)[1032]. This provides precursors needed for muscle repair and growth. In addition to mTOR, here are the other contributing factors to muscle hypertrophy:

- **Mitogen-activated protein kinase (MAPK)** regulates gene expression, redox status, and metabolism[1033]. It links the cellular stress with the adaptive response in myocytes induced by exercise[1034].
- **IGF-1** facilitates the anabolic reaction in response to mechanical loading[1035]. Insulin also promotes anabolism by inducing hyperplasia and satellite cell differentiation[1036].
- **Testosterone** has an anabolic effect on muscle tissue[1037]. Mechanical loading magnifies the effects of MPS and anabolism[1038]. Suppression of testosterone has been shown to inhibit the anabolic response to resistance exercise[1039]. Resistance training also upregulates androgen receptors in humans, which attracts testosterone in target tissues[1040].
- **Growth hormone** promotes fat mobilization while increasing the uptake of amino acids into proteins and muscle[1041]. Exercise-induced increase of GH is linked with muscle hypertrophy[1042] although it's not directly an anabolic hormone like testosterone.
- **Cell swelling or hydration** stimulates anabolic processes by increasing MPS and decreasing breakdown[1043,1044]. It's thought to be caused by the increased pressure against membranes, which gets perceived as a threat, leading to cell response and reinforcement. Cell swelling is maximized by heavy glycolytic exercise, which accumulates lactate[1045], as well as blood flow restriction training.

- **Hypoxia or states of low oxygen** contribute to muscle hypertrophy by increasing lactate accumulation[1046]. This has been also shown to raise GH by 290% and attenuate atrophy in patients of bed rest[1047]. Hypoxia also generates low-grade ROS, which is a part of the hypertrophy response after resistance training[1048,1049].

Muscle growth can happen in many ways and it manifests itself differently based on the type of stimulus received. The two contributing factors are sarcoplasmic hypertrophy and myofibrillar hypertrophy.

- **Sarcoplasmic hypertrophy** focuses on increasing muscle glycogen storage. It increases the volume of sarcoplasmic fluid in the muscle cells and isn't accompanied by significant strength gain. The amount of potential blood being stored in the muscles increases. This is the bodybuilder approach. As much size as possible. It's caused by several sets of 8-12 reps against a submaximal load around 60-80% of your maximum.
- **Myofibrillar hypertrophy** increases the number of proteins necessary for adding muscular strength and will also cause small enhancements in size. This is the Olympic weightlifter and gymnast approach – as much strength with the least amount of weight. It's caused by muscle contractions against 80-90% of the one-repetition maximum for 2-6 reps.

Both powerlifters and bodybuilders are known to have exceptional muscle mass and strength. However, they train in completely different rep ranges and intensities. Which method is better or superior for hypertrophy is not clear[1050]. It's hypothesized that there are 3 primary factors that initiate the hypertrophic response to exercise: mechanical tension, muscle damage, and metabolic stress[1051,1052].

- **Mechanical tension** is essential for muscle growth and it can increase muscle mass while decreasing atrophy[1053]. However, some training modalities that incorporate high intensities like powerlifting or gymnastics can increase strength without significant hypertrophy.
- **Muscle damage** results from exercise localized damage to muscle tissue, which creates a hypertrophic response[1054]. It's similar to inflammation that the body then starts to heal and in so doing augments in size.
- **Metabolic stress** doesn't seem to be essential for muscle growth[1055] but it can still have a hypertrophic effect and promote cell swelling[1056]. This results from the accumulation of metabolites like lactate, phosphate, creatine, hydrogen ions, and others[1057].

Brad Schoenfeld is one of the most published researchers of muscle growth. In his 2010 paper, he poses the mechanisms of hypertrophy, which can be summarized as such[1058]:

Current research suggests that maximum gains in muscle hypertrophy are achieved by training regimens that produce significant metabolic stress while maintaining a moderate degree of muscle tension.

A hypertrophy-oriented program should employ a repetition range of 6–12 reps per set with rest intervals of 60–90 seconds between sets. Exercises should be varied in a multiplanar, multiangled fashion to ensure maximal stimulation of all muscle fibers.

Multiple sets should be employed in the context of a split training routine to heighten the anabolic milieu. At least some of the sets should be carried out to the point of concentric muscular failure, perhaps alternating micro cycles of sets to failure with those not performed to failure to minimize the potential for overtraining. Concentric repetitions should be performed at fast to moderate speeds (1–3 seconds) while eccentric repetitions should be performed at slightly slower speeds (2–4 seconds).

Training should be periodized so that the hypertrophy phase culminates in a brief period of higher volume overreaching followed by a taper to allow for optimal supercompensation of muscle tissue.

This paragraph alone can sum up the most optimal way of training for most individuals based on our current knowledge. It's optimized for muscle hypertrophy, which what the majority of people are looking for anyway – be strong, look good naked, and live longer.

Protein Intake, Meal Timing, and Muscle Growth

Studies have found that fasting lowers the expression of mTOR and IGF-1, which are both needed for cellular growth by increasing one of their inhibiting proteins called IGFBP1[1059]. Within 12-14 hours of fasting, SIRT1 gene regulation starts rising which will begin to suppress mTOR and AKT[1060,1061], thus down-regulating mTOR mediated protein synthesis.

mTOR has quite a detrimental role in anabolism. Inhibiting mTOR blocks the anabolic effects of resistance training and prevents muscle growth[1062]. mTOR is clearly anabolic but also anti-catabolic. It's going to protect the body against the harmful effects of cortisol and glucocorticoids on muscle tissue[1063]. Fasting is one of the most anti-TOR things there is because of upregulating AMPK and depleted amino acid availability. Although fasting increases growth hormone exponentially, it also decreases serum IGF-1 levels, which again decreases the body's anabolic state[1064]. Simply put, you're not building muscle when in a fasted state.

Muscle growth results from the positive balance between muscle protein breakdown (catabolism) and muscle protein synthesis (anabolism) (See figure below). You can stay in a highly catabolic state most of the day (fasting) if you compensate for it with enough anabolic stimuli (eating). It's possible to build muscle with a time-restricted feeding schedule, even with eating just one meal a day. Whether or not it's the most optimal thing for muscle growth remains to be up for debate.

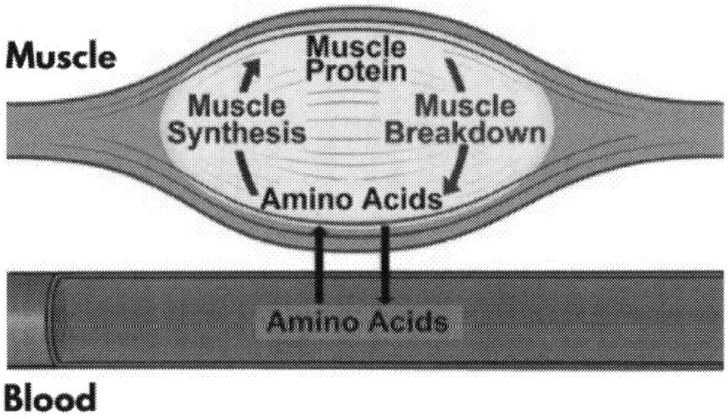

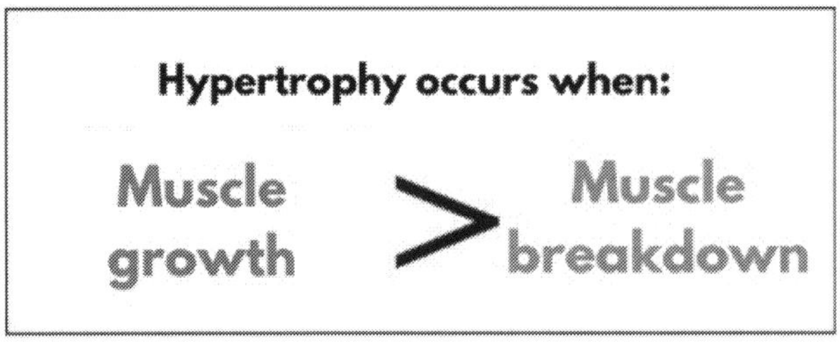

If you're trying to maximize your genetic potential for lean muscle mass, then it's naturally going to be easier to do so with a more frequent eating schedule. Likewise, it would be possible to go through periods of more anabolism followed by brief periods of higher catabolism that can lead to a positive muscle building condition because of practicing intermittent fasting.

I have been doing intermittent fasting since high school and have gained lean muscle mass and virtually zero fat. Throughout this period, I've never eaten breakfast, mostly skip lunch, eat the majority of my food in a single meal, have built an impressive physique, gotten stronger, and have never lost my six-pack abs. With deep keto-adaptation and proper nutrient timing, I'm able to build muscle at a steady linear progression.

From my own experience, intermittent fasting on a low carb ketogenic diet is much easier than on a high carb diet. Because of switching into ketosis, you start using fat as your primary fuel source and thus have little to no hunger, stable energy throughout the day, zero cravings, your brain gets less distracted and you reduce muscle catabolism in the fast. Fasting for such a long time eating carbs will make you go catabolic much faster because your body is inefficient at burning its body fat during the fast. However, you can mitigate that by eating a high protein diet.

The recommended dietary allowance (RDA) for protein is 0.36g/lbs of bodyweight which for an average individual weighing between 150-180 pounds would be 55-70 grams of protein per day[1065]. However, this is not ideal for most of the population and most people need more, especially if you're exercising.

In general, the optimal amount of protein tends to be somewhere between 0.7-1.0 g/lbs or 1.5-2.0 g/kgs of lean body mass (LBM), which for the same average individual weighing between 150-180 pounds would be 110-160 grams of protein at a minimum. Optimal ranges are between 0.7-1.2 g/lb of lean body mass. If you're not exercising, then you can safely stick to 0.6-0.7 grams but if you're working out you should stick to 0.8-1.2 grams. A higher protein intake is more important while doing intermittent fasting because you have to go for a longer time without eating and thus need the additional amino acids to protect against catabolism.

It's thought that your body can absorb only 30-40 grams of protein per meal and everything else gets wasted away. However, that's based on the idea that you only need about 20-40 grams of protein to maximize muscle protein synthesis[1066]. Any more than 40 grams doesn't stimulate MPS further. However, this doesn't really tell you much about how much protein you can end up absorbing in one meal. It just tells you that if you want to keep the muscle-building signal activated more frequently then you'd have to spike muscle protein synthesis more frequently as well. There's no indication of how it affects muscle protein synthesis over the 24-hour period.

The stimulation for muscle growth after resistance training will remain elevated for a long period of time. Studies have found that the potentiation of exercise-induced increases in myofibrillar protein synthesis and Akt/mTOR signaling by protein consumption is sustained for at least 24 hours post-workout[1067]. Even if you stimulate MPS twice within those 24 hours compared to 6 times you can still build muscle if you eat enough protein within that time frame.

Eating fewer meals and consuming more than 30 grams of protein in one sitting with intermittent fasting has not been shown to have any negative consequences in terms of lean tissue maintenance. One study done on women who ate their daily protein requirements of 79g of protein in either a single meal or 4 meals saw no difference in terms of protein metabolism and absorption[1068]. Several intermittent fasting studies have also shown that eating your entire days' protein in a 4-hour eating window has had no negative effects on muscle preservation[1069,1070,1071,1072]. Recent research indicates that, when fasted, we can use up to 3.5g/kg/day of protein and breakdown and metabolize up to 4.3g/kg/day[1073].

When it comes to body composition and fat loss, then meal timing has been shown to be irrelevant as intermittent fasting doesn't slow down your metabolism or make you lose muscle. If you're doing intermittent fasting with 2 meals a day or even just one, you can spike muscle protein synthesis and that's going to be more than enough to force your body to grow. You just have to make sure you hit your overall daily requirement for protein and calories.

The biggest advantage of building muscle with intermittent fasting is that it adds additional benefits to your longevity and anti-aging. It's not a good idea to be constantly stimulating mTOR and protein synthesis as discussed earlier. Furthermore, eating too much overall, or much rather too frequently, can directly accelerate aging as discussed earlier.

Methionine Restriction

It's thought that methionine restriction (MR) can increase lifespan and slow down aging[1074]. Methionine is an essential amino acid and a substrate of other important amino acids like taurine, cysteine, and glutathione. It's the most abundant amino acid in the human diet and can be found in almost all foods. Proteins that are low in methionine are considered to be incomplete proteins[1075]. Plant-based proteins that are high in protein are lower in methionine than animal proteins.

The increased lifespan seen in methionine restriction is considered to be caused by the downregulation of anabolic pathways like IGF-1, mTOR, and insulin[1076]. These complexes in excess can accelerate aging and promote disease. It's even possible that the longevity effects of MR can happen without caloric restriction, which so far is the only known way of lengthening life in many species.

Low protein intake reduces IGF-1 levels and incidence of cancer in 65 and younger but not older people[1077]. However...both high, as well as low levels of IGF-1, are associated with cancer mortality in older men[1078]. The same applies to risk of dying from all causes[1079]. In fact, a meta-analysis of 12 studies with over 14 000 participants found that people with low IGF-1 were at a 1.27x risk of dying and those with higher levels were at a 1.18x risk. Lower levels of IGF-1 may actually be more detrimental as you age as you'll be more predisposed to muscle loss and bone fractures. Studies find an association with low IGF-1 and sarcopenia in older people[1080,1081].

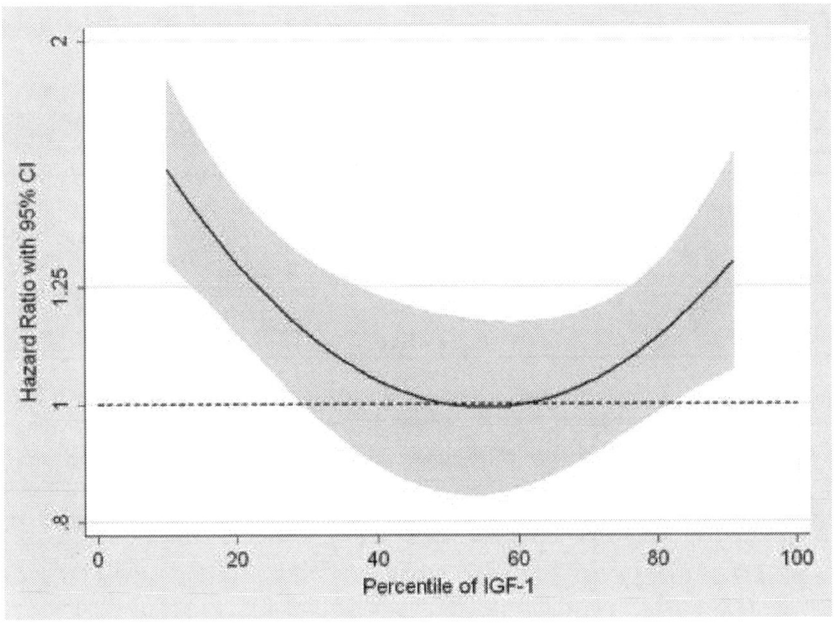

Source: Burgers et al (2011)

The Journal of clinical endocrinology and metabolism, 96(9), 2912–2920.

In healthy individuals, IGF-1 expression would be balanced by the IGF-1 binding protein (IGFBP), which blocks IGF-1s effects. That's why IGF-1 is bad only if you have too much free serum IGF-1 in the blood. Things that keep IGF-1 in the blood for longer are high eating frequency, high levels of insulin, combining carbs with fats and excessive protein, and physical inactivity. Exercise, intermittent fasting, and restricting either carbs or protein clears IGF-1 from the blood faster.

Calorie Restriction VS Methionine Restriction

It's thought that restricting just protein can give the same effect on life-extension than caloric restriction without needing to restrict calories. The link is lower methionine. However, a large 2016 meta-analysis demonstrated that CR combined with PR had an identical effect on life-extension as just CR without restricting protein[1082]. Therefore, eating less protein isn't the most important part of increased longevity when it comes to caloric restriction.

There are also many studies in which rodents under CR are fed a higher percentage of protein, including methionine, to keep it at the same level as those who eat without caloric restriction[1083,1084]. With protein being matched, the rodents under CR still gain the benefits of CR and do better than the ones eating with no restriction. Compared to caloric restriction, protein restriction does not contribute to the benefits of caloric restriction. The benefits of reduced protein intake are caused primarily by decreased calorie intake.

A 2008 study found that a low protein diet was more effective in increasing lifespan in fruit flies than caloric restriction[1085]. However, the researchers found that the carbohydrate to protein ratio was also quite important. Lifespan in fruit flies was the longest on diets with a 1:16 ratio of protein to carbohydrate. As this ratio decreased, lifespans also got shorter.

Protein isn't the only thing that raises IGF-1 and mTOR. Insulin that can be stimulated by carbohydrates also raises IGF-1 levels and turns on mTOR. In a low carb state protein doesn't spike insulin almost at all because there are no carbs. The reason has to do with a lower insulin to glucagon ratio. If glucagon is higher than insulin, which happens during fasting, eating a low carb diet or with low blood sugar, then consuming pure protein won't spike insulin but actually raises glucagon. This turns this idea that protein restriction shortens lifespan completely upside down.

Methionine Restriction and Autophagy

Another critical component to life-span extension seen in MR and caloric restriction is autophagy. Restricting methionine in Baker's yeast extends their lifespan, which was accompanied by an increase in autophagy[1086]. However, when autophagy genes were deleted, the extension in lifespan was also prevented. So, even if you restrict protein and calories but don't activate autophagy, then you're probably not going to gain the lifespan extension benefits of methionine restriction. Compared to caloric restriction or low protein diets, intermittent fasting is probably by far the most powerful strategies for lowering IGF-1 and promoting longevity.

In my example, I eat only once a day and I'm getting significantly more protein than the average person. Despite eating around 2.0-2.5 g/kg of protein, my serum IGF-1 levels were on the lower end at 103 ng/ml with the reference range being 90-357 ng/ml. That's lower than any person who's eating a protein-restricted diet or someone who doesn't eat meat. The reason I'm able to eat more than enough protein while keeping my IGF-1 low is because of intermittent fasting, resistance training, and being keto-adapted.

My body simply uses that protein and IGF-1 to support lean muscle growth and maintenance.

The argument I'm trying to make is that you shouldn't blame it all on protein. It doesn't matter what kind of a diet or how little protein you eat if your energy pathways don't induce a medium to low IGF-1 level and low levels of insulin. You have to realize that insulin and carbohydrates play a massive role in IGF-1 levels and how methionine is going to affect longevity depends on that particular context.

Furthermore, glycine supplementation has been found to have the same effects on life-extension as methionine restriction[1087]. Glycine is found in organ meats, ligaments, drumsticks, and all of these bone broth parts. If you're eating meat, then you would want to have less steak and more of these glycine-rich foods.

If the argument is that methionine restriction promotes longevity by suppressing IGF-1, then you would want to adopt a fasting focused lifestyle instead that incorporates a very low eating frequency, resistance training, and controls for carbohydrates in a flexible way. This will balance mTOR and autophagy, which determines the overall rate of your longevity.

Balancing Autophagy and mTOR

The Protein Kinase Triad (mTOR/Autophagy/AMPK) senses the energy status of the body and determines whether your cells will be favoring anabolic processes of growth or catabolic processes of self-devouring and preservation. By regulating and controlling the expression of certain nutritional factors you can get drastic results in your body composition as well as expected lifespan.

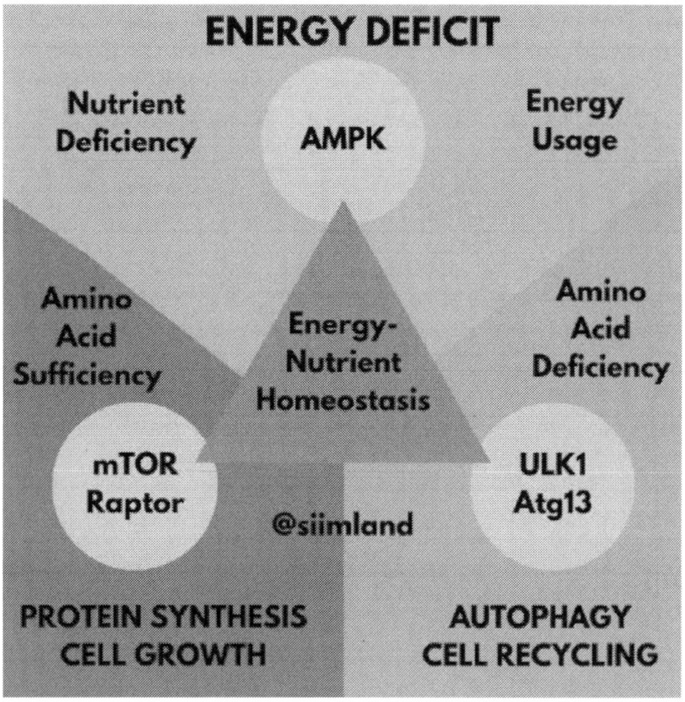

Both mTOR and autophagy are amazing. Fortunately, the human body has evolved with these metabolic pathways that promote survival and growth. However, they're not always optimal for our health and longevity in the context of modern society.

There's always an evolutionary trade-off between anabolism and catabolism as well as growth and repair. No free lunches in nature.

- Being too anabolic and growing rapidly may speed up your biological clock so to say by causing oxidative stress to the mitochondria and simply making your other organs work harder. If you grow fast you'll inevitably age faster as well.

- Being too catabolic and degrading at a quicker rate than you can repair will also damage vital cells and other processes in the body. In this case, you'll die sooner just by virtue of physical deterioration.

Being at a balance, on the other hand, is a myth because you'll never be completely stagnant and balanced. In fact, staying balanced in the wrong state can actually keep you stressed out or age faster. That's why it's important to rotate through different parts of the cycle and achieve equilibrium through motion. That's why intermittent fasting is still the most powerful component of this. Honestly, it's the easiest and most effective way to balance mTOR with autophagy – you just don't eat for a while.

Sarcopenia and Autophagy

To prevent sarcopenia, maintaining physical activity and doing resistance training is incredibly important. However, there are a few overlooked things like autophagy and calorie restriction that can also have the same effect. Common sense would say that autophagy promotes muscle wasting and sarcopenia because of breaking down proteins. However, research shows that autophagy actually maintains muscle and protects against age-related muscle dysfunction[1088].

Increased basal autophagy protects against sarcopenia by degrading misfolded proteins and organelles. It eliminates inflammatory cytokines and ROS that can turn your muscles resistant to MPS and anabolism. Systemic low-grade inflammation is the root cause of many diseases like sarcopenia, aging, metabolic syndrome, insulin resistance, atherosclerosis, and Alzheimer's[1089]. One of the main effects of autophagy is reduced inflammation. Inhibiting autophagy predisposes you to lose more muscle when under stress or while exercising because the body is less adapted, especially as you get older.

Autophagy plays a role in skeletal muscle mitochondrial biogenesis[1090]. Dysregulated autophagic flux inhibits lysosomal storage processes involved in muscle biogenesis[1091]. Dysfunctional mitochondria ignite a cascade of signaling events that lead to motor neuron and muscle fiber death[1092]. Mitophagy selectively degrades these damaged mitochondria and prevents them from becoming pathological.

Exercise activates the AMPK pathway, which stimulates autophagy. To activate the activation of autophagy during exercise, the activation of AMPK is also needed[1093]. AMPK regulates both protein synthesis and breakdown pathways. AMPK has a vital role in skeletal muscle homeostasis. Additionally, AMPK has been recently shown to be a critical regulator of skeletal muscle protein turnover or anabolic balance throughout the day[1094].

Caloric restriction (CR) without malnutrition has also been shown to reduce age-related muscle loss[1095]. The mechanisms by which this happens include improved mitochondrial functioning and biogenesis, increased basal autophagy, and reduced oxidative stress. CR reverses the decline in NAD metabolism and circadian misalignment, which enables the body to maintain a higher level of energy production and efficiency. Unfortunately, long-term calorie restriction isn't optimal nor sustainable as we discussed in chapter two. To not lose muscle from calorie restriction, you'd want to get a higher percentage of protein, focus on nutrient-dense foods, and incorporate periods of strategic refeeding. Becoming malnourished and frail because of CR is very common and it would speed up age-related sarcopenia.

Intermittent fasting is the more powerful and effective version of CR where you're essentially eating close to zero calories or nothing at all. You can get the longevity benefits of autophagy and NAD without necessarily restricting your calories or becoming malnourished. Instead, you confine your daily eating window and thus leverage the benefits of time-restricted eating. Of course, time-restricted eating can also lead to sarcopenia if you're not eating enough protein or are too catabolic but the other lifestyle benefits are far greater. You can completely avoid these negative side-effects by just making sure you meet your daily protein requirements.

Excessive exercise or too much fasting can cause excessive autophagy activation and result in muscle atrophy and catabolism[1096]. Chronic energy deprivation prevents recovery and suppresses anabolic hormones. To maintain better muscle, you need appropriate autophagy and mitochondrial quality control in the right amounts and at the right time[1097]. That optimal dose is very context-dependent and varies greatly between individuals.

Signs of Too Much Autophagy

There is no viable method of measuring autophagy in humans unless you're at a special medical laboratory. What's more, science doesn't currently know how much autophagy is good and when is the best time to activate it. What you can do right now is just look at your symptoms and make changes based on that.

Here are the signs of too much autophagy:

- **Muscle Loss** – Although autophagy can protect against sarcopenia, fasting for too long or undereating will eventually make you lose muscle. If you notice yourself losing lean tissue or are starting to look skinny fat, then you need to bump up your anabolism with either eating more often, getting more protein, or increasing total calorie intake slightly.
- **You're Getting Weaker** – If you're getting weaker at the gym, then it's probably because of losing muscle or not having enough energy to perform. Truth be told, you can't reach your physical potential by being in autophagy all the time. You need to eat enough protein and calories to keep making progress in any exercise.
- **Low Insulin and IGF-1 Signalling** – High fasting insulin is associated with metabolic syndrome and diabetes[1098], which happen due to energy excess. Furthermore, lower insulin/IGF-1 indicates more autophagy because of decreased nutrient signaling. Fasting, low carb diets, and exercise all stimulate autophagy a bit and they also reduce

insulin/IGF-1. The reference ranges are again very subjective. Generally, a fasting insulin < 25 mIU/L (< 174 pmol/L) is considered normal. Chronically low IGF-1 below 100 ng/ml can lead to sarcopenia, frailty, hair loss, hormonal imbalances, autoimmune issues, and slower healing. If you don't suffer these symptoms, then an IGF-1 below 100 can actually be good for longevity because you're not overstimulating growth. Fasting insulin and IGF-1 are also the most viable ways of measuring mTOR. Bodybuilders who are trying to build muscle would aim for IGF-1 above 500 ng/ml but that's not healthy for the general population. In my opinion, IGF-1 above 300 ng/ml would accelerate aging. It would at least keep you aging at the standard pace as the rest of the population does.

- **Hair Loss -** To prevent hair loss, you need primarily zinc, protein, vitamin A, B vitamins, vitamin C, vitamin D, vitamin E, collagen, iron, selenium, magnesium, and a few others. Depleting these nutrients like protein and zinc promote autophagy as a way of recycling them from your tissues, including hair and nails. You're basically using autophagy to get the nutrients you need from your tissues because you're not getting them from diet. The most common ones include zinc, B vitamins, D, and magnesium, which is why eating some oysters and organ meats as a supplement can be useful. Fasting too much too often will also make things worse.
- **Neuropathy** – Nerve damage in your fingers, hands, and feet can cause hypoxia, resulting in impaired blood flow,

not enough oxygen, and cold or numb limbs[1099]. Hypoxia promotes autophagy and angiogenesis but it can also create neuropathy. If your fingers and toes are cold or blue all the time then it's because of too much fasting and autophagy impairing blood circulation.

- **Heart Failure** – Autophagy gets activated in response to cellular stress that's happening in heart disease[1100]. In some cases, it's beneficial for reducing inflammation and oxidative stress whereas at others maladaptive and promotes disease progression. Defective autophagy or the accumulation of autophagosomes has reportedly caused cardiac dysfunction and heart failure[1101]. However, more often than not, heart failure results from hypoxia and ischemia. Autophagy is just there because it's trying to repair the damage. At the same time, too much autophagy can also cause atrophy of heart muscle cells, leaving it vulnerable to arrhythmia and irregular heartbeat. That's why increased basal autophagy like with time-restricted eating can be beneficial but prolonged fasting may not be ideal.
- **Weight Loss** – The process of autophagy is involved in fat oxidation as well, in which it's called lipophagy[1102]. When you're fasting you will lose weight and probably experience some autophagy. However, when you're losing weight without fasting, then you may experience slight increased basal cell turnover from the calorie restriction, which also promotes some autophagy. Generally, weight loss isn't a negative side-effect for most people because

they want it to happen but if you're losing weight uncontrollably then you can stop it by feeding yourself more calories and decreasing autophagy.

- **Weight Loss Plateaus** – Hitting a weight loss plateau or seeing metabolic slowdown can be partially be caused by too much autophagy. If you lose muscle because of autophagy your metabolism will inevitably drop because muscle burns more calories. This will also reduce thyroid functioning. To overcome that you need to stop autophagy and kick yourself out of ketosis by eating more protein and some carbs. This will stimulate IGF-1, mTOR, and insulin to increase metabolic rate, muscle growth, glycogen replenishment, and leptin.
- **Chronic Fatigue** – Being tired all the time and needing stimulants aren't that much caused by too much autophagy but more so because of the energy depletion that's required for it. If you're under-eating, overexercising, missing some key nutrients, and fasting too much, you'll eventually start dragging your feet. Autophagy is involved in removing old and dysfunctional mitochondria, which can increase your energy efficiency and vitality but too much of anything is still bad[1103]. Chronic fatigue may be just a sign of energy shortage that leads to increased autophagy.
- **Diabetic Ketoacidosis** – Ketoacidosis is defined as a dangerously high level of both ketones and glucose in the blood, causing the blood to become more acidic and promote oxidative stress[1104]. It happens primarily in

alcohol poisoning or diabetics. Autophagy is another protective mechanism trying to repair the damage so it's not causing the issue but it's a sign of its presence[1105]. If you're fasting or eating a keto diet your ketones will probably rise but your blood sugar will also drop, preventing ketoacidosis and maintaining healthy levels of ketones. If you're experiencing ketones above 5-10 mmol/l and your glucose is also above 120 mg/dl, then you should contact a medical professional and break ketosis. Elevated ketones are generally not an issue as long as your blood sugar is low to accommodate that shift in fuel sources.

Keep in mind that some of these symptoms don't have to overlap. Meaning, you can have low IGF-1, which is generally a good thing, as long as you're not losing muscle and/or are building it. In that case, low IGF-1 isn't bad but actually good because your body seems to be doing just fine without it and is healthier. If you're losing muscle even with elevated IGF-1 then that's much worse because your body is aging faster and at the same time not gaining the metabolic benefits of increased lean tissue. So, all the biomarkers themselves have to be taken within the context of your entire health and physique.

Signs of Not Enough Autophagy

There is always some autophagy happening somewhere in your body and to different degrees. This basal autophagy is just regulating the immune system for pathogens, helping with burning fat, and trying to wipe the floor so to say. However, for deeper cellular cleanup that would fight disease and remove senescent cells, you need to ramp up autophagy a lot more by increasing energy stress whether through exercise or fasting.

Here are the signs of not enough autophagy that tell you should try to increase autophagy:

- **High Insulin and IGF-1 Signalling** – High fasting insulin and IGF-1 reflect energy excess. Insulin and IGF-1 stimulate mTOR the anabolic switch that suppresses autophagy. IGF-1 levels above 300 ng/ml are associated with increased mortality but the same can apply to levels below 100[1106]. Ideally, you want to be somewhere between 100-300. A lower insulin IGF-1 signal indicates more autophagy because of decreased nutrient signaling. Fasting, low carb diets, and exercise all stimulate autophagy a bit and they also reduce insulin/IGF-1. You basically want to reach optimal body composition and the lower end of the normal fasting blood sugar/insulin range.
- **Insulin Resistance** – Hepatic autophagy is suppressed in the presence of insulin resistance and hyperinsulinemia[1107]. With high insulin, autophagy won't begin effectively because insulin inhibits the formation of key autophagy genes. Generally, a fasting insulin < 25 mIU/L (< 174 pmol/L) is considered normal. Anything above that may hint towards insulin resistance or pre-diabetes. Defective hepatic autophagy in obesity promotes endoplasmic

reticulum stress and causes insulin resistance[1108]. So, too much autophagy can also become detrimental if expressed in excess.

- **Heart Disease** - Suppressing cardiac autophagy in the heart is associated with age-related cardiac function and aging of the heart[1109,1110]. Insulin regulates myocardial autophagy. Glycogen content in the heart is regulated by extracellular insulin and glucose in cardiomyocytes. Inhibiting glycogen-specific autophagy (glycophagy) contributes to glycogen accumulation and cardiac failure[1111]. Hypertension, insulin resistance, high blood sugar, and being overweight all promote heart disease. Getting healthier by any means will improve the situation but autophagy is causally linked to mediating some of the benefits.

- **Atherosclerosis** - Autophagy is present in atherosclerosis as shown in both *in vitro* and *in vivo*[1112]. This is so because the autophagic process is trying to clear out the plaques and re-establish homeostasis. It's been shown that knocking out ATG5, the key gene for autophagy regulation in macrophages, can cause the aggravation of atherosclerosis lipid plaques by increasing apoptosis and oxidative stress[1113]. Macrophage autophagy plays a protective role in advanced atherosclerosis. Autophagy plays a vital role in protecting cells from oxidative stress, reducing apoptosis, and growing lesion stability. Impaired autophagy contributes to arterial aging[1114]. Inhibiting autophagy accelerates plaque necrosis[1115]. However, excessive autophagy can lead to autophagic cell death in smooth muscle and endothelial cells, resulting in decreased collagen synthesis and thinner fibrous cap. That can cause plaque rupture and cardiovascular events. Generally, the presence of excessive atherosclerosis is due to poor lifestyle and health but also because of not enough autophagy that would eat up the plaque.

- **Inflammation** - Autophagy removes aggregated inflammasome structures, thus dampening the pro-inflammatory response[1116]. Autophagy reduces NFKB (nuclear factor kappa-light-chain-enhancer of activated B cells) a protein complex that produces inflammation and cytokines. Autophagy regulates inflammation in adipocytes. Suppressing autophagy increases inflammatory responses via endoplasmic reticulum stress and regulation of insulin resistance[1117]. Autophagy is important in treating Crohn's disease[1118]. Inhibiting autophagy causes lung inflammation in cystic fibrosis[1119]. Biomarkers of too much inflammation are CRP (above 3.0 mg/L is high) and elevated white blood cell count.
- **Weak Immune System** - Autophagy plays a role in shaping immune system development, fueling the host's immune responses, and directly controlling intracellular microbes as a cell-autonomous innate defense[1120]. In recent years, a lot of studies have shown how fasting affects the immune system in a positive way[1121]. Not only does it protect against immune system damage but also helps to induce its regeneration by activating stem cells. Signs of a

weak immune system include getting the cold easily, coughing frequently, constantly runny nose, fibromyalgia, sore throat, allergies, autoimmune issues, and chronic fatigue.

- **Neurodegeneration** - Loss of autophagy promotes the formation of protein aggregates and neurodegeneration in mice[1122]. Autophagy is like a clearance system that eliminates these aggregates from the nervous system[1123]. Early stages of Alzheimer's show deficient autophagy and autophagy is important for metabolizing beta-amyloid[1124]. The most important thing for preventing neurodegeneration is sleep but primarily because that's the time period autophagy gets processed especially in the brain. If you tend to forget things easily, can't memorize information, and lack concentration then you should try to improve the quality of your sleep.
- **Gut Health** - Lysosomal disorders and deficient autophagy may lead to irritable bowel disease (IBD)[1125]. Certain probiotic bacteria that have been shown to reduce inflammation such as Bacteroides fragilis activates autophagy and requires autophagy for its therapeutic effects[1126]. Autophagy also eliminates pathogens and inflammatory cytokines from the gut.
- **Liver Spots** - Lipofuscin is an age-related pigment. It's a breakdown product of proteins and lipids formed by the oxidation and peroxidation of lysosomal contents. They can cause the accumulation of defective organelles and

aggregates because of reducing autophagy[1127]. Inhibiting autophagy results in an impaired turnover of lysosomes and accumulation of lipofuscin-like material[1128]. Liver spots and dark pigment on people's hands, face, and skin is oxidized lipofuscin. Sufficient autophagy eliminates the source of lipofuscin, which is oxidized iron and polyunsaturated fats.

- **Bad Skin Health** – Wrinkles and skin damage happens as a result of aging and oxidative stress. Autophagy in the skin fibroblasts promotes skin health and keeps it more elastic[1129]. Acne, pimples, or other spots are caused by over-expression of mTOR and IGF-1[1130]. Your skin reflects the health of your gut and digestion. With enough autophagy, you can keep your skin clear and not inflamed.
- **Fatty Liver** - It's been shown that impaired autophagy prevents the clearance of excessive lipid droplets, damaged mitochondria, and toxic proteins, which get generated during the progression of liver disease, thus contributing to the development of liver disease, fibrosis, and tumors[1131]. That's why regular autophagy will help to keep things in check. Signs of a fatty liver include high amounts of visceral fat that pops out and insulin resistance.
- **Muscle Loss** –Regular basal autophagy actually strengthens your body against the catabolic stress from exercise as well as fasting. So, if you're the kind of guy who immediately starts losing their gains whenever they

skip a meal, then it's because of inadequate autophagic adaptation.

All of these signs aren't solely signs of not enough autophagy. Some of them are more so than others but they're primarily signs of disease and imbalances in the body's energy homeostasis. If you're obese and insulin resistant, then it means you've spent too much time in a state of over-abundance and not enough time in self-recycling. To bring yourself back into balance, you have to increase autophagy whether with exercise, fasting, or calorie restriction.

How to Know If You're In Autophagy

The closest you can get is to look at your glucose ketone index because it reflects the relationship between the main regulators of autophagy – mTOR and AMPK.

Here's the Glucose Ketone Index Formula: (Your Glucose Level / 18) / Your Ketones Level = Your Glucose Ketone Index

- Measure your blood glucose by pricking your finger and all that. Write down the number you got.
- Measure your blood ketones by pricking your finger again *(ouch).* Write down the number you got.
- Divide your blood glucose number by 18.

    - If your device is using mg/dl, then dividing that with 18 converts it over to mmol/l
    - If your device is already showing mmol/l, then you don't need to divide anything and can skip this step
- Divide your result from the previous step by your ketone numbers.
- The end result is your GKI.

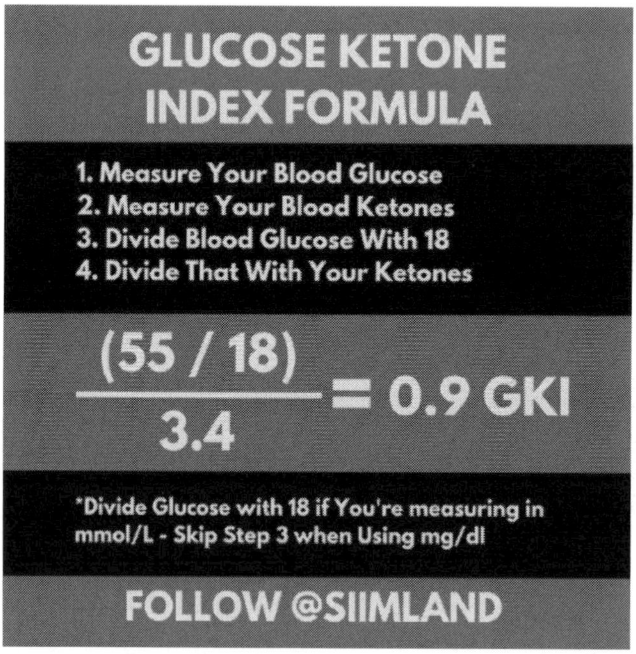

GLUCOSE KETONE INDEX FORMULA
1. Measure Your Blood Glucose
2. Measure Your Blood Ketones
3. Divide Blood Glucose With 18
4. Divide That With Your Ketones
(55 / 18)
3.4
= 0.9 GKI
*Divide Glucose with 18 if You're measuring in mmol/L - Skip Step 3 when Using mg/dl
FOLLOW @SIIMLAND

In general, having a GKI below 3.0 indicates high levels of ketosis in relation to low levels of glucose; 3-6 shows moderate ketosis, and 6-9 is mild ketosis. Anything above 9 and 10 is no ketosis.

The GKI indicates an estimated relationship between insulin and glucagon with a lower score being more ketotic. Ketosis and autophagy aren't mutually inclusive as you can be in ketosis without autophagy but it can at least give some insight into your nutrient status.

mTOR and IGF-1 Protocol

- Maintaining muscle mass is incredibly vital for longevity and anti-aging because of age-related sarcopenia. It becomes increasingly more difficult to build muscle the older you get, which is why it's best to build it throughout your lifetime. Your resiliency against stressors and hardship will also be much better if you're physically stronger.
- The key switch for muscle growth is mTOR, which gets turned on primarily by resistance training and consuming protein. It takes about 2.6-3 grams of leucine to maximize muscle protein synthesis in one sitting. You can reach that threshold by eating 30-40 grams of protein per meal.
- Higher protein diets tend to be better for muscle growth, to prevent sarcopenia, and improve body composition. You should aim for 0.6-0.8 grams per pound of lean body mass on rest days and 0.8-1.2 grams on workout days. This ensures adequate MPS. Restricting protein because of methionine or IGF-1 is not worth it and misleading because a higher eating frequency and spiking insulin multiple times a day is the determining factor to how high your mTOR and IGF-1 are. Protein is nothing to worry about if you're eating only once or twice a day. In fact, with intermittent fasting, you should be eating a higher protein

diet to compensate for the catabolic periods with increased anabolism.

- Resistance training stimulates mTOR and keeps it elevated for up to 48 hours. That's why you should workout optimally 3-4 times a week. The minimum effective dose is 2 times but it results in less muscle hypertrophy than working out more frequently. Based on current research, maximum growth can be achieved by working with 6-12 reps per set with 60-90 seconds for rest. In total it takes about 3-4 sets per exercise.
- When doing intermittent fasting, it's important to keep your protein intake high to compensate for the lower eating frequency. For nutrient partitioning, it's better to consume the majority of your calories after working out.
- Too much autophagy can cause muscle loss but not enough of it will also make the body more vulnerable to catabolic stressors. The optimal amount depends on your results and biomarkers. If you're losing muscle and strength, then you should dial down on how much fasting you're doing. Likewise, if your fasting insulin and IGF-1 are elevated, then you should lower it by eating fewer carbs, losing weight, and doing intermittent fasting.
- It's not recommended for older people above 60 to fast over 24 hours because of losing muscle. The cost to benefit ratio isn't that good and you'd be better off by maintaining lean tissue. They should increase their protein intake and add an additional meal to stimulate MPS more often.

Chapter Nine:
Exercises for Longevity and Hormesis

"Not less than two hours a day should be devoted to exercise, and the weather should be little regarded."
— Thomas Jefferson

So far we've talked about packing on solid muscle and getting strong because it's going to slow down aging and make us tougher. However, we ought to keep in mind what kind of physical parameters and measurements specifically are associated with longevity and increased lifespan. That's why I wanted to dedicate a separate chapter to this topic, although it's a shorter one.

It's thought that strength, not muscle mass *per se*, is associated with mortality[1132]. Although losing muscle is a contributor to increased atrophy in old age, the elderly and obese people tend to have less strength and force production than healthier ones[1133]. Maintaining muscle mass alone has not been shown to be enough to prevent loss of muscle strength, which suggests that muscle quality is equally as important for longevity.

Factors that underpin muscle quality include insulin sensitivity, motor unit control, body composition, metabolic status, aerobic capacity, fibrosis, and neural activation. All of them determine how well your muscle functions as well as the rate of decline with age.

Muscle quality is intricately connected with muscle strength and power. With age, you tend to see a progressive decrease of type IIb fibers also known as fast-twitch muscle fibers. Part of this may be because of disuse and sedentism but another reason is probably the pure atrophy of lean tissue. Therefore, having simply bulkier muscles like a bodybuilder may not necessarily correlate with increased longevity and lifespan. Even though you may exhibit higher insulin sensitivity and growth hormone, the body will still wean off because of the excess weight it has to carry. The key is to have quality muscle that's able to carry heavier loads and protect itself against muscular dystrophy. Being able to exert more force with less muscle size indicates higher muscle quality and mitochondrial density. That's why it's beneficial to focus on functional exercises that incorporate power and strength, not just hypertrophy.

It's suggested that overall grip strength may act as a biomarker of aging across the entire lifespan[1134]. Low grip strength is associated with all-cause mortality, cardiovascular events, myocardial infarction, and stroke. It's thought to be a stronger predictor of dying than systolic blood pressure[1135]. A recent UK study on over 400 000 participants tested the association between grip strength, obesity, and mortality[1136]. It was found that a stronger grip was associated with an 8% decreased risk of mortality. Adiposity measurements were inconsistent with mortality but a BMI over 35 and abdominal obesity were strong predictors of mortality, independent of grip strength. Therefore, even if you are a strong-but-fat person with good grip strength, it's not justified nor optimal to be obese. Strong hands require you to be working and lifting things consistently.

Grip strength is best trained with frequent use. Ever seen the hands of a plumber, construction worker or a farmer? They're not doing any fancy gym workouts and get their exercise solely from daily work. Picking things up, carrying around bricks, and working with different tools keeps the forearms constantly engaged and ultimately makes them stronger. Hypertrophy and increased size are just a side-effect.

Another potential indicator of muscle quality is leg strength. In 2011, researchers found that leg strength was one of the most important factors for determining physical function and mortality[1137]. The link was purely based on strength and longevity regardless of the amount of muscle mass. Glutes and hamstrings are one of the most functional muscles because you need them for running away or chasing game, thus they provide an incredible survival benefit.

What are the best exercises for building grip and leg strength? Compound lifts like squats, deadlifts, barbell rows, and bench press are amazing for developing full-body strength and muscle. Furthermore, resistance training increases bone mass and lowers the risk of osteoporosis, which a lot of people suffer from. Elderly folks who don't exercise and have lost their muscle are prone to hip fractures and joint pains just because of lacking bone density. In fact, one of the causes of death is that they get an injury or they fall on the ground and they get hospitalized or die altogether.

Fortunately, it's never too late to start working out. It's been shown that elderly sedentary people can gain more than 50%

strength after 6 weeks of resistance training. They even had to exercise only 2-3 times per week with about 70-80% of maximal strength[1138]. Age is not an excuse to not train or be un-fit.

How to Do Resistance Training

Lifting weights and bodybuilding are incredible for injury prevention and rehabilitation. Strength training's been found to be more effective in treating painful muscles and mobility issues[1139]. On the flip side, things like Pilates do not seem to improve symptoms of back pain or functionality[1140].

Compared to aerobic exercise, strength-based resistance training protects against aging a lot more. It's been found that strength training can have a lot of the same health benefits as cardio, such as reduced chronic inflammation and improved cardiovascular health[1141].

Here's a list of functional exercises you should do to improve strength, power, and muscle mass: bench press, barbell squat, barbell row, deadlift, pull-ups, push-ups, handstands, dips, walking lunges, sprints, shoulder press, kettlebell swings, farmer's carries. The body doesn't care where the stimulus is coming from so calisthenics as well as weights both are great. Adaptation results from perceived exertion and necessity. In fact, it's even found that blood flow restriction (BFR) training combined with low-load resistance training enhances muscle hypertrophy and strength[1142].

BFR can give your body an effective stimulus for muscle growth and maintenance without overloading the joints and cardiovascular system[1143]. Traditionally, heavy strength training between 60-80% of your 1RM has been the go-to method for increasing muscle and strength[1144]. However, it's been shown that BFR can be effective even at 20-30% of 1RM[1145]. You can basically trick your body into thinking it's lifting a much larger amount of weight than it actually is. This is especially useful for the aging population or for someone who can't exercise at their full capacity due to injuries or rehab.

The 3 main variables of effective training you should know about are:

- **Volume** – the total quantity of movement performed during each exercise, training session, and training cycle.
- **Intensity** – the amount of load, weight lifted, speed attained. Intensity is the qualitative component of training. It's about doing more work per unit of time.
- **Frequency** – the amount of density, how often you train, how many times per week you execute certain movements. This is another quantitative aspect, but it is more concerned with recovery.

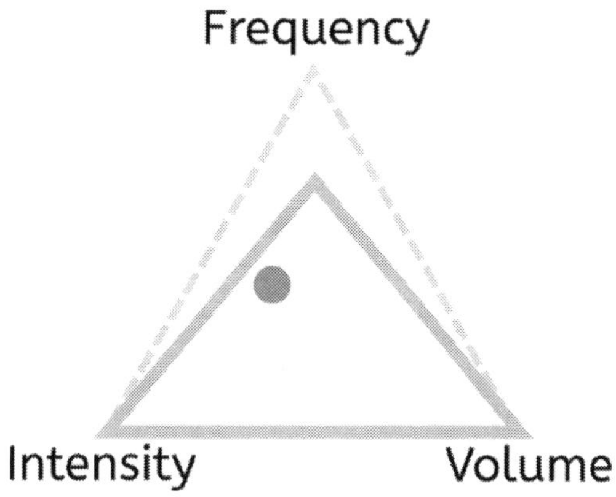

Together, these are the triad of training variables. The maximum capacity of each aspect is at the endpoint. For instance, for volume, it would be two hours of training, for intensity 80-90% of near-maximum effort, for frequency it would be training twice every day. You should strive for creating a balance between these extremes. Otherwise, you will reach burnout. If you were to go for a high amount of volume every single day, then you can't be doing it intensely. Training only 2 times per week (low frequency) allows you to do a lot of work at greater loads as well.

We can manipulate intensity by (1) increasing the difficulty of our exercise, by adding more weight to the bar or progressing in bodyweight movements. Volume can be modulated by (2) the number of reps done per set, (3) the number of sets per exercise, and (4) the total amount of exercises we do. To continually get stronger and build muscle, we can't be repeating the same workout over and over again, because the body will adapt to the stress. Therefore, improving at least in some of those 4 metrics indicates progress.

Current research shows that training a muscle twice a week leads to superior hypertrophy than once a week[1146]. In 2015, Schoenfeld *et al* showed how training a muscle group 3 times a week with full-body workouts was better for muscle growth than training it once a week on a split routine[1147]. Therefore, you'd want to be targeting the main muscle groups like legs, chest, back, shoulders, and arms at least 2-3 times per week.

Higher training frequencies are also seen to be more effective in highly resistance-trained men[1148], which makes sense from a physiological perspective as well. The more you train and the stronger you get the more stimulation you need to facilitate further growth. Advanced trainees are more resistant to muscle damage and neuromuscular fatigue. They also show a blunted hormonal-anabolic response to training volume[1149]. That's a good example of a hormetic adaptation outside of muscle hypertrophy and a reason you may want to take breaks every once in a while.

Exercising too frequently may actually inhibit mTOR a little bit by upregulating myostatin. That's why consistently working out frequently may be more beneficial for longevity whereas for pure hypertrophy purposes you'd benefit from taking breaks more often. Not exercising desensitizes the body to the anabolic stimulus again and potentiates further growth.

Increasing training frequency also lowers rates of perceived exertion (RPE)[1150], reduces delayed onset muscle soreness (DOMS), and increases the testosterone to cortisol ratio[1151]. Training more frequently requires that you scale down the volume or intensity of each workout, which enables you to train more often. However, doing high-intensity, high-volume, and high-frequency as in working out for 2 hours every day with heavy loads will lead to burnout, unless you have really optimized your recovery or are on some performance enhancers.

It's not necessary to workout 4-6 times a week if it doesn't fit your schedule. As long as you put enough volume and intensity into your program, you can probably see as good results with a lower frequency and training only 2-3 times a week. In that case, you'd have to simply make your single workout sessions matter a lot more by increasing the amount of volume done with each exercise.

What routine you choose to follow depends on your preference. You can either do a full-body routine, upper and lower body splits, or the push/pull/legs split. Train about 3-4 days a week and leave at least 1 day for complete recovery. On days when you're not doing resistance training, you can do some light aerobic cardio to keep yourself moving.

Cardiovascular Exercise and Fitness

However, there are still some benefits to doing endurance-based exercise and cardio. To protect yourself against aging and disease, you'd want to maintain aerobic fitness regardless of your physique goals.

A large 2011 meta-analysis found a direct correlation between gait speed and survival in older people[1152]. Walking requires energy, movement control, fitness, and puts mild hormesis on the body. Slow gait may reflect either damaged systems or a high-energy cost[1153]. It could also mean you're pushing the envelope too much and are compensating for an impeding allostatic overload.

Aerobic capacity, which is your maximal ability to use oxygen during physical activity, tends to decline after the age of 50[1154]. Having better aerobic capacity helps muscles to use oxygen more efficiently, which will improve physical performance. Enhanced endurance exercise includes increased capillary supply, mitochondrial biogenesis, and enriched transport of electrons in the mitochondrial transport chain. With higher aerobic fitness you'll also use less muscle glycogen and produce less lactate, which allows you to perform at greater intensities.

Both aerobic and anaerobic training contributes to the development of cardiovascular fitness. It means you can improve your endurance even without necessarily training it. There are much more effective and healthier ways of training your aerobic system than just cardio. Excess aerobic training increases the risk of oxidative damage in the muscles which may speed up sarcopenia, especially when dieting. On top of that, the prolonged stress hormones and free radicals in the blood may damage the mitochondria.

The worst thing about too much cardio is that it's not even the best way to lose fat or become fit. How many people you know who've been exercising for a long time doing different runs, spin classes, and cardio sessions but they haven't gotten any fitter? They're exercising like maniacs but they're still fat or worse skinny fat. That's because the body receives an inefficient stimulus that would enforce proper adaptation. This is called *Black Hole Training* – a nightmare exercise zone between a piece of cake and a Navy SEAL workout.

The Black Hole is a heart rate zone that exceeds your aerobic capacity just a tiny bit. Once you can't hold a conversation anymore and have to breathe through your mouth, then you're using more glycogen and less fat for fuel. For a few minutes, that's fine, but most people never go running for 10. They hit the runner's flow because of the adrenaline rush and can easily empty their glycogen tank. Once this happens, the body still needs glucose to perform at such intensity. As a result, it begins to break down the protein.

Intensity Zone	VO_2 (%max)	Heart Rate (%max)	Lactate ($mmol \cdot L^{-1}$)	Duration	
1	45-65	55-75	0.8-1.5	1-6 h	'Zone 1'
2	66-80	75-85	1.5-2.5	1-3 h	
3	81-87	85-90	2.5-4	50-90 min	'Zone 2'
4	88-93	90-95	4-6	30-60 min	'Zone 3'
5	94-100	95-100	6-10	15-30 min	

While cardio is beneficial and has its place, you shouldn't make it the focus of your training. At least you'd want to avoid the Black Hole as much as possible. If you're doing cardio for 30+ minutes, then you should stay aerobic for the majority of the time. That's when your heart rate is below 60-70% of your VO2 max. At that intensity, you're using fat not glucose as fuel. Going higher than that will simply make you more glycolytic. When you burn through your glycogen stores and you keep exercising, then at that point you may end up sacrificing a bit of the lean tissue by converting some protein into glucose. With keto-adaptation, you can protect yourself against that to a certain extent but it's simply not worth it.

The 80/20 to exercise is to focus on resistance training and strength because it's what gives you the most results in terms of longevity. Endurance work can also be good but not if done excessively or for too long. Of course, you don't want to get winded because of climbing a flight of stairs. There are other more effective and time-efficient ways of doing aerobic conditioning than just cardio.

High-Intensity Interval Training

In 1996, a Japanese researcher Dr. Izumi Tabata conducted a study on two groups to assess the differences between training intensities on aerobic fitness[1155]. All the subjects used cycling ergometers to train on for 6 weeks.

- The control group did 60 minutes of moderate-intensity exercise 5 times per week at 70% of their VO2 max. It's called low-intensity steady-state cardio (LISS).
- The other group did high-intensity interval training (HIIT) with 20/10 sessions repeated 8 times at 170% VO2 max. They basically pushed themselves to the absolute limit. 20 seconds beyond-maximum exertion followed by 10 seconds of rest. In total, the workout lasted for 4 minutes.

The results were quite astonishing. Over the course of 6 weeks, the control group trained for 1800 minutes vs the 120 minutes of the HIIT group. The subjects who did LISS increased their VO2 max from 53 +/- 5 ml.kg-1 min-1 to 58 +/- 3 ml.kg-1.min-1 but their anaerobic capacity didn't improve significantly. The individuals who did HIIT increased their VO2 max by 7 ml.kg-1.min-1 while their anaerobic fitness also improved by 28%!

The Tabata study shows that high-intensity training is much more effective in inducing physiological adaptations and is a more time-efficient way of exercising. It also makes sense from the perspective of the body. You won't get stronger by lifting a pillow off the ground for millions of times. Lift a 400-pound barbell just twice and you'll get a much bigger response from the body.

Strength and fitness are the result of adapting to stressors. There needs to be a very good reason for your body to build muscle or get faster. Making a hole in the wall with a spoon will take an eternity. Hit it with a sledgehammer and you'll get out of jail in no time.

However, there are some implications about the Tabata study that need to be covered.

- **First, the aerobic group did improve their VO2 max quite significantly**. They did get fitter because the body became more efficient at running on lower intensities and thus they could ramp it up a notch.
- **Secondly, although the HIIT-ers increased their mean VO2 max slightly more, they started off with a lower value and may have had more room for "newbie-gains".** The LISS group was already fitter and thus it was more difficult for them to see improvements.

- **Thirdly, the HIIT group still ended up at a lower VO2 point than the LISS group.** They didn't become exponentially fitter although their relative fitness increased more.

Despite that, the HIIT group did improve their anaerobic fitness as well as their aerobic, whereas the LISS cyclers didn't. Unless you're training specifically for endurance, then you don't have to do hours of cardio. For an average person trying to be fit and healthy, doing high-intensity training is still a much more effective and time-efficient way of exercising. It saves time, preserves joints, and maintains a better metabolic rate.

High-intensity interval and Tabata training also coincide with resistance training. Both of them are more intense, they burn muscle glycogen, improve insulin sensitivity, and stimulate the sympathetic nervous system. HIIT may lead to similar physiological adaptations in terms of muscle growth than resistance training. Not entirely but it's not going to hinder the beneficial muscle-building signals as much as cardio would.

Doing cardio alongside resistance training decreases the positive effects of both. Concurrent strength and endurance training decrease gains in cardiorespiratory fitness, explosiveness, strength, and muscle mass[1156]. Confucius a Chinese philosopher said: "*The man who chases two rabbits, catches neither.*" That's why it's not necessarily a good idea to combine a bunch of endurance exercises with hypertrophic ones. You can do your endurance work and aerobics as much as you like. However, whenever you're doing resistance training to stimulate muscle growth you have to focus on primarily power and strength.

Next to fasting, exercise can also increase mitochondrial biogenesis and function through the same pathways of AMPK and FOXO proteins. High-intensity interval training increases the ability of mitochondria to produce energy by 69% in older people and 49% in younger[1157]. Even just acute exercise increases FOXO1 phosphorylation, improves insulin sensitivity, and promotes mitochondrial biogenesis[1158]. Exercise also increases NAD+ and sirtuins[1159,1160].

There are many ways to improve your aerobic fitness but the main tenet is increased muscle quality. The best way to protect against sarcopenia and muscle-wasting is to combine both resistance training with HIIT cardio and then get a minimalistic dose on endurance. That's the main way to improve the quality of your life with exercise.

Intermittent Hypoxia Training

In addition to the countless stressors our bodies can adapt to, low levels of oxygen or hypoxia can pose another benefit through hormesis. It's often associated with sleep-disordered breathing and other pathologies but can also have exercise performance results[1161]. For example, altitude training has been used by athletes in many sports for decades to improve endurance and speed[1162]. Living at higher altitudes is also associated with lower mortality from cardiovascular disease, stroke, and certain cancers[1163]. Although protective against many ailments, high altitudes can also promote disease progression once it sets in. More on this shortly.

Hypoxia is defined as breathing less than 21% oxygen or oxygen saturation below 80-90%. Complete oxygen deprivation is called anoxia. To cope with oxygen deprivation, the cells adjust their metabolic requirements. Adaptation to hypoxia requires the activation and coordination of many pathways like hypoxia-inducible factors (HIFs), the mTOR complex, unfolded protein response (UPR), and autophagy[1164,1165].

Here are the effects of hypoxia on the body:

- **Hypoxic exposure increases the oxygen-carrying capacity of red blood cells**. Hypoxia-inducible factor 1 (HIF-1) tells the body to produce erythropoietin (EPO), which in turn makes bone marrow create more red blood cells[1166]. EPO is the infamous doping used by athletes, especially long-distance cyclists. Your body can make its own as well.
    - HIF-1 regulates vascular endothelial growth factor (VEGF), which is important in angiogenesis and cell permeability[1167]. This expression appears to be more robust in intermittent instead of sustained hypoxia[1168].
    - Prolonged hypoxia can lead to the accumulation of lactic acid, which is associated with cancer metabolism via the Warburg Effect and cell death[1169]. It can also cause adrenal stress, heart palpitations, and fatigue.

- **HIF mediates the reduced cardiovascular disease mortality seen in altitude[1170].** People living higher above the ground tend to have lower blood pressure and C-LDL[1171]. Intermittent hypoxia increases nitric oxide production, which promotes vasoprotection, neuroprotection, cardioprotection, and defense against stress[1172].
    - Symptoms of altitude sickness include fatigue, nausea, numbness, tingling of extremities, and cerebral anoxia. In severe cases, it can also cause headaches, confusion, breathlessness, hypertension, and even heart failure[1173].
- **Intermittent Hypoxia has been shown to promote BDNF-dependent neurogenesis and antidepressant-like effects[1174].** A shorter dose of 4 hours/day for 14 days appears to be beneficial and hormetic in rats whereas chronic exposure (8 h/day for 14 days) can decrease hippocampal BDNF levels[1175]. Excessive oxygen deprivation will damage the brain and impair cognitive functioning.
- **Exercising in hypoxia can promote ischemic preconditioning of the heart[1176].** This can provide subsequent protection against more severe hypoxic situations and other stressors. Even just a single night of sleeping at moderate altitude before exercising may reduce the risk of sudden cardiac death in men over 34 with a history of coronary artery disease and/or prior infarction[1177].

- The response to hypoxia in the lungs is vasoconstriction or narrowing of the blood vessels also called hypoxic pulmonary vasoconstriction (HPV)[1178]. When pulmonary capillary pressure remains elevated for at least 2 weeks, the lungs become more resistant to edema because of lymph vessel expansion. This increases their ability to carry away fluids by as much as 10-fold.

- **Hypoxia-Inducible Factor (HIF) is a major factor in cell survival to hypoxia that induces autophagy[1179].** HIF gets activated when oxygen tension decreases[1180]. It then guides many genes that promote survival instead of cell death under hypoxic stress to restore O2 homeostasis[1181].
    - Unfortunately, tumor cells and cancers can also rely on autophagy to self-proliferate and survive nutritional stress. HIF can promote tumor expansion, metastasis, and drug resistance by supplying cells with more oxygen. Hypoxia increases blood vessel growth that supports angiogenesis[1182]. Hypoxia-induced autophagy promotes tumor cell survival and adaptation to antiangiogenic treatment[1183]. This is the point I was trying to make earlier – some oxygen deprivation in a hormetic fashion is great but, in some cases, it can also supply malignancies.

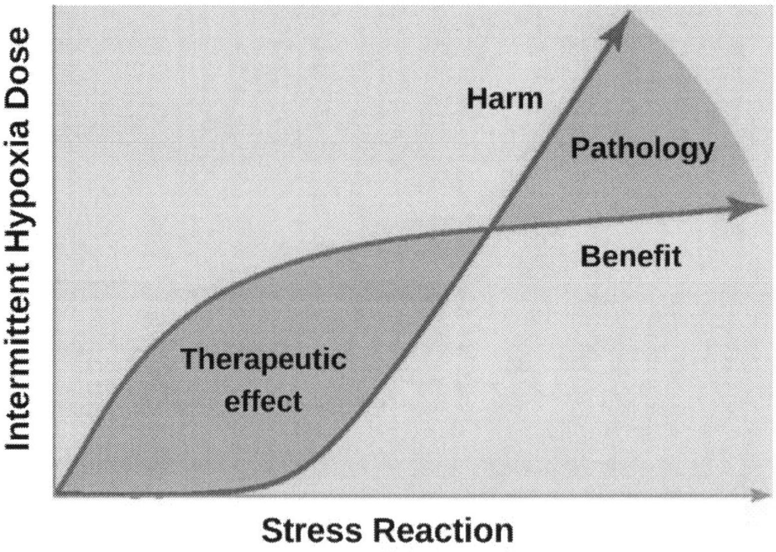

Although chronic hypoxia is a pathological condition, it's a part of normal physiology during exercise, being in altitude or hypoventilation. Exercising or even just spending time in higher altitudes with less oxygen can pose a hormetic response but not if done chronically. It could be especially harmful in people with pre-existing heart problems or brain injury.

Here are some ways of training hypoxia besides living in higher altitudes or going for a skiing trip in the Alps:

- **Oxygen Saturation Monitor** – Using a fingertip pulse oximeter you can accurately measure your blood oxygen saturation and pulse rate. It's going to tell you whether or not you're indeed in hypoxia, especially while trying to train for it.
- **Altitude Training Tents** – There are altitude training tents or chambers that create the same environment as if you're

being in altitude. Companies like Hypoxico have tents and masks for both exercise as well as sleeping. Imagine, you can workout and sleep as if you're in the Alps without ever actually going there.

- **Voluntary Hypoventilation** – It's been shown that swimmers can train in hypoxia at sea level through voluntary hypoventilation[1184]. They're deliberately trying to slow down their breathing rate while continuing to exercise. You can try to do the same during your workouts. Instead of gasping for air, make a conscious effort to slow yourself down and not breathe uncontrollably. On the flip side, doing an insane HIIT workout that leaves you breathless itself already achieves hypoxia. The difference is you're just in such high demand for oxygen that the hypoxic environment is just the result.
- **Buteyko Breathing** – This technique is named after the Ukrainian physician Konstantin Pavlovich Buteyko who developed it in the 50s. The main idea is to reduce hyperventilation and normalize breathing through nasal breathing, reduced breathing rate, and relaxation. You're trying to consciously reduce either your breathing rate or volume. By using a measurement called the Control Pause (CP), you can increase the amount of time you can comfortably hold your breath in between breaths. A good CP is about 30 seconds and 45-60 seconds is amazing. Being able to CP for only 15 seconds or less may indicate respiratory problems, stress, and disordered breathing patterns.

- **Breath-Hold Walking** – I started doing these during my commutes to school in college. There was this one street I always had to go through and to make better use of my time I decided to practice hypoxia and meditation at the same time. I locked eyes with a signpost a few dozen feet away and held my breath while walking towards it. After my eyes couldn't see it anymore, I took a small inhale and picked another target. It was actually quite fun and made the walk slightly more challenging. Looking at a single point while your mind is trying to escape enhances self-control and mindfulness but the breath-hold makes it that much harder. You can do this even without anything to look at – just hold your breath for about 10 seconds while walking and repeat for a few minutes.
- **Box Breathing** - Think of a box with 4 facets. They are stages of your breathing with each phase lasting for 4 seconds. By the end of it, you will have reached a semi-meditative state which makes you completely centered within your body and calm. It feels amazing and you are entirely in the rest and digest mode.
    - Breath in for 4
    - Hold it for 4
    - Exhale for 4
    - Pause for 4
    - Repeat for 4 minutes in total

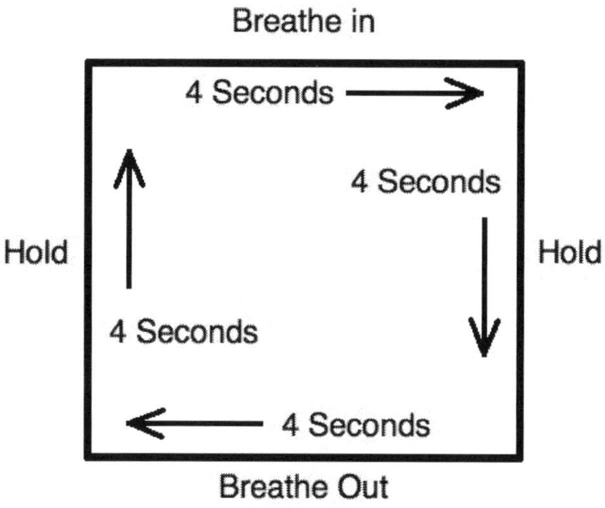

- **Alternate Nostril Breathing** – In Kundalini *Yoga it's called "Nadi Shodhana PranayamaI"* and it's great for reducing stress, becoming mindful, and cleaning the nasal pathways. Take your right thumb and press it on your right nostril. Breathe out through the left nostril. Close the left nostril with your right ring finger and release your right thumb from the right nostril. Breathe in through the right nostril and close it with your right thumb again. Open the left nostril and breathe out. Repeat this alternating pattern for about 5 minutes. You can use your left hand as well but just switch the fingers.
- **Hyperbaric Oxygen Therapy (HBOT)** – HBOT involves giving the body pure oxygen in a small airtight chamber or tank. It's used in medical clinics, to enhance athletic performance, recovery, and longevity. HBOT promotes new blood vessel formation like hypoxia, bone growth, wound healing, and improved oxygenation of tissues[1185]. Having your own HBOT chamber is very expensive for

most people as it can cost well over 15k. However, there are hyperbaric clinics in most large cities of the world now and you can get an hour-long session for relatively cheap. Generally, it's more effective for people who live in highly polluted areas, they have respiratory problems or some other co-morbidities.

Hypoxia during sleep causes sleep abnormalities, sleep apnea, hemodynamic stress, and other health problems related to poor sleep quality[1186]. It's also a factor in cognitive decline and neurodegeneration[1187]. Generally, hypertension, obesity, breathing problems, asthma, and smoking causes this. However, environmental factors like air travel, high altitudes, and air quality also play a role.

How to Breathe Better

Sleep apnea or apnoea is a sleep disorder characterized by dysfunctional breathing, shallow breath, and paused breathing during sleep[1188]. You may stop breathing for a few seconds up to a few minutes several times throughout the night.

Symptoms of sleep apnea include excessive daytime sleepiness, chronic fatigue, and impaired alertness. The biggest predictor is probably snoring and breathing problems during the night. So, you'd want to pay attention to how your family members sound like when they're sleeping. Some people may occasionally snore more and louder when they're drunk, taking a nap, or being in an armchair.

To treat sleep apnea, you'd want to start with losing weight, improving your biomarkers, becoming more insulin sensitive, and eating at the right time. Overeating, especially at night, may bring about snoring and difficulties breathing in bed. You would also stop other afflictive habits like smoking, drinking alcohol, and eating processed food.

There are also different mouthpieces, breathing devices, and machines that help you to breathe properly. However, you'd want to consult with your doctor beforehand as you may need surgery instead. I'd say it's better to fix your nose and nasal pathways rather than live dysfunctionally for the rest of your life. Crooked noses or jawbones may put you at a higher risk of sleep apnea. If not right now, then later in life.

To breathe better at night, you want to teach yourself to breathe better during the day. It's something most people don't think about at all yet it's one of the most powerful things that can control the state of your nervous system.

Humans can breathe through their nose as well as mouth. You may breathe through the mouth during exercise to supply your body with more oxygen but during rest you should stick to nasal breathing as much as possible. It maintains parasympathetic dominance, reduces stress, and promotes digestion.

You're breathing automatically basically 95% of the time without thinking about it. If you do it through the mouth, then your body has just adapted to compensating for some issues.

Mouth breathing even at rest is considered abnormal because it hints towards subconscious adaptations to improperly developed nasal airways, chronic stress, and just breathing wrong[1189]. It affects facial development in children and is associated with Down's syndrome, sleep apnea, and snoring.

Oxygen is probably the most essential nutrient our bodies need. You'd die in minutes without enough oxygen and that's why breathing is so crucial to health. Nasal breathing improves the delivery of oxygen throughout the body by producing nitric oxide in the nasal sinuses. This promotes blood flow and respiration. Mouth breathing, however, doesn't create this effect and is thus an inferior breathing method.

Here's how to treat mouth breathing and start breathing properly:

- **Go See a Doctor** – Maybe your nose or jawline is poorly developer and clogged. For some people, surgery is the only thing that fixes their snoring and breathing.
- **See a Therapist** – You may also want to consult a myofunctional therapist who could help you to retrain your facial muscles and breathing patterns.
- **Practice Nasal Breathing** – Consciously try to breathe through your nose as much as possible. Keep your mouth shut when you're at rest.
- **Proper Tongue Position** – Your tongue position is a good indicator of how well you breathe. It promotes better delivery of oxygen, keeps the teeth straight, and opens up

nasal airways. A low downward tongue position is a sign of mouth breathing as it opens up breathing space in the mouth. You should keep your tongue high up against the palate to seal the oral cavity. If you need to retrain yourself, then it might be necessary to consciously push it up the roof of your mouth and stick it there for practice.

- **Stress Reduction** – When you're stressed out whether because of exercise, work, sleep deprivation, or overeating, then you're keeping yourself in a sympathetic state. That promotes mouth breathing and shallow breaths. All of this exacerbates the issue and leads to habitual mouth breathing. To avoid this cycle, incorporate active stress management into your life.
- **Deep Belly Breathing** – Shallow breahts into your chest are also a reason why people tend to be sympathetic dominant and breathe through their mouths. To stay parasympathetic, you want to breathe through your belly. Deep breaths into the diaphragm also massage your intestines, improve digestion, and relax the body. Holding yourself upward and tight is just stressful. I recommend practicing deep belly breathing through the diaphragm every day for at least 5 minutes.
- **Nasal Breathing and Mouth Taping** - If you think you're suffering from sleep apnea, snoring, or breathing problems during sleep, then you should try mouth taping. While this may sound bizarre, it's quite effective and not at all painful or risky. This will encourage breathing through your nose

throughout the night, which has several health benefits aside from regulating sleep-disordered breathing that can progress to sleep apnea. Simply place a small piece of medical tape (please do not use industrial types of tape which can damage your skin) across your lips.

- **Fix Nutrient Deficiencies** – To prevent poor facial development and jawline abnormalities, you want to get enough fat-soluble vitamins into your diet. Vitamin A, D, E, K2, and omega-3s are crucial for dental health, especially in growing children. Eat organ meats, liver, eggs, fish, meat, and vegetables.

But are you going to do 5 minutes of box breathing during military firefights or, in our case, HIIT? No, the purpose of it is to simply practice the skill of taking control of your breathing. During the actual event of when you would be lifting heavy weights, you simply use the same technique to calm down your nervous system. In between sets you will be gasping for air and can use mindful breathing to bring yourself down from being too stressed out. Training stimulates the fight or flight hormone which we then will be able to dissipate immediately for increased performance and faster recovery.

Periodization and Adaptation

Immediately after a workout we get slightly weaker. However, as our body heals itself and the nervous system recovers, we'll go through super-compensation and come back stronger. The training stimulus must exceed a certain threshold that causes good adaptations (to not undertrain), and it mustn't be too much that causes excessive damage from which we can't recover from. More is not always better and we should be mindful of under which conditions our body is at.

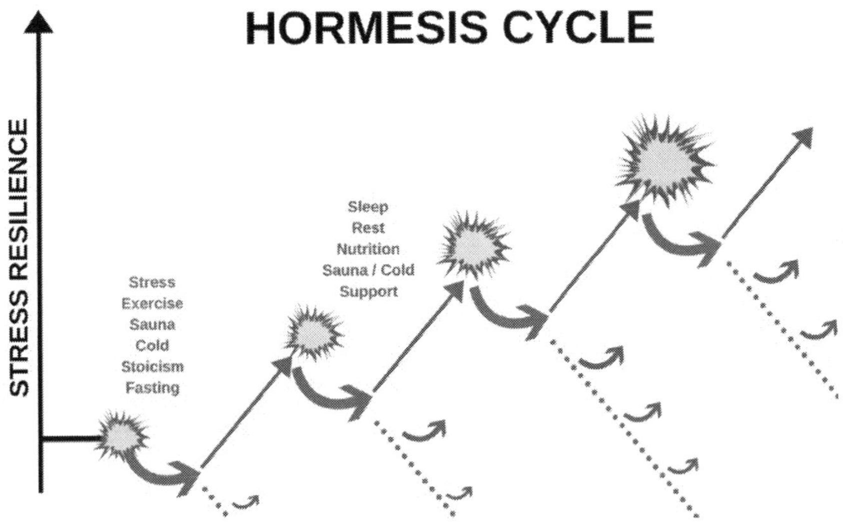

Generally speaking, optimal recovery from workouts takes about 48-72 hours. If you didn't push yourself through the dirt like a maniac, then your muscles should be repaired by that time. However, the central nervous system may need up to 6-7 days of rest. That's why it's important to not overdo the intensity and volume. You'll feel it when you've fried your CNS. Your motor control and balance will decrease and you'll be more tired.

Here are signs of over-training and/or under-recovery:

- Troubles maintaining balance and motor control
- Chronic muscle soreness or nagging injuries

- Brain fog, forgetfulness, getting distracted, and dullness
- Problems falling asleep and getting up in the morning
- Higher basal heart rate than normal
- Losing muscle strength and power
- Lack of motivation to train and move around
- Decreased performance and plateaus in progress
- Increased perceived effort above what's normal
- Mood swings, agitation, and mild depression
- Low thyroid and high cortisol

Periodization is about strategically designing our workouts by systematically manipulating variations in training specificity, intensity, and volume. The goal is to maximize your gains while reducing the risk of injury, staleness, and overtraining. It will also address peak performance for competitions or events. An intelligently structured design will include several different chunks or periods across the entire year that have their own priorities.

In terms of recovery and adaptation, then the most critical component of it is sleep. That's the time your body is repairing itself and facilitating growth. Sleep deprivation releases cortisol, which is the catabolic stress hormone. This will promote tissue breakdown and the accumulation of fat. Given that stress alters the immune system, it's also going to affect the speed of healing. Biopsy wounds heal 25% slower in chronically stressed out people or those caring for a person with Alzheimer's disease[1190].

Exercise and Hormetic Recovery

Historically, antioxidant supplements, non-steroidal anti-inflammatory drugs, and cryotherapy post-workout have been shown to decrease exercise-induced oxidative stress. This may slow down or completely nullify the adaptive response, especially in terms of muscle hypertrophy. ROS and inflammation created during exercise are key signaling molecules for adaptation and immediately blocking them may mitigate the responsiveness to exercise[1191]. However, it's not so with all types of exercise.

- Pomegranate extract, blackcurrant powder, and beetroot juice enhance blood flow and time to exhaustion thanks to their nitrate and flavonoid content[1192][1193][1194]. Taking them before working out can improve performance and give a better cell swelling effect.
- Antioxidant supplementation before exercise has been shown to interfere with mitochondrial biogenesis, which is a key adaptation to endurance performance capacity[1195]. Multiple studies have also found antioxidant supplements to impair anabolic signaling and muscle hypertrophy[1196]. However, it doesn't seem to affect muscle strength[1197].
- Cocoa polyphenol consumption can reduce exercise-induced oxidative stress but not inflammation[1198]. Tart-cherry juice lowers inflammation, muscle soreness, and pain levels[1199].
- Vitamin C doesn't appear to benefit exercise performance and it may impair exercise performance at doses above 1000 mg/day[1200]. Getting about 200 mg-s from 5 or more

servings of vegetables a day appears to be enough to manage oxidative stress without impairing training adaptations.

- Vitamin E supplementation for several weeks doesn't show performance benefits and tends to actually lean towards some impairment[1201]. The only exception is training at altitude because vitamin E prevents the breakdown of red blood cells that happens at altitude exposure.
- Carotenoids like astaxanthin may attenuate exercise-induced muscle damage and free radicals[1202]. Polyphenols can have the same effect[1203], which is why it's more optimal to take them before working out and preferably in a semi-fasted state.
- N-acetylcysteine (NAC) enhances skeletal muscle function and endurance[1204]. One study found that 1800 mg/day of NAC for four days increased sub-maximal knee extensor endurance[1205]. NAC at doses of 150 mg/kg enhanced submaximal handgrip strength endurance[1206]. Taking 1200 mg/day resulted in improved repeated sprint performance[1207]. It appears to not affect exercise adaptation but in high amounts can cause vomiting, nausea, and diarrhea[1208].

Although there is some evidence showing that antioxidants may blunt exercise adaptations, it's not fully conclusive and applicable to all forms of training. There's also a big difference between the modest amount of antioxidants you get from real food and the large quantities in supplements.

Generally, eating whole foods and vegetables probably won't have a negative effect on hormesis or exercise adaptations. They may promote recovery by modulating inflammation but not shutting it down completely. However, taking large doses of antioxidant supplements around your training could shut down the beneficial response, especially hypertrophy. Endurance and strength are less affected than muscle growth.

The critical part is the timing of when you go through these interventions both hormetic and anti-inflammatory. Prepare with recovery, experience stress, allow time for adaptation, start recovery but don't over-do it.

Exercise Hormesis Protocol

- Muscle quality is determined by its strength and power, which can be trained with resistance training. Grip strength and leg strength both are associated with better longevity because of their functional benefits. You should train with compound lifts, kettlebells, calisthenics, and anything that can be applied in the real world. This will preserve antifragility and reduces the down-side of unpredictable fitness situations.
- Cardiovascular fitness is great for improving heart health, increasing aerobic capacity, and general vitality. Being able to walk fast and long is a good sign. You can train cardio with both low as well as high-intensity exercise albeit HIIT is more time-efficient and effective.

- Avoid training in the Black Hole. This will cause too much oxidative stress and free radical damage to the body, promoting sarcopenia, frailty, and over-training. Instead, when you do cardio, do it below your anaerobic threshold and save your glycogen for short but heavy HIIT sessions.
- Antioxidants around exercise may have a negative effect on training adaptations, especially in terms of muscle hypertrophy and bodybuilding. Getting antioxidants from real food and vegetables is probably safe but taking supplements is not advisable because they'll blunt hormesis.

Chapter Ten:
Of Electromagnetic Frequencies and Radiation

"An absence of natural radiation may be as harmful as an abnormally large exposure of radiation."

Steven Magee

In addition to all the environmental pollutants and toxins, our health is also under attack by the technology we all so like to use. Artificial light, electromagnetic frequencies, radiation, and invisible wavelengths affect our biology on the cellular level without us being aware of it. Most people are fine and don't experience any negative side-effects but others are seriously affected.

What's worse, we're not planning to stop the use of technology but will start implementing it into our lives even more. There's not much you as an individual can do about it because society yearns for more bandwidth and faster connections. It's only a matter of time before we're completely enmeshed and surrounded by so much tech that we completely forget about what's it like to be unplugged.

This chapter talks about the effects of artificial light, EMF, magnetism, and radiation on your health. More importantly, how to not be wiped out but instead leverage it for hormetic adaptation and antifragility.

Electromagnetic Frequencies (EMFs)

The human eye can detect only a narrow range of wavelengths in the electromagnetic spectrum beaming down from the sun. It's dubbed visible light and contains all the colors of the rainbow – purple, indigo, blue, green, yellow, orange, and red. There are many types of energy humans can't see such as gamma rays, X-rays, far infrared, microwaves, and radio waves. However, this doesn't deny the fact that these wavelengths can still affect our bodies, albeit they're invisible.

There are two types of EMFs. Low-level radiation or non-ionizing radiation emanates from cellphones, Wi-Fi, MRIs, 5G towers, etc. High-level radiation or ionizing radiation comes from the sun's UV lights and X-ray machines. Naturally, the only part of the electromagnetic spectrum humans would get exposed to was visible light from the sun. Cosmic rays are reached only in space and during air travel whereas radio and microwaves originate from the invention of electricity. Nowadays, we're all pretty much constantly surrounded by different EMFs and radio frequencies (RFs).

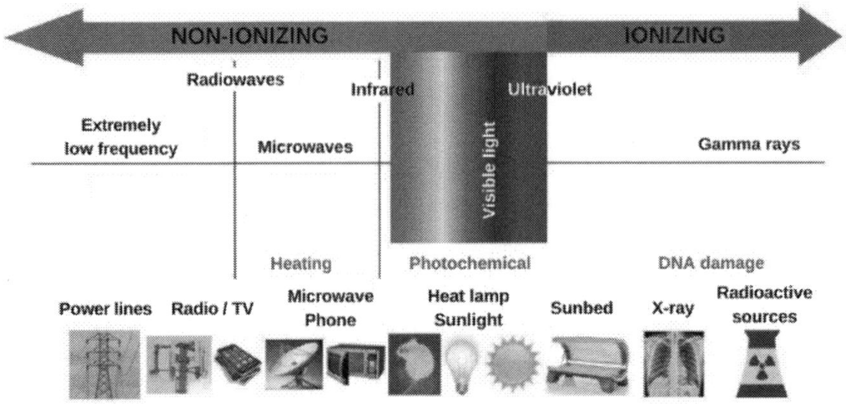

Electricity can be described as the movement of electrons through a conductor like a wire or cable. It can flow in two ways: direct current (DC) or alternating current (AC). In DC, electrons move in a single direction or only forward and in AC they switch between going forward and backward. You can essentially transfer energy much farther away with AC, which is why it's being used in electricity infrastructure. In nature, we would be exposed to primarily DC but after the invention of AC in the 19th century, we're gradually being exposed to it more. Interestingly, magnetic storms and geomagnetic pulsations also follow the AC current and they're associated with higher rates of heart attacks, psychiatric disease, anxiety, crime, and overall mortality[1209]. How would you feel about being constantly in the middle of a magnetic storm? What if you knew that you pretty much already are?

The Earth has an electromagnetic field surrounding it. They have various peaks in extremely low frequencies (ELFs) called the Schumann Resonance. Our bodies also emit a frequency that resonates with the earth's 7-13 Hz-s. Funny enough, that corresponds with your brain waves when in a relaxed state of alpha waves. Man-made EMFs and RFs disrupt this homeostasis and cause cellular dis-ease. There's a lot of controversy about the safety of EMFs and their impact on health. Intuitively, you'd think it's harmful and all bad but the jury is still out there.

According to the World Health Organization, radiofrequency EMFs are a possible carcinogen to humans[1210], based on an increased incidence of breast cancer, associated with wireless phone use. However, some studies don't see a higher risk except for only the highest levels of exposure[1211]. More long-term research is needed.

Exposure to EMF for years is linked with leukemia, especially in children[1212,1213]. EMF has also shown to cause neurological, sleep, and nerve functioning problems[1214]. In addition to that, higher exposure to EMF during pregnancy increases the chances of the child getting asthma by 3 times[1215]. Mobile phone use reduces sperm motility and viability[1216].

Since the 2000s, mobile networks have been using 4G connections, which increased internet speed by 500 times compared to 3G and allowed HD streaming. We're now entering the 2020s with 5G and it's going to change the game drastically. It's estimated that 5G will be 100 times faster than 4G with speeds up to 100 gigabits per second. For reference, I recall having 500-megabyte connection during primary school, which was around 2007 for me. One gigabyte equals 1024 megabytes. So, nowadays we can have over 200 times faster connections than even just a decade ago.

5G connection is based on the millimeter wave (MMW) bandwidth, which is between 30-300 GHz and has been shown to penetrate human skin[1217]. MMWs are even used by the US Department of Defense in crowd control weapons called Active Denial Systems (look up the Non-Lethal Weapons Program FAQ at https://jnlwp.defense.gov/About/Frequently-Asked-Questions/Active-Denial-System-FAQs/). They're not lethal but it comes to show the potential of these frequencies. MMWs are also linked to eye problems, cataracts, impaired heart rate variability, pain, suppressed immunity, antibiotic resistance, and reproductive problems in rats[1218,1219,1220,1221]. In 2015, more than 230 scientists who study the effects of non-ionizing EMF in 41 countries signed an international appeal to the United Nations regarding the safety and protection from EMFs due to their effects on health even at low levels[1222]. Two years later, in 2017, over 180 doctors and scientists from 35 countries signed a petition to delay the rollout of 5G because of the potential health risks on wildlife and humans[1223].

Living on Earth exposes us to certain magnetic frequencies, especially from the Sun's UV rays but it's not as intense or predominant as it is in modern society. Since the end of World War I at 1918, we're exposed to 1 quintillion times more EMFs than 100 years ago[1224]. In case you're wondering, 1 quintillion is 100 billion billions. That's a 10 with 18 zeros. It would be naive to think that this kind of exposure doesn't affect us in any way. Most people aren't aware of this nor do they experience the harm until their body is already somewhat broken. At that point, it'll be much harder to reverse the damage and an antifragile approach would be to prepare for it in advance by practicing hormesis.

A lot of research that finds EMF to be safe is funded by the telecommunications industry to create biased results or design flaws. I think that being chronically surrounded by a ton of EMFs is probably not a good idea and you don't want to keep your devices near your body all the time. However, humans can probably develop some level of resistance through hormesis, which we'll be talking about soon.

How EMF Affects Your Body

There are many ways EMF and RF affect our biology. The main mechanism is through disrupting the cell's redox status and homeostasis, causing more oxidative stress, imbalances, and disease.

Low-frequency radiation like smartphones and Wi-Fi routers activate the voltage-gated calcium channels (VGCCs), also known as voltage-dependent calcium channels (VDCCs), which are a group of ion channels located in the outer membrane of cells[1225,1226]. This produces excessive in-flux of intracellular calcium that can lead to pathophysiological and in some cases therapeutic effects[1227]. Nitric oxide and superoxide get increased and they react instantly to create a powerful oxidant called peroxynitrate[1228]. Its ionic formula is $ONOO^-$ and chemical one HNO_3^-.

Peroxynitrate promotes massive amounts of oxidative stress, mitochondrial dysfunction, inflammation, DNA damage, apoptosis, necrosis, lipid peroxidation, protein oxidation, inactivation of enzymes, and ROS production (See figure). It also modifies tyrosine into nitrotyrosines, which are associated with atherosclerosis, myocardial ischemia, and IBD[1229]. For a full overview of how peroxynitrate and NO can be harmful, check out the 140-page paper with 1500 references by Pacher et al (2007) titled 'Nitric Oxide and Peroxynitrite in Health and Disease'[1230]. It's implicated in every pathology, starting with hypertension, heart failure, and ending with diabetes and vascular aging[1231].

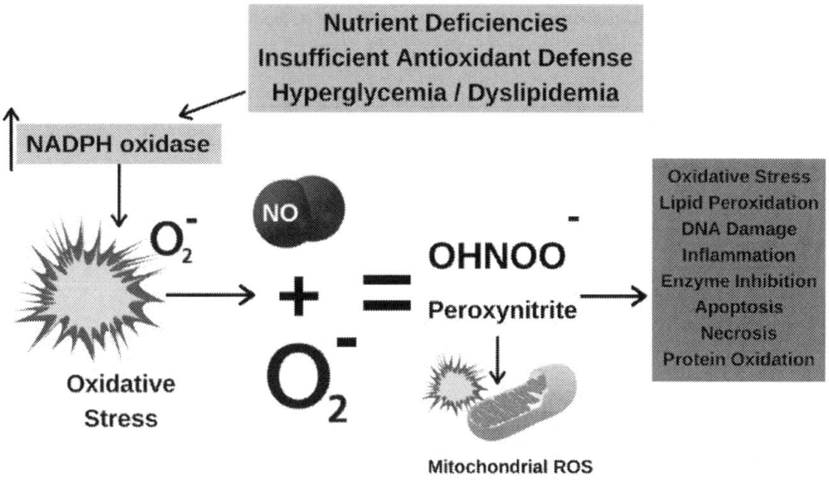

Ionizing radiation has been found to accelerate the development of atherosclerosis or plaque formation in mice[1232]. This can cause CVD, stroke, or other forms of heart failure. Whether or not non-ionizing radiation has the same effect is not clear. The International Commission on Non-Ionizing Radiation Protection (ICNIRP) has stated that the link between CVD and magnetic frequencies is weak and it remains speculative until we get more evidence[1233]. However, it doesn't deny the fact that being exposed to EMF both ionizing and non-ionizing, creates oxidative stress by increasing intracellular calcium. It's an extra stressor we have to deal with. The increased amount of free radicals also damage the mitochondria and spread inflammation. Oxidative stress and inflammation play a crucial role in the development of atherosclerosis and CVD[1234,1235].

The main way your body deals with oxidative stress and DNA damage is by activating sirtuins and a group of enzymes called PARPs (poly ADP ribose polymerases). Both of them consume

NAD+ and can lead to energy depletion. When that happens, you're left with chronic fatigue, getting sick easily, lackluster performance, brain fog, and decelerated healing. NAD deficiency causes tissue damage in response to ionizing radiation and administrating it protects against these effects[1236]. NADPH is also essential for lowering oxidative stress and inflammation[1237].

EMF and RF exposure are slowly depleting your NAD stores because it's being used for DNA repair. That can explain why people experience negative symptoms only in their later life. However, how many people are stuck in large compartmental buildings or offices where there's a ton of EMF, yet they don't know why they're tired. The reason has to do with how their NAD is secretly being depleted by this low-grade exposure.

Does Autophagy Protect Against EMF

EMF and exposure to radiation cause subliminal stress to all cells of your body. This, in turn, will also promote autophagy to protect against the damage. Autophagy is a defense mechanism against physiological stress such as starvation, exercise, heat, or concussions, including magnetism.

In one study, 6 mice were exposed to 835 MHz RF-EMF at a specific absorption rate of 4,0 W/kg for 5 hours a day over 12 weeks. Frequencies around 830-840 MHz are used in radiotelephone services. The results showed that long-term exposure to 835 MHz induces hyperactivity, autophagy, and demyelination in the cortical neurons of those mice.[1238] Myelin is the insulating sheath around nerves that transmits electrical impulses. Basically, the mice's brains started to break down and the body activated autophagy to slow down the degeneration and repair the damage. Of course, 835 MHz to rodents is far more deadly than it is for humans but it does show that small amounts of EMF and radiowaves can act as hormetic stressors with beneficial effects if it leads to increased autophagy or DNA repair.

It's been found that autophagy protects against DNA damage in mice exposed to radioactive electromagnetic frequencies[1239]. Autophagy plays a vital role in protecting cells from oxidative stress, reducing apoptosis, and growing lesion stability. Impaired autophagy contributes to arterial aging[1240]. Inhibiting autophagy accelerates plaque necrosis[1241]. It's been shown that knocking out ATG5, the key gene for autophagy regulation in macrophages, can cause the aggravation of atherosclerosis lipid plaques by increasing apoptosis and oxidative stress[1242]. So, you can find at least some solace in the fact that your body does a pretty good job of protecting itself against these things thanks to autophagy. This further supports the idea that being exposed to these kinds of environmental stressors in a state of heightened autophagy is protective and reduces the negative side-effects.

Poor sleep, jet lag, shift work, and even EMF exposure resemble mild forms of traumatic brain injury or TBI. It's been shown that the brain activates autophagy in response to TBI as to deal with the stress and start healing the damage[1243]. Ketone bodies and ketogenic diets are also now being used to treat brain injuries and concussions[1244]. Neurological stress like sleep deprivation or EMF is a form of TBI or concussion that leads to mild autophagy in the brain. You need to clear out the toxic proteins and aggregates that accumulate there via this thing called aggrephagy, which is a sub-category of autophagy[1245]. Autophagy gets activated in response to concussions because that's the least your body can do to cope with it.

Banging your head against a wall or shooting EMF up there isn't worth it even if it does activate autophagy. Don't do it. Just know that whenever you do hit your head, you should promote the healing process by stimulating both ketosis and autophagy by consuming some exogenous ketones or MCT oil. I would also say that getting hit by something or being exposed to EMF while fasting would also result in less injury because the body is already in ketosis and autophagy, thus its self-defense system is more robust and eager to deal with stress. You'd also have to be keto-adapted and metabolically flexible.

Hypothetically, EMF exposure is a hormetic response that should increase your body's resiliency against it in the future. Hopefully, it's true because when you look the world around you where there's a cell-tower almost everywhere in sight. If you were to drop a hunter-gatherer from the wild in the middle of Times Square, their head would explode because they're not used to all the artificial light and EMF, whereas people living there barely notice anything. Your tolerance to these things is already much higher than that of your grandparents or ancestors and it'll probably keep increasing in the future as long as you grow into it gradually. It's going to be a bigger problem if you're immediately thrown into an ocean of radiation and EMF that you haven't experienced before.

The critical part of adapting to EMF is being able to recover from it and allowing the body to heal. That's how the hormesis cycle also works – exhaustion followed by recuperation. You can potentially become hormetically stronger and more resilient against EMF but only if you get periods of no exposure to it.

EMF and Sleep

Your body heals inflammation primarily during sleep. Sleeping in a high EMF environment keeps the body more wakeful and under low-grade stress. This prevents you from going into deeper stages of rest and recovery, which requires a relaxed state. Cellular recycling processes like autophagy also need you to be in a lower redox state of less electrical activity.

Daily EMF exposure is associated with disturbances in NREM sleep and overall poor sleep quality[1246]. Your pineal gland and thus melatonin production can also be affected by EMFs in a negative way[1247]. This effect is even more pronounced in EMF exposure during the night. EMF from cell phones and Wi-Fi routers can increase EEG brain activity, which increases high-frequency beta and gamma waves and less slow delta waves that are associated with deep sleep[1248].

Here's what to protect yourself against EMF and become more resilient:

- **Scan Your Bedroom** – To know how much radiation you're getting, use an EMF and EMC detector. You'd be surprised that even in a room with no electronics there's still this low-grade pulsation of radiowaves that inevitably has an effect on your physiology. The closer you are to the emitting device the bigger effect it has. That's why keeping your phone on airplane mode when you're not using it is paramount. Dimmer switches are a source of dirty electricity so consider using regular on/off ones.
- **Turn Off Your WiFi** – It's a good idea to minimize the amount of EMFs in your surroundings. You can't avoid all the radio waves coming from your neighbors but the least you can and should do is turn off your WiFi. If you have a router right next to the bed, then it's going to be quite detrimental to your sleep quality. With the exponential increases in bandwidth and speed, it's basically a mini cell tower. Ideally, you'd want to swap out all WiFi in your house for an Ethernet cable connection. It's a lot faster and safer than wireless. The same applies to all the Bluetooth smart TVs and refrigerators. Do you really need those sorts of things? It's much smarter to make your household as EMF-proof as possible so that you'd have a better time dealing with it elsewhere.
- **Use Battery-Powered Alarm Clocks** – If you happen to be still using alarm clocks, then stick to the old-school

battery-powered ones. It would be quite unfortunate to have a small electric EMF generator buzzing right next to your head for the entire night. They should also have no artificial light because that's going to further disrupt melatonin production. If you have children, don't keep wireless baby monitors or other devices near their cradle while they're sleeping. It's crazy if you think about it. Instead, use a hard-wired monitor or be a good parent and bring the child into your room.

- **Keep Your Phone on Airplane Mode** – If you're carrying your phone in your pocket or on your body, it should always be on airplane mode. Otherwise, you're blasting your organs and cells with EMF. When having a call, use the speaker function or the microphone from headphones. Putting it next to your head is one of the stupidest ideas I can think of. I take it back…putting your smartphone under your pillow while you sleep with 5G turned on is much worse. It should be off or on airplane mode during the night, especially if you keep it next to your bed.
- **Get an EMF Shield** – There are some devices out there like the BlueShield that protect against EMF exposure and minimizes it. You can place a unit in your room and it will cover the majority of your household. It's pretty amazing and well worth the investment. You should also use laptop pads, phone pockets, and air tube headphones like ShieldYourBody. Using a Faraday bag will also shield some of the radiation from smartphones. However, don't fall for feeling some sort of safety or complete protection

from these devices. They're not going to work better than the fundamentals already discussed, such as airplane mode and avoiding wireless.

- **Get Grounded** - Walking outside on the grass with your bare feet can lower the excitation from EMF. When you're grounded, electrons move from the Earth to the body and vice versa. This will maintain the body's negative charge electrical potential similar to the Schuman Resonance. You will then experience less oxidative stress and inflammation because of putting yourself into a state of less excitation. Unfortunately, grounding can be dangerous in most urban areas, especially in North America, because of underground wiring and dirty electricity. That's hyper-exposure. The safest and most effective place for grounding is in beach waters where you get full-spectrum sunlight, you're grounded, and are exposed to the negative ions from the sea.
- **Use a Grounding Mat** – To get the same grounding effect indoors, you'd have to drive a metal rod into the ground outside and have a wire run from the rod inside. That's pretty complicated not to mention having to be in contact with the wire itself all the time. Fortunately, there are technological alternatives like grounding mats, earthing sheets, bands, and patches. Check out Groundology. I'm sleeping on a grounding mat all night.
- **Make a Faraday Cage** – If you really want to put on your tinfoil hat and get your hands dirty, then you can also

create your own Faraday Cage for your bed DYI style. A Faraday Cage by definition is an enclosure created with conductive materials that block both static and non-static external electrical fields. It was invented and named after Michael Faraday who built a room with metal foil in 1836 and used an electrostatic generator to produce high-voltage discharges on the outside of the room. He then proved with an electroscope that the inside walls had no electric charge because they were blocked by the cage.

Examples of everyday Faraday Cages include the microwave oven – the metal exterior traps the radiation within itself and doesn't spread outward – as well as airplanes and cars that if struck by lightning won't directly affect the people inside. The metal walls of vehicles and cages absorb the electric charge but the current won't pass through to the inside. That can be useful for sleeping in a reduced EMF setting. I'm not going to teach you how to make a Faraday cage because I think it's not necessary. You can get most of the benefits of improved sleep by simply using simpler EMF shields and grounding mats. However, you might need extra-strong protection if you live in a highly radiated environment or are very sensitive to EMF.

Radiation Hormesis

Let's also talk about radiation hormesis. Although it's something most people aren't exposed to on a daily basis, we should still understand some of the general principles and mechanisms. It was actually this phenomenon that ignited the research of hormetic stressors in the last century and led to the development of several models of stress adaptation.

Radiation hormesis involves low-dose ionizing radiation and it's hypothetically beneficial in small amounts by stimulating the body's repair mechanisms[1249]. It essentially causes DNA damage and oxidative stress, which turns on glutathione, Nrf2, and PARPs. This concept has replaced the linear no-threshold model used for decades with many laboratory studies proving it. For example, cells first exposed to a low dose of X rays (1 cGy) developed an adaptive response to the subsequent higher dose (1 Gy)[1250]. They suffered about 50% fewer chromosome breaks despite the 100 times larger second dose.

Regulatory agencies that assess the safety of different compounds have a zero-tolerance stance on carcinogens. Meaning, that the risk is directly proportional to the dose and there is no safe level of exposure. This is called the linear no-threshold model (LNT), which has become the standard for assessing the safety of carcinogenic and chemical agents. In the case of non-carcinogens, they assume that there is a threshold below which there is no harm.

The linear no-threshold (LNT) model was explained first by Hermann Mueller who showed that radiation can cause genetic mutations[1251]. He also said it leads to cancer[1252]. In his 1946 Nobel

Prize Lecture, Mueller said that mutations are "*directly and simply proportional to the dose of irradiation applied*" and that there is no threshold dose[1253]. This framework had been well-established in radiation safety regulations for the past few decades. However, there have been many disagreements about its efficacy and appropriateness.

Many controlled in vitro and animal studies contradict the LNT model. They show adaptive responses to low dose radiation, resulting in reduced mutations and cancers[1254]. It demonstrates the phenomenon of radiation hormesis[1255]. In 1995, Cohen published a study on lung cancer mortality where it was shown that human response to low-level radiation is not linear[1256].

Edward Calabrese published an article in 2011 where he highlighted how Herman J. Mueller had access to contradictory evidence but didn't disclose it[1257]. "*Apparently it was a matter of political expediency in an effort to ban above-ground atomic testing*[1258]". This stems from the impending fear of nuclear warfare that was created during World War II that still lurks around today.

After the 1986 Chernobyl accident in Ukraine, there was a Europe-wide concern of children being born with increased mutations due to the radiation[1259]. In Denmark alone, hundreds of abortions were performed on healthy unborn babies in fear of the no-threshold effect[1260]. By 1999, millions of births had been recorded in the EUROCAT database and divided into 'exposed' and the control group. The researchers concluded that "*in retrospect, the widespread fear in the population about the possible effects of exposure on the unborn was not justified*"[1261]. High dose radiation therapy is known to cause pregnancy anomalies and defects, but studies show this to have a threshold dose below which no increased cases of mutations are observed[1262].

Reports on the mortality of atomic bomb survivors have shown that they have a significantly increased risk of cancer throughout their life in proportion to the radiation dose[1263]. That's to be expected because the amount of hazard from a nuclear explosion is probably thousands if not millions of times higher than you'd experience anywhere else. Younger people may also be more sensitive to the initiation of carcinogenesis at the time of exposure. However, Russo et al (2012) show that:

> *In interventional cardiologists, chronic exposure to low-dose radiation is associated with an altered redox balance mirrored by an increase in hydrogen peroxide and with two possibly adaptive cellular responses: (i) an enhanced antioxidant defence (increase in GSH, counteracting increased oxyradical stress) and (ii) an increased susceptibility to apoptotic induction which might efficiently remove genetically damaged cells.* [1264]

One of the most known counter-examples to LNT is Ramsar, a city in Iran with about 33 000 inhabitants. It's considered to have the highest natural background radiation on Earth, several times higher than the International Commission on Radiological Protection (ICRP) recommends for radiation workers[1265]. Despite that, the population doesn't seem to suffer from any disease. Unfortunately, the amount of people in high-exposure areas is quite small with roughly 1800 inhabitants and they only receive on average 6 millisieverts per year[1266], so there's not enough data to draw conclusions about cancer epidemiology[1267]. However, there are other non-carcinogenic differences between residents living in high background radiation and normal exposure such as chromosomal abnormalities and infertility[1268,1269]. So, there is some evidence of a threshold response.

As of now, the United States National Research Council endorses the LNT model and claims that radiation hormesis is unwarranted[1270]. Since 2001, The National Council on Radiation Protection and Measurements supports the idea as well[1271]. However, several organizations oppose this such as the French Academy of Sciences and the National Academy of Medicine[1272]. They published a report in 2005 that favors the threshold dose response[1273]. The Health Physics Society's position, first adopted in 1996 and revised in 2019, states that *„underlying dose-response relationships at molecular levels appear mainly nonlinear"*[1274]. The American Nuclear Society recommends additional research on the linear no-threshold hypothesis before changing any current radiation protection guidelines due to the evidence that high dose exposure has been shown to have health risks[1275].

Tubiana (2009) writes that the linear no-threshold relationship is inconsistent with radiation biologic and experimental data. He says:

> *Life is characterized by the ability to build defenses against toxic agents, whether internal or environmental. The*

defenses are overwhelmed at high doses and are stimulated at low doses, which is incompatible with the LNT model. [1276]

The LNT model doesn't take into account the degree of extrapolation between human and animal data, which may set very low allowable doses for carcinogens. A typical risk assessment of a single additional case of cancer per million people over a 70-year lifetime can equal an assumption of more than 5–6 times from the high exposure animal studies to the low doses assumed to be safe for humans[1277]. Furthermore, the LNT model doesn't take into account how our bodies have developed mechanisms for detoxifying chemicals and adapting to them.

In toxicology and pharmacology, the threshold model has become the key model whereas the LNT model remains used in estimating cancer risks by virtually all regulatory agencies. Who knows, maybe it's for the better, as an ounce of prevention is worth a pound of cure. Governments and institutions have to be extremely conservative as to avoid potential errors and consequences. Even if radiation hormesis is real they can't and won't recommend it because most people don't know how to calculate their risks and safe dosage.

There's quite a lot of pre-clinical and clinical data showing that low dose radiation can be used for treating cancer patients either alone or as an adjunct to standard therapies[1278,1279]. In early-stage cancers, low dose radiation boosts the immune system, thus increasing the person's resiliency and decreasing the need for additional treatments such as chemotherapy or radiation therapy which have many negative side-effects[1280]. It's also observed that low dose radiation in between standard radiation therapy may improve primary tumor control and reduce metastasis in patients of non-Hodgkin's lymphoma[1281]. Furthermore, there's a reduced incidence of second cancers per kg in tissues exposed to low dose radiation at ~20 cGy compared to unexposed tissues[1282]. Albeit more clinical trials are needed to determine whether or not low dose radiation can be effective for contributing to cancer therapy, it's still exemplary of hormesis in action.

I'm not advising anyone to engage in radiation therapy of any doses because the modern environment is already quite radiated. Most people have microwaves, they're living in cities and they're surrounded by technology all the time. You don't need additional exposure. Instead, you ought to focus on keeping your body strong with other safer hormetic interventions and getting enough downtime.

Fasting has also been shown to improve the effectiveness of chemotherapy in rats as well as humans by protecting against the radiation and killing off more cancer cells[1283,1284]. It essentially puts the body into a more robust state wherein its stress adaptation and resilience are higher. Fasting furthermore protects against the radiation and UV rays while flying high in the air. It's going to reduce the negative effects of traveling across timezones and circadian rhythm mismatches. That's why you shouldn't really eat on an airplane because the food they serve there will definitely kick you out of ketosis and autophagy. The processed carbs and vegetable oils will also create additional inflammation and oxidative stress, making the hazard of flying that much greater.

In conclusion, you probably can't avoid all EMFs for the rest of your life and you probably shouldn't. If you want to keep living in the modern world which includes using wireless technology, occasional flight travel, and 5G, then you should focus more on increasing your body's ability to deal with these stressors rather than avoiding it. A stronger immune system inside a healthy body will probably adapt to mild EMF and radiation exposure. You should be grounded and in low-EMF environments, especially when you're sleeping or taking a sauna. Nevertheless, adding intermittent fasting and promoting some ketosis are still the top ways to making your body hormetically more robust against EMF.

Red Light Therapy

One beneficial hormetic activity that does expose your body to mild radiation is red light therapy. The body responds very positively to specific wavelengths between the 600-900 nm range. When red light is shined on the skin, it can penetrate several millimeters into the skin where it has a beneficial biochemical effect. The mitochondria absorb the red light, which triggers a release in nitric oxide, an increase in ATP production, and a decrease in oxidative stress[1285].

Red light wavelengths specifically have many rejuvenating and longevity-boosting benefits. Here are the benefits of red light therapy or photobiomodulation as it's called.

- Relieves muscle pain, aches, and can reduce arthritis[1286]. It can also reduce pain symptoms in cancer patients[1287].

- Can promote recovery of skeletal muscle injuries[1288,1289] and speeds up wound healing[1290,1291].
- Enhances physical performance and muscle recovery time[1292, 1293]. It's been even questioned as to whether or not red light therapy should be allowed in athletic events because of its effectiveness[1294].
- Increases mitochondrial functioning and energy levels by releasing ATP from cells and stimulating DNA/RNA synthesis[1295,1296]
- Reduces inflammation and oxidative stress[1297,1298].
- Better blood circulation, vasodilation, and lymph flow[1299]
- Normalizes thyroid hormones and can help with low thyroid[1300]
- Improves skin condition and can be used to treat acne[1301,1302]. Protects your skin against wrinkles, stretch marks and sun-induced aging[1303]
- Fights various skin and teeth infections by killing the bad bacteria[1304,1305]
- Stimulates the production of collagen and elastin that promote skin complexion, joint strength, and hair growth[1306]
- Supports hair growth and alleviates hair loss[1307,1308]
- Prevents age-related macular degeneration in the eyes[1309]
- Boosts cognitive function as well by increasing mitochondria in the head, improving memory, attention, and overall brain health[1310]

There are no reported side effects to the therapeutic use of photobiomodulation. The most common side effect of red light therapy is tiredness and redness of the skin if you use too much[1311].

Amber lights at 590 nm-s induce the breakdown of lipid droplets through autophagy-related lysosomal degradation[1312]. Lipophagy is the degradation of lipid particles, cholesterol, and triglycerides by autophagy. Red light laser therapy at 635 nm has been used to trigger lipolysis or oxidation of fat in subcutaneous adipose tissue[1313]. So, this increased fat burning happens in conjunction with lipophagy and the effects of red light wavelengths.

Red light between 550-670 nm has also been shown to accelerate skin recovery whereas blue light slows it down[1314]. Autophagy plays a crucial role in skin health by protecting it against environmental stressors, UV radiation, infections, and the aging of fibroblasts[1315,1316]. Fibroblasts create collagen in the skin and as they get older they show less autophagy. Not enough autophagy speeds up skin aging and formation of wrinkles because of poor elasticity[1317].

Moderate amber and red light in the evening will also promote melatonin production, thus resulting in more autophagy during sleep[1318]. Blue light at night has the opposite effect. Observational studies have shown a correlation between exposure to light at night with obesity and type-2 diabetes[1319,1320]. This is primarily due to the suppression of melatonin and autophagy that will result in sub-optimal sleep quality.

When it comes to red light dosage, then you have to know your devices light's power density (in mW/cm^2) by measuring it at different distances with a solar power meter.

Here's a chart of the main power densities for most devices:

	Infrared Light Device	Red Light Device	Red / IR Combo
Distance (cm)	Power Density (mw/cm^2)	Power Density (mw/cm^2)	Power Density (mw/cm^2)
0	1200	1100	1000
5	500	450	600
10	200	180	450
15	100	90	200
20	70	65	100
25	45	40	85
30	35	30	70
35	20	18	55
40	18	15	40

The therapeutic range tends to be between 10 - 200mW/cm^2. The optimal dose can be calculated with the following formula: **Power Density (in mW/cm^2) x Time (in seconds) = Dose (in J/cm^2).** With higher intensities, you need less time to reach the minimal effective dose. Most studies find benefits in the range of 0.1 J/cm^2 to 6 J/cm^2 but some go to even as high as 70 J/cm^2. The formula for calculating the time is this: **Time = Dose ÷ (Power density x 0.001).**

In general, you can get a therapeutic benefit by exposing yourself to red light for 10-15 minutes while standing about 20-30 centimeters away from it once a day. I sit in front of my red light panel in the afternoon for about 10-15 minutes. Sometimes I might even exercise or meditate.

You can get red light from sunsets, infrared saunas, in medical clinics, functional fitness facilities like Upgrade Labs, or by using your own red light therapy device at home such as the Joovv, Red Light Rising or Red Light Man.

PEMF Therapy

We've already talked about grounding, earthing, and EMF protection. However, some magnetic waves such as PEMFs can also be beneficial for healing our bodies.

Pulsed Electromagnetic Frequencies (PEMF) are low to medium intensity magnetic frequencies that are pulsating at a particular frequency. These magnetic frequencies can have a profound impact on the health and energies of our bodies. Your mitochondria and all other cells in the body are being influenced by electromagnetic frequencies and PEMF can optimize your cellular functioning.

A study done in 1989 on rats found that long-term PEMF therapy of 10 Hz at 10 hours a day increased the cellular respiration of two important enzymes in the Krebs cycle by 3 times[1321]. This directly improves oxygen utilization and how well the mitochondria can produce energy.

PEMF therapy is most used by NASA astronauts because they have to spend a lot of time out in space away from the Earth's natural magnetic field. One NASA study found that PEMF waves can promote the proliferation and growth of neuronal cells and neural tissue regeneration by 2.5-4 times more than non-wave cells.

Here are some scientifically proven benefits of PEMF therapy:

- Improved symptoms of arthritis[1322], chronic pain[1323], fibromyalgia[1324], osteoporosis[1325], PTSD, Parkinson's[1326] and even cancer.
- One study found that PEMF supported bone formation in 85% of the participants who'd endured failed posterior lumbar interbody fusion [1327].
- It's going to improve the quality of your sleep and help with insomnia.
- Better sleep and mitochondria will help you to lose fat, reverse insulin resistance, and diabetes.
- Increased athletic performance and faster recovery from exercise. A lot of professional athletes are using PEMF devices after their workouts to speed up the body's repair processes.
- PEMF can fight depression by alleviating pain in the body and replacing medications that often have negative side effects on neurological disorders.
- PEMF waves can heal damaged tissues and cells to relieve pain and stimulate repair processes.

- Because your mitochondria get a boost, you'll have more energy, vitality and you'll feel a lot better.

PEMF is safe for humans and it doesn't have any negative side effects. For thousands of years, people have been using different stones, metals, and other conductors to manipulate these electromagnetic frequencies to the best of their ability. With modern science, we have access to different PEMF devices that can be used to optimize our cellular functioning and overall health.

Here's a list of some PEMF devices you can use:

- PEMF mattresses can be used to sleep upon. They'll create a magnetic field around you while sleeping, which improves the quality of your slumber
- PEMF devices such as Earthpulse can be anchored in your bed to create a similar field around your bed
- Some PEMF devices help to induce deep sleep by emanating a magnetic field on your chest
- PEMF wristbands and other wearable gadgets can be applied on your body during the day at various times to ground yourself again
- PEMF micro pulse devices have long coils that you can lay on a particular region of the body. Flux Health has convenient micro pulse machines.

Most PEMF systems are quite expensive, ranging from 500-10 000 dollars. The most cost-effective devices I know come from

Flux Health. They're being developed by one of the leading researchers of PEMF Bob Dennis, PhD who worked with NASA.

There's also this smaller device that you can attach onto your chest and create a similar PEMF field grounding effect on your body. It's called the Somniresonance SR1 Delta Sleeper and it's perfect for traveling. You just strap it on your chest and leave it there for the night. The device will keep working for about 20 minutes and then shuts off. During that time, you may find it easier to gradually transition into more delta waves and fall asleep faster.

Jet Lag Prevention Plan

To be honest, I don't like traveling that much. Don't get me wrong, I love to see historical places, meet new people, and experience different cultures, but I hate the aspect of having to actually go there. Long-distance travels whether, by car, bus, airplane, or ship are uncomfortable and boring. It also disrupts your circadian rhythms, which may lead to a decrease in performance and well-being for many days after the fact.

How to Prepare for Jet Lag and Air Travel

- **Before your trip, move your bedtime more towards the time zone you're flying into**. I tend to sleep deprive myself just a little bit prior to the flight so I could fall asleep during the trip and then wake up in the morning.
- **Eat a meal with a lot of DHA and omega-3 fatty acids before traveling.** This will give your cells enough healthy

fats needed to protect the membrane of the cells and reduce inflammation. Great foods for this are salmon, mackerel, sardines, eggs, and algae.

- **Load up on sulfur foods.** Sulfur-rich foods like eggs, cruciferous vegetables, and meat have more sulfur, which helps with the production of glutathione. Glutathione is the most powerful antioxidant system your body has.
- **Don't take antioxidant supplements.** I wouldn't take additional antioxidant supplements like vitamin C or E because they might make the body weaker prior to going into an airplane. Don't use them during the flight either. Instead, have them after landing when you're grounded and in a new environment. That's the time you want to focus on recovery and allow your body to turn its guard down. Lowering your guard while in a soup of EMF can leave you vulnerable to more damage by blocking autophagy and hormesis. That low-grade stress up in the air will actually protect you.
- **Drink molecular hydrogen.** Hydrogen-rich water is infused with extra hydrogen molecules. Molecular hydrogen protects against free radical damage and oxidative stress[1328]. Loading up on molecular hydrogen water like Trusii can help to increase your body's capacity to deal with jet lag and the stress flying causes.
- **Consume some baking soda with water before getting on the plane.** Baking soda promotes alkalinity and balances the pH levels of the blood. It also decreases the

build-up of lactic acid in the muscles and prevents muscle tightness. Staying in ketosis will also lower the accumulation of lactic acid and supports your DHA levels.

- **Take activated charcoal before and after the flight.** This helps to eliminate the germs and toxins you got exposed to. It also improves kidney function and filtration of various substances[1329].
- **Not eating and doing intermittent fasting while flying is another great hack.** Not only is the food served on airplanes unhealthy and not fresh but staying in a fasted state will also reset your circadian rhythm by suspending your metabolism. Fasting will also increase autophagy and reduce inflammation, which tends to suffer while up in the air.

Jet Lag Treatment During the Flight

- **Set the time of your computer and smartphone to the time zone you're heading into so you could subconsciously start living in that time zone.** Structure your sleeping times and napping around the time of your destination.
- **Don't sit next to a window with an open curtain**. Because you're so high up in the air, you'll get exposed to excessive UVA light and cosmic radiation. Pull down the curtain when you can and try to sit in the aisle, so you could move around more easily as well.

- **Doing mobility exercises is also a must**. You should walk around at least once every 30 minutes to prevent muscle cramps and immobility. Every hour or two you can have a slightly longer break where you do some Yoga poses like the downward dog, warrior pose, deep squat, and Sun salutation. I like to also sit on a tennis ball and put it behind my back as to do roll out the fascia and make me squirm around in the chair.
- **Wearing compression clothes and socks will promote blood flow and stimulates lymph flow**. This prevents clotting and cramps as well.
- **Drink lots of fluoride-free water, preferably with sodium and minerals** so you would stay hydrated, keep yourself moving between the bathroom, and protecting your cells from drying out from the air pressure. As I said, don't take vitamin C or any antioxidants.
- **Cover yourself from EMF**. Wear long-sleeve clothing to protect your skin from blue light, wear blue-blocking glasses during the flight to preserve your DHA loss in your eye, and alleviate the circadian mismatch.
- **Practice polyphasic sleeping**. It means you'll be having several short naps rather than one big block of sleep throughout the night. This helps you to adjust to the time zone better and makes the time go by faster. I tend to have 20-30 minute power naps several times when I'm traveling. In the case of a 10-hour flight, you'd want to have 3-4 of

these power naps every 2-3 hours. Use a sleeping mask and noise-canceling airphones while you sleep.

- **Use blue light exposure to strategically adjust yourself to the target destination time zone.** Blue light in the morning until the afternoon sets off a proper circadian rhythm and increases your alertness. Using blue screens or a device like the Human Charger when it would be morning at the destination can help you to start living in that time zone before landing already. The Human Charger also has an app that tells you exactly when to use the device for the best effect.

What to Do After the Flight

- **After landing your first priority is to ground yourself.** The Earth's negatively charged electrons will balance out the positive charge you've accumulated during the flight. Standing on the grass with your bare feet will help to eliminate free radicals and it has an antioxidant effect. You should also go through a more thorough mobility sequence where you loosen up your joints and get the blood flowing.
- **Taking some more activated charcoal will also flush out the germs you've accumulated.** This is a good time to also take some antioxidant supplements like vitamin C and glutathione to re-strengthen your immune system.
- **Your first meal after flying should be ketogenic and with plenty of DHA.** This will lower inflammation and

replenishes the cells with essential fatty acids. This is the time you can also take antioxidants like vitamin C to lower oxidative stress and turn the body's guard down.

- **Reducing overall stress will lower cortisol and alleviates the negative side effects of jet lag.** You should try to do some foam rolling, take more magnesium, have an Epson bath, or meditate.
- **When you're in the destination time zone, you should wait until it's the evening before going to sleep.** Try to get into bed earlier around 9-10 PM and take a high dose melatonin supplement of about 0.5-1 grams to really knock yourself out and get a good night's sleep.

Nutritional strategies will also condition your body to handle EMFs and radiation. **Here's a list of foods and supplements that can alleviate the negative side-effects of air travel, CT scans, 5G, and the like:**

- **Magnesium** is a calcium channel blocker that reduces the effects of VGCCs and EMF. It's also essential for reducing stress, improving cardiovascular health, heart rate variability, and overall cellular function. Most people are deficient in magnesium.
- **Molecular Hydrogen** protects against oxidative stress caused by peroxynitrate[1330]. It's a powerful antioxidant that I would take in preparation of a flight or afterward. You can take molecular hydrogen tablets with you and mix them with water very easily.

- **Spices and Polyphenols** have been shown to prevent and repair the damage from peroxynitrate[1331]. Everything from tea, rosemary, thyme, and turmeric can have a protective effect by stimulating Nrf2 as well as fighting AGEs.
    - Genistein is an isoflavone found in soybeans that's been shown to protect mice against ionizing radiation just one hour before exposure[1332]. It works by inhibiting lipid peroxidation and stimulating red and white blood cells[1333].
    - Curcumin treatment prior to irradiation restores antioxidant levels and has radioprotective effects[1334]. It also reduces DNA damage and tumor formation in rats and human lymphocytes exposed to gamma-rays[1335,1336]. Ginseng and ginger have been shown to have similar benefits[1337].
    - Garlic extract protects red blood cells from radiation damage through glutathione[1338]. Pre-treating mice with garlic extract also prevents radiation damage to chromosomes[1339].
    - Ginkgo biloba has been shown to reduce markers of DNA damage in people and animals exposed to radiation[1340,1341]. It was even used to treat workers at the Chernobyl nuclear disaster[1342].
- **ACE Vitamins** – Vitamins A, C, and E are the main dietary antioxidants that can protect against radiation-induced free radical damage and DNA damage. They're used by airplane pilots and astronauts[1343,1344].

- Supplemental beta-carotene the precursor of vitamin A reduced lipid oxidation caused by radiation during the Chernobyl nuclear accident[1345]. Animal studies show vitamin A can reverse radiation-induced gene expression abnormalities[1346]. It might be a good idea to eat some liver and carrots before a flight or afterward.
- Combining high-dose vitamin C with conventional cancer treatment slows down disease progression and reduces some of the side-effects of chemotherapy[1347]. It's failed to benefit patients with advanced cancer and malignancies[1348] but can still alleviate radiation-induced damage. Workers at the Fukushima nuclear plant who supplemented with vitamin C had a much greater reduction in DNA damage and cancer risk than those who didn't[1349]. Vitamin C is also radioprotective against iodine-131 in vivo[1350]. I wouldn't still mega-dose vitamin C before going on an airplane or getting exposed to radiation. Instead, I'd eat more antioxidant-rich foods to get my vitamin C that way and then maybe take larger doses as a supplement after the fact.
- Vitamin E protects against lipid peroxidation and toxicity that will inevitably happen if your cells get exposed to radiation[1351]. Studies in mice also show that vitamin E protects them from dying to otherwise lethal levels of gamma rays[1352]. Taking

100 mg-s of vitamin E before exposure can be an effective strategy.

- **N-Acetyl-Cysteine (NAC)** is a potent supporter of glutathione and intracellular antioxidant activity. It reduces damage from radiation, oxidative stress, cytokines, and DNA damage[1353]. Because NAC works through a different mechanism than vitamin C, it's safer to take it prior to EMF or radiation exposure. Instead of reducing oxidation like vitamin C, it revs up the body's defense systems like glutathione, thus making it stronger beforehand.

If these examples don't illustrate the power of nutrition and food then I don't know what will. I mean, certain nutrients and ingredients can greatly protect you against the negative side-effects of radiation, which is supposed to be lethal in too large amounts. It comes to show how powerful our bodies can be if it receives the right resources and proper conditioning.

EMF and Radiation Hormesis Protocol

- EMFs and exposure to radiation will only become more predominant. There's not much you can do about it, unless you become an activist or move to a forest cabin above the Arctic Circle. To thrive in the modern technological environment, you should condition your body to handle these wavelengths through hormesis.
    - Don't freak out about EMFs and 5G because stress makes you more vulnerable.

    - Don't bring in unnecessary sources of EMF into your house, such as smart TVs, refrigerators, Alexa, etc.
    - Swap out your Wi-Fi for Ethernet connection and turn it off during the night.
    - Sleep in as low of an EMF environment as possible with your phone turned on airplane mode and no Wi-Fi. Use grounding mats and sheets to provide additional protection.
    - Keep your phone on airplane mode most of the time, especially when carrying it in your pocket. Ditch the Bluetooth headphones and wireless gadgets.
- Practice time-restricted eating and intermittent fasting to promote cellular autophagy, Nrf2, and glutathione for fighting the oxidative stress created by EMFs.
    - When on airplanes, stay in ketosis and fast throughout the flight to keep your body in a protected state.
    - Don't eat a bunch of carbs, sugar, or inflammatory fats when surrounded by radiation or high amounts of EMFs because they'll promote inflammation.
- Pick up red light therapy to promote hormesis to radiation and gain other rejuvenating benefits at the same time. Don't over-do it though because it'll be too stimulating. Generally, 10-15 minutes a day is more than enough.

Chapter Eleven:

Of NAD+ and Methylation

"If you want to find the secrets of the universe, think in terms of energy, frequency and vibration."

Nikola Tesla

In his 1930 Nobel Prize speech, Euler-Chelpin said that NAD+ is: *„one of the most widespread and biologically most important activators within the plant and animal world.*[1354]*"* NAD+ or nicotinamide adenine dinucleotide is a co-enzyme involved in virtually all energetic processes inside your body without which you'd die[1355]. What's more, your ability to cope and adapt to stress is very much limited to how well your body can use NAD. Here's why it's essential:

- **NAD+ is needed for energy production and skeletal muscle homeostasis**[1356]. Not enough NAD+ causes chronic fatigue, exhaustion, and atrophy. Giving NAD+ precursors to mice with muscle dystrophy improves their muscle function[1357].
- **NAD+ supports mitochondrial functioning** during youth and restores it in later life[1358]. It's essential for shuffling around energy.
- **Low NAD+ impairs brain function, increases brain degeneration, inflammation, body fat, fatigue, and**

muscle loss. NAD+ in mice improves cognitive functioning and protects against Alzheimer's[1359].

- **NAD+ inhibition causes cellular death because of not being able to produce energy**[1360]. Low NAD+ also reduces the availability of oxygen to cells, which lowers thyroid and metabolism.
- **NAD+ protects the cells against oxidative stress with the help of sirtuins**[1361]**.** NAD+ activates sirtuins which then help to grow blood vessels and muscle[1362].
- **NAD+ replenishment improves lifespan and healthspan through mitophagy and DNA repair**[1363]**.** NAD+ supplementation can promote DNA repair in mice[1364].

NAD is critical for converting food into energy, repairing DNA damage, strengthening the immune system, burning fat, and regulating the body's circadian clock[1365]. Unfortunately, as you get older your NAD levels drop and this accelerates aging and disease[1366]. NAD levels drop to less than half of what you had in your 20s after the age of 60[1367]. Declining NAD is linked to age-related metabolic dysfunction and is a key contributor to aging[1368]. Research has found that raising NAD levels with one of its precursors nicotinamide riboside (NR) can reverse signs of aging and lowers the risk of many diseases[1369]. However, there aren't many clinical trials in humans proving it yet.

NAD is intrinsically involved in the Krebs cycle and ATP production. It has two forms: NAD+ and NADH which both govern electron transfer reactions (See figure below):

- NAD+ is an oxidizing agent that picks up electrons from other molecules and thus becomes reduced.
- NADH is a reducing agent that forms from reduced NAD+ and it can then be used to donate electrons to other molecules, thus becoming NAD+ again.

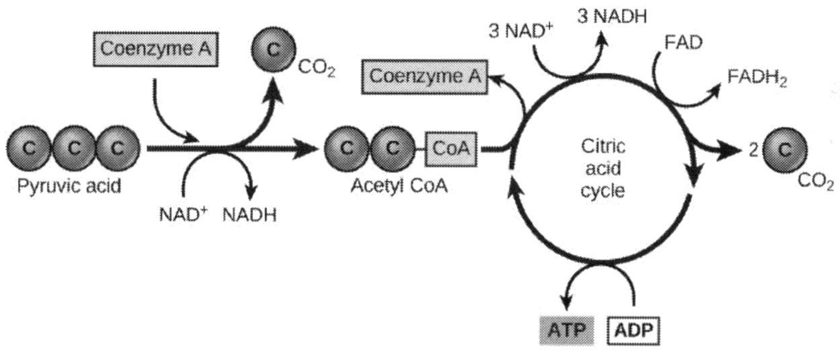

Electrons of NADH can store energy that gets converted into ATP in the mitochondria during a process called „oxidative phosphorylation". A higher NAD+ to NADH ratio creates more readily available energy around and makes the cells produce energy more efficiently. An abundance of energy from food raises NADH and lowers NAD+ levels. Fasting, exercise, and any form of energy depletion deplete NADH production, thus resulting in an increased NAD/NADH ratio. It works by not giving you a massive amount of energy like drinking sugar would but it does so by releasing the stored energy of NADH.

It's hypothesized that NAD levels decline with age because it's being destroyed by the overactivity of an NAD-consuming enzyme called CD38[1370]. CD38 is a membrane-bound hydrolase that regulates metabolism and is regulated by an inflammatory cytokine called NF-kB[1371]. Basically, more inflammation results in higher CD38, which depletes NAD. This happens through decreased SIRT3 in the mitochondria[1372]. CD38 is also the main enzyme that degrades NMN – another NAD precursor.

Inflammation is one of the main NAD consumers that occurs as a by-product of aging[1373]. With age, your inflammation will increase because the body is less robust and frailer. Unfortunately, we're also exposed to a lot more additional inflammatory sources such as air pollution, heavy metals, pesticides, EMF, and poor quality food, which depletes our NAD even further. With lower oxidative stress, achieved by either fasting or exercise, your NAD levels will be thus higher because it's not being consumed to fight inflammation.

NAD+ and Immune System Function

Here's how NAD affects the immune system:

- **NAD-biosynthetic pathways regulate immune cells and innate immunity[1374].** During an immune response, macrophages upregulate NAMPT, which governs the NAD salvage pathway, to control inflammation and cell survival. NAD also regulates cytokines, blood lymphocytes, and monocytes.[1375] Injecting NAD into mice protected them against autoimmune diseases and prolonged survival after skin transplantation.[1376,1377]

- **NAD is involved in the body's anti-viral defense systems like interferon.[1378]** Coronavirus infection dysregulates NAD metabolism by over-expressing PARP activity, which inhibits anti-viral activity.[1379] Supplementing with nicotinamide or nicotinamide riboside can restore the anti-viral effects of PARPs whereas inhibiting them is less likely to do so.

- **Low NAD decreases cellular antioxidant activity.** This will promote inflammation and oxidative stress that weaken the immune system. Lipopolysaccharide infection causes a pro-inflammatory state and reduces NAD within hours.[1380] This will lead to the creation of too many reactive oxygen species that begin to ravage even more. NAD is essential for cells to reduce stress-induced damage.[1381]

- **NAD helps to protect against and detoxify heavy metals.** One study on roundworms found that NAD+ supplementation protects against methylmercury toxicity[1382]. So, keeping your NAD elevated with lifestyle or supplements could be protective against environmental toxins. Cadmium elevates inflammatory IL-4 levels and alters metabolites associated with fatty acid metabolism, leading to increased pulmonary inflammation during a viral infection[1383].

- **Low NAD accelerates aging and speeds up immunosenescence.** NAD-consuming enzymes like

CD38, ADP-ribosyltransferases (ARTs), poly-ADP-ribose-polymerases (PARPs), and sirtuins are involved in the aging process as well as immunity[1384].

NAD/NADH homeostasis determines the survival of pathogens and various bacteria, including *Mycobacterium tuberculosis* (Mtb) – the one that causes tuberculosis[1385]. Several viruses hijack the enzymes involved with NAD homeostasis and use it to survive[1386]. It's estimated that about 17% of the enzymatic reactions in Mtb use NADPH as a cofactor[1387]. Mtb can use both the de novo as well as the salvage pathway[1388]. NAD+ starvation is bactericidal for many strains of bacteria and could be a potential target for certain diseases.

At the same time, NAD-boosting compounds like NR and pterostilbene have been shown to have anti-bacterial properties.[1389] Topical NAM has anti-mycobacterial abilities found in two studies on tuberculosis and HIV patients.[1390] This might be the cause of increasing sirtuins and SIRT1 that improve the function of cells.[1391]

Accumulation of NADH regulates biofilm formation and improves Mtb survival in macrophages.[1392] NADH can also bind to isoniazid to create isonicotinic-NADH that has powerful antimycobacterial properties.[1393] The NAD/NADH ratio thus works like a double-edged sword that can both fuel a healthy immune response as well as promote the survival of pathogens. The determining factor probably depends on many other processes in the body and overall energy balance.

For example, if you're generally a healthy person with high NAD then supplementing extra NAD is probably not necessary. On the other hand, someone who's suffering from chronic fatigue or mitochondrial dysfunction could benefit from getting supplemental NAD as a means of getting out of the slump. Low NAD breeds lower NAD and keeps you in the vicious cycle. The particular stage of the infection will also have a major impact on the final outcome. Raising NAD in certain phases of a viral infection could help to prevent it from becoming severe and suppressing it at others will help to starve it out.

Viruses and infected cells are battling in a constant tug of war by trying to drag NAD to their side. Infected healthy cells use NAD to defend themselves against the virus whereas the virus tries to hijack NAD and prevent itself from dying. Recent research has shown infections deplete NAD by activating PARPs. Furthermore, the highly inflammatory cytokine storm that accompanies COVID19 and other infections are also elevated, thus depleting NAD further.[1394] Almost all the pathologies of COVID19 result from low NAD+.

NAD Biosynthesis Pathway

NAD can be created via several pathways – (1) the De Novo Biosynthesis Pathway that uses precursors like tryptophan or vitamin B3 from diet and (2) the Preiss-Handler Pathway that converts nicotinic acid (NA) into NAD+. There's also the Salvage Pathway that cycles NR and Nam. (See figure below)

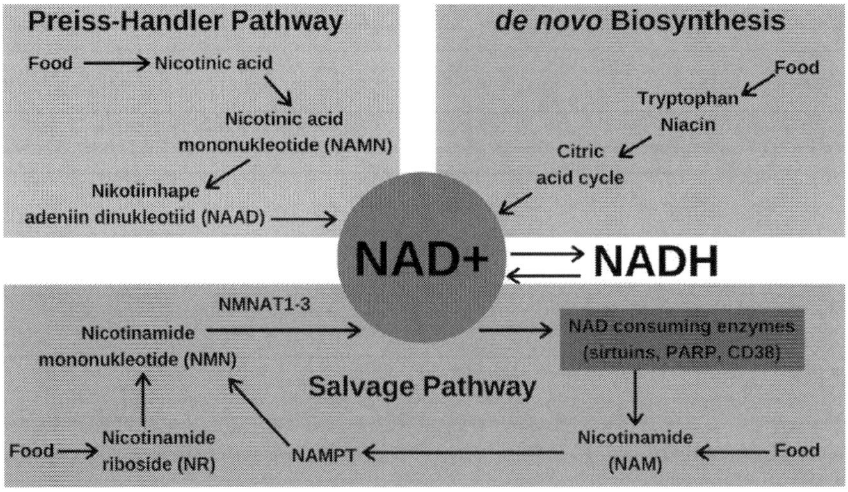

NAD is consumed by NAD-consuming enzymes, sirtuins, PARPs, and CD38[1395]. This will decrease intracellular levels of NAD+ biosynthesis but produces nicotinamide (NAM). NAM gets recycled into NAD by the NAD salvage pathway. Via this same pathway, NAM is converted into nicotinamide mononucleotide (NMN) by NAMPT (nicotinamide phosphoribosyltransferase), and NMN is converted into NAD by nicotinamide mononucleotide adenylyltransferases 1–3 (NMNAT1–3)[1396]. This is a phenomenal way your body can produce and recycle its energy.

When NAD+-dependent enzymes like sirtuins use NAD, they break off the components of NAD they need and send the rest to be recycled back into NAD+. This happens to the original NAM, NA, and tryptophan as well when the Press-Handler Pathway merges with the Salvage Pathway. To overcome the rate-limiting step of NAD biosynthesis via the salvage pathway you need proper circadian alignment and SIRT1 expression[1397]. Disrupting circadian rhythms suppresses SIRT1, which then lowers the availability of NAD+[1398]. Chromatin is a mass of genetic material composed of DNA and proteins that form chromosomes during eukaryotic cell division. The structure of chromatin changes in a circadian manner over the 24-hour period and NAD+ is a key metabolite for circadian chromatin remodeling[1399].

Circadian rhythms of NAD+ are driven by the circadian clock called CLOCK, which is a histone acetyltransferase. CLOCK:BMAL1 regulates the circadian expression of NAMPT, the enzyme that provides a rate-limiting step in the NAD+ salvage pathway[1400]. SIRT1 promotes NAMPT and contributes to the circadian synthesis of its own enzyme. This regulatory network is called the 'NAD World'.

Aging also weakens the expression of circadian gene expression[1401], causing sleep fragmentation and age-related diseases[1402]. Mice with circadian gene knockouts show accelerated aging, shortened lifespan, cancer, and other ailments[1403]. Fortunately, it's been found that calorie restriction can reverse the rewiring of disrupted circadian rhythms, thus alleviating the side-effects of aging[1404]. It does so by increased sirtuin, autophagy, and NAD. However, as we've already found out in chapter two, intermittent fasting mimics calorie restriction and time-restricted eating is an intricate part of proper circadian rhythm alignment.

It's been shown that mice fed a fattening diet are protected against obesity, hypertension, inflammation, and circadian clock gene expression patterns if they do time-restricted eating[1405]. Mice without time restrictions get obese and sick quite fast. This suggests that the timing of when you eat has a profound effect on your metabolic health and circadian alignment. It also means that if you are misaligned with the circadian rhythms, then eating within a smaller time window may help you to get away with it even if it's going to disrupt your circadian rhythms. Probably because you compensate for it by increasing NAD, sirtuins, and autophagy in the fasted state.

Sirtuins and NAD+

Sirtuins are a family of proteins that act as metabolic sensors. They deacetylase the coenzyme NAD+ into free nicotinamide. Simply put, they break down acetyl from proteins to maintain their functioning for longer. The NAD/NADH ratio determines the nutritional status of the cell and sirtuins are there to respond to the availability of NAD+ in the body.

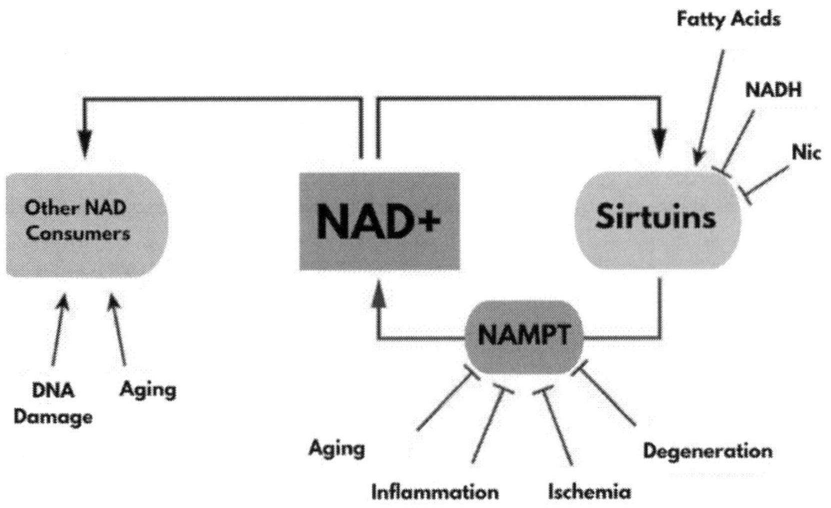

SIRT6 overexpression has been found to lengthen the lifespan of male mice by as much as 15,8%[1406]. SIRT6 deficiencies in mice accelerate their aging[1407]. Cellular deterioration and senescence are thought to be caused mostly by the accumulation of unrepairable DNA damage[1408]. SIRT1 plays an important role in activating DNA repair proteins[1409]. It's specifically involved with repairing the double helix of DNA[1410]. SIRT1 can also induce cellular autophagy by directly deacetylating AuTophaGy (ATG) proteins such as Atg5, Atg7, and Atg8[1411]. This then promotes mitophagy and helps to eliminate old worn-out cell parts.

A lot of research shows that NAD+ homeostasis is very connected with autophagy and fasting[1412]. Metabolizing NAD+ by PARP or SIRT1 activates autophagy. It does so by creating ATP depletion and an energy deficiency, which enforces self-eating. So, NAD+ by itself won't increase autophagy. Only consuming it will. Creating more NAD+ will not lead to accelerated autophagy because it increases the cells' energy charge. Intracellular NAD+ depletion leads to the inhibition of cell growth and autophagy induction[1413]. The goal should not be to deplete NAD+ because that will interfere with proper cellular functioning. Fortunately, lower NAD+ can be compensated for by an increase in autophagy.

How to Increase NAD+ Levels

To produce NAD, it needs to be created from one of its precursors or through endogenous bioenergetics. Here's how to increase NAD:

- **NAD can be synthesized from the amino acid tryptophan or aspartic acid.** Animal protein tends to be a more bioavailable source of NAD precursors as well as tryptophan. Poultry, meat, fish, algae, nuts, and seeds are the top sources.
- **Vitamin B3 or niacin supplementation can also increase NAD+ levels**. Daily requirements for NAD+ biosynthesis can be met with 20 mg of niacin a day. However, there's growing evidence saying that substantially higher rates of NAD+ are beneficial against aging and neurodegeneration[1414]. Niacin can be found in animal foods and proteins. The niacin in plants or maize is less bioavailable and can still cause deficiencies in B3[1415].
- **Fermented foods like sauerkraut and yogurt have precursors that increase NAD+.** During fermentation, pyruvate is converted into lactic acid. NADH accumulates and is oxidized into NAD+ to eliminate the excess[1416]. Many bacteria use this process to complete the NAD/NADH cycle. Alcohol fermentation can have the same effect, but it'll also promote more oxidative stress and inflammation.

- **Ketone bodies lower the production of reactive oxygen species in the mitochondria by increasing NADH oxidation into NAD+[1417].** A ketogenic diet promotes NAD+ levels because of fatty acid oxidation and glucose depletion[1418]. However, fruit can also activate enzymes that help to convert NADH into NAD+[1419].
- **Fasting and calorie restriction increase NAD+ and SIRT1 levels** which have many anti-aging benefits and it dictates cell survival[1420,1421]. However, it does so by raising the NAD ratio in relation to NADH by lowering oxidative stress and inflammation that would otherwise consume NAD. Another method is through the salvage pathway involving NAMPT but for that, you need circadian alignment as was discussed earlier.
- **Oxaloacetate supplementation can increase lifespan** in roundworms by increasing NAD and FOXO proteins[1422]. Oxaloacetate is a compound that supports energy production in the citric acid cycle. It can be taken as a supplement.
- **Exercise also increases NAD+, NAMPT, and sirtuins[1423].** It promotes the shuffling and redistribution of NAD through the salvage pathway similar to fasting. More on that shortly.
- **Heat shock proteins increase NAD+ and sirtuins[1424].** Taking a sauna feels like exercise and has similar benefits so go at it!

There is a small amount of NMN in vegetables like broccoli, tomatoes, avocadoes, cabbage, etc. Unfortunately, it would be very difficult to eat enough of them to reach the effective dose of NAD[1425] because you get like 2 mg/kg.

Do Fasting and Keto Increase NAD+

Nutritional ketosis has been shown to increase NAD/NADH ratio in healthy human brains. A study found that consuming MCT oil increased NAD by 3,4% and reduced NADH by 13%, resulting in an 18% increase in the NAD/NADH ratio[1426].

Ketones provide an alternative fuel source to the brain but they also lean the body towards more acetyl-CoA generation, which requires less NAD+ than glucose[1427]. To produce two acetyl-CoA molecules, one glucose molecule requires four molecules of NAD+ to be converted into NADH whereas synthesis of acetoacetate requires zero NAD+ and beta-hydroxybutyrate requires one NAD molecule[1428].

Ketone bodies work like signaling molecules that have anti-inflammatory effects by reducing oxidative stress and inhibiting histone deacetylases (HDACs)[1429,1430]. Beta-hydroxybutyrate blocks NLRP3 inflammasome-mediated inflammatory disease[1431]. This prevents the loss of NAD to inflammation and CD38.

Diabetes and high blood sugar offset the redox balance of NADH/NAD+ by overproducing NADH and thus lowering NAD+. Such imbalances if maintained for too long can lead to oxidative damage and contribute to the development of diabetes[1432].

Ketogenic diets are low carb, thus they prevent this off-set redox balance between NADH and NAD.

When in ketosis or any other form of energy stress like exercise or fasting, you activate a sensor called AMPK. With low ATP, AMPK signals the creation of NAD through NAMPT, which is one of the rate-limiting enzymes in the NAD salvage pathway[1433]. AMPK also protects against oxidative stress. So, the main way ketosis and fasting help with NAD is by preventing its loss and making you consume less of it but ketosis also leads to an increase in NAD through NAMPT and AMPK.

Does Exercise Increase NAD+

Resting human muscle is estimated to have NAD+ concentrations of ~1.5–1.9 mmol/kg and NADH ~0.08–0.20 mmol/kg[1434,1435]. NAD+ concentrations are positively correlated with the percentage of slow-twitch muscle fibers[1436]. They get trained primarily during aerobic endurance training but so does anaerobic HIIT.

NAD levels fluctuate in response to exercise and nutrition[1437]. It's being constantly consumed by DNA repairing enzymes, oxidative stress, inflammation, and ATP. Muscle contraction or heat production decrease NADH levels and thus result in an increase of NAD+[1438,1439]. The elevation of NAD happens primarily in the mitochondria[1440].

In mice, swimming increases NAD+ in the muscles[1441] but in rats it's endurance exercise[1442]. Swimming to exhaustion in rats decreases NAD but it gets restored during rest[1443]. Injections of nicotinamide elongate how long the rats were able to swim. Therefore, the rate-limiting factor of how long rats can exercise at maximum intensity could be the amount of NAD.

In humans, exercise at 65% and 100% VO2 max reduces muscle NAD levels[1444]. This might have to do with the more oxidative and anaerobic environment created by higher intensity exertion. During maximal exercise, NADH can increase up to 140% because your body is producing a lot of lactic acid[1445]. No change has been noted throughout exercise at 75% of VO2 max[1446], whereas the NAD+/NADH ratio decreased at 50% VO2 max[1447]. This elevation of NADH drops down to resting levels during recovery and it might play an important role in mitochondrial adaptations to exercise[1448]. NADH will also be eventually oxidized into NAD after the exertion once lactic acid production stops.

To me, it seems that sub-maximal exercise like cardio or hiking elevates NAD thanks to oxidative phosphorylation i.e. nasal breathing. High-intensity exercise like weightlifting or HIIT depletes NAD because of increased NADH and lactic acid. However, the intense nature of this leads to future adaptations that improve mitochondrial density and oxygen efficiency at rest. It's just that you can't sustain such high intensities for too long because your NAD gets depleted. In any case, exercise is good for increasing your basal levels of energy and NAD+.

NAD Supplements

There are many NAD precursors like nicotinic acid (NA) or niacin, nicotinamide (NAM), nicotinamide riboside (NR), and nicotinamide mononucleotide (NMN)[1449]. All of them have unique effects on the metabolism and NR seems to elevate NAD the most in a dose-dependent manner[1450]. This can restore more youthfulness in older people who already suffer from low NAD and can't do much to re-salvage it.

Safety assessment studies on NR have found that oral doses of 5000 mg/kg/day don't cause mortality but they have toxic side-effects on the liver, kidney, and other organs. The lowest observed adverse effects happened at 1000 mg/kg/day and 300 mg/kg/day had no side-effects[1451]. Just 100, 300, and 1000 mg-s of NR has been shown to raise NAD levels in humans[1452] and most supplements have serving sizes of 300 mg-s.

Supplementing 400 mg/kg/day of NR for 1 week in mice increases their liver as well as muscle NAD levels[1453]. They also were protected against high fat over-feeding thanks to an increase in energy expenditure. NR-fed mice also had improved endurance, and enhanced thermogenesis in response to cold exposure, suggesting better muscle oxidative performance and brown adipose tissue functioning.

One study found that giving rats NR at a dose of 300 mg/kg for 21 days reduced their swimming performance by 35% compared to the control group[1454]. The researchers hypothesized that this was due to using rats instead of mice, swimming instead of treadmill-running, and NR decreasing fatty acid oxidation during exercise, leading to early fatigue. It might have also been due to NR off-setting the redox homeostasis of cells and inhibiting the hormetic stress from exercise similar to antioxidants[1455]. If anything, then taking NAD precursors around exercise may not be the best idea and they should be kept at least a few hours apart.

Tissue-specific NAD biosynthesis in only skeletal muscle doesn't seem to be enough to trigger muscle mitochondrial biogenesis or change their function[1456]. This isn't sufficient to promote oxidative metabolism and protect mice against high-fat feeding. So, just muscle-specific elevation of NAD may not be enough to gain its benefits and you also want to raise it systemically in the liver and other tissues.

Compared to others, NR elevates NAD the most and drives more NAD consuming activities[1457]. NAM gets to NAD in 3 steps but NR in just 2. In fact, NR by-passes the rate-limiting step that NAM has to go through, which would limit how much NAD gets produced. NMN has the least side-effects compared to others. For instance, niacin causes the infamous flush and itchiness when taken at high doses. A protein called Slc12a8 helps to absorb NMN in the gut, which makes oral supplementation a potentially effective way of boosting NAD[1458].

In one long-term study, mice were given NMN. The group who received a high dose (300 mg/kg a day) lost 18% of their body weight with no added exercise[1459]. Young mice who got the treatment didn't see any benefits, but the older ones did. In 2013, David Sinclair's lab reported that injecting NMN to old mice for just a week restored their mitochondria to a youthful state[1460].

They noted that the muscle strength of NMN-injected mice didn't improve and proposed it was due to the short treatment period of just a single week, which wasn't long enough to see a significant change in muscle mitochondria.

Although both can create NAD, NMN and NR have key differences and unique characteristics. Here are the most significant ones:

- **NR is a form of vitamin B3 but NMN is not**. NMN is a nucleotide derived from ribose and nicotinamide used for DNA synthesis and repair[1461].
- **NMN can't enter the cell easily**. It can transform into NR before entering the cell where it can transform back to NMN and become NAD+[1462].
- **NMN is larger than NR**. This makes it less efficient in entering cells. However, new research shows that NMN has to become NR for only certain cells. The Slc12a8 transporter allows NMN to go into cells but it's only located in the gut cells of mice and requires sodium ions[1463].

- **NMN creates NAD in 3 steps but NR in 2.** This goes back to the larger size of NMN as well as the biosynthesis pathway of NAD. Due to being larger, NMN takes a longer time.
- **NR has more human clinical trials than NMN.** NR has been deemed to be safe even in large doses but there are no real studies about the safety of NMN[1464]. NMN hasn't been

evaluated by the FDA yet although it's available on the market for a few years already[1465]. Nevertheless, high dose studies of NMN show no adverse health effects or toxicity.

Intravenous NAD infusions are by far the most effective and fastest way of boosting intracellular NAD. They're being used in anti-aging clinics and medical facilities.

I've experimented with both NMN and NR. They both seem to raise your homeostatic energy levels a little bit but you notice their effects primarily when you're more tired or sleep-deprived. If you're already young, fully optimized, and aligned with the circadian rhythms it'll have less of an effect. I'm still using either NR or NMN after a bad night's sleep or when jet-lagged.

The problem with taking NR or NMN is that it can potentially be methylated and lost through urine. This not only makes you excrete the NAD precursor but we also lose the methyl donor, which could be put into use for other important purposes, such as overall cellular functioning, athletic performance, digestion, and cognition[1466]. To prevent that, you can take NR with meals to direct it into NAD biosynthesis as opposed to methylation. Combining it with additional methyl donors like trimethylglycine, creatine, B12 or glycine will also prevent the loss of methyl groups. More on methylation shortly.

NADPH

The NAD/NADH ratio is primarily involved in balancing catabolic reactions. However, for anabolism, there needs to be an additional phosphate group.

Nicotinamide Adenine Dinucleotide Phosphate (NADP+) is a form of NAD that promotes anabolic reactions such as photosynthesis, and nucleic acid synthesis. It's used by all forms of cellular life[1467]. NADPH is the reduced form of NADP+. The extra phosphate gets added during the de-novo or salvage pathway by the NAD+ kinase. This allows NAM synthesis and conversion of NADPH back into NADH to maintain balance[1468].

NADPH protects against the oxidative stress from excessive reactive oxygen species (ROS). It also allows for the regeneration of glutathione (GSH) a master antioxidant pathway[1469]. NADPH is important because it will reduce and neutralize oxidized antioxidants and free radicals. It's also used for synthesizing cholesterol, which itself is also a major antioxidant in the body.

Glutathione is your body's main antioxidant system that protects against reactive oxygen species and free radicals like peroxides, lipid peroxides, and heavy metals[1470]. It not only keeps your body working optimally but also alleviates a lot of the side-effects of aging and modern life. Without enough NADPH, your body can't recharge glutathione after it becomes oxidized. This will put breaks on all the detoxification systems.

Dietary glutathione is poorly absorbed. However, people who eat glutathione-rich foods have a lower risk of cancer[1471]. Dairy, cereal, and grains are low in glutathione; fruit and veggies are moderate, and fresh meat is higher[1472]. It can also be obtained from cruciferous and allium vegetables[1473,1474]. Supplements that promote glutathione production are N-acetylcysteine, Alpha-Lipoic Acid, and reduced glutathione[1475,1476]. The most effective method is to probably get glutathione intravenously, however, more data is needed to support its bioavailability[1477]. Most glutathione supplements are destroyed by the digestive tract and they're poorly absorbed when taken orally.

NADPH oxidase or NOX is a complex of enzymes bound to the cellular membrane. It senses the presence of oxygen and nutrients to balance the body's ROS[1478]. Inhibiting NOX increases NADPH and combats oxidative stress. NOX proteins are involved in the inflammation of the vascular adventitia[1479]. However, NOX also generates free radicals that destroy pathogens through a process called the respiratory burst[1480]. So, some small amounts of oxidative stress is still needed for optimal cellular homeostasis.

Glycine lowers oxidative stress by lowering NOX. It brings chloride into endothelial cells, which reduces the cell's ability to push out chloride ions. Chloride ions are needed for generating superoxide and free radicals.

Glycine is one of the main amino acids that comprises glutathione next to glutamine and cysteine[1481]. It's a powerful antioxidant needed for GHS as well as liver detoxification pathways. A 2019 paper by Dr. James DiNicolantonio writes:

> *Supplemental glycine may be useful for the prevention and control of atherosclerosis, heart failure, angiogenesis*

associated with cancer or retinal disorders and a range of inflammation-driven syndromes, including metabolic syndrome[1482].

Glycine supplementation has been found to have the same effects on life-extension as methionine restriction[1483]. It also promotes vascular health, lowers inflammation, improves sleep, maintains cartilage integrity, and reduces oxidative stress. You can get glycine from animal protein, specifically collagen-rich foods like tendons, ligaments, skin, bone broth, etc. Taking it as a supplement is also powerful. Glycine alone doesn't spike insulin or blood sugar because it's an inhibitory amino acid. Combining glycine with 25 grams of glucose lowers the blood sugar response by about >50%[1484]. You can consume 3-5 grams of glycine 1-3 times per day. It’s an amazing alternative to sugar and artificial sweeteners because of its sweet flavor.

Methylation

Methylation is the process of adding methyl groups to other molecules and transferring them around. Methyl groups are atoms with a single carbon unit that attaches to hydrogen (CH3).

The role of methylation is to control almost every reaction inside your body, starting with digestion, your brain plasticity, and ending with genetic expression of DNA repair. Methylation also affects liver health, detox pathways, neurotransmitter balance, fatty acid metabolism, inflammation, cellular repair, and fighting infections. Improper methylation is linked with autoimmune conditions[1485].

The role of methylation in the immune system is to maintain the function of immune cells like T-cells and activate the immune response.[1486] DNA methylation is a transcriptional modulator of the immune system by turning on and off certain genes as well as preserving host defenses.[1487] Improper methylation is linked with autoimmune conditions[1488]. Patients of lupus show abnormal methylation. Methylation also promotes detoxification pathways such as glutathione and autophagy. They're important for removing pathogens and neutralizing free radicals.[1489]

The methylation cycle is directly connected to the transsulfuration pathway, which is where glutathione is produced. Homocysteine, a by-product of methylation, creates cysteine, which forms glutathione. Sulfur coming from amino acids like methionine is one of the precursors for glutathione production. Hypomethylation results in low glutathione because sulfur is being excreted and low glutathione leads to slow methylation. This eventually leads to the accumulation of toxins and oxidative stress that causes disease and chronic fatigue.

There are many things that deplete your body's glutathione levels, such as environmental toxins, poor lifestyle habits, sleep deprivation, drinking alcohol, chronic stress, getting older, inflammatory foods, and nutrient deficiencies[1490,1491]. However, not getting enough amino acids and sulfur also results in low glutathione because of insufficient methylation.

NAD and methylation are very inter-dependent of each other and you need both for carrying out healthy cellular processes. The key difference is in acetylation. While methylation is the process of adding and transferring methyl groups, acetylation is the process of adding acetyl groups instead. An acetyl group binds together an additional carbon atom and a methyl group with an oxygen atom – C(O)CH3 – whereas a methyl group is just a single carbon attached to hydrogen (CH3).

- Acetyl groups can replace reactive hydrogen atoms in molecules and they work via substitution. Most commonly, acetylation happens in proteins and is called protein acetylation.
- Methyl groups can also replace reactive atoms but they also work by introducing methyl donors to molecules. Methylation is involved with DNA and protein methylation.

Methylation occurs inside the methylation cycle where it methylates histones and DNA. This will regulate gene expression or epigenetics. Acetylation happens in the mitochondria via the production of acetyl-CoA. Acetyl-CoA is one of the main energy currencies that gets created from metabolizing food and will be transferred into the citric acid cycle (CAC). NAD is being shuffled around the CAC to carry electrons and produce ATP.

Sirtuins and NAD seem to work independently of DNA methylation as you can create acetyl groups and HDACs without methyl donors[1492]. You don't need NAD for methylation either as

NAD metabolism happens inside the citric acid cycle, not the methylation cycle. However, imbalanced NAD metabolism can surely disrupt methylation and in turn result in affecting NAD itself. Too high NAD+ can actually lead to mitochondrial dysfunction and hypermethylation. Maintaining the oxidation of NADH, which is the reduced form of NAD+, prevents these ill effects, preserves methionine salvage, and prevents changes in DNA methylation and gene expression[1493].

For methylation to work properly, you need enough methyl groups from diet. One of the most important ones is folate also known as 5-MTHF. Folate (vitamin B9) turns on a methyl donor called SAMe (S-adenosylmethionine), which donates one of its methyls to form a methyl group. Folic acid is the artificial supplemental form of folate, which can promote cancer in high amounts[1494]. So, it's better to not take supplements that have folic acid and get folate from real food. Other methyl donors are all the B vitamins, methionine from protein, choline, creatine, betaine HCL, trimethylglycine, N-acetyl L-cysteine (NAC), and minerals like zinc.

You do need proper methylation for NAD homeostasis and energy production. Imbalances in methylation can increase inflammation and oxidative stress, which deplete NAD and contribute to aging[1495].

- **Not enough methionine, folate, choline, or betaine** can cause methyl deficiencies and result in high homocysteine and inflammation[1496]. Too many methyl donors can also have negative side-effects[1497].

- **Too much methionine in relation to glycine** can raise homocysteine, which promotes inflammation and oxidation[1498]. Higher intakes of TMG, glycine, and choline can lower circulating homocysteine[1499].
- **High intake of niacin or B3 can deplete methyl groups[1500]**. It also disrupts glucose metabolism and liver functioning[1501]. A study compared 300 mg/day of niacin and nicotinamide in healthy adults[1502]. They found that nicotinamide raised homocysteine more significantly than niacin. Taking 1000 mg/day of nicotinamide riboside hasn't shown to deplete methyl donors or cause liver problems[1503]. However, you would still want to combine NR with some additional methyl donors as discussed before.
- **NAD precursors like NMN and niacin deplete methyl donors.** If you're taking a lot of NAD supplements, then it's important to also consume additional methyl donors whether that be TMG, creatine, or folate.

Here are a few ways of assessing your methylation status and NAD metabolism:

- **Homocysteine** – The healthy range is said to be between 5-15 μmol/L but every 5 μmol/L could increase the risk of coronary heart disease risk by about 20%. Populations with low B vitamin intake can reach up to 20 μmol/L[1504]. To lower homocysteine get more methyl donors like TMG, glycine, or folate.

- **Vitamin B12** – Low levels of B12 or B6 cause poor methylation. The healthy range is between 350-950 pg/mL.
- **Folate** – Folate is a key methyl donor and low levels of it can raise homocysteine and inflammation. The healthy range is between 2,2-20 ng/mL.

Some people have a genetic predisposition for high homocysteine levels if they have the MTHFR gene, which is an enzyme that helps to regulate homocysteine in the blood. High homocysteine is a risk factor for cardiovascular disease[1505]. We all have 2 MTHFR genes from both of our parents. If one of the MTHFR genes is mutated, you're 'heterozygous'. With two you're 'homozygous'. MTHFR gene mutations can cause errors in homocysteine regulation and methylation. A single mutation isn't a medical concern and not everyone with two mutations develops hyperhomocysteinemia. Even further, getting more folate from the diet can also compensate for the mutation.

NAD and Methylation Hormesis Protocol

- Most of your daily NAD is created via the salvage pathway thanks to NAMPT and AMPK. To overcome the rate-limiting step in this synthesis, you need proper circadian alignment. That's why fatigue and brain fog are very much caused by poor sleep and desynchronized diurnal rhythms.
- Things that promote NAD re-synthesis are most of the hormetic activities such as fasting, exercise, ketosis, and

sauna. You can also create it from eating fermented foods, tryptophan, and methyl donors.

- Don't take NAD precursors around exercise because they may negate the adaptations similar to antioxidants. It's a safe bet to have them a few hours before working out. You should also combine them with some methyl donors like trimethylglycine, B12, glycine, or creatine to prevent decreased methylation.
- Both NR and NMN seem to be effective for boosting NAD but make sure you combine them with additional methyl donors. I also use them only when I had bad sleep or experience circadian mismatches to alleviate the imbalance. At other times it might just create a negative feedback loop and dependency.

Chapter Twelve: Of Barriers and Membranes

"The only defense against the world is a thorough knowledge of it."

— John Locke

Over the past few decades, certain food groups have been constantly criticized for causing cardiovascular disease and obesity. Since the 80s it's been saturated fat and cholesterol but after the wake of paleo and ketogenic diets in the 2000s, it's sugar and carbs getting their share of the blame. What gives?

In the previous chapter about metabolic flexibility and keto-adaptation I discussed how it's not a particular food that's causing the problem but more so the combination of them all. If you combine high amounts of fats and carbs you disrupt the Randel cycle, in which fatty acids and carbohydrates are competing for oxidation. This is thought to cause insulin resistance, dyslipidemia, overconsumption of calories, and atherosclerosis. Eating either a low carb ketogenic or a high carb low fat diet can be equally as healthy because they're lacking one of those macronutrients.

Although sugar and carbohydrates are often considered to be unhealthy, they're not as harmful as bad fats. In fact, I would eat a large amount of carbs over rancid or inflammatory oils every day of the week. While we wait for die-hard keto zealots to crawl back into their chairs, let me explain.

You can easily burn off the sugar you ate or even store it as glycogen. With proper timing, you can leverage carbs for performance, muscle protein synthesis, increased metabolic rate, thyroid functioning, and higher insulin sensitivity. It's much harder to get rid of bad fats and proteins because they literally become a part of your cells. Cell membranes consist of a lipid bilayer, cholesterol, triglycerides, and various proteins that control the movement of substances moving in and out[1506]. It's very difficult to eliminate them after consumption. What's worse, if your cells are made of inflammatory substances and fats, then they're going to be causing subliminal inflammation all the time. The effect is minute but imagine being poison-dripped over years or decades. That's surely going to cut down on your longevity and healthspan.

When I say 'bad fats and protein', then I refer to anything inflammatory, oxidized or rancid. Things like high omega-6 vegetable oils, trans fats, fake-butter margarine spreads are all too common in the food supply but even things like sun-damaged olive oil, fried fish, and moldy nuts can be equally harmful. Unfortunately, that's what the Standard American Diet (SAD) is mostly composed of.

There's no hormetic effect in eating bad fats – you're not making yourself stronger or more resilient but the opposite. If we went back in time, then the linear no-threshold (LNT) model would be perfectly applicable – pathogenesis is directly proportional to the dose and amount of exposure with no threshold of benefits. That's what we'll cover in the rest of this chapter – the anti-hormesis substances.

Bad Fats and Healthy Fats

In addition to stable blood sugar and body temperature, homeostasis also includes the balance between various fatty acids. The most important ones for survival are omega-9, omega-6, and omega-3 fats. They can't be produced by the body itself, thus need to be derived from diet.

- **Omega-3 Fatty Acids are an integral part of the cell membrane** and they regulate many other hormonal processes. They have great anti-inflammatory benefits that protect against heart disease, eczema, arthritis, and cancer[1507]. Dietary omega-3s help with modulating inflammation and certain aspects of immunity[1508]. Omega-3s are polyunsaturated fatty acids (PUFAs), which refers to their multiple unsaturated double bonds. That makes them very unstable and easily oxidized. Great sources of omega-3s are salmon, grass-fed beef, sardines, krill oil, algae, and some nuts. There are 3 types of omega-3s:
    - **EPA (EicosaPentaenoic Acid) and DHA (DocosaHexaenoic Acid)** are animal-sourced long-chain omega-3 fats both essential for the nervous system, brain, and general health. They're found especially in seafood.
    - **ALA (Alpha-Linolenic Acid)** is mostly a plant-based short-chain omega-3 fatty acid. Most animals, including humans, can't directly use ALA so it gets converted into DHA first. Humans can convert only about 8% of ALA into DHA[1509],

which is why animal foods like salmon and oysters are much better sources of omega-3s and DHA.

- **Omega-6 Fatty Acids are also essential polyunsaturated fats.** They differ from omega-3s by having 6 carbon atoms at the last double bond instead of 3. Omega-6s are primarily used for energy and they have to be derived from diet. Unfortunately, most people are getting too many omega-6s. The most common omega-6 is linolenic acid (LA). Another omega-6 is conjugated linoleic acid (CLA) with some health benefits. Vegetable oils, processed foods, salad dressings, and some nuts are the highest sources of omega-6 fatty acids and PUFAs.
- **Omega-9 Fatty acids are monounsaturated fats with a single double bond.** These ones aren't necessarily essential as the body can produce its own. In fact, omega-9s are the most abundant fats in cells. They've been found to lower triglycerides and VLDL[1510] as well as improve insulin sensitivity[1511]. You can get omega-9s from olive oil, and some nuts.

Anthropological research suggests that our hunter-gatherer ancestors consumed omega-6 and omega-3 fats in a ratio of roughly 1:1[1512]. After the industrial revolution about 140 years ago, the ratios between omega-6s and omega-3s shifted heavily towards the consumption of omega-6s[1513]. Today, the average American eats a ratio of 10:1, 20:1, or even as high as 25:1[1514], which is definitely not optimal. No wonder there are so many chronic diseases related to inflammation and cardiac events. The worst you could get away with, in my opinion, is a balance of 4:1 in favor of omega-6 but only if it's coming from whole food sources. Any amount of oxidized fat whether from grass-fed steak cooked in canola oil or charred salmon off-sets this balance further by turning the PUFAs inflammatory.

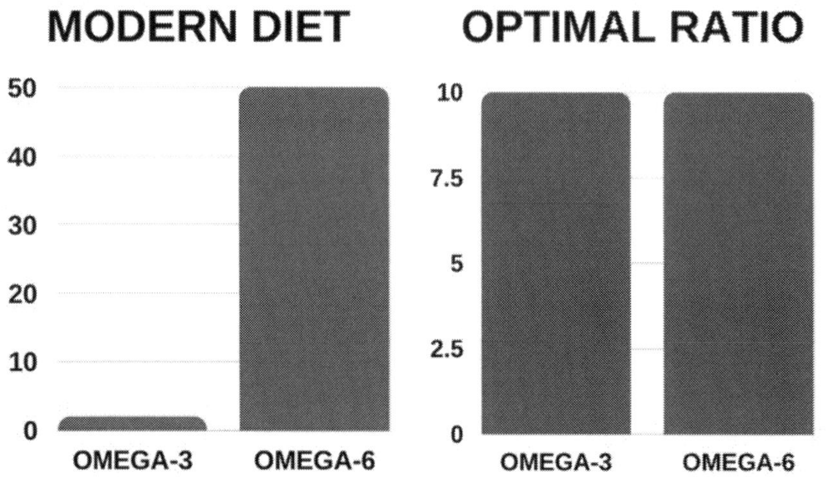

Lipid Peroxidation

Lipid peroxidation is the oxidation of lipids. It's the process by which reactive oxygen species and free radicals steal electrons from the lipids in cell membranes, causing cellular damage. Cell membranes consist of primarily lipids, which is why exposure to dietary fatty acids plays a key part in determining the composition of your cells.

Oxidation of lipids is deemed to be a crucial step in the development of several diseases. Lipid peroxidation can cause DNA damage, lipofuscinogenesis, mutagenesis, and carcinogenesis[1515]. It damages the skin and can cause inflammatory acne[1516]. In mice, T-cell lipid peroxidation causes ferroptosis or programmed cell death by iron and prevents immunity to infection[1517]. Supplemental vitamin E was able to negate those effects. Because most people are eating too many inflammatory fats they experience lipid peroxidation even when exposed to the sunlight[1518]. Their cell membranes become oxidized due to the heat, which will make them get sunburnt more easily as well as promote skin cancer. Inhibiting lipid peroxidation restores impaired VEGF expression and stimulates wound healing[1519]. This protects against cardiovascular disease as well as keeps your tissue healthy. PUFAs are easy targets for lipid peroxidation because of their chain bonds. That's why you'd want to limit your intake of these fats, especially if they're processed.

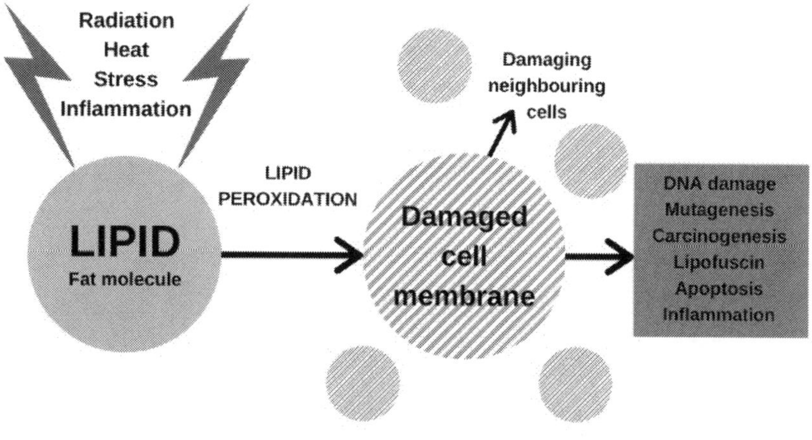

Vegetable oils are extracted from various seeds, such as rapeseed (canola oil), soybean, corn, sunflower, peanut, and safflower. To get the oil from these plants, they have to go through processing and high amounts of heat, which damages their fatty acid composition and causes lipid peroxidation. In addition to that, before the oils get put onto store shelves, they get processed even more with different acids and solvents to improve the composition of the product. As a bonus, they get deodorized and mixed with chemical colorings to mask the horrible residue of the processing process.

If you take it a step further and hydrogenize the vegetable oil, then it will eventually become more solid and intact. Meet margarine and all of the other trans-fatty acid type vegetable spreads that I can't believe aren't butter... A Medical Research Council survey showed that men eating butter ran half the risk of developing heart disease as those using margarine[1520]. Consumption of trans fats has been linked to obesity, metabolic syndrome, increased oxidative stress, heart disease, cancer, and Alzheimer's[1521,1522]. *Friends don't let friends eat margarine...*

To keep your ratios in balance, eliminate all vegetable oils from your house like canola oil, sunflower oil, rapeseed oil, etc. Avoid high-temperature cooking of fats, prefer grass-fed meat, pastured eggs, wild fish, and don't eat packaged foods. It's not that unoxidized omega-6 fats are bad. It's just that more often than not they tend to be oxidized already or are processed in a way that makes them so. Too many nuts and seeds in the diet can also be pro-inflammatory due to their omega-6 content, especially if you're not consuming enough omega-3s to counterbalance them.

Here's a list of the different oils and their fatty acid content:

FOOD	OMEGA-6 (g)	OMEGA-3 (g)	RATIO 6:3
FISH			
Salmon (4 oz/113 g)	0.2	2.3	1:12
Mackerel (4 oz/113 g)	0.2	2.2	1:11
Swordfish (4 oz/113 g)	0.3.	1.7	1:6
Sardines (4 oz/113 g)	4.0	1.8	2.2:1
Canned Tuna (4 oz/113 g)	3.0	0.2	15:1
Lobster (4 oz/113 g)	0.006	0.12	1:20
Cod (4 oz/113 g)	0.1	0.6	1:6
VEGETABLES			
Spinach (1 cup/110 g)	30.6	166	1:5.4
Kale (1 cup/110 g)	0.1.	0.1	1:1
Collards (1 cup/110 g)	133	177	1:1.3
Chard (1 cup/110 g)	43.7	5.3	8.2:1
Sauerkraut (1 cup/110 g)	37	36	1:1
Brussels Sprouts (1 cup/110 g)	123	270	1:1.3
NUTS AND SEEDS			
Walnuts (1 oz/28 g)	10.8	2.6	4.2:1
Flaxseeds (1 oz/28 g)	1.6	6.3	1:4
Pecans (1 oz/28 g)	5.7	0.3	21:1

Poppy Seeds (1 oz/28 g)	7.9	0.1	104:1
Pumpkin Seeds (1 oz/28 g)	2.5	0.1	114:1
Sesame Seeds (1 oz/28 g)	6	0.1	57:1
Almonds (1 oz/28 g)	3.3	0.002	1987:1
Cashews (1 oz/28 g)	2.1	0.017	125:1
Chia Seeds (1 oz/28 g)	1.6	4.9	1:3
Pistachios (1 oz/28 g)	3.7	0.071	52:1
Sunflower Seeds (1 oz/28 g)	6.5	0.021	312:1
Lentils (1 oz/28 g)	0.0384	0.0104	3.7:1
OILS AND FATS			
Butter (1 Tbsp)	0.18	0.83	1:1.5
Lard (1 Tbsp)	1.0	0.1	10:1
Cod Liver Oil (1 Tbsp)	2.8	1.3	2.2:1
Grain-Fed Tallow (1 Tbsp)	3.35	0.2	16.8:1
Grass-Fed Tallow (1 Tbsp)	1.2	0.8	1.5:1
Peanut Oil (1 Tbsp)	4.95	Trace	1:0.0
Soybean Oil (1 Tbsp)	7.0	0.9	7.8:1
Canola Oil (1 Tbsp)	2.8	1.3	2.2:1
Walnut Oil (1 Tbsp)	7.2	1.4	5.1:1
Sunflower Oil (1 Tbsp)	6	0.0	6:1

Margarine (1 Tbsp)	2.4	0.04	6:1
Peanut Butter (1 Tbsp)	1.4	0.008	17:1
Almond Butter (1 Tbsp)	1.2	0.04	2.8:1
Flaxseed Oil (1 Tbsp)	2.0	6.9	1:3.5
Olive Oil (1 Tbsp)	1.1	0.1	11:1
MEAT			
Ground Pork (6 oz/170 g)	2.83	0.119	23.8:1
Chicken	2.2	0.16	13.8:1
Grain-Fed Beef	0.73	0.08	9:1
Grass-Fed Beef	0.72	0.15	4.9:1
Domestic Lamb	1.9	0.6	3.3:1
Grass-Fed Lamb	1.7	2.2	0.7:1
Farmed Salmon	1.7	4.5	0.39:1
Wild Salmon	0.3	3.6	0.08:1

Most people aren't deficient in omega-3s or DHA. They're just out of balance in terms of their fatty acid ratios – too much omega-6s in relation to the small amount of omega-3s in their diet. High amounts of inflammation and oxidative stress from other sources is also a contributing factor to the reason why some people may need to increase their omega-3 intake.

The end-products of lipid peroxidation are reactive aldehydes such as malondialdehyde (MDA) and 4-hydroxynonenal (HNE). They are also called lipid peroxides or lipid oxidation products (LOPs). There are diagnostic tests like TBARS Assay that assess lipid peroxidation by quantifying MDA. However, this test is not specific for lipid peroxidation because there are other sources of malondialdehyde that cause oxidative stress[1523]. Additional biomarkers for lipid oxidation are F2 isoprostanes and isofurans.

Generally, if your blood test for inflammation and lipids is fine then you probably don't need to deliberately increase your fish oil or omega-3 consumption. It's more important to avoid the pro-oxidative foods like vegetable oils and rancid fats that would create inflammation in the first place. Instead of taking fish oil, you should want to get your omega-3s from eating fish every once in a while, grass-fed meat, and maybe algae.

There's quite a long history of epidemiological studies linking fish consumption with reduced risk of cardiovascular disease, lower inflammation, and general metabolic health[1524]. People who eat fish once or twice a week have been shown to have 50% fewer strokes, 50% lower CVD risk, and 34% lower CVD mortality risk compared to those eating no fish[1525,1526]. Supplemental fish oil reduces risk factors for CVD but hasn't definitively shown to prevent it[1527]. Unfortunately, rancid or spoiled fish oil causes lipid peroxidation and will promote inflammation. The amount of spoilage depends on extraction, processing, and containment practices[1528]. High heat and humidity promote oxidation quite easily. Most conventional fish oil supplements sit on store shelves exposed to light and heat for months even years. According to consumerlabs.com rancid fish oil supplements are quite common but not all of them are ruined[1529].

How do you know if your fish oil or olive oil are spoiled and rancid? If it smells bad and tastes awful, then it's probably oxidized in some way. A lot of companies also use flavorings and odors like lemon or strawberry to mask the smell. Fats should be kept away from heat, sunlight, and oxygen. The best place for storage is the fridge and it should be consumed within a few weeks or months.

Compared to fish oil, krill oil is safer, more sustainable, and with higher potency. A 2011 study found that the two have essentially the same metabolic effects despite krill oil containing less EPA and DHA[1530]. However, krill oil is absorbed 10-15 times better. Another benefit of krill oil is that the omega-3s are bound to phospholipids, which don't get destroyed in the gut and can cross the blood-brain barrier[1531]. They're also a more sustainable food source.

In my opinion, algae supplements like spirulina are much safer than fish oil because they're not oxidized and rancid. The potential for heavy metal toxicity is also relatively low if you keep in mind the fact that most fish is already polluted to a certain extent. Spirulina lowers serum MDA, which is a marker of lipid peroxidation[1532]. Chlorella is also able to detoxify metals and other toxins so it can overcome some of its shortcomings[1533].

Cholesterol Good or Bad

The same applies to cholesterol, which has been receiving a bad rep for decades due to being implicated in CVD. Is it really that cholesterol itself is harmful or is the oxidized environment turning it dangerous, thus increasing your risk of heart disease?

It turns out, half of heart disease patients have normal cholesterol levels yet they have an underlying risk of plaque build-up in the arteries[1534]. According to a 2009 study, nearly 75% of hospitalized patients for a heart attack had cholesterol levels that would not put them into the high-risk category[1535]. In fact, low cholesterol is associated with mortality from heart disease, strokes, and cancer, whereas higher cholesterol has not been seen to be a bigger risk factor[1536,1537]. It's not cholesterol itself that's causing the problems but its oxidation and lipid peroxidation.

Cholesterol is an organic molecule produced by all animal cells. It's an essential component of the cell membrane and helps with its functioning. It's also a steroid hormone that promotes hormone, bile, and vitamin D synthesis. Because of its molecular substance, cholesterol doesn't mix with blood and it's carried around the body by lipoproteins. There are different types of them:

- **Very Low Density Lipoprotein (VLDL)** – VLDL delivers triglycerides and cholesterol throughout the body to be stripped off from energy or for storage.

- **Intermediate Density Lipoprotein (IDL)** – IDL helps the transport of cholesterol and fats but its density is between that of LDL and VLDL.
- **Low Density Lipoprotein (LDL)** – LDL carries energy through the bloodstream and directs nutrients into the cells.
- **High Density Lipoprotein (HDL)** – HDL collects unused cholesterol from the blood and brings it back to the liver for recycling.

HDL is considered the healthy one and LDL the bad one because it prevents the clogging of arteries and accumulation of cholesterol.

The build-up of plaque in the arteries is mostly driven by inflammation. If you suffer from higher inflammation and oxidative stress, then you'll also have more free radicals in the blood which can oxidize LDL. Oxidation of LDL by free radicals is associated with an increased risk of cardiovascular disease[1538]. It'll also promote lipid peroxidation because you're essentially creating heat that will damage vulnerable fats.

Free radicals and high inflammation damages the arterial lining called the endothelium, which the cholesterol then gets stuck to. Over the long run, this leads to the accumulation of plaques and increased risk of heart disease. Additionally, higher VLDL is considered a better predictor of heart disease risk than LDL[1539]. I'm not telling you that crazy high cholesterol is justified or completely safe. It's just important to recognize that some of it is more harmful. In most cases, it's not a problem as long as you don't suffer from additional inflammation or oxidative stress.

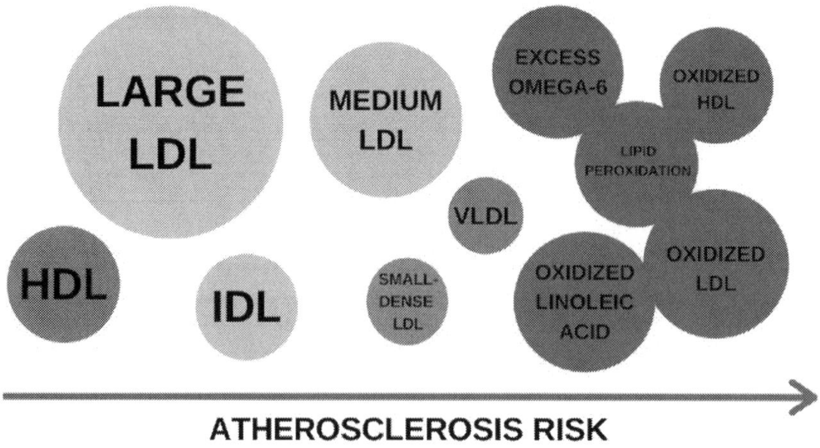

As a protective measure, you'd want to take care of your endothelial functioning, which describes how the endothelial regulates vascular tone and oxidative stress. It's the interior cell lining of blood vessels and arteries. Endothelial dysfunction is involved in the development of atherosclerosis and predicts vascular pathology because it impairs blood flow and decreases the ability of arteries to dilate[1540,1541]. Hypertension and hypercholesterolemia can contribute to endothelial dysfunction[1542]. Nitric oxide (NO) suppresses platelet aggregation, inflammation, oxidative stress, and improves the transport of lipids and cholesterol. It'll essentially promote blood flow and reduces the time these particles stay in your bloodstream. That's why regular exercise, sauna, and eating things like beetroot can help with better endothelial function.

A review paper on the heart-related health effects of cholesterol concluded: *"Epidemiological data do not support a link between dietary cholesterol and cardiovascular disease."*[1543] You could make the argument that high cholesterol leads to atherosclerosis because the plaques are created by cholesterol build-up. However, the root cause of the issue is inflammation and arterial scarring in the first place. If you'd have lower C-reactive protein, then cholesterol would simply be transported around the body by VLDL and if it's not needed for nutrition it'd be transited back to the liver by HDL.

Unfortunately, no matter how hard you try or how clean unoxidized food you eat, we're all susceptible to higher lipid peroxidation and inflammation than we were in the past. This is due to decreased nutrient quality, chronic stress, short sleep, air pollution, environmental toxins, and even invisible electromagnetic frequencies that also cause subliminal oxidative stress. That's why you shouldn't think you can just eat an unlimited amount of cholesterol and be fine because the higher your cholesterol the greater the chance of it becoming oxidized eventually. To prevent that, you should actively take some protective measures against lipid peroxidation and oxidative stress, in general, to prevent your cholesterol from becoming mutinous.

Chronic stress has been shown to raise serum lipids and cholesterol[1544]. We also know by now that hypercholesteremia can be caused by low thyroid functioning because it's not being converted into steroid hormones. That's why staying in a state of metabolic winter and starvation all the time can be bad for your longevity by predisposing you to weight gain and atherosclerosis. Even if you embrace the intermittent fasting or low carb lifestyle.

Here's how to reduce lipid peroxidation as well as protect against it:

- **Avoid Vegetable Oils and Trans Fats** – Canola oil, sunflower oil, rapeseed oil, and margarine are remarkably high in pro-inflammatory omega-6 fatty acids that cause insulin resistance and oxidative stress[1545]. They're high in PUFAs and almost guaranteed to oxidize. Olive oil is an exception as it reduces low-density lipoprotein uptake by macrophages and decreases the potential of lipid peroxidation[1546]. The antioxidants and polyphenols also protect it against the heat.
- **Vitamin C and E** – antioxidants can inhibit lipid peroxidation. Supplementing with vitamin C or vitamin E alone can reduce lipid peroxidation to a similar extent but combining them together seems to have no benefit beyond that of either vitamin alone[1547]. If you know you're about to be exposed to inflammatory or oxidized food, then you can mitigate the damage by taking vitamin E beforehand. It's found that a diet with more fruits and vegetables can increase the body's antioxidant status compared to a control group[1548]. But your body can also increase its antioxidant activity with things like exercise, sauna, and fasting.
- **Exercise** – Because intense exercise creates oxidative stress it will also increase lipid peroxidation and MDA

levels[1549]. However, it's going to result in lower levels of inflammation afterward. As long as you're not overtraining or combining it with other forms of allostatic overload.

- **Don't Over-Cook Food** – High-temperature cooking, deep-frying, grilling, and sauteing destroys the nutrients and antioxidants in food thus making them more pro-inflammatory. Repeatedly heated vegetable oils create lipid peroxidation[1550]. Most frying oils in restaurants are also inflammatory and made from vegetable oils. Taking a vitamin E capsule beforehand can mitigate some of the oxidation and heat-damage.

- **Carotenoids** – the compounds that give vegetables their bright color can inhibit lipid peroxidation as well as hemoglobin oxidation[1551,1552]. Another study found that carotenoids actually increased lipid hydroperoxide in membranes enriched with PUFAs[1553]. So, if you've been surviving on a diet of a lot of PUFAs like the majority of people have for years, then your membranes are already made of PUFAs and eating a lot of carotenoids may make things worse. Astaxanthin, on the other hand, had the opposite effect and reduced lipid peroxidation. It's probably due to the oxidation of PUFAs, which astaxanthin can counteract but carotenoids can't. You can get astaxanthin from algae, salmon, and pink seafood. Just make sure you're not going to overheat it. Eating more saturated fats as opposed to PUFAs will also re-compose your cell membranes towards being made of saturated fats, which are more heat-stable and not oxidized. Depending on your past dietary history, it can take several months and years to fully rid yourself of oxidized PUFAs stuck in your cell membranes.

- **Spirulina** – Spirulina lowers serum MDA, which is a marker of lipid peroxidation[1554]. This is partly because of the fatty acid profile of algae as well as its ability to chelate and detoxify compounds. Taking 2-3 grams of spirulina with meals would lower its lipid oxidation potential. I'm taking it with every meal.
- **Soymilk** – Consumption of soymilk for 28 days has been shown to reduce markers of lipid peroxidation like MDA in apparently healthy individuals[1555]. However, it also lowered micronutrient status and resulted in some micronutrient deficiencies. Soy is a known endocrine disruptor and can lower testosterone in men. So, I would focus on other compounds with fewer side-effects. In my opinion, it's not worth it and things like algae are already more effective.
- **Crushed Garlic** – Garlic contains a compound called allicin that gets activated when you crush it. One study found that swallowing whole garlic without damaging it had no effect on serum lipid levels but crushed garlic reduced cholesterol, triglycerides, blood pressure, and MDA[1556]. It also has anti-bacterial and anti-fungal properties.

- **Turmeric** – the active compound of curcumin or turmeric can lower lipid peroxidation by enhancing antioxidant enzymes like superoxide dismutase and catalse[1557]. This decreases reactive oxygen species and can repair the DNA damage that occurs because of lipid peroxides[1558]. Curcumin also chelates iron and thus reduces its potential for oxidation.
- **Carbon 60** or Buckminster Fullerine is about 100 times more powerful than vitamin C and it works as an anti-oxidant. It's been found that C60 protects against oxidative stress and lipid peroxidation[1559].

The danger with fats is that they tend to oxidize if used improperly. That's why heating some of them is out of the question. To not cause inflammation we need to be wise with how we use our fats. Here's a chart of the smoking points for different fats:

Fat Source	Smoke Point °C/F	Omega-6: Omega-3 Ratio
Unrefined Flaxseed Oil	107°C / 225 F	1:4
Unrefined Safflower Oil	107°C / 225°F	133:1
Unrefined Sunflower	107°C / 225°F	40:1

Oil		
Unrefined Corn Oil	160°C / 320°F	83:1
Extra Virgin Olive Oil	160°C / 320°F	73% monounsaturated, high in Omega 9
Unrefined Peanut Oil	160°C / 320°F	32:1
Semi Refined Safflower Oil	160°C / 320°F	133:1, (75% Omega 9)
Unrefined Soy Oil	160°C / 320°F	8:1 (most are GMO)
Unrefined Walnut Oil	160°C / 320°F	5:1
Hemp Seed Oil	165°C / 330°F	3:1
Butter	177°C / 350°F	9:1, Mostly saturated & monosaturated
Coconut Oil	177°C / 350°F	86% healthy saturated, lauric acid (has antibacterial, antioxidant, and antiviral properties). Contains 66% medium chain triglycerides (MCTs).
Unrefined Sesame Oil	177°C / 350°F	138:1
Lard	182°C / 370°F	11:1 high in saturated

Macadamia Nut Oil	199°C / 390°F	1:1, 80% monounsaturated, (83% Omega-9)
Refined Canola Oil	204°C / 400°F	3:1, 80% of Canola in the US in GMO
Semi Refined Walnut Oil	204°C / 400°F	5:1
Sesame Oil	210°C / 410°F	42:1
Cottonseed Oil	216°C / 420°F	54:1
Grapeseed Oil	216°C / 420°F	676:1, (12% saturated, 17% monounsaturated)
Virgin Olive Oil	216°C / 420°F	13:1, 74% monosaturated (71.3% Omega 9)
Almond Oil	216°C / 420°F	Omega-6 only
Hazelnut Oil	221°C / 430°F	75% monosaturated (no Omega 3, 78% Omega 9)
Peanut Oil	227°C / 440°F	32:1
Sunflower Oil	227°C / 440°F	40:1
Refined Corn Oil	232°C / 450°F	83:1
Palm Oil	232°C / 450°F	46:1, mostly saturated and monosaturated

Palm Kernel Oil	232°C / 450°F	82% saturated (No Omega 3)
Ghee (Clarified Butter)	252°C /485°F	0:0, 62% saturated fat
Rice Bran Oil	254°C / 490°F	21:1, Good source of vitamin E & antioxidants
Refined Safflower Oil	266°C / 510°F	133:1 (74% Omega 9)
Avocado Oil	271°C / 520°F	12:1, 70% monosaturated, (68% Omega-9 fatty acids) High in vitamin E.

Instead of using vegetable and seed oils, a much safer alternative is to cook with animal fats like lard or tallow. They're comprised of mostly saturated fat, which is much more heat-stable and won't become oxidized. Compared to something like avocado oil, they're also much cheaper and easier to come by. Cooking with olive oil at moderate temperatures is also fine because the antioxidants and polyphenols protect against lipid peroxidation. However, you'd have to know that the olive oil you do use actually has those polyphenols because most conventional products don't. Some brands mix olive oil with canola oil 50-50 so you have to read the labels and do your research. It's well worth it to be pre-cautionary because you're literally made of the fats you eat.

Lipofuscin

One of the main 'wear-and-tear' aging pigments is called lipofuscin. Lipofuscin is a golden brown-looking granular pigment that contains protein, lipid, and lysosomal accumulates[1560]. It's found in the liver, heart, muscle, and nerve cells where it deranges the processing of healthy cells.

Lipofuscin (LF) forms inside lysosomes that are organelles used to break down macromolecules and other particles via autophagy. Lipofuscin contains proteins, sugars, and fats but also metals, iron, copper, and zinc. It consists of oxidized proteins and unsaturated fatty acids[1561]. Your body does a pretty good job in cleaning house with regular autophagy. Unfortunately, all of the parts can't be broken down completely and that leads to the accumulation of lipofuscin. Why this happens is still unclear but the main reason seems to be oxidative stress[1562].

LF is cytotoxic because of its ability to incorporate redox-active transition metals, resulting in a redox-active surface[1563]. This promotes free radical stress and reactive oxygen species formation. Lipofuscin in heart cells is associated with heart disease and sudden cardiac death. However, that accumulation reflects more chronological aging rather than direct pathology[1564].

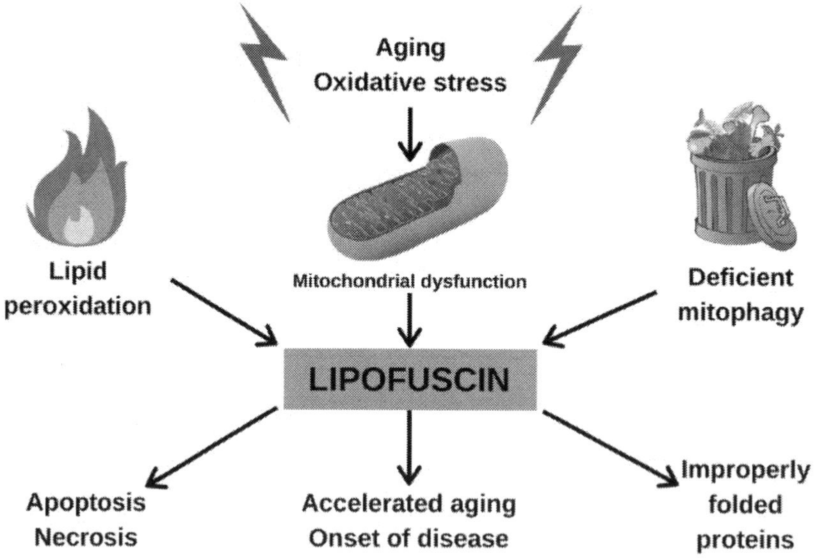

Melanin and lipofuscin are considered to be hallmarks of skin aging[1565]. Skin discoloration and spots indicate liver dysfunction and lipofuscin accumulation. Those yellowish-brown spots on people's hands and face are called liver spots but it's lipofuscin. LF also promotes neurodegeneration and other pathologies[1566]. It's a major risk factor for macular degeneration[1567].

Lipofuscin accumulates primarily in the lysosomes. Autophagosomes and lysosomes are not needed for forming LF but they constitute storage for LF aggregates[1568]. Therefore, a dysfunctional autophagy-lysosomal system plays a big role in lipofuscinogenesis[1569]. Decreased and fawlty autophagy promotes aging in many animals and cell models[1570][1571], which inevitably supports lipofuscin accumulation. Impaired mitochondrial autophagy or mitophagy also contributes to lipofuscinogenesis[1572]. Peroxidation of polyunsaturated fatty acids of cellular membranes by free radicals creates lipofuscin[1573].

Lipid peroxidation plays an important role in apoptosis or programmed cell death. The by-products of lipid peroxidation interact with membrane receptors to signal for apoptosis[1574,1575]. This can kill off or damage healthy cells in addition to the old ones. Oxidized fats have a complex relationship with autophagy. The products of lipid peroxidation can drive cellular dysfunction in an autophagic cell death way[1576]. Lipid peroxides can also induce lysosomal dysfunction and lipofuscinogenesis, which reduces autophagic activity[1577]. During myocardial ischemia and reperfusion, lipid peroxidation compromises autophagy signaling of AMPK and mTOR[1578]. Lipid peroxidation can activate mTORC1 by conjugating liver kinase B1 (LKB1), which inhibits AMPK, thus suppressing autophagy[1579].

It's another chicken and an egg type of problem – as autophagy slows down you form lipofuscin, which accelerates aging and disease, thus impairing autophagy even further, and accumulating even more LF. Lipofuscin inhibits the degradation of oxidized proteins by binding to proteolytic enzymes, leading to more oxidation. Age-related diseases can be very much alleviated with sufficient clearance and autophagy. It's one of the best preventative tools you have – to not get sick in the first place and to avoid the accumulation of lipofuscin that begins to cause more and more issues.

How to Avoid Lipofuscin

Lipofuscin causes aging and aging promotes the accumulation of lipofuscin. As your metabolism becomes less efficient with age, your ability to remove lipofuscin also diminishes. Thus you accumulate LF and accelerate aging.

Here are the main causes and sources of lipofuscin:

- **Oxidative Stress** - Free radical damage of any kind injures the cells and impedes their functioning. Being stressed out and inflamed promotes oxidative stress and excess cortisol, which can lead to disease and burnout[1580]. Chronically elevated cortisol can promote the creation of reactive oxygen species because your body is under stress.
- **Advanced Glycation End-Products (AGEs)**[1581] – Glycated proteins and fats that promote oxidative stress, aging, inflammation, and atherosclerosis[1582]. There's also ALEs or advanced lipid peroxidation end-products that damage proteins[1583].
- **Cell Senescence** – Senescent cells are zombie cells that spread inflammation and oxidative stress. They're essentially dead but they still stick around to injure everything around. This is a reflection of inadequate autophagy, oxidative stress, and aging[1584]. Although there's an increase in autophagy during senescence[1585], inhibiting autophagy can induce senescence through metabolic and proteostatic dysfunction. Studies also find that re-establishing autophagy may reverse senescence and restore

regenerative functions in geriatric satellite cells[1586]. So, autophagy gets activated during cellular senescence to clear out the junk.

- **High Iron Levels** – Too much ferritin promotes lipofuscin formation because it reacts with other macromolecules inside lysosomes[1587]. More of it will also be left in the lysosomes after degradation. Production of ROS during iron metabolism causes lipid peroxidation[1588]. Consuming PUFAs or oxidized fats in conjunction with high iron levels or an iron-rich meal exacerbates lipofuscinogenesis and additional oxidative stress. It essentially accelerates the rusting process, creating even more oxidation end-products.
- **Zinc Deficiency** – Zinc is important for immune and endocrine system functioning thus it supports lysosomal degradation and autophagy. Insufficient zinc leads to the accumulation of lipofuscin in the retinal pigment epithelium of pigmented rats[1589]. A large amount of zinc gets used by the immune system to fight infections as well as prevent immunosenescence[1590]. Zinc also has potential protective effects in iron overload[1591].
- **Excess Estrogen** – Estrogen is a sex hormone both men and women have but it's more relevant during female puberty and ovulation. Too much estrogen may amplify lipofuscin formation[1592]. It increases iron absorption as well. Birth control pills also increase estrogen. Estrogen is involved in the stress response and can amplify its effects

in the brain[1593]. That's why you may be more susceptible to stress-related conditions with too high estrogen, which itself will become an additional stressor.

- **Plastic Containers** – Environmental plastics are known xenoestrogens or compounds that imitate estrogen inside your body. This has been shown to cause precocious puberty at an unusually early age and other hormonal imbalances[1594]. Water bottles, plastic boxes, Tupperware cause oxidative stress by activating cytochrome P450 in the liver. Bisphenol A or BPA is a major contributor to the creation of free radicals and ROS[1595]. If you're still using conventional plastic containers, then I highly recommend you swap them out completely to either stainless steel, glass or BPA free options. Additional xenoestrogens include DDT, dioxin, endosulfan, PBB, PCBs, phthalates, zeranol, and countless others.
- **Pesticides and Insecticides** – Conventional fruit and vegetables that are sprayed with pesticides increase oxidative stress by activating cytochrome P450 enzymes[1596]. They also induce lipid peroxidation, stimulate free radical production, and disturb the body's antioxidant capabilities[1597]. In an animal model, exposure to gestational pesticides induces oxidative stress and lipid peroxidation in offspring that persists at adult age[1598].

Managing lipofuscin is relevant in the context of stress and inflammation because in most cases it's the consequence of overbearing environmental stressors and toxins. What's more, the accumulation of LF is going to inhibit every other cellular process and accelerates aging. Imagine how little work you can do if your desk is a complete mess – you might get something done but it's drastically slower than what you could do if everything were to be organized.

Here's how to avoid lipofuscin and reduce its allostatic load on your system:

- **Keep Your Iron Levels Optimal** – Normal ferritin levels range from 12-300 ng/ml for males and 12-150 ng/ml for women. You may get too much iron from supplements and animal protein. Alcohol increases the absorption of iron. Regular blood donations and menstruation can also lower iron. Things that chelate iron include coffee, green tea, spirulina, curcumin, and plant polyphenols[1599].
    - Copper helps to balance iron and hemoglobin metabolism. A protein called caeruloplasmin is essential for iron transportation and it needs copper to function properly[1600]. Symptoms of anemia could also then result from a copper deficiency even if the body has adequate ferritin levels. Copper is also important for the functioning of organs and the metabolism. You can get it from organ meats, oysters, spirulina, mushrooms, dark chocolate, and leafy greens.
- **Calorie Restriction** – Eating fewer calories causes less oxidative stress, lowers iron, and promotes autophagy[1601].

Intermittent fasting is also a potent stimulus for cell clean-up via autophagy. Mitophagy clears out pro-inflammatory mitochondria and organelles that spread senescence[1602].

- **Consume Autophagy Boosters** – There are certain autophagy boosting foods that can fight lipofuscin and incinerate visceral fat like berberine, turmeric, cayenne pepper, coffee, green tea, and ginger[1603]. For a full list, refer to the chapter about xenohormesis and plant hormetics.
- **Don't Smoke** – Smoking promotes oxidative stress, accelerates aging, and promotes atherosclerosis. Malondialdehyde is a compound that indicates oxidative stress by smoking but it's also found in rancid and overcooked food. Malondialdehyde causes autoxidation of PUFAs[1604]. So, even eating healthy fats like salmon can cause lipid peroxidation and lipofuscinogenesis if you're inflamed from the inside out.
- **Just Enough Sunlight** – Direct UV light creates oxidative stress in the skin and raises body temperature. This will also stimulate autophagy and lipofuscinogenesis if done in excess, not to mention skin aging. Generally, you don't have to be afraid of the sun. Just don't get sunburnt or spend the entire day out in the open. However, if you get sunburnt very easily, then it's probably because of the PUFAs in your fat tissue getting oxidized. In that case, I would be more conservative with exposure to heat and the sun because you're causing mild lipid peroxidation. It's

going to take a few months if not years of no rancid oils to clean things up but it's well worth it.

- **Cyclodextrin** – A compound used in drugs that stimulates autophagy and clears cellular junk, including lipofuscin[1605,1606]. Oral α-cyclodextrin has been shown to be safe in humans and it lowers small LDL particles[1607].
- **Vitamin E** – Deficient vitamin E promotes LF accumulation[1608] and it can reduce LF as well[1609]. Vitamin E also protects against the oxidation of vitamin A, which reduces lipofuscin[1610]. Get your vitamin E from low PUFA foods like vegetables not nuts or seeds. Taking a vitamin E capsule before eating PUFAs or oxidized fats would also effectively reduce the negative side-effects of lipid peroxidation as well. Just make sure you're not using a supplement that's made of sunflower oil or the like because that would kind of defeat the whole purpose.
- **Glutathione** – Reduced glutathione leads to lipofuscin formation[1611]. You stimulate glutathione with sulfur-rich foods like cruciferous vegetables, eggs, meat, and liposomal glutathione supplements. Fasting and exercise do so as well.
- **Coenzyme Q10** – CoQ10 decreases oxidative stress and improves mitochondrial functioning[1612,1613]. CoQ10 is a major nutrient for energy production and cellular homeostasis. You can get it from organ meats, dark chocolate, raw cacao, beef, and supplements.

- **Creatine** – Creatine has been shown to reduce free radicals and lipofuscin in the brains of rats[1614]. It also promotes cognition and muscle function. Taking 3-5 grams of creatine a day is great for methylation support, exercise performance, recovery, hypertrophy, and anti-aging.

You can't avoid lipofuscin and lipid peroxidation completely because it'll happen as a by-product of aging and living itself. What you can do is just reduce your exposure to the unnecessary aspects of it that yield no hormetic benefit.

Advanced Glycation End-Products

When you bite into an apple and leave it out in the air then you'll eventually see it turning brown and rotten. That's glycation in action as the sugar and fructose in the fruit becomes oxidized when exposed to oxygen.

Advanced Glycation End Products (AGEs) are proteins that become glycated after exposure to sugar. Advanced Lipoxidation End Products (ALEs) are glycated lipids and fats. They're the same thing with minor differences.

AGEs are thought to promote aging, inflammation, and worsen many diseases such as diabetes, atherosclerosis, chronic kidney disease, and Alzheimer's[1615]. They can be formed either internally or obtained from outside sources. Low levels of AGEs are generally fine and not an issue because the body has its defense mechanisms and antioxidant systems like glutathione to eliminate them. Chronic AGE accumulation, however, will promote inflammation and cellular damage[1616]. AGEs are found in the progression of many age-related diseases, such as Alzheimer's[1617], cardiovascular disease[1618], renal dysfunction[1619], diabetes, insulin resistance, and stroke[1620].

Factors that determine the formation of AGEs include how fast the proteins get glycated, the degree of hyperglycemia or how high your blood sugar is, and the extent of oxidative stress in the body[1621, 1622,1623]. AGEs essentially oxidize your healthy tissues, which leads to more inflammation and programmed cell death. This speeds up the aging of your skin, weakens tendons, damages the mitochondria, and jeopardizes all aspects of cellular functioning. Getting exposed to oxidative stress and AGEs can be seen in the form of dark pimples and spots on your skin. You'll also get more wrinkles and look older than you are.

AGEs directly cross-link certain proteins like collagen to promote vascular stiffness and thus affect vascular structure negatively[1624]. They also signal more oxidative stress and promote key proinflammatory cytokines. What's worse, AGEs cause glycation of LDL cholesterol, which promotes its oxidation[1625]. Oxidized LDL is a major risk factor in atherosclerosis[1626]. This is a serious danger and should be avoided at all costs. When AGEs engage with the receptor for AGEs called RAGE, it'll upregulate the transcription factor nuclear factor-kB (NF-kB), which causes endothelial permeability[1627]. This creates an inflamed environment inside the body that has serious implications on cardiovascular health.

The biggest source of AGEs tends to be diet because of the quality of ingredients and cooking methods used. Especially processed food that combines carbs and fats as well as high-temperature grilling, broiling, or deep-frying. Foods high in fat and protein are more susceptible to AGE formation during cooking than vegetables or plants[1628]. Animal foods, in particular, like meat, eggs, and fish tend to glycate quite easily if over-cooked. However, you absorb only a very small amount of AGEs from dietary sources. It's also been found that vegetarians have higher concentrations of AGEs compared to non-vegetarians[1629]. Therefore, the AGEs you get from food aren't nearly as important as the ones you create during internal metabolic reactions.

You'd think that processed food like french fries and grilled burgers create a lot of AGEs but this is not the case in studies. In fact, consumption of fruit, fruit juices, cereal, and sweets are much more associated with elevated AGEs. Probably due to eating those sugars with other foods that have protein or fats. The '*fructositis*' hypothesis states that a higher fructose to glucose ratio promotes an intestinal environment that promotes the formation of AGEs and other proinflammatory cytokines internally. Glycation of fructose in early research was also drastically underestimated because of inaccurate measuring techniques[1630].

A diet high in AGEs is considered to start above 15 000 kilo units of AGEs per day[1631]. Here's a list of the amount of AGEs in common foods:

- Almonds (30g): 1642 kU
- Avocado (30g): 473 kU
- Cashews (30g): 2019 kU
- Peanut Butter (30g): 2255 kU
- Beef (90g): 1468 kU
- Chicken, boiled in water (90g): 1011 kU
- Chicken, roasted then BBQ (90g): 7922 kU
- Fried Bacon (13g): 11905 kU
- Raw Lamb (90g): 743 kU
- Raw Salmon (90g): 472 kU
- Broiled Salmon (90g): 3012 kU
- White Cheese (30g): 2603 kU
- 1 fried egg: 1,237 kU
- Whole Wheat Bread (30g): 16 kU
- 1 tablespoon of cream: 325 kU
- ¼ cup (59 ml) of whole milk: 3 kU
- Celery (100g): 43 kU
- Grilled Vegetables (100g): 226 kU

Dietary AGEs will make things worse and affect AGEs negatively but other factors that create a glycating environment in the body like chronically elevated blood sugar and inflammation are much more important. That's another reason why to keep your glucose and insulin lower.

How to Reduce AGEs

Here's what to do to prevent and reduce AGEs:

- **Avoid High Fructose Corn Syrup (HFCS)** – Everything from sauces, sodas, juices, TV dinners, frozen pizzas, canned vegetables, and candy has some HFCS. It promotes hyperglycemia, insulin resistance, glycation, and obesity[1632]. Consuming fructose-sweetened beverages has been shown to increase visceral adiposity, high lipids, and decreased insulin sensitivity[1633]. Amongst adults between the ages of 45-59, HFCS, fruit drinks, and apple juice is associated with prevalent coronary heart disease[1634].
- **Don't Eat Too Much Fructose** – In rats, long-term fructose consumption accelerates glycation[1635]. Fruit that's higher in fructose and lower in glucose is more pro-AGE. Things like apples, pears, mangos, oranges, bananas are all quite high in sugar. A few servings of fruit every once in a while is fine but it really shouldn't be eaten in excess.
- **Don't Combine Fats and Carbs** – Eating fat together with carbs promotes lipoxidation and ALE formation, which are equally as bad as AGEs. It's plausible that dietary AGEs

are not that big of a deal if you have adequate glutathione levels to mop up the debris. However, peroxidized PUFAs are a major source of ALEs or and AGEs. Fish and chips just became even more harmful to your health. Packaged and processed food generally have added sugars and fats all of which will turn into AGEs when eaten. They're also prone to oxidation because of the processing and heating. TV dinners and almost everything in the frozen food section is quite horrible if you know the details.

- **Avoid High-Temperature Cooking** – Temperatures above 120 C (~248 F) speed up the formation of AGEs, especially if you cook with inflammatory vegetable oils. Slow cooking is generally a safer alternative but whenever you do choose to grill or sear something then do it only briefly.
- **Limit Charred Meat** – Overcooking or barbequing animal protein will create AGEs like heterocyclic amines (HAs), and polycyclic aromatic hydrocarbons (PAHs). High intake of HAs from meat is associated with oxidative stress and colon cancer[1636]. There's no evidence that it has a hormetic effect by stimulating Nrf2 but in low doses it can increase glutathione S-transferase in the liver[1637]. Fortunately, marinating meat before cooking reduces their AGE formation and HA content by up to 90%[1638]. Generally, I wouldn't recommend eating BBQ or grilled food on a regular basis and use it only as an occasional treat.

- **Keep Exercising** – Regular moderate exercise reduces AGEs and lowers oxidative stress[1639]. Too much exercise, however, can promote inflammation and stress. It'll become more damaging when combined with other stressors like sleep deprivation or an inflammatory diet.

Arguably, some AGEs can be a positive hormetic stressor by upregulating the body's antioxidant pathways like glutathione and superoxide dismutase. However, there are some things like HFCS and deep-fried food that just don't have a beneficial effect. It's just not worth it and the collateral damage would be much greater than the small benefit.

The clearance of AGEs is done through the proteolysis of AGEs, which is the breakdown of proteins. This produces AGE peptides and AGE free adducts that will be released into the plasma and excreted through urine. Compounds that help to clear AGEs are vitamin C, alpha-lipoic acid[1640], taurine[1641], aspirin[1642], carnosine[1643], metformin, resveratrol[1644], curcumin[1645], and polyphenols. Phytates found in grains, legumes, and nuts can chelate metal cautions and decrease the formation of AGEs[1646]. However, they can cause inflammatory issues and oxidative stress in other aspects. It's a double-edged sword. Ancestral grains and whole grains that actually have those phytates and other phytonutrients are fine in moderation but the generic white flour bread doesn't even have that. There's just the carbs and gluten – the worst combo.

Poor kidney functioning and dysfunctional mitochondria make it harder to excrete AGEs. This is a negative feedback loop as AGEs inhibit their work in the first place. To fix your kidneys and clear out the liver, you want to cure insulin resistance and the best way of doing that is to fast for extended periods. Daily time-restricted eating can help you to lose weight and lower blood sugar but to fix insulin problems completely you'd have to fast a bit longer.

Liver detox

Your liver is one of the most important organs of your body because it monitors nutrient status, regulates energy pathways, and also mediates the removal of toxins. Unfortunately, in the modern environment, our bodies are exposed to countless environmental toxins, chemicals, metals, and pesticides all of which damage our health and impair liver functioning.

Liver detoxification or cleansing is the process of releasing accumulated toxins from the body and eliminating them. There are two stages to this: liver detox phase 1 and phase 2 both of which have important roles.

- **Phase One consists of oxidation reduction and hydrolysis.** It's initiated by a group of enzymes called the cytochrome P450, which are located in the membrane system of liver cells called hepatocytes. They help to neutralize substances like alcohol and caffeine by converting toxins into less harmful ones.

- **Phase Two is called the conjugation pathway** where liver cells bind the toxin with another substance to make it water-soluble. This enables the toxin to be excreted via urine, sweat, or bile.

To not accumulate the same toxins you're trying to eliminate, you want to have both of your liver detox phases to work properly and in conjunction with each other. Phase 1 detox converts the toxins into less harmful ones. This process also produces free radicals, which can damage the liver if in excess. Some substances can also turn carcinogenic when in over-activation. If phase 2 doesn't kick in and eliminate the compounds that phase 1 produces, the toxins can start to build up. This will lead to DNA damage, further toxicity, and inflammation.

- **Substances that may overactivate phase 1 detox** are caffeine, alcohol, exhaust fumes, pesticides, dioxin, saturated fat, and sulphonamides.
- **Co-factors that support phase 1 detox** are niacin, NADH, riboflavin, magnesium, iron, sulforaphane from cruciferous vegetables.
- **Required nutrients for phase 1 detox** are folic acid, B3, B6, B12, vitamin A, C, E, D3, and N-acetyl cysteine.
- **Compounds that inhibit phase 1 detox** include grapefruit and curcumin because of downregulating P450 activity.

Phase 2 detox the conjugation phase includes many antioxidant pathways, such as glutathione, sulfate, glycine, and glucuronide conjugations. It enables the liver to turn toxins, drugs, and other harmful substances into a water-soluble form that then can be excreted.

Here's how to promote phase 2 detox:

- **Sulphur Foods** like cruciferous vegetables, garlic, onions, eggs, and leeks support phase two.
- **Glutathione** is the master antioxidant that can be stimulated by cruciferous vegetables, red meat, or taken as liposomal glutathione for most effect.
- **Required nutrients for phase 2 detox** include glutamine, glycine, carnitine, lysine HCL, taurine, MSM, N-acetyl cysteine, and sulfur.

If phase 1 causes an overload of toxins and phase 2 can't keep up, then the toxins may get accumulated and bound into your body's fat. This can cause severe hormonal imbalances, weight loss issues, brain dysfunction, and disease. That's why it can be better to go through a slower detox phase 1 as to not cause an accumulation.

Liver Detox Steps

Here are the steps of cleansing your liver:

- **Remove the Toxins** – Stop eating inflammatory foods and processed substances. It's also best to not get exposed to other chemicals, heavy metals, gas fumes, paint, etc. One study on roundworms found that NAD+ supplementation protects against methylmercury toxicity[1647]. So, keeping your NAD elevated with lifestyle or supplements can be protective against environmental toxins pro-actively.
- **Consume Phase One Detox Foods** – Initially, you want to convert the toxins you already have into less harmful ones. This means no alcohol, no coffee, no fatty foods, no pesticides. Instead, eat things like cruciferous vegetables, celery, bitter herbs, chlorella, artichoke, beetroot, lemons, leafy greens, green tea, milk thistle, and wild fish. Do this for about a few days up to one week.
- **Take Activated Charcoal in Between Meals** – This helps to bind the toxins and eliminate them. Other binders you can use are silica, chlorella, bentonite clay, and enterosgel. Be wary of not taking it with food because it'll decrease the absorption of other key nutrients you get.
- **Do Some Intermittent Fasting** – Skipping meals puts less stress on digestion and promotes healing. It can also support detox. Low levels of GADD45β in humans is accompanied by increased fatty liver and elevated blood

sugar. Fasting stimulates GADD45β production, which repairs the metabolism[1648].

- **Week Two Consume Phase Two Detox Foods** – After a day of phase 1 foods you want to introduce phase two foods like turmeric, cruciferous, garlic, onions, eggs, and other anti-inflammatory foods.
- **Monitor Your Symptoms** – If you start to feel worse and nauseous during the first few days of phase 1, then you should already move on with phase 2 foods to flush out the toxins.
- **Get a Sweat On** – Sweating helps to eliminate toxins and harmful substances. Take a frequent sauna, exercise daily, do yoga, and try to bring up your core temperature. It was thought that you can't sweat out toxins or heavy metals and it's essentially a bogus idea. However, recent research shows that you actually can and it's one of the most effective ways of doing so[1649].
- **Re-Introduce Nutrient Density** – After about a week or two of phase 1 and 2 detox, you'd want to re-introduce some of the other nutrient-dense foods that help with recovery. They include organ meats, fish, eggs, red meat, tubers, and some fruit.

Here are some foods that promote liver detox:

- **Artichoke** contains phenolic compounds that stimulate bile production and protect the liver[1650].
- **Beetroot** contains a pigment called betalains, which helps to lower chronic inflammation and repairs cells[1651]. It also has betaine and pectin that support the elimination of toxins[1652].
- **Bitter herbs** support phase 1 and phase 2 detox by improving digestion and releasing bile.
- **Broccoli Sprouts** have sulforaphane, which activates one of the main antioxidant systems Nrf2[1653]. One study found that broccoli sprouts can pick up pollutants from the bloodstream and flush them out via urine[1654]. Broccoli also protects against liver damage and fatty liver disease[1655]. Cabbage contains glucosinolates that have chemoprotective effects linked with phase 1 and phase 2 detoxification.
- **Chlorella** can bind to heavy metals and toxins that cause inflammation[1656].
- **Citrus fruit** like lemons, oranges, and tangerines contain a compound called D-limonene, which can help to reverse liver damage[1657]. There might be something there to drinking lemon water.
- **Vitamin C** is a water-soluble antioxidant that can lower oxidative stress and flush out some toxins during phase 2.

- **Dark Berries** like blueberries, elderberries, bilberries, cranberries, blackberries, currants, and cherries are full of antioxidants and anthocyanins that reduce oxidative stress.
- **Dandelion** root and leaf promote the removal of reactive oxygen species and free radicals.
- **Glutathione** is the master antioxidant in your body. It can be activated by sulfur-rich foods like cruciferous, eggs, bone broth, and garlic.
- **Green Tea** has many antioxidants and also boosts phase 1 and phase 2 detoxification.
- **Leafy Greens** like arugula, kale, spinach, collard greens, help to remove toxins from the blood because of their chlorophyll content[1658].
- **Milk Thistle** is a herb with many antioxidant and anti-inflammatory properties. It also boosts glutathione and liver detox[1659].
- **Turmeric Extract** has been found to protect against liver damage in animal studies[1660]. It also stimulates bile production that helps with digesting fats[1661].
- **Wild Fish** like salmon, trout, herring, mackerel, and sardines are rich in omega-3 fatty acids that lower inflammation and reduce fatty liver[1662]. Bigger fish like shark and tuna have more mercury and other heavy metals because they're larger and have spent more time in polluted waters whereas smaller ones like sardines and salmon have less[1663]. The sad truth is that virtually all fish

today is polluted to a certain extent aside from maybe a few places of pristine nature.

- **Be Wary of Fish Oil** - In 2012, 35 fish oil supplements were tested for PCBs and they all had trace amounts of PCBs with 2 samples exceeding the allowed limit of 3 picograms per gram of fish oil[1664]. Microalgae oil is an alternative to fish oil and it has a lower risk of pollutant exposure[1665]. Dioxins and polychlorinated biphenyls (PCBs) can be carcinogenic at low levels of exposure over time.

Liver cleansing and detox are something you should do every once in a while to maintain proper liver functioning and health. It's a good idea to have a 1-2 week liver cleanse a few times a year.

Gut-Brain Axis

Science has now discovered that microbial life inside living organisms has played a crucial role in shaping the evolution of said organism. In fact, the mitochondria are descendants of bacteria that millions of years ago developed a co-existence with our ancestral bodies. This has led to the suggestion of the concept of a Microbe-Gut-Brain (MGB) Axis[1666].

The MGB Axis is this network of biochemical signaling between the gastrointestinal tract (GI) and the central nervous system (CNS). It includes the enteric nervous system (ENS), the endocrine system, the hypothalamic-pituitary axis (HPA), the autonomic nervous system, the vagus nerve, the endocrine system, and the gut microbiome[1667].

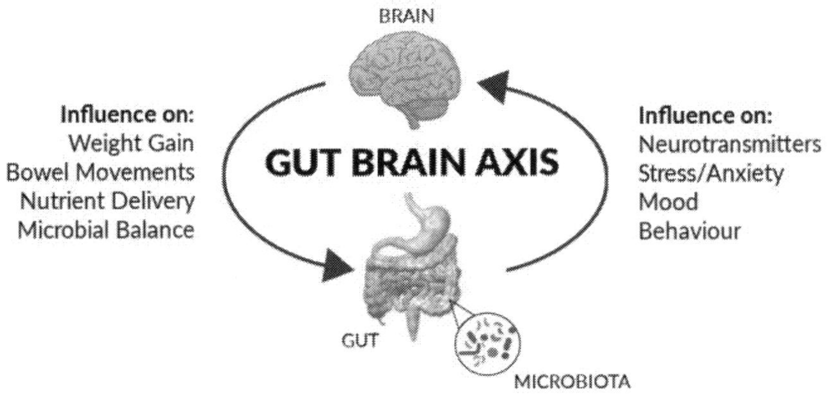

Your immune cells, muscle cells, cells of the gastrointestinal tract are all mediators of the neuro-immuno-endocrine system that are influenced by both the brain and the gut microbiome. In fact, it's been thought that the microbiome plays a much more influential role in the state of your being than the brain[1668]. It's like a super-complex ecosystem that consists of trillions of microorganisms and bacteria. The 'thing' you call 'I' is comprised of this collective consciousness of many living organisms and cells inside your body.

Your gut lining is the most intimate point of contact between you and the world, which is why it's important to keep away potential toxins and foreign intruders. Intestinal permeability or leaky gut was considered a made-up condition for years and a myth. However, research shows it might be implicated in many diseases like Crohn's, celiac, arthritis, diabetes, IBS, obesity, and allergies[1669, 1670]. Intestinal permeability describes food particles and other matter inside the gut passing through the gut lining into the bloodstream[1671]. This upregulates the body's antibodies and spreads inflammation, causing potential autoimmune issues.

The blood-brain barrier (BBB) is what separates the bloodstream from the brain. It's made of tight junctions similar to the ones in the gut. This network is supposed to select which particles are allowed into the brain because it's your central control room. A dysfunctional or permeable BBB can allow inflammatory substances and toxins to enter the brain, causing more inflammation, neuronal stress, and poor cognition. It often walks hand-in-hand with intestinal permeability, hence the term leaky brain.

If you've ever suffered from frequent brain fog, bloating, forgetfulness, chronic fatigue, mood swings, indigestion, sugar cravings, or poor concentration, then it's probably due to some gut problems. Dysfunctions in the BBB reflect intestinal permeability because research has shown a clear relationship between changes in the gut flora and brain function[1672].

How to Fix Leaky Gut and Brain

Here's how to keep your gut lining healthy to prevent unwanted substances from entering the blood:

- **Be careful with food sensitivities**. If you react negatively to some foods like lectins or grains, then it will cause more inflammation in the gut, promoting intestinal permeability. Potential allergens include gluten, soy, dairy, eggs, lectins, oxalates, and nuts. Whether or not you should hormetically microdose with these ingredients depends on your level of sensitivity and how you react. If you have no negative symptoms after consumption, then you're fine. Just pay attention to any potential downside. If you get wiped out or develop serious autoimmunity against any food, then it's better to just avoid it. Alternatively, consult with a gut specialist or functional medicine practitioner.

- **Avoid gluten and grains**. Gliadin, which is a protein found in wheat increases another protein called zonulin, which makes the gut more permeable[1673]. Zonulin is the only substance that regulates BBB and gut tight junctions[1674]. Serum zonulin has been found to be much higher in people with celiac disease compared to healthy controls[1675]. Conventional bread and grain products are pretty high in gluten and zonulin but you can reduce that amount with fermentation and sourdough practices. There's nothing wrong with eating a bit of homemade bread every once in a while, if you can tolerate it.

- **Eat probiotic foods that populate beneficial bacteria in the gut.** Sauerkraut, fermentations, kimchi, kombucha, raw kefir, and yogurt have billions of *Lactobacillus and Bifidobacterium* that can promote a healthy microbiome. *Akkermansia* has also shown to protect against obesity and type-2 diabetes[1676]. You can get them from polyphenol-rich foods. Prebiotics and dietary fiber can improve the integrity of the gut lining and reduce inflammation[1677]. Short-chain fatty acids like butyrate have anti-inflammatory properties[1678]. They repair the gut lining in the colon.

- **You should also eat foods that promote healing of the intestinal lining** such as bone broth, the amino acid glutamine, animal fats, and stews[1679]. A diet with vegetables, fish, and meat benefits the gut microbiome and its metabolome[1680]. The microbiome creates butyrate from digesting fiber and vegetables, but you can also obtain it from animal fats and amino acids to a certain degree. However, too much fat in the diet can also be problematic. A study on mice found that a diet composed of 40% saturated fat from cocoa butter increased BBB permeability[1681]. Fortunately, adding things like niacin, aged garlic extract, alpha lipoic acid (ALA), curcumin, astragalus, and cruciferous vegetables protected against that.
- **Focus on fixing dysbiosis.** You should work with a functional medicine practitioner to help you guide through the process and help you with the potential issues that may arise. I also recommend you to read Dr. Ruscio's book Healthy Gut Healthy You.
- **Avoid anti-biotics, chlorinated water, household chemicals, processed foods, detergents, and pesticides** because they wreak havoc to the microbiome. There's a lot of environmental toxins we get exposed to on a daily basis, which is why it's important to keep our bodies detox pathways active.

- **Add daily periods of time-restricted feeding.** Extend your overnight fast until at least 14-16 hours to give your digestion a break and help with the assimilation of nutrients. Fasting promotes the diversity and dynamics of the microbiome, which is determined by feeding and fasting cycles of the host[1682]. At the same time, it will still starve off some of the pathogens, viruses, and bad bacteria. Furthermore, ghrelin the hunger hormone has been shown to prevent BBB dysfunction after traumatic brain injury[1683].
- **Exercise can improve gut microbial composition independent of diet[1684]**. Part of it may have to do with promoting cellular clearance through autophagy but part of it may also have to do with increased neurotrophic factors in the brain.
- **Get enough sleep**. REM sleep regulates BBB function because that's the time most of neurological processing occurs[1685]. In mice, sleep restriction impairs BBB functioning by down-regulating electrical signaling and membrane potential[1686]. Short or poor sleep promotes oxidative stress, insulin resistance, and raises blood pressure, which is implicated in the pathophysiology of BBB function[1687].
- **Skip alcohol**. Alcohol-induced oxidative stress in brain endothelial cells causes BBB dysfunction[1688]. Drinking increases ROS and expression of CYPE1, which is an enzyme that converts ethanol into acetaldehyde. This damages the tight junctions in both the gut and the brain.

Instead of alcohol, consume coffee and teas as caffeine has been shown to prevent BBB dysfunction in a rabbit model of Alzheimer's[1689].

Your mind is only as sharp as the health of your body. With leaky tight junctions, rancid membranes, glycated blood vessels, liver spots, and oxidized cells you'll never reach your fullest potential. It'll also undermine any attempts to stay antifragile and resilient against stress.

Gut and Membrane Hormesis Protocol

- Avoid vegetable oils and trans fats. Even healthy fats like fish can become oxidized if heated. Don't cook with PUFAs and use saturated fats instead. When eating out in restaurants, take spirulina and vitamin E to protect against the lipid peroxidation.
- Fish oil supplements are a potential source of lipid peroxidation. Especially the conventional ones that sit on store shelves for months. Keep all your fats away from sunlight, heat, and oxygen because they'll become oxidized otherwise.
- High cholesterol in of itself isn't as bad as oxidized cholesterol. If you have high cholesterol but low inflammation, then the atherogenic potential of that cholesterol is much lower. However, given the amount of environmental stressors we're exposed to, it's only a matter

of time before anything in your blood suffers from some oxidation. If you have high cholesterol but low inflammation, then you may suffer from low thyroid. Breaking ketosis is a useful tool for reversing that.

- Manage your iron levels with curcumin, coffee, green tea, and spirulina to prevent lipofuscin accumulation. Don't create unnecessary oxidative stress either because it'll promote lipofuscinogenesis.
- To avoid AGEs don't eat sugar or too much fructose together with fat. Don't combine them with other macronutrients either. Charred or fried food will be much higher in AGEs than steamed or cooked food. Having your detox and glutathione pathways active will help to clear AGEs.
- Keep your gut lining sealed by consuming enough amino acids and protein. Figure out which foods you react negatively to and limit their consumption. Feeling bloated after meals, indigestion, undigested food particles in your stool, brain fog, joint pain, and inflammation may indicate intestinal permeability.

Chapter Thirteen: Of Circadian Rhythms and Sleep Optimization

"There is even rhythm in being empty."

Miyamoto Musashi

Maintaining homeostasis and allostasis requires the ability to predict and quickly respond to the ever-changing conditions of the environment. That's why virtually all lifeforms have developed internal diurnal rhythms that are connected to the day and night cycles. They're called circadian rhythms and they allow organisms to adapt to the environmental changes that happen on a frequent basis to optimize survival, reproduction, and calorie intake.

The term '*circadian*' was first coined by Franz Halberg in 1959 who derived it from the Latin words '*circa*' (about) and '*dies*' (day)[1690]. In 1977, the International Committee on Nomenclature of the International Society for Chronobiology adopted the official definition for circadian rhythms, which goes like this:

> *Circadian: relating to biologic variations or rhythms with a frequency of 1 cycle in 24 ± 4-h; circa (about, approximately) and dies (day or 24 h)*[1691].

In 2017, three scientists Jeffrey C. Hall, Michael Rosbash, and Michael W. Young were awarded the Nobel Prize in Physiology or Medicine for their discoveries of molecular mechanisms controlling the circadian rhythm[1692]. The men identified a gene in fruit flies that controls their circadian rhythms. They named this gene *period,* which encodes a protein called PER. PER accumulates during the night and degrades during the day, thus it oscillates over a 24-hour cycle, in synchrony with the circadian rhythm.

The mammalian circadian system is composed of a central pacemaker in the brain's hypothalamus called the suprachiasmatic nucleus (SCN) and multiple peripheral clocks in other tissues, such as the liver, pancreas, heart, and GI tract. By now we know that every cell and organ in the body has its own circadian clock and they're all coordinated to work together with the SCN[1693]. The main circadian clock consists of the transcription factor BMAL1-CLOCK that activates circadian genes like *Per 1-3*[1694]. Low levels of *Per1* and *2* are associated with gastric cancer[1695]. CLOCK also regulates the circadian rhythms with histone acetylation and deacetylation with BMAL1[1696]. It shows intrinsic histone acetyltransferase (HAT) activity which turns on or off genes.

Early research in humans speculated that most people's circadian rhythm is closer to 25 hours when isolated from external cues[1697]. However, these results were misleading because the participants were exposed to artificial light. In 1999, a Harvard study found that the human circadian rhythm is about 24 hours and 11 minutes, which is closer to the solar day[1698]. This 24-hour period is referred to as the free-running of the circadian rhythm. Decreasing RNA methylation has been shown to elongate circadian rhythms[1699].

Disruptions in circadian rhythms are linked to obesity, diabetes, cardiovascular disease, Alzheimer's, and cancer[1700]. There's also a lot of evidence to show that they affect aging and longevity in many ways[1701,1702]. CLOCK/BMAL1 regulates the transcription of antioxidant systems like Nrf2 as well as nutrient-sensing for the metabolism[1703]. Research in many species suggests that there's a negative relationship between lifespan and deviation of the free-run circadian period from 24-hours[1704,1705]. Basically, disrupted circadian rhythms can shorten your life and accelerate aging.

Circadian rhythms affect epigenetics and vice versa. Things like DNA methylation, RNA encoding, histone acetylation, and deacetylation all regulate clock genes in some shape or form[1706]. This leads to the epigenetic expression or shut down of particular genes that affect disease as well as longevity. Essentially, if your circadian clocks are working in sync, then your body will work like clockwork (pun intended) but if they're misaligned, it'll result in more oxidative stress and inflammation.

An organism whose circadian clock corresponds with its surroundings is *entrained*. Their established rhythm persists even when outside signals disappear and no stimulus is received. If your circadian rhythms are working properly, then isolating you in a bunker with either constant light or darkness wouldn't drastically disrupt your body's rhythmic physiological processes because the clocks have already been wired to function rhythmically. Naturally, you will experience off-sets due to traveling, jet lag, or staying up on some nights but they shouldn't become an issue as long as you've entrained your internal clocks to bare that allostatic load.

However, to be antifragile, you shouldn't be solely dependent on being always in sync with the circadian rhythms. In a perfect world scenario, we could just go to bed and wake up at the same time every day for the rest of our life. It'd be amazing and very healthy. Unfortunately, circadian mismatches are becoming a more frequent part of our lives whether that be due to binge-watching TV in the evening, traveling across timezones, or shift work. To not be wiped out by these disruptions, we ought to condition our bodies to adapt to them in advance. This doesn't mean staying up all night or becoming nocturnal. Instead, you should just teach yourself to sleep in unfavorable situations and at random times. It'll make you able to fall asleep and trigger the body's rejuvenating systems more efficiently anywhere you'd like. More on that later in this chapter.

Here are some key points in the typical 24-hour cycle:

- 6 A.M. Cortisol levels increase to wake you up
- 7 A.M. Melatonin production stops
- 9 A.M. Sex hormone production peaks
- 10 A.M. Mental alertness levels peak
- 2:30 P.M. Best motor coordination
- 3:30 P.M. Fastest reaction time
- 5 P.M. Greatest cardiovascular efficiency and muscle strength

- 7 P.M. Highest blood pressure and body temperature
- 9 P.M. Melatonin production begins to prepare the body for sleep
- 10 P.M. Bowel movements suppressed as the body quiets down
- 2 A.M. Deepest sleep
- 4 A.M. Lowest body temperature

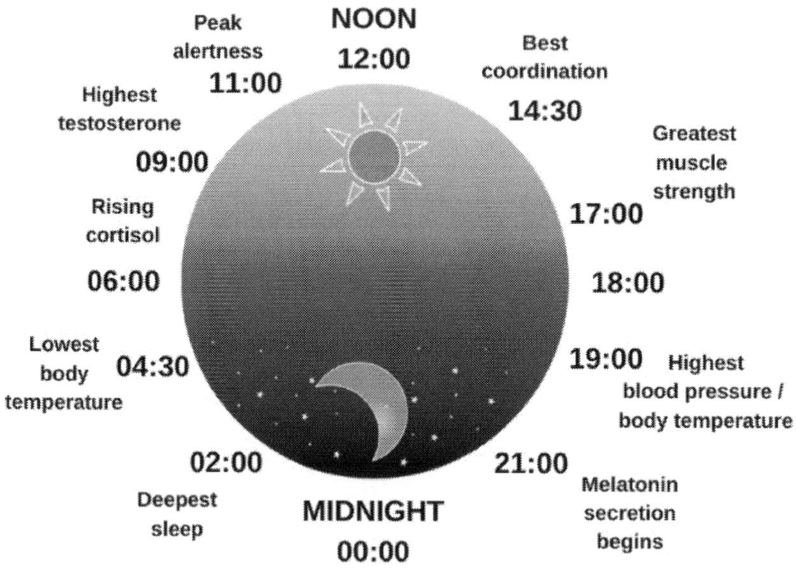

Of course, these numbers may be different from person to person and their environment. This is just a general overview of how and when the body does certain things.

What Affects the Circadian Rhythms

There are many signaling factors that regulate circadian rhythms such as light, temperatures, magnetic frequencies, movement, and food. Most of the circadian signaling is transmitted through your eyes. When light enters the retinas and gets transmitted into the brain it stimulates the suprachiasmatic nucleus (SCN). The SCN responds to light cues and darkness as well as magnetism. Other signaling factors include temperature, movement, and food, which is why there are multiple clocks in different organs.

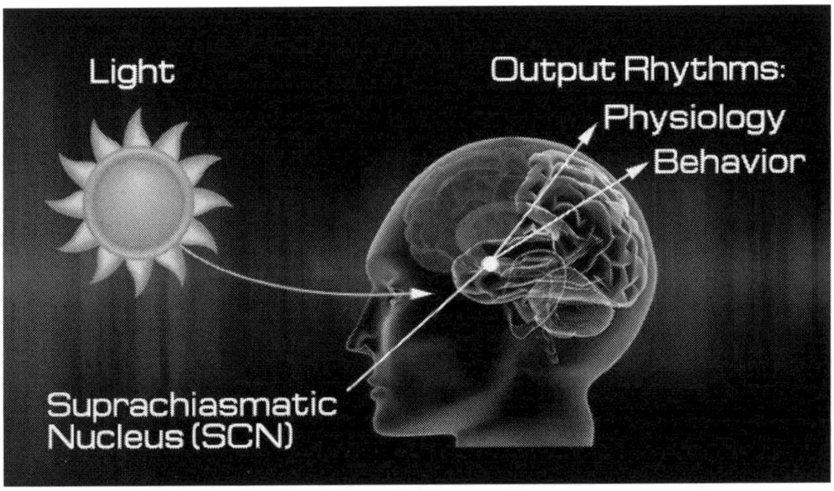

Light directly affects the production of melatonin 'the sleep hormone' also called 'hormone of darkness'[1707]. It has a critical role in the regulation of sleep-wakefulness cycles and most of it gets secreted during the night. A newborn baby doesn't produce their melatonin until three months old. From then on, its production peaks at adolescence and settles down in adulthood. As you get older, melatonin production starts to decrease. It's thought the elderly don't usually get as much sleep as younger people because of this drop in melatonin[1708]. That may be implicated in the development of neurodegeneration and other ailments.

Aging impairs circadian systems through pupillary myosis, reduced transmission of blue light by the crystalline lens[1709], a reduction in the density of intrinsically photosensitive retinal ganglion cells (ipRGCs) in the retina and the atrophy of ipRGCs dendritic trees[1710]. The SCN also becomes more susceptible to age-related impairment in the form of neuronal degeneration[1711].

Older people exhibit an earlier chronotype than youngsters by going to bed and waking up earlier[1712]. It seems that when over 60 you tend to become more of a morning person[1713]. This reflects in cognitive tests as well – the elderly are as sharp as young people in the early parts of the day but not so much later[1714]. Because of decreased melatonin, seniors wake up more frequently, fall asleep slower, and spend less time in deep and REM sleep[1715,1716,1717]. Similar observations are seen in rhesus monkeys[1718], hamsters[1719], and fruit flies[1720].

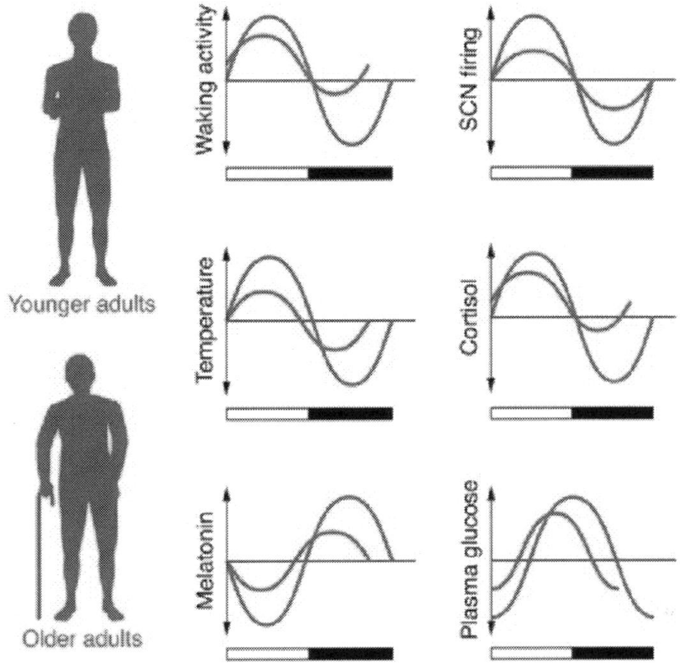

Source: Hood and Amir (2017) 'The aging clock: Circadian rhythms and later life',
Journal of Clinical Investigation 127 (2): 437-446

Fortunately, it's been found that calorie restriction or intermittent fasting can reverse the rewiring of disrupted circadian rhythms, thus alleviating the side-effects of aging[1721]. This effect is caused by the upregulation of autophagy, NAD+, and sirtuins, which affect the body's clocks. Sirtuins detect cellular energy balance and modulate the circadian epigenome[1722]. They basically provide information about the organism's nutrient status and therefore conduct either processes of growth or survival. SIRT1 is the main sirtuin gene that controls circadian rhythms and connects it with cellular metabolism[1723]. SIRT1 also delays aging and extends lifespan in mice[1724]. Enhanced SIRT1 activity can have widespread health benefits in humans as well[1725].

Calorie restriction and fasting increase lifespan because of autophagy and sirtuins. If you don't activate them either because of eating excess calories or circadian mismatches, then you may not gain their benefits[1726,1727]. Autophagy, melatonin, and growth hormone get released the most during deep sleep[1728]. That's why time-restricted eating and intermittent fasting can provide unique metabolic effects that slow down aging.

Melanopsin is a type of photopigment found in mammals that regulates the synchronization of circadian rhythms[1729]. It does so by relaying photoelectric and magnetic signals to the pineal gland and SCN. In humans, melanopsin is found in the ipRGCs in your eyes. White adipose tissue and the skin also have melanopsin receptors[1730]. So, your entire body is sensitive to light and is primed to read its cues.

Blue light exposure to the eyes plays a paricularly important role in regulating your circadian rhythms and diurnal cycles. It has antibacterial properties, boosts wakefulness, increases alertness, and can adjust the circadian clock. The reason why the sky is blue is that the blue light coming from the Sun collides with the air molecules and makes the blue light scatter everywhere. Too much blue light at the wrong time can damage your mitochondria, promote insulin resistance[1731], cause insomnia, depression, and increase inflammation.

Blue light has a short wavelength of 380-500 nanometers, which makes it produce higher amounts of energy. Naturally, you wouldn't get exposed to much blue light aside from the early to afternoon parts of the day. However, ever since the invention of the light bulb, our environment has many additional sources of blue light. Because of technology and new gadgets, we're getting exposed to more blue light for longer periods of time which can offset the circadian rhythm and cause damage to our health.

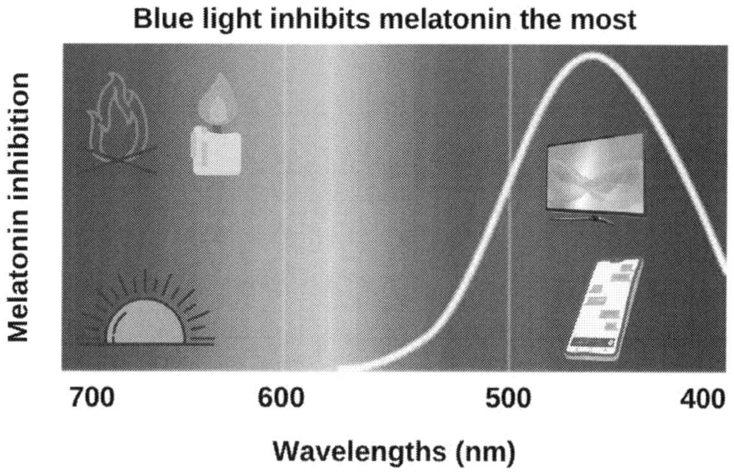

Blue light between 415-455 nm is linked with eye damage[1732]. The high energy blue light damages the retina, causing cataracts, age-related macular degeneration, and disrupts circadian rhythms. Autophagy protects the retina from light-induced degeneration[1733]. Mitophagy gets activated in response to photo-oxidative stress in photoreceptor cells and dysregulation of this process can cause retinopathy. However, this is just a defense mechanism to the oxidative stress similar to EMFs, not inherently beneficial.

Autophagy is increased in age-related macular degeneration but it's not clear whether or not autophagy is the cause or effect[1734]. Young and healthy eyes have basal autophagy that's balanced and regulated based on requirements. Old and unhealthy eyes have more aggregates and oxidative stress, which also triggers more autophagy. So, it's more likely that autophagy is just there to try to clean-up the damage instead of causing it. If you're staring at a computer screen or smartphone most of the day then you should use blue light blocking glasses with the see-through lens. This will prevent excessive autophagy and apoptosis in the retina, which will protect eyesight.

Blue light-mediated damage in retinal pigment epithelium cells activates autophagy via glucose-related protein 78 (GRP78) to alleviate the stress[1735]. Taking an antioxidant called N-acetyl-cysteine (NAC) has been shown to suppress blue light-induced autophagy activation[1736]. NAC inhibits the initiation of autophagy by scavenging the reactive oxygen species created by LED lights, thus signaling there's no need for additional autophagy. The same effect has been seen in taking bilberry and lingonberry extract[1737].

It's been found that white LED light induces autophagy in hippocampal neuron cells[1738]. Most LEDs include some blue or green light wavelengths around 400-500 nm-s. Blue light at those wavelengths causes oxidative stress in human corneal epithelial cells[1739]. You will also probably see an increase in autophagy to repair the oxidative stress but it won't benefit your health. It's just damage control that shouldn't be there in the first place.

On the flip side, amber lights at 590 nm-s induce the breakdown of lipid droplets through autophagy-related lysosomal degradation[1740]. Red light between 550-670 nm has also been shown to accelerate skin recovery whereas blue light slows it down[1741]. Moderate amber and red light in the evening will also promote melatonin production, thus resulting in more autophagy during sleep[1742]. Blue light at night has the opposite effect. Orange, reddish, and amber wavelengths mimic the natural sunset whereas blue/green resembles midday.

If you inhibit melatonin secretion because of blue light at night, then you're going to lower growth hormone, which makes it more difficult for you to burn fat and build muscle. You'll also prevent the brain from clearing out the toxins that get accumulated there during the day. Alzheimer's and neurodegeneration are characterized by the accumulation of beta-amyloid plaque. They get removed primarily during sleep in conjunction with melatonin and autophagy. So, if you're noticing brain fog, forgetfulness, and cognitive decline, then you should start taking your circadian rhythms and sleep more seriously.

There are some health benefits to blue light at the right time, such as increased alertness, circadian alignment, improved mood, and antimicrobial effects[1743]. Exposure to light and sunlight during the day is healthy and needed for autophagy production. Here's how to optimize it:

- **Morning AM light is needed for producing melatonin at night**. It does so by increasing a protein in the brain called POMC (Proopiomelanocortin)[1744]. The UV-A light also helps to lower cortisol by activating the HPA axis. UV light hitting your skin activates p53, which upregulates the gene encoding POMC[1745]. The precursor of POMC is processed into several by-products that regulate important hormones tied to leptin, satiety, and obesity.

- **Autophagy won't begin effectively in individuals with low vitamin D levels** because the autophagosomes won't be manufactured in enough quantity[1746]. The vitamin D receptor also regulates autophagy and activating it with vitamin D can induce autophagy[1747]. Getting sufficient sunlight optimizes cellular functioning as well as gives the blueprint on how to act based on the circadian rhythms.

So, blue light at the right time in the right amounts is healthy and necessary for optimal cellular functioning. It's just important to not let it disrupt your sleep quality and hormonal profile at night.

Effects of Sleep on Health

You spend nearly 1/3rd of your life being asleep. From the perspective of evolution, it's quite a paradoxical phenomenon. Although sleep has many restorative benefits on the body, it's still a gambit. In nature, while your sleeping, you're putting yourself at greater risk of being eaten, getting killed, or missing out on some feeding opportunities. The fact that virtually all animals have some sort of a sleep-wakefulness cycle shows how important it must've been for organisms to develop this process.

Not getting enough sleep is quite common in modern society. About 40% of people report getting less than recommended amounts of sleep, which is 7 hours at a minimum[1748].

Here are the negative side-effects of sleep deprivation on your body and physique:

- **Causes Serious Health Problems**, such as increased blood pressure, higher stress hormone levels, greater risk of diabetes, heart attacks, and irregular heartbeat[1749]. People with insomnia have a 90% chance of suffering from any other health condition[1750].
- **Promotes Diabetes.** Just 4 nights of sleeping about 4.5 hours reduce people's insulin sensitivity by 16% and make fat cells 30% more insulin sensitive[1751].
- **Increased Risk of Fibromyalgia**, which is a medical condition of chronic widespread pain and heightened pain response to pressure[1752].
- **Baggy Eyes, or Periorbital Puffiness**, is a hint towards minor sleep deprivation. Minor dark circles and some bags under your eyes.
- **Makes You Fat**. Sleep deprivation makes your body more susceptible to gaining weight. Poor sleep drains your willpower and decreases self-control[1753], which makes you more prone to overeating.
- **Muscle Loss**. Sleeping 5.5 hours instead of 8.5 per night results in a lower proportion of energy being burned from fat and more of it coming from carbohydrates and protein[1754]. This predisposes you to fat gain and muscle loss.

- **Increased Risk of Disease.** Sleep deprivation mirrors physical stress[1755]. It weakens the immune system which makes us more prone to colds and other illnesses[1756].
- **Hormonal Malfunctioning.** It decreases your testosterone and leads to lower libido in both men and women[1757]. Human growth hormone gets released during the first hours of our sleep which is incredibly important for building tissue and maintaining leanness.
- **Reduced Performance.** You won't have enough energy or power to perform at your peak both cognitively and physically[1758]. This also inhibits workout results and motivation.
- **Pre-Mature Aging.** Lack of sleep releases cortisol which is the catabolic stress hormone. Your body will begin to break down its muscle and accumulate fat. It also accelerates aging and makes your skin more wrinkled and drier.

There isn't a single health problem that isn't affected by sleep deprivation. It's the most fundamental thing for being healthy and vibrant.

Sleep Deprivation and the Brain

Here's how sleep deprivation affects your brain and cognition:

- **Confusion, Memory Lapses, or Losses[1759].** It can also create false memories with you getting random mental images as if you were dreaming while being awake.
- **Psychosis, bipolar disorder, schizophrenia**, and other mental illnesses are linked to sleep deprivation[1760]. If you start seeing random things or are super paranoid for some reason, then you might be just sleep-deprived.
- **Decreased Cognition[1761].** Your mental performance and ability to focus will be severely hindered. Poor sleep reduces brain activity in the thalamus and prefrontal cortex[1762]. Just a single night of sleep deprivation compromises frontal function and fluency[1763]. Sacrificing sleep time for working or studying actually makes us less productive. Even a few days of not getting enough sleep leads to the same performance drop-off as not having slept for 24 hours.
- **Impaired Judgement.** Lack of sleep interferes with how you make decisions and interpret outside events. Even just 24 hours of sleeplessness makes you 4.5 times more likely to sign a false confession[1764]. Especially about how good you sleep.
- **Leptin Resistance.** Short sleep duration is associated with reduced leptin, the satiety hormone, elevated ghrelin, the hunger hormone, and increased body mass index[1765].

You'll feel less satisfied with eating while simultaneously being hungrier for more junk food.

- **Accelerates Alzheimer's Diseases.** Poor sleep is linked with neurodegeneration and other ailments[1766]. It encourages the spreading of toxic Alzheimer's proteins[1767]. Sleep loss precedes symptoms of Alzheimer's. So, you may already be on track of getting the disease several years before showing the first symptoms. A lack of deep sleep, especially, could indicate the development of Alzheimer's[1768].
- **Sleep Loss Breaks Down the Brain.** Insufficient sleep promotes the activity of astrocytic phagocytosis and microglial activation, which break down the brain physically[1769]. I don't know about you but I don't want anything to destroy my brain.
- **Sleeping less than 5 hours a night increases your chances of dying by about 15%[1770].** Shorter sleep durations are associated with a greater risk of death[1771]. Not enough sleep has played quite a big role in many tragic accidents and events that involve airplanes, ships, and even nuclear reactor meltdowns[1772]. An estimate of 1 in 25 adult drivers report having fallen asleep while driving. Not in their entire lifetime, but in the last 30 days[1773]. How scary is that? Imagine how many tired drivers are constantly dozing off in traffic.

On a societal scale, sleep deprivation has major costs on all countries and their GDPs. The US loses up to $411 billion a year,

Japan $138 billion, Germany $60 billion, the UK $50 billion, and Canada $21 billion[1774]. That's just a massive inefficiency dragging down the entire economy.

Sleep deprivation also drags down productivity at the workplace quite massively. The US loses an equivalent of 1,2 million working days a year to poor sleep, Japan 600 000, the UK and Germany 200 000 days, and Canada 80 000 days. Just showing up to work and suboptimal cognition severely hinder overall performance. You're better off working fewer hours and getting more sleep as you'll be more productive and achieve more. Not to mention staying healthier.

Sleep deprivation is subjective, meaning the threshold at which you experience its effects vary between people. Some people have genetics that enables them to sleep less whereas others need more. You have to focus on the actual symptoms of sleep deprivation, which include chronic fatigue, caffeine dependency, frequent yawning, muscle soreness, irritation, depression, problems concentrating, forgetfulness, paranoia, food cravings, energy dips, and getting sick easily. Declining health and poor biomarkers like high blood sugar, triglycerides, and inflammation may also indicate that you're probably not fully optimized.

The signs of sleep deprivation accumulate over the course of many weeks and months. They're easy to miss and shrug off as temporary. Unfortunately, as you get more used to suboptimal sleep, your brain begins to compensate for the lack and it becomes your new normal. At that point, you've simply forgotten what it feels like to be fully recovered. To a certain extent, it's beneficial because your body is adapting to the restriction but it can hinder your performance or well-being.

What Drives Sleep Drive

The sleep-wake cycle is regulated by two separate mechanisms: sleep-wake homeostasis (Process S) and the circadian rhythm (Process C). This model was first posited by a Swiss researcher Alexander Borbély in the early 1980s.

- **Sleep-Wake Homeostasis (Process S)** is the accumulation of sleep-inducing chemicals in the brain. It's driven by glycogen depletion, accumulation of adenosine, and general fatigue[1775]. Proper balance is called homeostasis whereas an imbalance is sleep deprivation. The longer you've been awake, the stronger your drive to sleep becomes.
- **The Circadian Rhythm (Process C)** is the regulation of the body's internal clocks that affect biological processes and cognition. This determines the timing of sleep and it coordinates with the light and dark cycles of the day. Process C counteracts the homeostatic sleep drive during daytime and enhances it at night[1776], meaning with proper circadian entrainment you're supposed to be tired in the evening and energized in the morning.

Basically, sleep drive is a measure of your biological demand for sleep that builds up during wakefulness. It's the accumulation of fatigue that needs to be treated with proper shut-eye. Process S creates the pressure to sleep and Process C dictates the rhythm.

When you wake up in the morning your sleep drive is very low because you're just coming from sleeping. Of course, poor sleep has less of an effect on fixing sleep drive but you get the point. Sleep drive increases progressively throughout the day and drops when you fall asleep. The longer you stay awake the higher your sleep drive will be and vice versa. Someone engaged in more physical activities, mental gymnastics, or a lot of stress will also see a greater demand for sleep.

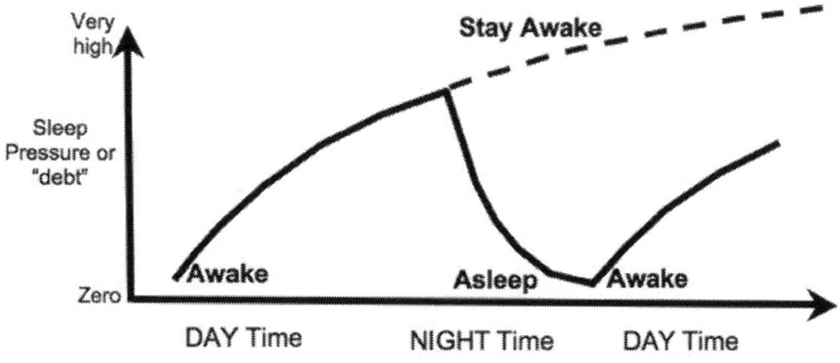

The strength of sleep drive is determined by the accumulation of a neurochemical in the brain called adenosine. It regulates the feeling of sleepiness, wakefulness, and sleep latency. Likewise, adenosine is low after waking up and builds up over the course of the day. Producing ATP the energy molecule also creates adenosine, which explains that good night's sleep after a hard days labor.

Caffeine blocks adenosine in the brain, which is why it boosts wakefulness and increases energy. However, it only masks the sleepiness because your body hasn't really fixed the rise of sleep drive. You're just blocking the accumulation of adenosine telling you're tired. And you feel as if you're not tired whereas, in reality, you're borrowing energy from your adrenals, glycogen, and cortisol.

To build up sleep drive and adenosine, you would have to stay awake for longer and increase your physical activity. That would drastically make you desire and need to sleep a lot more. Unfortunately, chronic sleep deprivation has a myriad of negative consequences.

Mild sleep restriction, however, may have a beneficial effect in making you fall asleep faster and improving the quality of your shut-eye. There's definitely a point of diminishing returns but making your sleep 30 minutes shorter can end up with better total sleep.

Sleep debt is shown to be cumulative and builds up over time. Humans seem to reach maximum subjective sleepiness after 30 hours of wakefulness[1777]. However, the physiological impact of not sleeping for that long will begin to accumulate and it's almost impossible to make up for lost sleep.

The longest scientifically documented period of wakefulness is held by Randy Gardner who intentionally went without sleep and no stimulants for 264 hours (11 days). He broke the previous record of 260 hours held by Tom Rounds from Honolulu[1778]. Gardner was said to have no long-term consequences after the fact although he experienced severely fragmented thinking, slurred speech, paranoia, confusion, and short-term memory loss[1779].

Sleep Anatomy 101

The most profound physiological changes during sleep happen in the brain. Compared to wakefulness, sleep reduces the brain's energy demands by decreasing cerebral metabolic rate of glucose by 44% and oxygen by 25%. Despite comprising only 5% of your total body weight, the brain uses about 20% of overall energy during the day[1780].

There are several brain structures involved with sleep. They are the (1) hypothalamus often called the emotional center of the brain that links the nervous system with the endocrine system via the pituitary gland, (2) the SCN, (3) the brain stem or reptilian brain that communicates with the hypothalamus to control sleep-wake transitions, (4) a mass of grey matter in the middle of the brain called the thalamus that transmits information from the senses to the cerebral cortex and regulates consciousness, and (5) the pineal gland that produces sleep-wake hormones like melatonin.

During sleep, the brain doesn't turn off but undergoes many neurological processes. There are 2 main types of sleep with characteristic brain wave patterns and activities.

- **Non-rapid-eye-movement (NREM)** sleep is anything that is not REM sleep. It consists of 3 separate stages.

- The first is NREM1 between drowsy wakefulness and sleep, in which your muscles are still quite active and you may occasionally open your eyes. During this time, the unsynchronized beta and gamma brain waves of a wakeful state (frequency of 12-30 Hz and 25-100 Hz) transition over to slower alpha waves (8-13 Hz) and then to theta (4-7 Hz). As it happens, your breathing down-regulates and heart rate gets slower. In this period, you're dosing off but you're not disconnected from your surroundings completely. You may experience sudden muscle twitches and can hear conversations. The duration is about 10 minutes. Overall, this stage should represent only 5% of your total sleep time.
- In NREM2 your muscle activity keeps decreasing and you start to slowly fade away into sleep. The majority of the brain activity is within the theta range but there are also quick bursts of brain activity in the 12-14 Hz region (sleep spindles) and K-complexes (a brief negative high voltage peak, followed by a slower positive complex, and a final negative peak, that occurs roughly every 1-2 minutes and is often followed by sleep spindles). They are spontaneous but usually happen in response to external stimuli such as sounds, touches, and internal breathing interruptions. Dreaming becomes possible. A 2002 study titled

Practice with Sleep Makes Perfect saw that getting enough NREM2 sleep improves motor skills[1781]. The function is to protect sleep and engage in information processing and memory consolidation. Duration 20-30 minutes.

  - Stage 3 (NREM3) is the deep or slow-wave sleep characterized by delta brain waves of 0.5-4 Hz. Here you are cut off from the conscious world around you and irresponsive to most sounds or other stimuli. During this time heart rate, blood pressure, breathing, neural activity, and brain temperature are all at their lowest. In addition to information processing and memory consolidation, dreaming, night terrors, sleepwalking, and talking occur as well. Growth Hormone gets secreted in small bursts and pulsations. This is the deepest stage and should represent around 15-20% of total sleep time. Duration 30-40 minutes.

- **Rapid-eye-movement (REM)** sleep is a phase characterized by random movement of the eyes, low muscle tone, and the possibility of lucid dreaming. It shares some physiological similarities to a wakeful state, such as rapid, low-voltage desynchronized brain waves. It allows us to learn complex tasks. Infants experience a lot of REM sleep and get roughly 50% of their total 16 hours of sleep as REM[1782]. One cycle of REM usually takes about 70-90 minutes depending on the length of overall sleep and comprises 20-25% of the total night's volume.

We usually awake from slow-wave sleep after the end of a REM phase or sometimes in the middle of it. This would happen gradually as the brain regulates its wave frequencies and activity accordingly to not cause any disturbances. If, however, we were to interrupt this process during a deep stage (NREM3) because of some external stimuli, such as an alarm clock, a crying baby, or someone shaking your body, then we would interrupt this natural pattern.

Studies have shown that the particular sleep stage at which you awake is important in amplifying sleep inertia or the feeling of drowsiness after waking up[1783]. As a result, we will suffer from grogginess and fatigue because of not having been able to go through it slowly. That's why it's optimal to wake up in either light sleep or at the end of REM. Waking up usually happens soon after a REM cycle or in the middle of it. Awakening is brought about by internal circadian clocks and reduced homeostatic need for sleep[1784]. This process involves heightened electrical activation in the brain, starting with the thalamus and spreading throughout the cortex.

One cycle of REM usually takes about 70-90 minutes depending on the length of overall NREM sleep and comprises 20-25% of the total night's volume. Light NREM1-2 takes about 50-60% and deep sleep NREM3 comprises up to 10-25%. The entire sleep cycle proceeds in that particular order: N1 → N2 → N3 → N2 → REM. You go into REM naturally as you come out of slow-wave sleep (N3) and are about to enter back into N1.

During the earlier parts of the night, you experience more deep sleep (N3) but the last two cycles before awakening have a greater proportion of REM[1785]. REM sleep usually requires a lower body temperature and occurs during the lowest point of the daily temperature cycles. NREM or deep sleep can occur independently of circadian time[1786]. So, you can get them from napping as well.

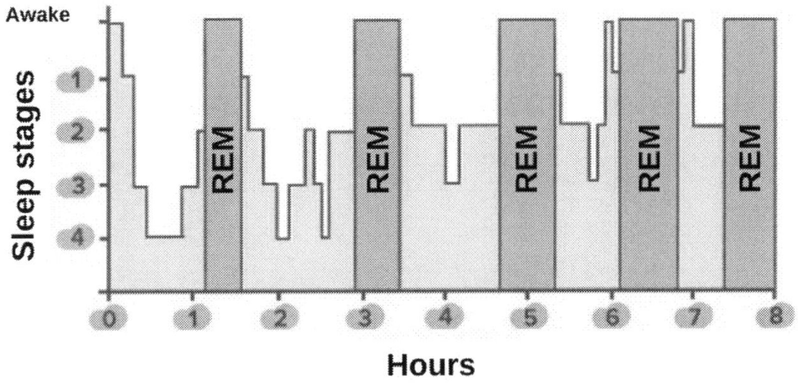

The optimal length of sleep varies between people and it's determined by genetics, lifestyle factors, circadian rhythms, and how much physical repair is needed. It's recommended that children should get about 10-12 hours a day and adults 7-9. Too much sleep, however, can also have negative side-effects, such as insomnia, restlessness, epilepsy, obesity, and other medical conditions.

A systemic review conducted at the University of Warwick saw that the risk of mortality amongst people who slept 6 hours or less increased by 12%[1787]. However, the risk of mortality amongst those who slept for 9 or more increased by as much as 30%. This was probably because there was a greater proportion of sick or hospitalized subjects who slept over 9 hours and they were already pre-disposed to dying any moment. Nevertheless, I would say the optimal length of sleep is still 6-9 hours.

According to a scientific review of more than 300 studies published between 2004 and 2014 to ascertain how many hours of sleep most people need to maintain their health, a panel of experts came up with the following recommendations.

Age Group	Hours of Sleep Needed for Health
Newborns (0 to 3 months)	14 to 17 hours
Infants (4 to 11 months)	12 to 15 hours
Toddlers (1 to 2 years)	11 to 14 hours
Preschoolers (3 to 5)	10 to 13 hours
School-Age Children (6 to 13)	9 to 11 hours
Teenagers (14 to 17)	8 to 10 hours
Adults (18 to 64)	7 to 9 hours
Seniors (65 and older)	7 to 8 hours

Physically, the most restorative stages are deep sleep and REM, which is why they should be prioritized much more. However, consistently over 25% of REM sleep may cause hyper brain activity, mood disorders, and other neurological issues.

It's a very good idea to use some sort of a sleep tracker so you'd have actual data about your progress. The best non-Bluetooth device out there is the OURA ring that can be kept in airplane mode. It also measures heart rate variability (HRV), body temperature, how fast you fall asleep, your physical activity, and how many times you wake up during the night.

Total Sleep Optimization

From the perspective of circadian rhythms, you're supposed to be sleeping when the sun is out or at least not be active. Once darkness falls, your body already tries to go into a deeper state of recovery and hibernation by releasing certain hormones such as melatonin. Likewise, at dawn, your brain should wake up as well by releasing cortisol.

An optimal circadian rhythm for sleeping cycles is somewhere between 9 PM – 8 AM. That's the longest timeframe wherein you should get the most of your sleep. This means waking up early at about 5-8 AM and going to bed between 9 and 11 PM. The actual timing may fluctuate between individuals, seasonality, and daily routines but the core is still the same.

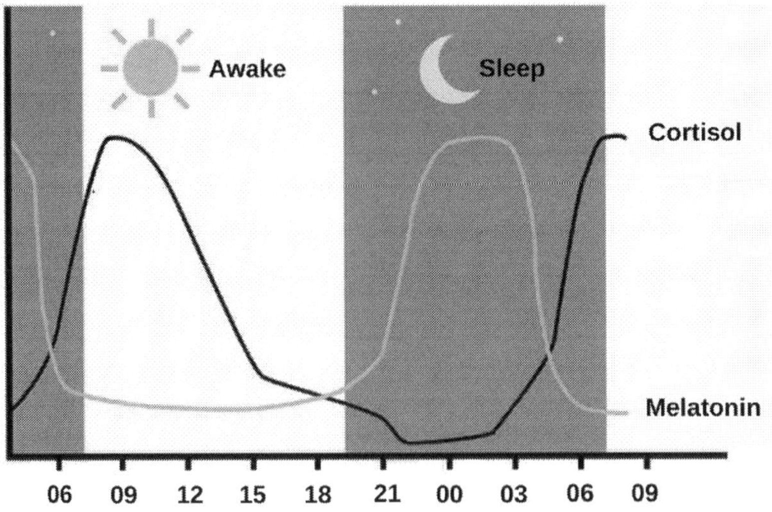

Although you may find yourself being a wolf or a night owl, it doesn't mean your optimal sleep time is between 2 AM and 10 AM. It's going to completely throw off your circadian rhythm and messes up your hormones. Chronotypes far outside the normal range of circadian rhythms are called sleep disorders[1788]. 'Night Owls' who justify being up all night as genetics are simply confused about how their bodies should work and have trashed circadian rhythms.

One of the most beneficial hormones for fat loss, muscle maintenance, and recovery is Human Growth Hormone (HGH). The majority of GH gets released during the initial hours of sleep between 11 PM and 2 AM when your body prioritizes physical repair. It has a circadian rhythm just as does the morning rise in cortisol. If you're not sleeping by that time, you're missing out on a lot of the benefits. Autophagy is also primarily processed in conjunction with that.

Likewise, sleeping in and waking up extremely late also throws off your body's circadian clock. If you wake up at 11 AM, then you've missed out on your natural release of cortisol, which makes you more tired and groggy. You've also woken up at noon with the sun reaching its zenith. Now you're sending very conflicting signals to your brain, which disrupts hormonal output and other metabolic processes. Granted, you can still be healthy and get synchronized with enough daylight but this kind of catching up places unnecessary extra stress on your body. Optimally, you still want to sleep in sync as possible.

Here's what to do to optimize your sleep and circadian rhythms:

- **Consistent Bed and Wake Up Time** – Going to bed and waking up around the same time entrains your circadian rhythms to follow certain patterns. It also improves sleep quality by enforcing the routine. A single night of sleep deprivation or disruption can be a hormetic stressor by making you antifragile, teaching how to function without optimal recovery, and increasing neurotrophic factors in the brain. You just have to make sure it won't become a bad habit and you're actually prepared to handle it.

- **Avoid Blue Light After 8 PM** – Research has shown that white LED lights are five times more efficient at blocking melatonin production than incandescent light bulbs[1789]. You don't want to be sitting under LED or fluorescent lights in the evening. Start blocking out blue light at least 2-3 hours before going to bed. This allows your body to start producing melatonin and make you more tired. I'm using blue light blocking glasses and screen filters like F.lux or Twilight.

- **Amber Lights at Night** – Natural sunset has a wavelength between 600-700 nanometers. It's the opposite of blue light, which is around 400-500 nm. Red, orange, and amber lights indicate the end of daytime and beginning of nighttime. Using amber lightbulbs, red light therapy devices, and orange screen filters in the evening can mimic the natural sunset, thus support the circadian rhythm as well as melatonin production.
- **Use an acupuncture mattress.** Purchase a small bedding that has little spikes on top of it. This is relatively cheap yet very effective. You can lay down before going to bed for 15 minutes or sleep on it throughout the night. I have used both options. At first, it feels like a lot of thorns are trying to penetrate your skin. After a while, the body relaxes and it becomes incredibly soothing. It creates a nice feeling of surging energy in the back. There's a lot of evidence for the health and stress management benefits of this. In China, needle therapy is a key component of traditional medicine. The Yogis of India have also been using nail beds for centuries.
- **Sleep in a Cooler Room** - Sleeping in too hot decreases REM and deep sleep[1790]. People with difficulties staying asleep often have elevated core body temperature at night[1791]. According to the National Sleep Foundation, the best

temperature for sleep is approximately 60–67°F (15–19°C)[1792]. Temperatures over 71°F (24°C) or below 53°F / 12°C are more likely to impair sleep quality. Both too cold as well as too hot can be detrimental. However, you have to find what works best for you.

- **Nasal Breathing and Mouth Taping** - If you think you're suffering from sleep apnea, snoring, or breathing problems during sleep, then you should try mouth taping. While this may sound bizarre, it's quite effective and not at all painful or risky. This will encourage breathing through your nose throughout the night, which has many health benefits aside from regulating sleep-disordered breathing that can progress to sleep apnea. Simply place a small piece of medical tape (please do not use industrial types of tape which can damage your skin) across your lips.

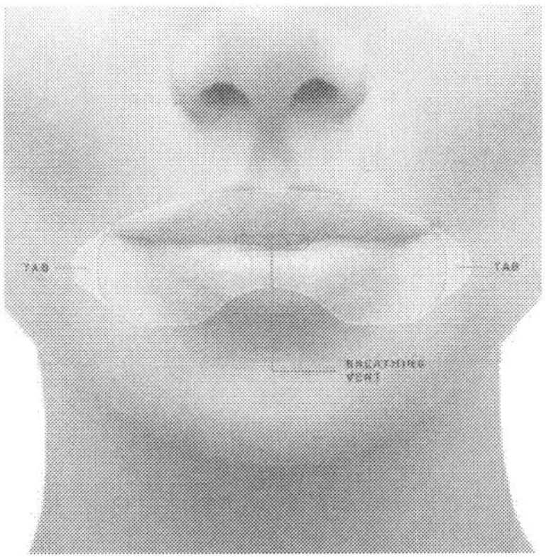

- **Block Out the Noise and Light** – Wearing a sleep mask is highly effective and will protect you from any potential disturbance by blue light sneaking in. Using regular inexpensive earplugs or noise-canceling headphones during the night is a simple way to block out the potential disturbing sounds. One study found that playing 'pink noise' synchronized to the subject's brain waves allowed them to stay in deep sleep for longer than when the sound was not played[1793]. They also saw 60% improved memory retention and were able to recall more words they had been shown before bed.

- **Improve Bedroom Air Quality** - Poor indoor air quality can cause sleeping problems and reduce deep sleep by affecting respiratory organs[1794]. Studies have found it's similar to second-hand smoking[1795]. That's why keep the house ventilated and open the windows frequently. A NASA study also found that different houseplants promote photosynthesis and turn CO2 into oxygen[1796]. Good ones would be devil's ivy, ferns, rubber plants, cactuses, snake plants, and weeping figs.
- **Expose Yourself to Daylight After Waking Up** – The first thing you should do after waking up is going outside for 10-15 minutes and get exposed to sunlight. That's going to immediately synchronize your body with the environment and sets off a proper circadian rhythm. You'll also feel more energized, wakeful, and happy throughout the day. Even if it's cloudy with no sun, some of the light waves will penetrate through the clouds and you'll still get the effect. Direct sunlight has a luminosity of about 32 000 to 130 000 lux compared to the 320-500 lux of typical indoor lighting.

- **Don't Drink Coffee After Noon** – Caffeine stays in your system for up to 10-12 hours. The half-life of caffeine is about 5.7 hours[1797], which means that if you drink coffee at noon, then 50% of it will still be in your system at 6 PM. That's why you should stop consuming everything that has caffeine after 2 PM. Ideally, you want to also postpone your first coffee by a few hours after waking up to allow cortisol to do its job. Between the hours of 8-9 AM, our cortisol levels are at their peak[1798]. The best time to drink coffee is between 9:30 AM and 11:30 AM. Cortisol peaks in the early morning, but also fluctuates during the day. Other times it rises are 12 PM – 1 PM and 5:30 PM and 6:30 PM, so avoid a cup of joe at those hours as well.
    - CYP1A2 is the main liver enzyme that breaks down caffeine. Variations and mutations in the CYP1A2 gene determine whether you're a fast or a slow metabolizer. People with a homozygous CYP1A2*1A allele are fast caffeine metabolizers, whereas those with CYP1A2*1F are slow caffeine metabolizers. Slow metabolizers may need more time to metabolize caffeine, thus it stays in their system for longer. There's also a link between slow metabolizers and increased risk of having non-fatal heart attacks and hypertension with caffeine[1799,1800]. If you have the slow metabolizer allele, then you may want to reduce your caffeine intake.
- **Go Outside Frequently** – You want to go outside and expose yourself to daylight as often as possible throughout

the day. It's going to keep you in sync with the circadian rhythms and also increases overall energy levels. One of the biggest reasons office workers drink so much coffee is that they're stuck inside under artificial lights that drain their energy and drowsy. Having short 10-15 minute walks spread across the day is an amazing way to not only burn more calories but also promote circadian alignment.

- **Experience Alterations in Temperature** – Circadian rhythms are affected by not only light but also temperature. Humans are not meant to be living in constant room temperature with central heating and air conditioning. Variations in temperatures are beneficial and causes positive stress. Do open the window to ventilate the air and let in a cold breeze as to shiver for just a few minutes. It will burn more calories and strengthens the immune system.

- **Keep Exercising During the Day** - According to a large 2018 meta-analysis, exercise may alleviate symptoms of insomnia without the use of hypnotics[1801]. Strength training has also been shown to increase deep sleep quality and sleep drive[1802]. Resistance training, in general, makes the body more efficient with falling asleep as well as staying in deeper stages of sleep. Based on the circadian rhythm, the best time to exercise is in the afternoon around 2-5 PM. That's when your nervous system has been warmed up and is ready to go. Your coordination, explosiveness, and strength are also the highest. Working out in the morning is fine as your cortisol levels are already the highest.

However, you should stop doing hard physical activities after 6-7 PM or at least 4 hours before going to bed.

- **Time-Restricted Eating** – Food is another circadian signaler that regulates metabolic processes and the endocrine system. Intermittent fasting has many health benefits but it's also supportive of circadian rhythms. Many researchers like Dr. Satchin Panda from the Salk Institute and Dr. Joseph Mercola think humans are meant to eat within 8-10 hours. That's the minimum amount of time to give our digestion enough time to rest as well as keep ourselves aligned with the daily cycle. You shouldn't eat when it's dark outside or too close to bedtime.
- **Eat a high protein meal for dinner** – the amino acid tryptophan gets converted into serotonin and then into melatonin[1803]. You can get it from poultry, meat, fish, nuts, and seeds. However, some carbohydrates can also enable that tryptophan to reach the brain thanks to insulin[1804]. Foods that disrupt sleep are spicy foods, caffeine, chocolate, fried foods, fatty foods, sugary foods, watery foods like watermelon because they may give you heartburn, indigestion, and make you wake up to go to the bathroom. Although some people say a glass of wine or a beer helps them fall asleep, alcohol is shown to decrease deep sleep quality[1805]. It also disrupts the natural REM cycle by shortening it in some instances and lengthening at others.

If you want to know how to sleep better, get more deep sleep, correct your circadian rhythms, sleep-wakefulness cycles, establish the optimal circadian rhythm, and improve overall recovery, then check out my Total Sleep Optimization Video Course at https://www.siimland.co/total-sleep-optimization-video-course/. It has over 8 hours of content, lectures, infographics, and walkthrough guides about getting the most out of your sleep.

Early Time Restricted Eating VS Late Time Restricted Eating

There are studies showing that eating breakfast and skipping dinner is better than the standard 3 meals a day[1806]. That makes sense because your body would be spending more time in a fasted state during which it can repair itself and clean house with autophagy.

A study published in May 2019 found that eating from 8 AM to 2 PM vs eating from 8 AM to 8 PM improved blood glucose control, lipid profile, and autophagy genes. However, that's compared to a non-fasting group eating between 8 AM and 8 PM. Of course, early time-restricted eating is better because it's actually going into a fasted state. Furthermore, this study was done on 11 overweight individuals over the course of just 4 days. An 8-hour window between 10 AM to 6 PM has also been shown to reduce blood pressure, lower body fatness, and improve cholesterol profile.

Another study in April 2019 – a few months earlier – found that a 9-hour eating window from 8 AM-5 PM and 12-9 PM both had the same results in terms of health and body composition[1807]. The subjects ate early in the day and skipped dinner for the first week and then switched it up by skipping breakfast and eating later. Their results were virtually identical in terms of weight loss, blood sugar, and insulin. However, eating the standard 3 meals a day and no fasting negated these benefits. So, the most important variable is eating less often and within a shorter time frame wherein you activate autophagy and suppress insulin.

How long it takes for you to experience the benefits of fasting depends on your metabolic status, energy balance, medical condition, physical activity, and metabolic flexibility. If your body is more fat-adapted thanks to already having practiced fasting before, then it will take less time to make these switches.

Restricting your carbohydrates in some shape or form and not overeating calories keeps you more ready to tap into autophagy, ketosis, and activate stem cells because there's less energy the body has to burn through. At the same time, the healthier you are, the better body composition you have, and the more physically active you are, the faster the transition will also be.

Extended fasting is more powerful but it's not something you can do all the time nor should. It's much more sustainable to stick to shorter fasts that are more frequent rather than super long fasts that require more time to recover from.

To get more benefits of extended fasting from shorter fasts you want to incorporate resistance training, some form of cardio, eat around your maintenance calories, lose fat, build muscle, and incorporate some autophagy boosting ingredients like coffee and medicinal mushrooms. This way you can still gain some of the longevity benefits from daily IF.

So far, the common theme of life-extension has shown that stress adaptation through mitohormesis as well as reduced energy expenditure promotes longevity. However, this is not the goal and it's not optimal in a lot of cases as we'll shortly see.

Benefits of Taking a Nap

Back in kindergarten, I remember the teachers making us go to bed and nap for about an hour. At that time I used to hate it but now I must say that having a short snooze is amazing. There's also research showing that there are a lot of health benefits to napping.

- **Reduces Stress** – Not getting enough sleep promotes cortisol production and taxes the adrenal glands. Excess leads to fat gain, weakened immune system, and muscle loss. In a study done on Greeks, people who had a 30-minute nap at least 3 times a week, had 37% less chance of dying from heart conditions and a 12% reduction in coronary mortality[1808].
- **Increases Alertness** – Naps can prevent fatigue, restore concentration, make you alert, and enhance performance. A NASA study found that a 40-minute nap increases alertness by 100%[1809].

- **Improves Learning and Working Memory** – Short mid-day naps have been found to improve sleep, cognitive tasks, and mental health in elderly Japanese[1810]. This is crucial for learning any skill, remembering information, or tackling complex tasks. You give your brain more downtime.
- **Enhances Relaxation and Mood** – Napping has great restorative effects on the body and brain. You get to deal with all the tension built up during the day and can hit a quick reset button for your mind. This can help you to fall asleep faster at night.
- **Promotes Creativity and Problem Solving** – The amount of cognitive work you can do depends on how much energy your brain has access to. Taking short naps in between studying or some other higher executive task is a great way to get a small boost of mental power. That's probably why Nikola Tesla and Leonardo da Vinci used it that much.

Sleeping at random times for indefinite time lengths can mess up your circadian rhythm and may keep you up at night. That's why you'd want to be strategic with your naps and have them with a purpose in mind.

The best time for a nap is mid-afternoon around 12 to 2 PM. Before hitting the sack, you should've been awake for about 5-7 hours. This way your body has a reason to benefit from sleeping. It all depends on what your sleep schedule is like but napping after 4

PM may keep you up at night. To get the most out of your naps, you'd want to keep them relatively short. If you wake up in the middle of a sleep cycle, you'll feel more tired than before. That's why you should aim for either a 10-30 minute power nap or a 90-minute post-lunch snooze.

Thanks to having practiced polyphasic sleeping in the past, I can doze off almost anywhere at any time if I want to. All I need is 20 minutes of my time, a good place to crawl up, and something to lay my head upon. From my experience in the military, I can also say that training yourself to nap in unconventional surfaces and environments like the bus, subway, floor, small couch, and bean bags will teach you how to fall asleep faster. It also makes you care less about these hard surfaces or bumps many people find difficult to sleep upon.

How to Recover from a Bad Night's Sleep

You're inevitably going to have sub-optimal nights of sleep, mornings you feel like being hit with a club, occasional episodes of insomnia, and days where you may only get a few hours of shut-eye. Make peace with it right now already in a stoic manner.

If you know you're about to experience a bad night's sleep or have to cut it short because of either traveling or something else, then the best thing you can do is to take some protective measures against that during the daytime already. This builds up your sleep bank so to say.

Here's what to do to mitigate and recover from sleep deprivation:

- **Drink Some Salted Water** – Hydration and a proper electrolyte balance is key for stress management. It'll lower cortisol and alleviates some of the insulin resistance that's caused by poor sleep.
- **Red Light Therapy** – Photobiomodulation or an infrared sauna even can lower inflammation and promote mitochondrial functioning. That's why I like to sit in front of my red light panel in the morning as well as in the evening. Using something like the Joovv or Red Light Rising after waking up from short sleep will help to deal with it better.
- **Take Exogenous Ketones** – When you feel like your brain and cognition have suffered from that sleep deprivation, then giving it high-quality fuel will help. Sugar and glucose can be damaging to brain cells, especially when you have higher oxidative stress than normal. That's why a much better option would be something like MCT oil, BHB salts, or ketone esters. Adding them to your coffee or making Bulletproof coffee would be a smarter and healthier option than straight-up black coffee.
- **Drink a Little Bit of Coffee** – If your mitochondria are optimized and you're getting consistently good sleep, then you shouldn't feel much worse from sleep deprivation. You should not need coffee to get out of bed or start the day. Nevertheless, drinking some coffee will help to mask sleep deprivation in a good way. You'll be able to start the day and stay active. To avoid over-stimulating cortisol, it's

still a good idea to wait at least a few hours after waking up before drinking coffee. This applies especially to situations where you wake up in the middle of the night. Even when you're up since 3 AM and are sitting at the airport at 4 AM, the best time to have coffee is between 9-12 AM. That's when your body's natural cortisol production begins to drop and you won't over-stimulate yourself. Adding healthy fats, cinnamon, turmeric, cacao, or MCTs would be slightly better than an espresso or Americano because you'll blunt the caffeine rush.

- **Circadian Entrainment** – Despite being short on sleep, you'd still want to try and stick to the same circadian rhythm. That means exposure to morning sunlight, daylight breaks, and blocking blue light at night. Just adjust to the current day and night environment of the timezone you find yourself in.
- **Powernaps** – Instead of napping before a short night's sleep, you'd want to do it afterward. This can help to recharge your batteries and still get some recovery. The recommended time for naps is about 6-7 hours after waking up and not any closer than 6-7 hours before going to bed. They should also be either 20-30 minutes or 70-90 minutes. This ensures you either get the full REM cycle or won't go into deep sleep and get shaken up from it at the 55-minute mark.
- **Exercise** – Cardiovascular training can help to mitigate sleep deprivation by boosting BDNF, boosting autophagy,

burning fat, and increasing your alertness. You shouldn't do HIIT or crazy Tabata when you're sleep deprived because it would put more stress on your body. Trying to hit personal records with resistance training is also very difficult when you're under-recovered. Long walks, cycling, or jogging are best for dealing with sleep deprivation.

'The shorter your sleep the shorter your life is going to be.' That might be true to a certain extent. However, I would much rather prefer saying *'the poorer the quality of your sleep the worse the quality of your waking life will be.'*

Circadian Rhythm and Sleep Protocol

- Circadian rhythms vary between individuals based on their genetics and lifestyle. However, trying to be in sync with the natural day and night cycles as much as possible will reduce the amount of oxidative stress and inflammation your body experiences on a daily basis. Consistency and sticking to the rhythm is the most important aspect of optimal circadian entrainment.
- The optimal amount of sleep is about 6-9 hours but the quality matters a lot more than the quantity. Athletes and children need more. Just because the elderly tend to sleep shorter doesn't mean it's healthy or that they wouldn't benefit from sleeping longer. Their body is just aging and becoming less robust. As we found out in chapter two, intermittent fasting and calorie restriction can reverse this negative rewiring through increased NAD metabolism.

- You need about 2-3 hours of deep and REM sleep per day for the best results. Light sleep is less important although you shouldn't want to restrict it. In total, it's said you need about 4-5 full sleep cycles per day but you could even dissect it into several chunks with polyphasic sleeping.
- Going to bed around 10-11 PM would be more productive because of the natural surge of growth hormone and melatonin that get secreted at that time. You should be asleep before midnight.
- Time restricted eating is an intrinsic part of the circadian rhythms and sleep cycles. A confined window for consuming food increases robustness. You should stop eating at least a few hours before bed and not eat when it's dark outside.
- Before going to bed, block out artificial light with filters and blue-blockers. Avoid bright screens and stimulation of any kind. Instead, wind down and try to relax.
- Sleep deprivation is less harmful and damaging in a state of ketosis whether through fasting or taking exogenous ketones. Caffeine can also help to mitigate the circadian mismatch if times properly. Taking a nap will speed up recovery and reduces the negative side-effects.

Chapter Fourteen: Of Self-Tracking and Monitoring

"Activity and rest are two vital aspects of life.
To find a balance in them is a skill in itself.
Wisdom is knowing when to have rest, when to have activity,
and how much of each to have.
Finding them in each other - activity in rest and rest in activity -
is the ultimate freedom."

Sri Sri Ravi Shankar

During the American Civil War in the late 19th century, a physician named Jacob Mendes Da Costa noticed a set of syndromes amongst soldiers that resembled cardiovascular disease. People called it 'soldier's heart' or Da Costa syndrome but the doctor himself called it irritable heart[1811]. Nowadays the term is no longer in use and has been replaced by more specific diagnoses.

Although physical exams did not show any abnormalities, the men had chest pains, palpitations, breathlessness, and chronic fatigue. It was thought to be caused by long-term physical exertion, carrying heavy objects all the time, lack of sleep, not enough food, and insufficient rest[1812]. Da Costa noted that these symptoms often developed after a fever, diarrhea, or extreme physical exertion. Soldiers would often get sick, go to the hospital to recover, and after returning to the frontlines could not keep up with others in terms of their daily activities. They would get out of breath faster, palpitations, their head went dizzy but other than that they appeared healthy.

The expression of soldier's heart stopped when patients ceased their strenuous activities and got enough recovery. Forced bed rest and appropriate levels of exercise were considered to be most effective. Other treatments included fixing posture, wearing loose clothes, and avoiding stooping.

In World War I, the term 'shell shock' was used to describe crippling anxiety and combat stress, which was thought to be caused by damage to the nerves. During World War II, it was called 'battle fatigue' and nowadays it's referred to as 'combat stress reaction (CRS)'. It's a prolonged fight or flight response that keeps the person in a constant state of sympathetic arousal. Experiments done by Selye showed what this kind of exposure does to the body in the long term.

Post-traumatic stress disorder (PTSD) is a condition characterized by negative thought loops, fearfulness, irritability, chronic fight or flight, nightmares, etc.[1813] It's common amongst soldiers and frontline fighters but can happen from things like traffic collisions, child abuse, sexual assault, and traumatic life events as well. People with PTSD are at a higher risk of suicide and self-harm[1814].

All these examples illustrate the fact that serious trauma, intense stress, or exertion even brief can leave a mark on the person. More often than not, physical ailments and damage can be repaired with adequate rest, nourishment, and relaxation. However, the psychological trauma may stick around after the fact and start affecting your behavior subconsciously.

A strong and hormetically adapted body ought not only endure stress but also bounce back from it fast. If you're going to be wiped out from exercise, working long hours, a single night of bad sleep, relationship turmoil, news about global emergencies, or anything the like, then you're being fragile. With antifragility, you can leverage these stressors in a way that you can return to the fray with new-found vigor.

In this chapter, I'm going to be talking about ways of monitoring stress, how to prevent it from becoming overwhelming, and the best recovery tactics. Hormesis works only when given enough time to rest. That's why a hardcore person who takes no days off may be doing themselves a disservice by burning out pre-emptively. A wiser approach would be to be proactive with stress management and thus live to die another day.

General Stress Management

First, let's start by recognizing how much stress is too much and how much is just enough. Here are the signs that you're being slowly overwhelmed.

- **Headaches and Pounding Head** – Pain in the head or neck region is often caused by stress. One study amongst 267 people found that in 45% of the cases chronic headaches were preceded by a stressful event[1815]. In military service members, headaches were triggered by environmental factors (74%), stress (67%), caloric consumption (60%), and fatigue (57%)[1816]. Other causes can be sleep deprivation, electrolyte imbalances, alcohol, and foods that contain tyramine, such as MSG, chocolate, and cheeses. It's also shown that being in ketosis and consuming ketones can be protective against migraines because the brain will get enough energy to stay functional[1817].

- **Chronic Soreness and Pain** – Working out may inevitably cause DOMS and fatigue, especially if you're new to a particular exercise. However, not being able to move your limbs properly or walking like a cripple is a sign of inadequate recovery. The problem is also that the perception of pain and discomfort will start raising your cortisol as has been shown by studies[1818]. Feeling sore and discomfort itself can cause stress. More often than not, it's caused by sleep deprivation but overtraining is an additional factor. Doing foam rolling, getting a massage, or using a massage gun will help to untie the fascia. Going to the sauna and stretching there may also provide relief. One of the best ways of numbing down all soreness in most cases is to take a 3-5 minute ice bath followed by some heat exposure.
- **Getting Sick Easily** – If you get a runny nose, coughing, shivers, or have a fever very often, then it can hint towards stress jeopardizing your immune system. People with chronic stress have been shown to experience a weakened immunity towards Influenza vaccination[1819]. Stressed out individuals also get 70% more respiratory infections and their symptoms last longer[1820]. To improve your resilience,

refer to the chapter about strengthening the immune system. However, when you're already sick, doing these kinds of hormetic activities like ice swimming or sauna may make things worse. At that point, bed rest and hydration are your best friend.

- **Sleep Problems and Insomnia** – When your body is in fight or flight, staying asleep is the least of your worries. Your brain would much rather keep you in the shallow stages of sleep and never let you go into deep and REM. That's why, if you're stressed out, you may find yourself waking up several times throughout the night and feeling tired in the morning. Work-related stress is associated with restlessness[1821]. Experiencing stressful events also increases the risk of insomnia and it precedes depression[1822].

    - To get out of fight or flight, you can lay on an acupuncture mattress. The small spikes will initially wake up the nervous system but eventually your body settles down into deep relaxation. It's quite phenomenal actually.
    - Falling asleep like a log quickly means your body is taxed and just wants to shut down. That's why appropriate exercise is great for improving sleep but not if it keeps you awake throughout the night.
    - If you're the kind of person who gets a lot of rumination and rushing thoughts before going to bed, then practice some meditation or breathing exercises before falling asleep. You should also be

wary of social media and the news. Watching motivational speeches or horror movies are all mentally stimulating. No wonder people can't sleep – they're just constantly wired up. To prevent that, make a rule to stop checking email and Instagram at least an hour or two before going to bed. Of course, blue light and EMF protection are also mandatory.

- **Depression and Lack of Motivation** – If you find yourself dragging your feet and not enjoying life, then don't start thinking there's something wrong with you. It might be that you're just stressed out or under-recovered. High stress is associated with depressive symptoms[1823]. Both acute and chronic stress can cause depression at least in the short term[1824]. It's just your body telling you to withdraw from life and social interactions so that you could spend less energy and "hibernate".
    - Deliberate social isolation and self-sabotage are also often signs of too much stress. To get out of this loop, you have to first take care of yourself and go through a short period of complete relaxation with minimal work. Once you've regained your bearings, it's important to get back into the world because loneliness can also turn into a stressor and cause depression. According to AARP, prolonged social isolation damages health and can be as harmful as smoking 15 cigarettes a day[1825]. Even introverts need some interaction with their peers but it's much less than extroverts. So, funnily,

introversion could be an evolutionary adaptation to loneliness. However, a more antifragile quality would be ambiversion. An ambivert can be both an introvert and an extrovert based on the situation.

- **Heartbeat Problems** – If you're showing signs of a soldier's heart or anything the like, then it can indicate high stress, which itself is a risk factor for cardiovascular disease[1826]. Stressful events raise your heartbeat and blood pressure[1827]. In a study, playing music during the exertion has been shown to reduce this elevation[1828]. That's why people might enjoy working out with their headphones on. Dehydration and electrolyte deficiencies can also cause palpitations, especially if you're sweating or moving around a lot. Drinking a glass of water with a tsp of sea salt can lower morning cortisol and supplementing magnesium is generally important because stress depletes magnesium and can cause scar tissue formation. Increased magnesium intake can alleviate cystic fibrosis[1829]. Similar stress management tactics like acupuncture therapy, massage, and spending more time in nature without social stimulation can be very effective.
- **Digestion Problems** – The sympathetic nervous system shuts down digestion and directs blood away from the gut towards the extremities. Constipation can be caused by chronic stress[1830]. This can lead to irritable bowel syndrome or disease and promote autoimmunity[1831]. With a slower transit time, food particles have a much higher chance of beginning to ferment in your gut whereas with a

fast thyroid your body will assimilate the nutrients much faster. Acute and intense stress may also cause diarrhea as the body tries to reduce the burden. In any case, you should never eat when you're stressed out. It can make you over-eat and gain weight[1832]. If you feel wired up, practice meditation, or go for a slow walk before eating. Thorough chewing and walking after meals will also improve digestion. Don't drink a lot of liquids with food because it'll dilute stomach acid, thus reducing your ability to break down the ingredients and making you more bloated.

- **Low Thyroid and Sex Hormones** – As discussed before, the body responds to chronic stress by lowering thyroid and energy expenditure. This, in turn, will have an impact on your testosterone and overall libido because there's not enough T3 to convert cholesterol into steroid hormones.
  - If your stress is related to over-training, fasting too much, low calorie dieting, or ketosis, then having a carb refeed is appropriate. An influx of glucose will raise insulin that then also replenishes muscle glycogen and thyroid hormones. Increasing iodine and selenium intake by eating sea vegetables, fish, and Brazil nuts can also help. Supplementing DHEA can give more assistance in increasing metabolic rate and alleviating the effects of cortisol.
  - If your stress is related to sleep and circadian mismatches, then you should catch up on the lost sleep first before implementing any other strategy. Eating more carbs while being sleep deprived

hinders glucose tolerance and insulin sensitivity. When sleep-deprived the safest option is to stay in ketosis until you've fixed yourself.

- If your stress is related to work, relationships, or responsibilities, then you should take some extra time for yourself and relax. You can't help others if you're not recovered yourself. In some cases that might mean taking a weekend vacation or just going to the spa. With the right communication, you should be able to explain the reasoning to your loved ones.

Stress is something we're all going to experience. Experiencing these symptoms isn't a bad thing. It's actually beneficial because you can identify where your limits are and what's making you stressed out. The problem is that most people lack enough self-awareness and inquiry to even go down that rabbit hole. They just start reacting and compensating for it with poor habits and decisions like binge-eating, social withdrawal, or self-abuse. What you need to do instead is recognize that you're starting to get taxed out and then take a proactive approach towards recovery and restoration. That's where all the hard work you've put in beforehand gets instated as hormesis.

Lab Work and Blood Tests

Here are some reference ranges for different biomarkers that can give insight into whether or not you're too stressed out. Keep in mind all these results have to be interpreted in the right context and with other factors kept in mind. That's why, if you don't have a lot of experience with these topics, you should work together with either a functional medicine doctor or someone well-versed with labs.

- **Cortisol Levels** – Cortisol being the main stress hormone can indicate how stressed out your body is overall. Because it fluctuates diurnally, fasting cortisol levels can be expected to be already elevated hours after waking up. Blood tests for cortisol are done in the early morning to establish a common standard for lab results. The normal cortisol range is between 6-23 mcg/dl. Higher than that may indicate either a bad night's sleep or just excessive stress. Testing later in the day can be misleading because of confounding variables like caffeine intake, work-related stress, exercise, or something else.
- **Thyroid Stimulating Hormone** – TSH regulates the amount of thyroid hormones in your blood. With low thyroid, you may see an increase in TSH and with hyperthyroidism the opposite. Here are the consensus references ranges for TSH:
    - 0.4 – 4.0 mU/L normal range
    - 4.7 - 10 mU/L hypothyroidism
    - Over 10 mU/L overt hypothyroidism

- 0.1 – 0.5 mU/L hyperthyroidism
- Less than 0.1 mU/L overt hyperthyroidism
- Normal T3 range 100-200 ng/dl
- Normal T4 range 4.5-11.2 mcg/dl

- **Testosterone and Free Testosterone** – With increasing stress, sex hormones like testosterone are one of the first to suffer. It's also noted that the average testosterone in men has been declining by about 1.2-1.3% every year since the 80s[1833,1834]. Men nowadays have about 20-30% less testosterone than their fathers and grandfathers. That's quite shocking and comes to show the negative side-effects of our modern environment and lifestyle. The bigger problem is that the reference ranges have also been decreased to accommodate that negative trend. So, what is considered normal by doctors nowadays is heavily underestimating what's optimal and you may want to aim for higher numbers. Here are the normal ranges for testosterone:

Age	Men Total Testosterone Average (ng/dl)	Men Free Testosterone Average (ng/ml)	Women Total Testosterone Average (ng/dl)	Women Free Testosterone Average (ng/dl)
15-18	300-1200	5.25-20.7	20-75	0.06-1.08
19-40	240-950	5-18	15-70	0.06-1.00
40-49	250-910	4.46-16.4	13-63	0.06-0.95
50-59	210-880	4.06-14.7	10-54	0.06-0.90
60+	200-600	3.67-10	9-46	0.06-0.71

- **Sex-Hormone Binding Globulin (SHBG)** - Sex hormone binding globulin is a glycoprotein that attaches to androgens and estrogens. It's synthesized in the liver and is involved with trafficking steroid hormones.[1835] Sex hormone binding globulin inhibits the function of sex hormones like testosterone. About 54% of testosterone is bound to serum albumin and 44% to SHBG. Only 1-2% of that is unbound or 'free' and thus biologically active and able to enter cells and turn on their receptors. Therefore, the bioavailability of sex hormones is affected by levels of SHBG. Normal SHBG levels for adult premenopausal females are 40–120 nmol/L, for postmenopausal women it's 28–112 nmol/L, and for adult males it's 20–60 nmol/L.
  - SHBG is regulated by thyroid hormones, insulin, androgen/estrogen balance, and diet. It decreases with high insulin, growth hormone, IGF-1, androgens, and prolactin. Anabolic steroids, PCOS, obesity, hypothyroidism, Cushing's syndrome, and acromegaly reduce SHBG as well[1836]. Low SHBG levels are associated with type 2 diabetes[1837].
  - Prolonged calorie restriction of more than 50% increases SHBG while lowering free and total testosterone. DHEA, another steroid hormone, is not affected by this because it lacks affinity for SHBG.[1838] Anorexia or malnutrition also increases SHBG[1839]. Low testosterone from excess SHBG can also lead to metabolic syndrome[1840].

- Things that increase SHBG are estrogen, pregnancy, cirrhosis, anorexia, and thyroxine (T4). Oral contraceptives or birth pills are associated with higher SHBG[1841]. This will make SHBG bind to sex hormones as it's supposed to. It might also explain why women on birth control tend to gain weight and have mood changes[1842].

- **Pregnenolone (P5)** – The mother of all hormones as it's called, pregnenolone is a precursor to all the other steroid hormones. Studies show it might help to combat fatigue[1843], improve memory[1844], protect against cognitive decline[1845], lower symptoms of depression[1846], and stress. Stress steals pregnenolone from other hormones and causes deficiencies. Here are the normal circulating levels of pregnenolone[1847]:
  - Men: 10 to 200 ng/dL
  - Women: 10 to 230 ng/dL
  - Children: 10 to 48 ng/dL
  - Adolescent boys: 10 to 50 ng/dL
  - Adolescent girls: 15 to 84 ng/dL
- **Progesterone (P4)** – Progesterone is also a steroid hormone crucial for the production of other steroid hormones. It improves HPA function[1848], reduces wakefulness in sleep[1849], lowers inflammation[1850], modulates immunity[1851], strengthens bone osteoblasts[1852], increases skin firmness[1853], and helps produce thyroid hormones. P4 is also depleted by cortisol through the

pregnenolone steal. With low progesterone, the balance with estrogen is off, resulting in higher estrogen levels in the body. This can cause bloating, PMS, breast tenderness, increased risk of fibroids, and ovarian cancer. Elevated progesterone reduces the sodium-retaining activity of aldosterone, which causes natriuresis and loss of extracellular fluid. Reduced progesterone, however, results in temporary sodium retention because of increased aldosterone[1854]. Progesterone also enhances serotonin receptors in the brain, which causes neurochemical imbalances. That's why some people with either too low or excess P4 may resort to other serotonin-boosting activities like alcohol, smoking, drugs, or other substance abuse[1855].

- In women, progesterone levels are relatively low before menstruation (less than 2 ng/mL), they rise after ovulation (above 5 ng/mL), and are elevated during the luteal phase. During the third trimester, P4 can go up to 200 ng/mL. The reference range for adult males is 0.27-0.9 ng/mL.
- To increase progesterone naturally, you can reduce estrogen, get enough zinc[1856], magnesium, and avoid alcohol[1857]. Vitamin B6 helps to lower stress and break down by-products of estrogen, reducing estrogen dominance[1858]. Cruciferous vegetables like broccoli, cauliflower, and cabbage are great for lowering estrogen dominance because of their

phytonutrient content[1859]. Mushrooms and plant polyphenols have the same effect[1860].

- **Oxidative Stress and Inflammation** – The best biomarker for oxidative stress and inflammation is C-reactive protein or CRP. CRP is produced by the liver in response to inflammation and oxidative stress. It can assess the risk of chronic stress-related diseases like CVD or atherosclerosis. Generally, lower is better but not completely zero because you do need minute amounts of inflammation for healthy redox balance.
    - 0.1-1.0 mg/l is low risk
    - 1-3 mg/l is moderate risk
    - 3-10 mg/dl is high risk
    - Over 10 may indicate an acute coronary process like a heart attack
- **Hemoglobin A1C** - HbA1c tells you your average blood sugar level over the past 2-3 months. It detects glycated hemoglobin, which is a protein found in red blood cells. You can use it to assess your overall glycemic health and blood glucose control as well. Here are the normal ranges for people without diabetes:
    - Normal 4-5.7%
    - Prediabetes 5.7-6.4%
    - Diabetes above 6.5%
- **Complete Blood Cell Count** – There are many different types of cells circulating your blood at any given moment.
    - White blood cells also called leukocytes help fight infections and bacteria. They may be elevated if

you're battling a virus, organ failure, or are inflamed. Low WBC can happen from an infection or taking certain medications. Children and adolescents have more WBCs than adults.

- Red blood cells or erythrocytes pick up oxygen and release it to the cells of the body. High RBC results from smoking, alcoholism, air pollution, and lung disease. Low RBC indicates anemia, bleeding, IBD, and malnutrition.
- Platelets are the ones that help your blood clot in response to injuries, cuts, or endothelial damage. High platelets may indicate progressing atherosclerosis. A low score can happen during pregnancy or autoimmunity.

	Red Cells per microliter (µL) of blood	**White Cells per microliter (µL) of blood**	**Platelets per microliter (µL) of blood**
Men	4.7 to 6.1 million	5,000 to 10,000	150,000 to 400,000
Women[2]	4.2 to 5.4 million	4,500 to 11,000	150,000 to 400,000
Children[3]	4.0 to 5.5 million	5,000 to 10,000	150,000 to 400,000

- **Cholesterol** – As said before, cholesterol alone isn't that dangerous as long as it doesn't oxidize. The problem is that higher cholesterol increases the likelihood that some of it will eventually become oxidized because of uncontrollable factors. That's why a precautionary measure would be to still try and keep your cholesterol in the normal range.

- **Total Cholesterol – 200 mg/dl (5 mmol/L) or less**. However, it's mostly irrelevant because of not giving enough detail about HDL and LDL, etc.
- **Triglycerides – 150 mg/dl (1.7 mmol/L) or less**. That's a good guestimate because you would expect triglycerides to go down when eating fewer carbs and burning more fat for energy.
- **HDL Cholesterol – 40 mg/dl (1.0 mmol/L) or more**. More HDL tends to be better and if you're eating a healthy diet you should see HDL at about 50-60 mg/dl.
- **LDL Cholesterol – 100 mg/dl (2.6 mmol/L) or less**. However, based on new research on low carb ketogenic diets, the amount of LDL for optimal health can vary greatly between people and is determined by many other things.
  - If your triglycerides are low and HDL is high, then it means you're using up those fatty acids. A higher LDL, in that case, isn't that relevant. Inflammation or C-Reactive Protein (CRP) would then also have to be low.
  - If your triglycerides are high and you have high CRP, then a higher LDL and total cholesterol aren't good because you'll have more inflammation, which will create more

scarring of arteries, which then can promote plaque formation.

  - For hunter-gatherers, free-living primates, healthy human neonates, and other wild animals, the normal LDL range is between 50-70 mg/dl, which prevents atherosclerosis[1861]. Based on current medical guidelines, you can have a slightly bigger buffer zone while staying below 100 mg/dl. More is not better nor protective against heart disease. Too much will still be detrimental, even if you think you're eating a healthy diet.
  - Some people, around 20%, have a genetic variation that makes them absorb or synthesize so much cholesterol that their diet does influence their blood cholesterol level[1862]. Even in these hyper-responders, however, a high cholesterol diet does not generally negatively influence their cholesterol profile[1863]. If total blood cholesterol increases at all during a high cholesterol diet, both 'good' HDL and 'bad' LDL cholesterol generally increase in the same proportion.

- **Liver Function** – The liver is one of the most important metabolic organs because it's the central ground for all nutrient signaling and assimilation. It's also subject to all the garbage people eat and are exposed to nowadays. That's why a healthy metabolism requires a healthy liver. There are several readings that indicate how well your liver is working. Here are the most relevant ones.

- **Serum Albumin** – Albumin is a type of protein produced by the liver that keeps fluid from leaking out of blood vessels. Serum albumin can tell how well your liver and kidneys are working. Low albumin can result from inflammation, liver disease, malnutrition, Crohn's, and celiac disease. Normal values range from 3.5-5.4 grams per decilitre.
- **Gamma Glutamyl Transpeptidase (GGT)** – GGT is an enzyme found in liver cells, the pancreas, and intestines. It helps with glutathione and transportation of peptides. Elevated GGT above 9-85 IU/L indicates cholestasis, non-alcoholic fatty liver (NAFLD), alcoholism, and acute pancreatitis.
- **Aspartate transaminase (AST)** – AST is an enzyme found in the liver, heart, muscles, kidneys, brain, and red blood cells. It catalyzes the conversion of different metabolites like aspartate, α-ketoglutarate to oxaloacetate, and glutamate. The reference range is 0-35 IU/L and AST is elevated in liver disease, cirrhosis, and mitochondrial dysfunction.
- **Alanine transaminase (ALT)** – ALT is another enzyme found in the liver, heart, kidneys, and muscle. Any kind of liver injury elevates ALT and it can happen because of hepatitis, cirrhosis, NAFLD, and exposure to toxins. The reference range is 7-56 IU/L.

    - **Alkaline phosphatase (ALP)** – ALP is an enzyme in the biliary ducts of the liver. It partakes in lipid metabolism and transportation. High levels indicate liver damage and hepatitis but low levels can mean hypothyroidism, anemia, and zinc deficiency. The reference range 41 to 133 IU/L.
- **Waist to Hip Ratio** – One of the easiest and most accurate measurements of general metabolic health and body composition is the waist to hip ratio. It measures the ratio between your waist circumference and hip circumference. People with more weight around their midsection are at a higher risk of heart disease, diabetes, and premature death than those who carry it around their hips and thighs[1864]. It also indicates higher stress-induced obesity and visceral fat. You can easily calculate your waist to hip ratio at home
    - First, measure the distance around the smallest part of your waist just above the belly button. That's your waist circumference.
    - Second, measure the distance around the widest part of your hips or buttocks. That's your hip circumference.
    - Third, divide your waist circumference by your hip circumference. That's your waist to hip ratio. A lower score indicates less obesity and is better.
    - According to the World Health Organization, obesity is defined as a waist to hip ratio above 0.90 for males and 0.85 for females[1865].

Health risk	Women	Men
Low	0.80 or lower	0.95 or lower
Moderate	0.81–0.85	0.96–1.0
High	0.86 or higher	1.0 or higher

- **DNA Methylation Age -** DNA methylation age of blood can predict all-cause mortality in later life, which strongly relates it to aging[1866]. It's based on Horvath's Epigenetic Clock Theory of Aging, which is considered currently one of the most accurate measurements of biological age[1867]. As of 2019, there is some, although inconsistent, evidence for an association between DNA methylation age and disease risk[1868]. Studies on over 41 000 participants have found that each 5-year increase in DNA methylation age was associated with an 8-15% increased risk of mortality. You can measure your DNA methylation age with at-home collection kits like myDNAge. It's a simple blood or urine test that can be taken at home and sent back for analysis. I took the myDNAge test at the beginning of 2020 at the age of 25. My DNA methylation age results say I'm 16 years old and 99% younger than other people at my age. Honestly, I wasn't surprised because I know what I'm doing and practice what I preach. It's the result of everything we've talked about in this book and my previous one Metabolic Autophagy.

All these tests might be overwhelming and confusing but they're still relevant for overall health and understanding where your body is at. If you're looking for a way to specifically assess stress and hormones, then check out the DUTCH test. It measures free cortisol, free cortisone, creatinine, DHEA, tetrahydrocortisone, progesterone metabolites, androgen metabolites, estrogen metabolites melatonin, and other relevant biomarkers. You're able to identify any circadian mismatches in your cortisol production as well as the by-products of metabolizing these hormones. Maybe your cortisol levels are fine but you're breaking it down too fast because of having high metabolites.

For a comprehensive overview of your nutritional status and metabolism, you can take the organic acid test. It measures the by-products of chemical reactions inside your body that get excreted through urine[1869]. Abnormalities occur usually from B-vitamin deficiencies[1870], insulin resistance, inflammation, oxidative stress, gut dysbiosis[1871], lactic acidosis[1872], impaired fatty acid metabolism, mitochondrial dysfunction[1873], and poor detoxification[1874].

Measuring Stress

You can also measure stress with psychological testing. The Holmes and Rahe Stress Scale (HRSS) rates 43 stressful life events on a scale of 100 and adds them all up. Depression Anxiety Stress Scales (DASS) is based on 42 self-reported stressors.

In 1970, Holmes and Rahe used their stress scale to keep track of the health of 2500 US sailors[1875]. They were asked to rate their life events over the next six months. There was a +0.118 correlation between higher stress and disease, which was significant enough to validate their hypothesis[1876]. Several other studies have supported these findings[1877] and it's been tested cross-culturally as well[1878].

Here's the Holmes and Rahe Stress Scale (HRSS):

Life Event	Life Change Units
Death of a spouse	100
Divorce	73
Marital separation	65
Imprisonment	63
Death of a close family member	63
Personal injury or illness	53
Marriage	50
Dismissal from work	47
Marital reconciliation	45
Retirement	45
Change in health of family member	44
Pregnancy	40

Sexual difficulties	39
Gain a new family member	39
Business readjustment	39
Change in the financial state	38
Death of a close friend	37
Change to a different line of work	36
Change in frequency of arguments	35
Major mortgage	32
Foreclosure of mortgage or loan	30
Change in responsibilities at work	29
Child leaving home	29
Trouble with in-laws	29
Outstanding personal achievement	28
Spouse starts or stops	26

work	
Beginning or end of school	26
Change in living conditions	25
Revision of personal habits	24
Trouble with boss	23
Change in working hours or conditions	20
Change in residence	20
Change in schools	20
Change in recreation	19
Change in church activities	19
Change in social activities	18
Minor mortgage or loan	17
Change in sleeping habits	16
Change in number of family reunions	15

Change in eating habits	15
Vacation	13
Major Holiday	12
Minor violation of law	11

Score of 300+: High risk of stress-related illness

Score of 150-299: Moderate risk of illness (30% less from the above risk)

Score <150: Only a slight risk of illness

DASS is a test to be taken over 5-10 minutes. It consists of 42 items that each reflect on negative emotional symptoms[1879]. They're rated on a four-point Likert scale of frequency over the last week. The purpose is to identify emotional disturbances and degree of depression. It's recommended to do this with a qualified psychologist or therapist combined with other assessments. That's why the HRSS is a more immediate and faster way to assess your current level of stress in the last few months or year.

Kidney Health and Stress

The body's homeostasis is very much linked to the functioning of the kidneys because they help to maintain the „internal milieu"[1880].

Part of this is due to the regulation of fluids, electrolyte balance, and the intricate connection with the HPA axis. Remember that the adrenal glands are located on top of the kidneys and they produce cortisol, including other steroid hormones. Therefore, your ability to cope with stress and regulate inflammation is very much dependent on your kidney and liver function as well.

Your kidneys are fist-sized organs located at the bottom of your rib cage. They filter waste, control fluid balance, activate vitamin D, regulate the pH, and electrolytes. Blood pressure regulating hormones are also produced by the kidneys. Unhealthy kidneys can lead to hypertension, cardiovascular disease, neuropathy, anemia, and kidney failure. Interestingly, those conditions are very much linked to excess stress on the body, which in turn can be caused by inadequate kidney functioning. Another chicken-and-egg problem.

To assess kidney health and functioning, you can get tested using various methods. You should consult with your physician first and do post-analysis.

- **Blood Pressure** – hypertension damages small blood vessels in the kidneys. Most people are below 140/90. If you have kidney failure, you'd want to be lower than 130/80 but 120/80 is the best.
- **Protein in Urine** – traces of albumin protein in urine is an early sign of kidney disease. Optimally, you want to have less than 30 mg-s of albumin per gram of urinary creatinine.

- **Serum Creatinine** – poor kidney functioning leads to the accumulation of creatinine in the blood, which is a normal waste product. A good score is 0.6-1.2 mg-s per deciliter of blood. However, it can vary, depending on other variables.
- **Glomerular Filtration Rate (GFR)** – this is a calculation of kidney functioning based on creatinine levels, age, race, and gender, etc. A score over 90 is good, 60-89 should be monitored, and less than 60 for three months indicates kidney damage.

To improve kidney health, you want to maintain an active lifestyle, try to lower your blood pressure, don't overeat excessive amounts of protein, stay hydrated, and manage your blood sugar. You should also avoid smoking, not take over-the-counter pain medication, and prevent fat gain.

Stress Management Tactics

Stress management involves a wide range of activities and techniques. It's supposed to be a proactive way for a person to control their stress levels and prevent it from becoming chronic. However, a lot of these tactics are also useful for managing already existing allostatic overload.

Here's a list of different stress management techniques you can try:

- **Autogenic Training** – The German psychiatrist Johannes Heinrich Schultz has developed this desensitization-relaxation technique to induce deep relaxation in 1932. It's based on repeated relaxing visualizations while concentrating on bodily perceptions such as heaviness, warmth, chills, or anything the like[1881]. You can perform this while sitting or staying in a horizontal position for a few minutes every day. Here are the six standard exercises prescribed by Schultz:
    - Muscular relaxation by repeating "my right arm is heavy". The goal is to quickly induce the feeling of heaviness in the particular limb that you're focusing on.
    - Passive concentration on warmth by repeating "my right arm is warm". Passiveness refers to allowing sensations to just happen and observing them without judgement.
    - Controlling cardiac activity by repeating "my heartbeat is calm and relaxed".
    - Passive concentration on respiration by repeating "my breath is calm and relaxed".
    - Passive concentration on the warmth of the abdomen by repeating "my stomach is warm".
    - Passive concentration on the coolness of the cranial region by repeating "my forehead is cool".

- **Emotional Freedom Technique** - If you're feeling anxious, stressed out, or just can't fall asleep, then try the Emotional Freedom Technique with tapping. It's acupuncture therapy without the needles. You can find tutorials on how to do it properly online. Essentially, you tap on acupressure meridians to release blockages. You start by acknowledging a certain problem or fear such as: *„Even though I'm afraid of being inadequate, I completely accept myself the way I am."* While repeating this statement three times, you keep tapping your finger on the karate chop point. Then you assess your feelings on a scale of 1-10.
- **Cognitive Therapy** – CT was developed by an American psychiatrist Aaron T. Beck in the 1960s. It's based on the idea that thoughts, emotions, and behavior are all connected. To overcome self-sabotage or other obstacles, one has to identify and fix dysfunctional thinking patterns and responses. This involves working with a therapist but with enough self-awareness and analysis you can also become aware of why you do certain things and how to overcome unwanted thought loops. But yea, if you want to really understand yourself better, then talking with a therapist would be pretty effective.
- **Cranial Release Techniques** – There are different types of massages but they can all help to relieve tension and stress that gets built up in the body. Muscles can become tight and sore from stress, over-training, and sedentary living. This, in turn, can become an additional source of pain and

stress. It's almost as if the body will be constantly perceiving that limited range of motion as a threat. That's why getting a massage from a professional at least once a month is a very effective strategy for preventing these imbalances. You can also use a regular foam roller, tennis ball, or massage gun at home daily.

- **Yoga Nidra** – This is a yogic sleep exercise that puts you into a state of consciousness in between wakefulness and sleeping. You're awake while simultaneously feeling as if you're about to slip into dreams. It includes aspects of autogenic training and muscular relaxation. Usually, it's done with the help of guided audio meditation whether that be from a music player or an instructor. What I do is search for a short 20-30 minute yoga nidra soundtrack on YouTube and take a nap while listening to it. It's like a power nap on steroids and I always feel incredibly rejuvenated afterward.
- **Mindfulness** – Trying to suppress thinking and emotions can itself become a big source of stress, especially if you're already stressed out. Mindfulness is about becoming aware of one's sensory experience in the present moment without judgement. It involves meditation and other similar techniques. Modern psychology has shown mindfulness to be beneficial in reducing depression, stress, anxiety, and even addiction[1882]. There are several exercises you can try:
    - **Seated focusing on one's breath.** Pay attention to the movements of the abdomen while breathing. You shouldn't try to control your respiration.

Instead, just be aware and notice it. The mind will inevitably start to wander and if it happens then simply return to the breath.

- **Body-scan meditation.** Either lay down or stay seated and direct your attention to different areas of the body such as the hands, feet, fingers, abdomen, or spine. You can also move from the top of your head down to the toes and vice versa.
- **Focus on the environment.** You start focusing on any kind of sensory perception such as sounds, smells, feelings, or sensations. Just observe it with no judgement. Mindful eating can be categorized here as well.
- **Yoga exercises.** Physical movements and yoga positions can be used to practice mindfulness whilst engaged in the exercise. Walking meditation, running, qi gong, tai chi, or hand-balancing are all quite meditative. If they help your mind stop thinking about work or responsibilities, then it'll also lower your stress.

- **Dynamic Meditation** – This is a technique I learned from my friend Elliott Hulse who's a strength coach, entrepreneur, and philosopher. I highly recommend checking out his legendary YouTube channel where Elliott showcases similar bioenergetics exercises and generally good advice for life. Dynamic meditation involves spontaneous dancing, shaking, shouting, hopping, and anything else that would blow off steam so to say. Elliott says the opposite of depression is expression and more often than not we tend to bury our stressors within the body. By letting go physically, it's also possible to release this tension. Elliott describes this process in 5 stages, where each can last anywhere from 1-2 or up to 10 minutes:

1. Chaotic breathing – you're charging your body with chaotic breaths, focusing on the exhalation for.
2. Catharsis or purification of emotions – you start to move your body completely irrationally and in whatever way comes to mind.
3. Grounding – raise your hands in the air and hop with your feet flat to the ground while letting out some vocals.
4. Tranquility – drop to the floor or stand up but stay motionless.
5. Dance – play some peaceful music or hum in your head while slowly dancing in rejoice.

- **Acupuncture Therapy** – If you have the opportunity, then going to an acupuncture therapist is a potential strategy for dealing with stress. It's one of the oldest healing practices in the world and has been shown to promote recovery from muscle fatigue, physical exertion, and pain[1883]. Symptoms of depression, anxiety, and headaches can also be relieved[1884]. You're inserting very thin needles into certain points throughout the body, which stimulates blood flow and releases endorphins. In movies, you might have seen big Chinese or Japanese mafia bosses getting acupuncture for stress relief and management. Maybe they're on to something. I mean, they probably have incredibly stressful jobs.

- **Forest Bathing** – Spending more time in nature is in most cases the best way to lower stress. This term was coined by the Japanese Ministry of Agriculture in the 1980s and it translates from "Shinrin-yoku". Although hiking and trail-running can be a form of forest bathing, the goal is to focus on fully immersing oneself in nature and taking a slow walk. Forest bathing has been shown to boost immunity, lower inflammation[1885], reduce stress, promote anti-cancer killer cells[1886], improve CVD risk factors[1887], and decrease blood glucose[1888]. I highly value my daily hour-long walks in nature because they expose me to all the essential oils and other natural particles but also help with creativity and productivity. There's a definite difference between people living in cities and those at the countryside. The high-speed environment has a subconscious effect on your general mental stimulation and stress levels. That's why city dwellers especially would benefit from daily walks in parks or forest paths.
- **Whole Body Vibration Therapy** – A small vibration platform can be an amazing recovery tool from workouts. It's been shown to promote bone density, reduce inflammation, and benefit the organ systems[1889]. I like to use it during the workday to break up long periods of sitting or stagnation with a bit of lymph flow. You can also do push-ups or squats on it for some extra challenge.

- **Pulsed Electromagnetic Frequencies (PEMF)** are low to medium intensity magnetic frequencies that are pulsating at a particular frequency. Studies have shown PEMF devices to improve symptoms of arthritis[1890], chronic pain[1891], fibromyalgia[1892], osteoporosis [1893], PTSD, and Parkinson's[1894]. PEMF therapy is most used by NASA astronauts because they have to spend a lot of time out in space away from the Earth's natural magnetic field. You can use PEMF mattresses to sleep upon or wear as a wristband.

It's important to note that what's relaxing for one person may not be so for someone else. Otherwise, it can become an additional stress. That's why you have to find what works for you in terms of relaxation and lowering allostatic load.

A critical thing to remember as well is to not eat when you're stressed out. The reason is that when you're in a sympathetic state, it's more difficult to digest food and feel satiated from it. When in fight or flight, most of the blood gets directed from the gut into the extremities, which lowers stomach acid and can cause bloating, constipation, and dysbiosis. That's why you always want to eat in a calm and relaxed environment to promote digestive enzymes and enjoy the meal fully.

Here are the foods you want to avoid if you're under stress:

- **Caffeine** – coffee and other caffeinated beverages raise cortisol and activate the sympathetic nervous system. Chronic stress combined with too much coffee is a recipe for chronic fatigue and adrenal overload.
- **Inflammatory foods** such as pastries, chips, cookies, candy, refined carbohydrates, trans fats, vegetable oils, etc. They promote oxidative stress and make you more inflamed.
- **A high-fat high carb diet** induces insulin resistance and inhibits hippocampal synaptic plasticity[1895]. People with HPA dysfunction often crave sugar and carbohydrates because it'll pick them up and gives a short increase in energy. A lot of addictive behavior is also caused by chronic stress.
- **High dose zinc supplementation** impairs memory and inhibits BDNF signaling[1896]. You should focus on getting zinc from whole foods like organ meats, meat, fish, pumpkin seeds, and legumes.

Some people will also resort to alcohol or comfort foods when they're stressed. Maybe 1-2 drinks for relaxation can be useful for calming your nerves and getting your mind out of its own rut. However, it's only a quick fix that doesn't address the underlying cause of why you're stressed out. If you have a stressful job, then first assess whether or not you want to keep on doing it. Secondly, get good at implementing stoicism and strengthening your psychological resilience. Part of it is the result of conditioning and succeeding in overcoming stressors but it's very much a mindset thing and determined by how you choose to respond to negative events. Reading stoic philosophers and practicing gratitude can help to put things into perspective by realizing that most of our problems are self-created and not that big of a deal.

Eating comfort foods in response to stress or emotional turmoil is a slippery slope because although you may find some brief relief it's only temporary. You're more than likely to get additional cravings and dissatisfaction after the fact. What's worse, you're teaching your brain to start wanting more junk whenever you're stressed. That's how neuroplasticity works and if your hippocampus equates donuts with happiness, then you'll never be happy without them. Of course, a flexible diet plan can fit any kind of food as long as you're not over-eating calories. It's just that it's much harder to stop eating hyperpalatable foods once you've started to eat them, especially if you're also loaded to the gills. I would much rather save my comfort foods for some special events like birthdays or something instead of letting stress hijack my self-control.

Adrenal Fatigue Fact or Fiction

A lot of people who suffer from stress also report chronic fatigue, exhaustion, and depression. The term used to describe it is

„adrenal fatigue“ but some people say it's a made-up condition that doesn't really exist.

Adrenal glands are tiny organs on top of your kidneys that produce a lot of hormones related to the fight or flight response such as cortisol. They're also a part of the HPA axis, thus they're important for balancing the nervous system.

Symptoms of adrenal dysfunction include anxiety, depression, muscle soreness, sleep disturbances, constipation, bloating, and circulatory problems. However, those symptoms can be caused by a myriad of things and Chronic Fatigue Syndrome (CFS) or adrenal fatigue are just too vague to pin-point the exact issue. Some doctors test for adrenal fatigue by measuring cortisol, thyroid-stimulating hormone (TSH), free T3, free T4, DHEA, and ACTH hormones. You can also order test kits and take them at home.

Most people who think they have adrenal fatigue are just sleep-deprived, their circadian rhythms are misaligned, they're overweight, they eat the wrong food, they don't get enough sunlight, they're not exercising, and or too stressed out. It's both a chicken and an egg type of problem. Adrenal insufficiency, on the other hand, is something a lot more common than full-blown adrenal fatigue. The two are different conditions although very similar.

If you feel tired and fatigued in the morning, then you have to fix your circadian rhythm and optimize your sleep. You must dig deep into optimizing everything and re-entraining your circadian rhythm. That's the only way you regain the vigor you should have after waking up. Your body might also be addicted to caffeine or sugar and needs time to re-habituate itself to function without stimulants.

Heart Rate Variability

Let's move on with actual physiological measurements, starting with Heart Rate Variability (HRV). It's the physical phenomenon of variation in the time interval between heartbeats. The measurements are observed by the variation in the beat-to-beat interval of heartbeats.

An average person's heart rate is about 60-100 beats per minute. However, it doesn't mean that your heart is beating once every second like a clock. There's variation between the heartbeats and HRV measures those intervals within a certain timeframe. Different devices use different units of measurement like milliseconds for up to 5 minutes. If the intervals between your heartbeat are consistent and repetitive, your HRV is low. With a greater variation between lengths, your HRV is higher.

Your HRV tells what state your nervous system is in, how recovered you are, and the amount of vagal tone. Generally, a higher HRV indicates being more recovered, less stressed, and higher parasympathetic activity. Decreased HRV has been shown to be a predictor of mortality after myocardial infarction[1897], cancer[1898], and sudden cardiac death[1899]. Lower HRV is also associated with heart failure, diabetic neuropathy[1900], and liver cirrhosis[1901].

Structuring your workouts and physical events around days when your HRV is highest is a great way to cause less stress and damage to your body. It can also be used to predict catching a cold or a virus. Usually, when you see an elevation in heart rate, body temperature, and a drop in HRV, you can expect to get sick unless you focus more on recovery and healing.

There are several methods used to track your HRV, depending on what technology you're using. Electrocardiogram (ECG) detects the R wave in the QRS complex and calculates the time between R waves (R-R interval). Something like the OURA ring uses photoplethysmography (PPG) to analyze your heartbeats. With PPG, the steepest increase in the signal prior to the peak marks a heartbeat. Instead of R-R intervals, PPG measures interbeat intervals (IBIs). See figure below to see what's the difference between ECG and PPG.

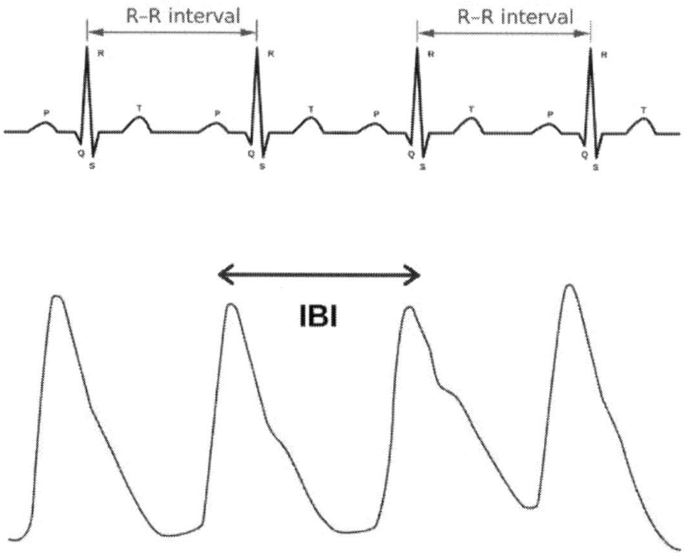

Based on my OURA ring data (See figure below), my average HRV is 110-180 and I've never seen it drop below 100. That's astronomically high compared to other biohackers I know and they're always amazed by this. Usually, most people stay around 30-60. Maybe I'm just very parasympathetically toned and fit but maybe it's just my ring telling me lies – it doesn't matter. What matters is just your individual baseline HRV status and how you deviate from it. All your decisions have to be based on what's normal for you.

You shouldn't compare your HRV with other people, because it's affected by many factors such as age, sex, hormones, physical condition, genetics, and lifestyle. It also fluctuates based on what you did the previous day etc. Using HRV should be thought of as recognizing the direction your nervous system is heading towards. First, you must monitor your HRV consistently for a few weeks to establish a baseline for your condition. Then you want to measure it on and off to see what's the trend like – are you improving, maintaining homeostasis, or getting worse.

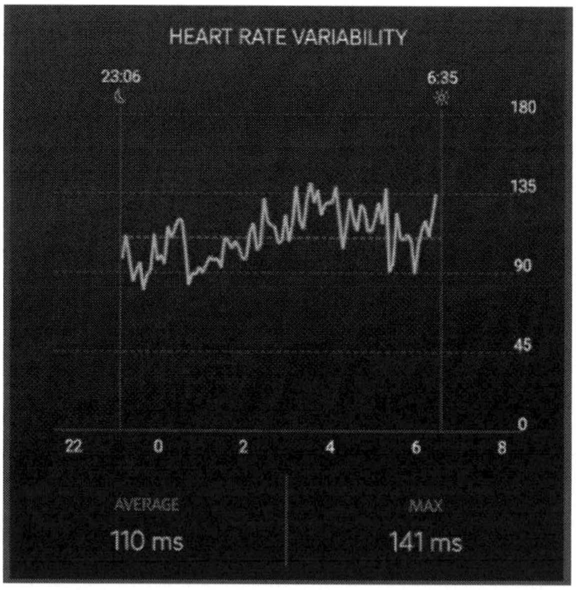

There are different ways to calculate HRV, but they all have to do with the amount of variation in the intervals between heartbeats. Oura uses rMSSD (Root Mean Square of the Successive Differences), which is the most commonly used HRV formula.

What Lowers HRV

Here are the things that lower your HRV and recovery:

- **Worrying** - Daily worry is related to low heart rate variability during waking and the subsequent nocturnal sleep period[1902]. If you're the worrier type, then I recommend getting more into stoicism and meditation.
- **Emotional Strain** – Time pressure, social anxiety, negative emotions, trauma, and stressful interactions

decrease HRV and make you more sympathetic dominant[1903]. To prevent this from having a negative effect, you have to increase your psychological resilience.

- **Post-Traumatic Stress Disorder** – PTSD victims experience more autonomic hyperactivity during rest and they cope with stress much worse[1904]. Working with a cognitive therapist or a psychologist might be necessary to fix these underlying issues.
- **Blood Sugar Problems and Insulin Resistance** – Big fluctuations in blood glucose cause stress on the pancreas and brain, which produces cardiac vagal withdrawal, thus lowering HRV[1905]. Diabetics have lower HRV and signs of early cardiac neuropathy[1906]. Eating too much sugar and processed food would also affect psychological resilience negatively.
- **Inflammation** – Higher pro-inflammatory cytokines decrease HRV, damage healthy cells, and create oxidative stress[1907]. Heart rate variability also predicts levels of inflammatory markers like CRP, IL-6, and glycated hemoglobin[1908]. With chronic low-grade inflammation, your body never has the chance to rest.
- **Eating Too Close to Bedtime** – Digesting food at night is a hassle and a big mistake. It's going to raise your body temperature, keeps the gut working, makes it harder to fall asleep, interferes with deep sleep, promotes acid reflux, bloating, and also decreases HRV. Stop eating at least 2-4 hours before going to bed. Alcohol also lowers HRV[1909].

- **Sleep Deprivation** - Poor sleep has been shown to decrease HRV during working hours[1910]. In a study on young healthy males, 24 hours of total sleep deprivation didn't reduce HRV but it did abolish the relationship between attention performance and HRV[1911]. So, if you're fully optimized and recovered on a regular basis, then short-term sleeplessness or disruptions shouldn't become a detriment. Chronic short sleep is worse than sleeping consistently all the time and then experiencing a massive shortcoming. HRV can be potentially used to predict an individual's decrements in psychomotor vigilance and attention caused by sleepiness[1912]. It can also identify sleeping disorders and other related ailments[1913].

How to Increase HRV

Here are the things that increase your Heart Rate Variability and parasympathetic tone:

- **Exercise** – Physical training decreases the risk of cardiovascular mortality and disease. Individuals who exercise regularly have a lower resting heart rate and higher HRV than sedentary people. During submaximal exercise, HRV tends to be lower than at rest because of the shifts in the parasympathetic and sympathetic nervous system[1914].
- **Fasting** – Being in a fasted state tends to lower your heart rate and increase HRV because the body tries to conserve

more energy[1915]. However, if you go hypoglycemic or become stressed out, then it'll lower HRV because of the increased stress[1916]. That's why fasting in ketosis and while being keto-adapted causes less harm compared to doing so on a sugar-burning metabolism. A 48 hour fast has been shown to cause parasympathetic withdrawal, thus lowering HRV[1917]. That's why you should do it intermittently though to avoid slowing down metabolic rate or thyroid functioning. I attribute my own high HRV scores mostly to time-restricted eating. Maybe it just requires you to get used to it.

- **Saunas** – Taking a sauna mimics a cardiovascular workout to a certain extent that improves aerobic fitness. It will also increase HRV and reduces risk of all-cause mortality[1918]. Just pay attention to your blood pressure and other biomarkers. If the heat makes you too stressed out afterward, your HRV will actually drop. The right dose is critical.
- **Cold Exposure** – Ice baths, cold showers, winter swimming, and polar plunges will also improve parasympathetic tone by lowering inflammation, making you more tired, and promoting lymph flow. Sympathetic activity rises briefly during the shock but lowers down afterward[1919]. You don't want to over-do it, however. In diabetics, cardiac sympathetic activity measured by HRV has been found to indicate myocardial ischemia in response to cold exposure[1920]. Thankfully HRV can be used to

predict these kinds of things by assessing your nervous system's tone at a particular moment.

- **Biofeedback** – Some form of neuro- or biofeedback reduces stress, anxiety, depression, and increases HRV[1921]. Neurofeedback can be trained at the right facilities like the Peak Brain Institute in LA led by my friend Dr. Andrew Hill. Programs like 40 Years of Zen or Biocybernaut also offer more thorough training that lasts for a few days.
- **Music Therapy** – One study found that playing Native American flutes increased HRV and reduced stress[1922]. Listening to music is great for relaxation and therapy. Humming and singing have been found to benefit HRV because it requires guided breath control[1923]. Just playing around and taking time to do the things you love is also amazing for relieving stress and getting out of fight or flight.
- **Meditation** – Sitting down to meditate can be a great way to lower stress and promote parasympathetic activity. It will calm you down, relaxes the organs, and thus increases HRV[1924]. The same applies to practices like yoga or Tai Chi. Different meditation practices have been shown to improve HRV, reduce heart rate and blood pressure[1925]. A 2016 review of 59 studies with 2358 participants found that yoga increases HRV and regular yoga practitioners had higher vagal tone[1926].
- **Sleep and Circadian Optimization** – Parasympathetic activity and HRV are supposed to be higher during

nighttime, especially in REM sleep[1927]. At least if your circadian rhythms are synchronized. Myocardial infarction has been shown to decrease this expression[1928]. Sleep is the time repair hormones like melatonin are supposed to repair and heal the body.

You don't want to live a completely stress-free life because it'll make you less prone to dealing with unexpected stressors but you don't want to be stressed out long-term either. Think of it as elusive cycling between stress exposure, recuperation, recovery, and adaptation. This way you'll have enough time to deal with it and get better.

Vagus Nerve

The parasympathetic control of the heart, lungs, and digestive tract is controlled by the vagus nerve (VN). In Latin, 'vagus' means 'wandering' and it refers to how you have many cranial nerves branching out throughout the body. The VN is the tenth cranial nerve or CN X. Being the longest autonomic nervous system's nerve, it runs from the brain down to the abdomen. It regulates heart rate, digestion, relaxation, mood, and immunity. What people call 'gut feeling' is very much mediated by the vagus nerve.

Poor vagal functioning is associated with poor sleep, fatigue, and digestive disorders[1929]. Stress inhibits the vagus nerve and has negative effects on the microbiota in the gastrointestinal tract[1930]. Because of how it regulates your appetite, a damaged or dysfunctional VN can lead to obesity[1931]. Furthermore, several studies show that vagal tone can be passed from mother to child, meaning if a mother experiences anxiety or depression during pregnancy, their child is more likely to experience the same effects[1932].

Stimulating the vagus nerve has been used to treat epilepsy, depression, neurological disorders, multiple sclerosis, Alzheimer's, and inflammation[1933,1934,1935,1936]. It's also been used for anxiety, addiction, chronic pain conditions, arrhythmias, and autoimmunity[1937,1938,1939]. Devices for this kind of treatment may also have negative side-effects, such as shortness of breath, nausea, headaches, and cardiac arrest[1940].

Functioning of the vagus nerve is called vagal tone and it reflects the parasympathetic nervous system. In most cases, it's not measured directly but as an assay of heart rate and heart rate variability[1941]. Improved RHR and HRV indicate you're more recovered and parasympathetically toned. According to Grossman and Taylor (2007), higher vagal tone *„reflects a functional energy reserve capacity from which the organism can draw during more active states"*[1942]. It essentially increases adaptive energy and can probably restore it.

You can also stimulate the vagus nerve at home without any fancy gadgets or expensive equipment. It's great for assessing your recovery and state of the nervous system as well as promoting relaxation.

- **Slow-Paced Breathing** – Deep breaths into the abdomen increase vagal tone[1943]. Coupling that with positive emotions is linked with greater coherence and better outcomes in personal health[1944]. Lengthening the exhale has also been shown to slow down heart rate and blood pressure[1945]. So, you can induce a deeper state of relaxation by breathing out longer than you inhale. For example, inhale for 3 seconds and exhale for 6. What I like to do is dedicate about 5-10 minutes of slow breathing in the mornings whenever I feel under recovered, tired, or stressed out. It's great for automatically lowering sympathetic arousal and promoting relaxation. Just sit or lay down with your hands across your diaphragm. Start breathing slowly as if a balloon is blowing up in your stomach.

- **Improve Gut Health** – A 2018 study in the journal *Frontiers in Neuroscience* stated that the microbiome affects psychiatric and inflammatory disorders via vagal afferent fibers in the gut[1946]. The effects of gut bacteria on mood and inflammation may be partly mediated by vagal tone. Histamine intolerance and low stomach acid can result from poor vagus nerve functioning[1947]. To keep your gut healthy, avoid processed carbs, rancid fats, and allergens. Instead, eat pre- and probiotic foods like fermented vegetables, sauerkraut, asparagus, onions, and some legumes. The *Lactobacillus* strain has been found to promote gut health via the VN[1948]. You can get it from kefir, yogurt, sauerkraut, miso, tempeh, sour cream, and cheese, granted it's unpasteurized. Stress management that involves meditation, hypnosis, or cognitive therapy can also improve both gut health and vagal tone[1949].

- **Cold Exposure** – Apparently, colder temperatures increase VN activity by releasing certain neurotransmitters in the gut[1950]. It's been found that if you're not used to the cold it can be a stressor but after acclimatization it will boost parasympathetic tone[1951]. So, the key is to condition yourself with cold showers and not freak out while doing it before jumping into harder plunges.
- **Gargling** – Gargling water at the back of your throat like during washing your teeth can stimulate the vagus nerve. It works similar to humming or throat singing.
- **Mild Exercise** – Exercising stimulates the vagus nerve and improves digestion[1952]. Different yoga techniques have also had a similar effect[1953]. Yet again, overtraining probably diminishes it.
- **Neck and Foot Massage** – Massaging the carotid sinus located in your neck stimulates the vagus nerve and lowers blood pressure[1954]. Foot massage has found to have similar benefits[1955]. It's recommended to get a massage from a professional but you can easily relieve stress by using either a foam roller or a regular tennis ball. Just lay on your back and put the ball behind your neck. Slowly apply slight pressure and move it around the neck. It's great to do at the end of the day to wind down and de-stress. You can do the same with the balls of your feet and calves.

- **Intermittent Fasting** – Yet another benefit to fasting is vagal stimulation[1956]. This can also improve satiety and hunger signaling through the VN.

Sleep Tracking 101

The most important physiological parameter to track is the quality of your sleep. That's where you'll get the biggest return on investment.

There are 4-5 stages of sleep, depending on how many you count. One full cycle lasts for 70-90 minutes and you need about 4-6 cycles per 24 hours. It includes light, deep, and REM sleep. Compared to deep and REM sleep, light sleep doesn't have that much benefit. I'm sure it has a role in recovery and facilitating better REM sleep but most of the physical repair, memory consolidation, and actual healing occurs during deep sleep/REM. That's what I'd always pay more attention to.

It doesn't matter how long you sleep per night or how much light sleep you got if you still feel tired, under-recovered, brain fogged, and didn't get enough deep/REM sleep. Light sleep itself is a superficial stage that's supposed to keep you between full wakefulness and complete deep sleep as to not get accidentally eaten by predators. It's like a fail-safe system for making sure you don't fall asleep in dangerous environments.

With enough practice and total sleep optimization, you can cut down on the total hours of sleep you get per night at the expense of light sleep without sacrificing deep sleep. You'll just teach your body to spend less time in the drowsy peripheral state wherein you aren't really sleeping nor recovering but you can't control your body or thoughts either. It's almost like an empty time that gets wasted.

It's recommended you have some sort of a sleep tracker like the OURA ring, Whoop, Fitbit, or Eight Sleep Mattress. This can tell you how much of each stage are you actually getting and how fast you fall asleep. I recommend the OURA ring because it can be put on airplane mode, you don't feel it while wearing, it's very accurate, and gives data about other important variables like heart rate variability, core temperature, heart rate, and sleep latency.

Here's what to pay attention to when tracking sleep:

- **Light Sleep** (NREM 1-2)– You don't need to get 50% of your total night's sleep as light sleep because that would entail 3-4 hours of light sleep. That's not optimal. The least light sleep you need is about 1-2 hours, which would comprise about 20-30% of the total night's sleep volume.
- **Deep Sleep** (NREM 3) – The minimal amount of deep sleep you should get is 20-30%, which entails about 2-3 hours of deep sleep. More isn't necessarily better but getting up to 35% would be more beneficial rather than harmful. Reaching 4 hours of deep sleep is quite difficult and chances are your sleep tracker is probably categorizing some of your REM sleep as deep sleep.
- **REM Sleep** – The average amount of REM sleep you should get per night is 20-25% of the total night's volume, which entails about 2 hours of REM. Consistently more

than 30% of REM causes hyper brain activity, mood disorders, and other neurological issues.

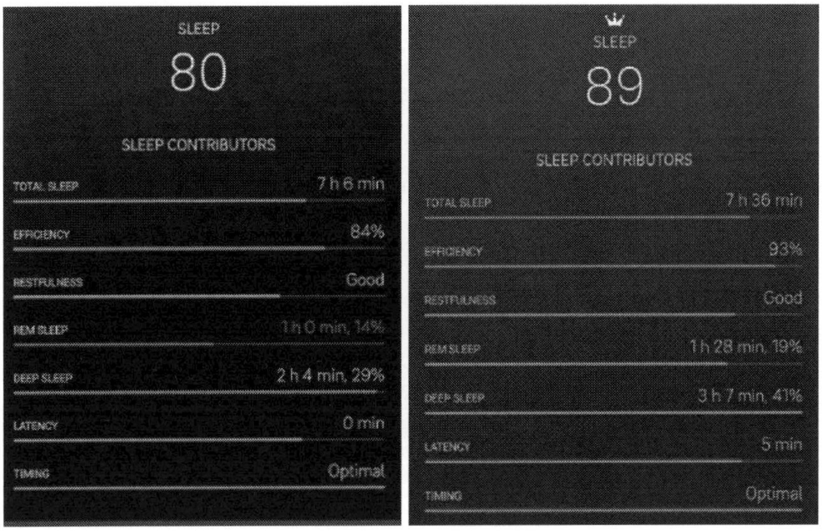

If your sleep tracker is telling you're getting very little REM sleep and a lot of deep sleep or vice versa, then chances are it's just misinterpreting the difference between the two and is categorizing them together. Using some other sleep tracking devices like the Dreem Headband, I've seen that I get a much higher proportion of REM and slightly less deep sleep, which is the opposite of my OURA ring. That's fine and it won't negate the usefulness of your data. I would much rather focus on looking at deep sleep and REM together rather than nothing at all.

A good night's sleep would be between 7-8 hours with about 2 hours of light sleep, 3 hours of deep sleep, and 2-3 hours of REM sleep. For a really efficient biohacker's sleep, it would be 6-7 hours total sleep, 1 hour of light sleep, 3 hours of deep sleep, and

2.5 hours of REM sleep. You never sacrifice deep sleep but try to cut down on light sleep.

If you're consistently getting very sub-optimal deep and REM sleep but too much light sleep, then you should look into Total Sleep Optimization. The biggest culprits are eating too close to bedtime, blue light sneaking in, screen time in the evening, watching movies, playing video games, over-hydration, electrolyte imbalances, and the wrong temperatures.

Your sleep quality can be evaluated both objectively and subjectively – with data and your intuition. Both are somewhat relevant and you shouldn't rely solely on one or the other. Subjective sleep quality refers to a sense of well-being, restfulness, and wakefulness after waking up in the morning. Objective sleep quality would be looking at your OURA score or some other datapoint. Insomniacs tend to be more demanding of their sleep quality than people with no sleeping problems. That's a pretty good point to remember as you may just end up creating false narratives in your head, making you sleep worse because of the anxiety and worrying.

You can also gauge why you sleep a certain way by looking at your other biomarkers like heart rate, heart rate variability, core temperature, and how often you wake up. Then you can create correlations i.e. the bed was too hot or I'm waking up too frequently to go to the bathroom.

In general, a slightly lower core temperature and slower heart rate will facilitate more restorative sleep because the body is trying to preserve energy. That's why being in a colder and chillier room not only makes it easier to fall asleep but also maintains deeper stages of sleep for longer.

- The perfect temperature is 20 degrees Celsius (65 Fahrenheit) +/- 5 degrees
- The normal resting heart rate for adults is 60-100. However, I'd say for healthy people who train and do other biohacks it should be below 60, at least during the night. My own heart rate while sleeping is 36-40 consistently. If I go above 40, I can immediately trace it back to either working out too close to bedtime, eating something bad, or too close to bedtime.

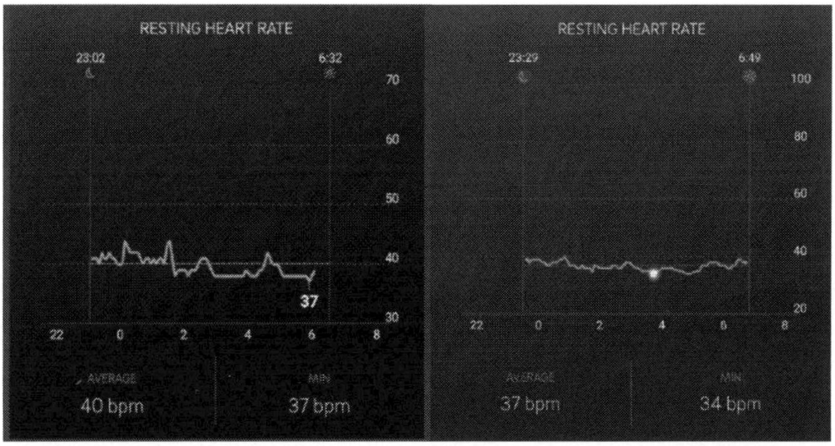

To lower your basal heart rate, you'd want to improve cardiovascular fitness and mitochondrial functioning. This will enable you to produce more energy with less effort and without wasting it. Regular aerobic conditioning is useful but I've seen the biggest effect come from intermittent fasting and becoming keto-adapted. This gives my body and brain a consistent supply of energy during the night and prevents waking up because of going hypoglycemic.

Hormesis Self-Tracking Protocol

- Assess your stress levels and adaptation with both intuitive as well as quantifiable metrics. If you feel sick and have a runny nose, it's probably not worthwhile to do HIIT. Instead, focus on recovery. Likewise, if you feel amazing but your OURA ring showed a poor score, follow your intuition. Active stress management will improve your ability to recognize the right signals and know what's the best way to act. Being stressed out dampens these signals.
- Signs that you're becoming overwhelmed include depression, chronic fatigue, muscle soreness, problems maintaining balance, confusion, procrastination, low thyroid, high blood pressure, insomnia, and constipation. Low HRV and elevated heart rate indicate you're edging on the side of overtraining or infection.
- Signs that you're in the zone of hormesis include motivation, rapid recovery from exercise, sharp cognition, muscle growth, fat loss, increased strength and speed, high HRV, sex drive, and general vigor. When in this state, you can keep pushing your adaptation and fitness further but increase it gradually.
- When tracking sleep, the most important variables are deep and REM sleep. Total sleep quantity isn't that relevant but it's optimally you'd still want to get 7-8 hours per night.

Your sleep demand depends on your level of physical activity, fitness, metabolic health, and circadian habits. More stress and higher exertion require longer sleeping times as well. Modern life already imposes us with frequent sleep disruptions and irregular sleeping patterns so there's no need to deliberately restrict your sleep on a regular basis. You're already bound to experience it eventually.

Chapter Fifteen: Of Supplementation

"The art of medicine consists of amusing the patient while nature cures the disease."

Voltaire

Humans have engaged in toxicology, pharmacology, alchemy, and medicine for thousands of years. Certain plants and compounds can be used to treat some illnesses or change the physiology in other ways. Even animals are aware of it as they self-medicate themselves by identifying the effects of a particular batch of grass by its taste. First, they take a small bite to assess the dose and then adjust it based on their response or particular needs.

Supplementation is a controversial topic because a lot of the claims certain companies make about their products are just blatantly false or at least half-truths. They mislead gullible people into buying expensive vitamin powder or snake oil. It doesn't apply to everyone but it happens quite often. Despite that, there's still plenty of great supplements out there that actually do work and they improve people's lives.

Supplementation should be done to fix underlying nutritional deficiencies, support a particular system in the body, getting an additional boost in performance, as a protective measure, or for just convenience. It's not supposed to replace good food and a nutrient-rich diet. However, most people will end up using some supplements at least every once in a while whether because of reasons discussed or due to the reduced nutritional value of our produce.

In this chapter, I'm going to outline a lot of the nutrients our bodies need to function optimally as well as whether or not you should supplement them. Before taking any supplements or meal replacements, you have to do your own research to understand the potential side-effects and abnormalities that may or may not occur. This is not professional medical advice and the responsibility yet again is solely on you. Furthermore, it's advisable to take blood tests to see what your deficiencies are before adding random supplements to your diet.

Essential Nutrients

An essential nutrient is defined as something that's needed for normal physiological functioning. It can't be synthesized inside the body and thus has to be obtained from a dietary source. For humans, there are 9 amino acids, 2 fatty acids, 13 vitamins, and 15 minerals that are considered essential[1957].

- **The minimum daily protein requirements** are set at 0.8g/kg or 0.36g/lb of bodyweight, which for an average adult who weighs around 100-200 pounds is roughly 40-80 grams of protein[1958]. However, that's for covering your bare nitrogen balance and I don't think it's optimal. For muscle growth and healthy aging, you would need more than that. Higher than minimal protein intake has many benefits such as increased preservation of lean body mass and weight loss. For physically active people, the maximum benefit is achieved at 0.7-0.82 g/lb of lean body mass but 1.0 might be better for older individuals[1959]. Out of the 20 amino acids, nine cannot be synthesized by the body itself and thus need to be obtained from diet. They are phenylalanine, valine, threonine, tryptophan, methionine, leucine, isoleucine, lysine, and histidine.

- Animal protein is considered complete protein because it contains all the essential amino acids whereas plant protein is incomplete. You can alleviate this shortcoming by combining different sources like rice and beans or just consuming more. Meat and eggs are not harmful in the context of a whole foods diet and as long as you don't combine fats with carbs. It's also worth reminding you to not oxidize fish or other PUFAs because they'll become very inflammatory.

- **Daily dietary fat intake** is suggested to be around 15%, including 2.5% as Linoleic Acid (LA) and 0.5% as Alpha-Linolenic Acid (ALA)[1960]. The reference intake value for LA is said to be 10g and for ALA 2g, which on a 2000 calorie diet can be covered with about 20-30 grams of dietary fat. DHA and EPA are conditionally essential for development and growth, which practically makes them essential. Although LA and ALA can be converted into DHA, it's not that effective of a process. It's recommended to get a minimum of 250-500 mg-s combined EPA and DHA per day[1961].

- Optimally, you'd want to keep your vegetable oil intake as low as possible and use more heat-stable fats like lard, tallow, coconut oil or avocado oil. Getting most of your fats from whole foods like meat and eggs is healthier than slabbing a ton of extra butter, or oil to your food. It'll keep the calories lower and burns more body fat. Also, you can also get your omega-3s from grass-fed beef, lamb, yolks, and fatty fish.

- **Essential vitamins are** Vitamin A, B1, B2, B3, B5, B6, B7, B9, B12, Vitamin C, Vitamin E, Vitamin K, and Choline.
- **Essential minerals for humans are** Calcium, Cobalt, Chloride, Chromium, Copper, Iodine, Iron, Magnesium, Manganese, Molybdenum, Phosphorus, Potassium, Selenium, Sodium, and Zinc.

Conditionally essential nutrients can be synthesized endogenously but they're in most cases insufficient. They become more essential in conditions such as pregnancy, premature birth, malnourishment, childhood growth, healing, and certain diseases. That's why you don't really want to put kids or old people on nutrient-deficient diets in hopes of making their body do extra work by converting some foods into essential nutrients.

Non-essential nutrients are not necessary for survival and they can either have a beneficial or a toxic effect. It doesn't mean you shouldn't consume them. They're just not inherently mandatory for normal physiological functioning. Here are some examples.

- **Carbohydrates and glucose aren't essential because the body can shift into ketosis and use ketones instead.** The brain and other vital organs do need a very minuscule amount of glucose for optimal functioning even after becoming keto-adapted. However, as we've mentioned before, the process of gluconeogenesis can create that glucose from dietary fat and protein intake so carbs aren't needed. Nevertheless, this may not be optimal all the time, which is why I prefer using a more cyclical approach as discussed earlier.
- **Dietary fiber isn't essential because humans can't digest it.** The emphasis on eating a lot of fiber comes from the idea that it helps with bowel movements, feeds the gut bacteria, and lowers cholesterol. Although fiber isn't essential, it's still advisable to eat some vegetables for gut diversity and antifragility. However, too much fiber and vegetables can cause digestive issues, bloating, and constipation, which is why you want to self-regulate your intake. Fortunately, butyrate can be gained from animal fats as well, such as butter, tallow, and meat but to get the other pre-biotic SCFAs like acetate and formate you'd want to eat some plants as well.

- **Phytochemicals and phytonutrients are non-nutritional parts of plants.** They're not essential for survival in humans but they help the plants survive harsh conditions and protect against predation. However, those same compounds can have a beneficial hormetic effect by making our own bodies more resilient. Different polyphenols, flavonoids, resveratrol, lignans, and catechins have all been shown to have great benefits on longevity and metabolic disorders as discussed in chapter four. What makes a poison deadly is the dose. That's why too much of anything will still be bad. You shouldn't take supplemental polyphenols or other plant phytonutrients because that might be too big of a stressor. Just stick to eating whole foods and use different spices and herbs as seasoning.
- **Alcohol is a non-essential non-nutrient that still has calories.** It means the body can't get any nutritional value from alcohol other than the empty calories. Now, the hormetic effect is something you may benefit from as certain spirits can fight off infections and promote ketone production even. However, the dose is probably quite small and you shouldn't be drinking every day. You definitely don't want to get hammered or even seriously intoxicated because you'll do more damage than good. Instead, one shot of vodka or a glass of red wine a few times per week is probably the minimal effective dose. Different kinds of vinegar also have small amounts of acetic acid and alcohol but they're great for blood sugar control and insulin regulation.

Here's a chart for the Recommended Daily Allowances for all the essential vitamins and minerals:

NUTRIENT	Average RDA	Upper Limit
Vitamin A	700-900 mcg	3000 mcg
Vitamin C	75-90 mg	2000 mg
Vitamin D	600-800 IU	4000 IU
Vitamin K	120 mcg	Not Established
Vitamin E	15-22 mg	1000 mg
Vitamin B1 (Thiamin)	1.2 mg	Not Established
Vitamin B2 (Riboflavin)	1.3 mg	Not Established
Vitamin B3 (Niacin)	14-16 mg	35 mg
Vitamin B5 (Pantothenic acid)	5 mg	Not Established
Vitamin B6 (Pyridoxine)	1.3-1.7 mg	100 mg
Vitamin B7 (Biotin)	30 mcg	Not Established
Vitamin B9 (Folate)	400 mcg	1000 mcg
Vitamin B12 (Cyanocobalamin)	2.4 mcg	Not Established
Calcium	1000-1200 mg	2000-2500 mg
Choline	425-550 mg	3500 mg
Chloride	1800-2300 mg	3600 mg
Chromium	35 mcg	Not Established
Copper	900 mcg	10 000 mcg
Fluoride	3-4 mg	10 mg
Iodine	150 mcg	1100 mcg
Iron	8-18 mg	45 mg
Magnesium	300-450 mg	500 mg
Manganese	1.8-2.3 mg	11 mg
Molybdenum	45 mcg	2000 mcg
Phosphorus	700-1250 mg	3000-4000 mg
Potassium	4700 mg	Not Established
Selenium	55 mcg	400 mcg

Sodium	1200-1500 mg	2500 mg
Zinc	8-11 mg	40 mg

Keep in mind that these are your average values for the average person. There are huge variations between populations, ethnic heritage, lifestyle habits, and dietary preferences. For instance, there's some evidence showing that the need for vitamin C increases only in diets of glucose-based metabolism. In fact, ascorbic acid and glucose compete for cellular transport[1962]. High levels of blood glucose inhibit the uptake of vitamin C because both use the same membrane transport chain and because glucose is a much more prioritized nutrient. Secondly, a person with an MTHFR mutation may also require extra folate and methyl donors.

The required intake of sodium will also fluctuate depending on your levels of physical activity, blood pressure levels, hydration levels, and what phase of the diet you're at. On the ketogenic diet, your need for sodium tends to increase because low insulin makes you hold onto less water and potentially excrete more electrolytes. What's more, higher sodium intake in a low carb state is harmless because there's not enough insulin around to raise blood pressure. Therefore, when eating keto low carb, you can increase your sodium up to 3000-5000 mg/day but when eating carbs it should stay within the RDA because of the higher insulin.

Here are the Essential Minerals to Cover:

- **Calcium** – A study found that fixing a calcium deficiency helped to regain normal REM sleep[1963]. It can also assist tryptophan to be converted into melatonin. RDA is 1000-1200 mg. Calcium deficiencies are common in older people or those who don't consume a lot of dairy. Before supplementing, you should know whether or not you're actually deficient because too much calcium promotes atherosclerosis and plaque formation. Especially if you're not getting enough vitamin K2. Calcium and Magnesium absorption compete with each other in doses higher than 250 mg-s so you shouldn't supplement them together. Consuming more dairy and calcium isn't healthier and won't strengthen your bones. Regions with the highest dairy consumption also have the highest rates of bone fractures and osteoporosis because they're not getting enough vitamin K. It's not recommended to supplement calcium if you're already eating meat and veggies or some dairy. The healthiest sources of calcium are fermented cheeses, clams, peas, and squash. Dairy can be okay for some people but it's not ideal because of increasing IGF-1. Vitamin K2 from cheese, on the other hand, is much more preferable because the K2 will direct the calcium into the right places and prevents accumulation.
- **Choline** – RDA is 425-550 mg. Choline is a precursor to acetylcholine – a neurotransmitter responsible for cognitive functioning and attention. It's also vital for cell

membranes, methyl metabolism, and cholesterol transportation[1964]. You need choline to prevent fatty liver and support methylation[1965]. Foods rich in choline are eggs, meat, and fish. If you eat these foods, then you don't have to supplement with choline. Eating 3-4 egg yolks a day is enough to cover your need for that. On a plant-based diet, it may be a good idea to take choline and inositol. If not choline, then trimethylglycine can be an alternative.

- **Iron** – RDA is 8-18 mg with most people getting about 10-15 mg/day from food and other supplements. Iron is essential for hemoglobin transportation, which helps to transfer oxygen to muscles and cells. Excess iron is a risk factor for cardiovascular disease and can be toxic so consult your doctor first before supplementation[1966]. Iron deficiencies are more common in diets with little or no meat. If your iron levels are high, then the surest and most effective way to reduce it is to donate blood. Women already lose some iron during menstruation but men don't. Adding iron chelators to your diet like curcumin or coffee will also help.
- **Iodine** – RDA is 150 mcg, but a lot of people are still deficient. Iodine is important for thyroid functioning and the metabolism[1967]. If you're not eating a lot of seafood, like oysters, salmon, algae, sea kelp, and lobster, then you may want to supplement iodine. Taking about 300-400 mcg can be good for fixing symptoms of low thyroid. Raw vegetables will also inhibit iodine absorption so if you feel like having hypothyroidism, then make sure you cook your

veggies or replace them with starchy tubers. Kale smoothies and pounds of broccoli a day isn't a good idea.

- **Potassium** - Potassium supplementation has been shown to have a positive effect on sleep quality[1968] and slow-wave-sleep. Deficiencies in potassium can weaken muscle contraction, cause arrhythmia, and impair insulin production[1969,1970,1971]. The estimated daily minimum for potassium is 2000 mg/day and the RDA 4700 mg/day. You shouldn't worry about eating too much potassium as long as you're eating some vegetables unless you're taking supplements. Even fresh meat has some potassium but only if it's not overcooked. Using potassium chloride salts with reduced sodium like NuSalt or taking potassium gluconate can be useful.
- **Magnesium** - Magnesium is important for managing stress, including all the other biological processes and reactions in your body. Although not conclusive, studies on humans as well as animals show that magnesium supplementation can alleviate many of the negative side-effects of stress like anxiety, depression, sleeping problems, etc[1972]. It also promotes sleep efficiency, onset, and total quality[1973].
  - Stress also depletes magnesium by activating the sympathetic nervous system and supplementation helps to reduce this effect[1974]. Stressful events like exercising, fasting, high blood sugar, insulin resistance, sleep deprivation or even feeling

anxious makes you burn through magnesium at a higher rate. That's why the more stressed out the more magnesium you need. Unfortunately, the less magnesium you have and the more of it you diminish.

- USDA data from 1950-1999 shows reliable declines in many vitamins and minerals for 43 common crops. Since 1975-1999 average calcium in vegetables has dropped by about 27%, iron by 37%, vitamin A 21%, vitamin C 30%[1975]. Between 1940 and 1991, magnesium content in vegetables has decreased by 24%, fruit by 17%, meat by 15%, and cheese by 26%[1976]. In the UK, it's approximately 35%[1977].
- The most absorbable forms of magnesium are citrate, glycinate, taurate, and aspartate. Avoid magnesium carbonate, sulfate, gluconate, and oxide because they're poorly absorbed and mostly used as fillers. The RDA is 350-450 mg a day, which ¾ of the population isn't meeting[1978]. Stress, insulin resistance, and coffee make you burn through magnesium so you might need up to 500 mg.

- **Selenium** – RDA is 55 mcg, but optimal doses are somewhere between 100-300 mcg. Selenium is important for hormones and energy production, especially testosterone. It also helps to detoxify mercury, heavy metals, and protects against toxicity in the first place.

Taking 200 mcg of selenium has been shown to increase glutathione peroxidase in patients with chronic kidney failure[1979]. Over 400 mcg of selenium, however, can be toxic and cause nausea.

- Mark Mattson wrote in his 2015 article that: *„Eating too many Brazil nuts can poison the liver and lungs because of the presence of the trace element selenium. Yet eating just a few supplies an essential nutrient that is incorporated into an enzyme that may help protect against heart disease and cancer"*[1980]. That's xenohormesis. The daily requirements for selenium can be met with eating only 2-3 Brazil nuts. Other foods include seafood, meat, organs, chicken, nuts, seeds, and carrots. However, the actual selenium content of a food greatly varies and depends upon soil quality and mineral content of the feed.

- **Zinc -** RDA for zinc is 8-12 mg/day. Zinc is an essential mineral involved in cell growth, protein synthesis, and protecting the immune system[1981]. The upper limit for zinc a day should be under 100 mg because you may get nausea, vomiting, and reduced immune functioning. Oysters are the most abundant sources of zinc with a massive 74 mg per serving. Other sources are beef, poultry, and some nuts. If you're a male, then you should pay close attention to your zinc consumption because it's one of the crucial minerals for testosterone production. But if you're eating a lot of seafood or red meat then you don't need to supplement it either.

Here are the Essential Vitamins to Cover:

- **Vitamin A** – RDA is 700-900 mcg with an upper limit of 3000 mcg-s. Vitamin A or retinol is important for nerve functioning, growth development, building new cells, and improving eyesight. The best sources of vitamin A are organ meats with liver giving you about 5000-7000 mcg from just 100 grams compared to the 700-800 of carrots. That's why it's better to eat liver only a few times per week. Higher doses of vitamin A like 12 000 mcg can become toxic and cause drowsiness and coma. The Inuit are known for developing hypervitaminosis A because of sometimes eating polar bear liver. Polar bears feed exclusively on seals and fish, which is why their liver contains extremely high amounts of vitamin A. Even just a mouthful has nearly 9000 mcg and you'd probably die if you ate the entire thing. If you eat some meat, organs, and tubers, you don't need to supplement vitamin A.
- **B vitamins are also essential and they can be found in animal foods**. If you're already eating a whole foods based diet that includes some meat then you really don't need to supplement this. Vegans, however, are commonly deficient in B-vitamins so you'd have to look into taking a B-complex supplement. As an omnivore, supplementing can be counter-productive because you may get the wrong ratios of vitamins. Instead, focus on eating some red meat

and fermented foods consistently. The optimal doses for B-vitamins are also slightly higher than the RDA. Here's what you should aim for.

- Vitamin B1 (Thiamin) —1.5 mg/day
- Vitamin B2 (Riboflavin) —1.7 mg/day
- Vitamin B3 (Niacin) —20 mg/day
- Vitamin B5 (Pantothenic Acid) —10 mg/day
- Vitamin B6 (Pyridoxine) —2 mg/day
- Vitamin B7 (Biotin) —300 mcg/day
- Vitamin B9 (Folic Acid) —400 mcg/day
- Vitamin B12 (Cobalamin) — 10 mcg/day

- **Vitamin B12** – Most of the population is low in B12, including omnivores, vegetarians, and vegans[1982]. Deficient B12 can cause neurological issues, dementia, and disturbed sleep[1983]. Foods higher in B12 are organ meats, fish, and fermented foods but supplementing 250-500 mcg a day is useful.

- **Vitamin C** – RDA is about 75-90 mg with an upper limit of 2000 mg. The function of vitamin C or ascorbic acid is to reduce oxidative stress by increasing antioxidants.
    - In the 18th century, sailors who went on long sea explorations developed ulcers, rotten teeth, and hair loss. This "*plague of the seas*" was caused by a medical condition called scurvy, which is a deficiency in many vitamins, mostly vitamin C, and B vitamins. The conventional treatment was eating lemons, oranges, sauerkraut, malt, marmalade, and

lemon juice. If you're eating fresh cruciferous vegetables and cabbage, then you probably don't have to worry about getting scurvy.

- Low vitamin C in the blood has been shown to cause trouble falling asleep and have more frequent interruptions in sleep[1984]. It also lowers histamine and may decrease sensitivity to FODMAPS[1985]. Vitamin C reduces the loss of carnitine through urine[1986] and increases its utilization in muscle[1987], which is beneficial for physical performance. Lastly, small amounts of vitamin C can reduce the risk of catching the cold[1988]. However, too much can actually predispose you to get sick because of blocking the beneficial ROS needed for optimal cellular functioning.
- Foods higher in vitamin C include vegetables, sauerkraut, fruit, and berries. The daily recommended dose would be 300-400 mg and perhaps 1000 mg during sickness but exceeding that can have a negative effect on your body's capacity to deal with sickness naturally. If you're eating some vegetables, then you probably don't need to take additional supplements.

- **Vitamin D** –Deficiencies in vitamin D are associated with autoimmunity and weakened immune system, which increases the susceptibility of getting an infection[1989]. Low levels of vitamin D are linked to poor sleep quality[1990].

Vitamin D levels between 60-80 ng/mL have been shown to improve sleep[1991]. An average adult should take at least 2000 IU of vitamin D but it would also depend on how much exposure you get to natural sunlight. The upper limit for adults is 4000-5000 IU/day. Vitamin D may become toxic with high levels of calcium or if you take 10 000-40 000 IU/day consistently.

- **Vitamin E** – RDA for Alpha-Tocopherol is 15 mcg with a 1000 mcg upper limit. It's a potent antioxidant and a fat-soluble vitamin. Vitamin E deficiencies are quite rare as it's found in vegetables, fish, and nuts. Instead of taking dietary vitamin E supplements, you can use vitamin E oils on your face and skin to reduce wrinkles, lighten dark spots, and promote anti-aging. Taking about 100-300 mg of vitamin E can protect against lipid peroxidation when exposed to bad fats[1992].
- **Vitamin K** - In the context of atherosclerosis, Vitamin K2 also directs calcium into the right place, namely the bones and teeth, instead of keeping it in the bloodstream to cause plaque formation. The RDA for vitamin K is roughly 60-120 mcg, and the optimal level is about 200 mcg. This optimal level is mostly the same for both vitamin K1 and K2. Although humans can convert some K1 into K2, the biggest effect comes from MK-4 utilization, which is most bioavailable in animal foods. It should be noted that while many sources may claim to hit the RDA for vitamin K, they have poor bioavailability - your body is unable to extract the full amount of said foods. That's why you

should eat a lot of organ meats, fermented foods, a bunch of cruciferous vegetables, and a bit of cheese. Supplementing vitamin K should be secondary. The top foods for K2 are natto, liver paste, egg yolks, cheeses, dark poultry, organ meats, and sauerkraut.

What about a multivitamin? - There are a lot of vitamins to be covered for our body to not only be healthy but function at its peak. It would be unreasonable to take too many tablets or pills while neglecting the importance of real food. Plus, there is the potential of interfering with vitamin absorption if you get it all together.

Adaptogens and Herbal Compounds

Folk medicine includes a lot of different herbs and adaptogens to bolster the immune system, fight sickness, and manage stress. Many studies have found them to be almost as effective as pharmaceuticals but with fewer side-effects. It's probably because we can get the right hormetic dose from these compounds with a much smaller risk of overdosing or creating imbalances in the nervous system.

Adaptogens are plant substances that promote homeostasis inside the body[1993]. If you're low in energy they will stimulate you up but if you're over-hyped they'll bring you down. Such compounds have been used in traditional medicine for centuries to promote stress resistance. By definition, a herb has to be non-toxic, non-specific, and affect the organism's physiology before it can be called adaptogenic. Currently, the EU and FDA don't accept the term in pharmacology because more data is needed[1994].

Compared to stimulants like caffeine, adaptogens create a more sustainable increase in energy without addiction or a crash. Stimulants "borrow" energy from your adrenal glands by raising adrenaline and cortisol but this can have many side-effects. Adaptogens, on the other hand, release energy gradually and based on your requirements. In a way, they will auto-regulate or adapt to the body's current homeostasis and state.

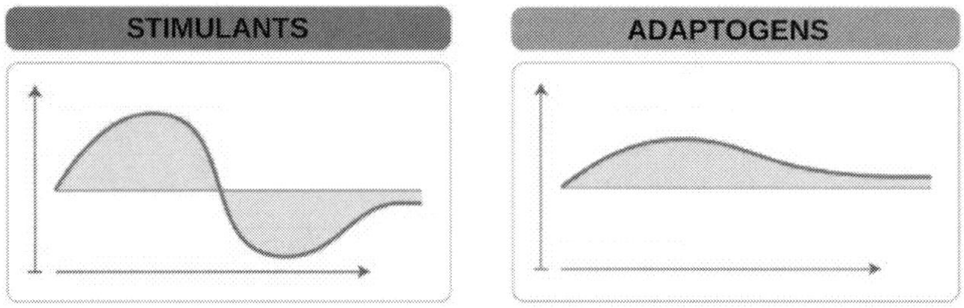

Adaptogens can be from different categories of plants or fungi. Most herbs have at least some sort of an adaptogenic effect by upregulating certain processes inside the body and creating mild hormesis. They can be used for general stress management, reducing inflammation, promoting energy production, fixing thyroid, or just exercise performance. For example, medicinal mushrooms stimulate the production of macrophages that eat identified pathogens[1995,1996].

Here's a list of the most common and effective adaptogens:

- **Chaga Mushroom (*Inonotus obliquus*).** Chaga is a mushroom that grows on birch trees that lowers cholesterol, triglycerides, inflammation, and oxidative stress[1997]. It has the highest ORAC value (a measure of antioxidant capacity) of any other food in the world. Polysaccharides from Chaga mushroom's fruiting body (PFIO) have been shown to effectively promote macrophage activation through the MAPK and NF-κB signaling pathways, which regulate the immune system function[1998]. Chaga will promote the health and integrity of the adrenal glands. This mushroom can be consumed as a powder, made into tinctures, or boiled into teas. Water or alcohol-based extracts are the most potent options. You can also harvest and grind it yourself. I consume about ½-1 teaspoons of Chaga throughout the day and love to add it to my coffee. Don't take too much though because Chaga is also high in oxalates. Any more than 4-5 tsps taken consistently can damage the liver and kidneys. Consuming Chaga with other high-oxalate foods like spinach or beets has been shown to cause zinc deficiency and neuropathy[1999].

- **Reishi Mushroom (*Ganoderma lucidum*).** Reishi or Lingzhi mushroom is a fungus that grows in humid regions. It improves the immune system and red blood cell functioning[2000], which makes the body more capable of fighting disease. In fact, a study of over 4000 breast cancer survivors found that 59% of them were consuming reishi[2001].

Reishi contains a huge variety of bioactive polysaccharides, beta-glucans, and over 120 different *triterpenoid* compounds[2002]. It increases overall well-being, raises HDL-cholesterol, decreases TNF-alpha, and reduces fatigue[2003,2004]. This mushroom has a more relaxing feeling to it and is best taken for relaxation and stress reduction. Daily dose would be similar to Chaga.

- **Shiitake Mushroom (*Lentinula edodes*)** - a Dark brown fungus that grows on decaying trees. It contains polysaccharides, terpenoids, and sterols that boost the immune system, lower cholesterol, and fight cancer[2005]. Regular shiitake mushroom consumption has been shown to improve immunity in young adults[2006].

- **Turkey Tail (*Coriolus/Trametes versicolor*).** Looks like a turkey tail but doesn't act like one – it's a mushroom, for god's sake! Turkey tail has been shown to fight against leukemia cells *in vitro*[2007] and improve the immune system of people getting chemotherapy[2008]. It contains 35 different phenolic compounds and flavonoid antioxidants quercetin and baicalein, which are strong antioxidants[2009]. Turkey tail also contains other substances, such as krestin and polysaccharide peptide (PSP), that activate macrophages and modulate immune response[2010,2011]. Turkey tail extract has been found to inhibit the growth of Staphylococcus aureus and Salmonella enterica *in vitro*[2012].

- **Lion's Mane** – A white mushroom that looks like a lion's mane. It's incredible for growing new brain cells and preventing cognitive decline[2013]. Supplementing just 3 grams a day has been shown to improve mental functioning of people with cognitive impairment[2014]. However, some people may be allergic to them. Generally, take ½-1 teaspoons a day.

- **Ashwagandha.** Based on animal studies, an adaptogenic herb ashwagandha has been shown to possess immunomodulatory effects by upregulating Th1 and macrophages[2015,2016]. In human studies, ashwagandha has been shown to lower stress and balance the immune system[2017, 2018]. In a small study of 5 people, ashwagandha was shown to upregulate the expression of CD4 on CD3+ T cells after 96 hours of consumption[2019].
- **Ginseng.** American as well as Asian ginseng, regulate immune cells such as macrophages, T-cells, and natural killer cells. It also has antimicrobial properties[2020]. Fermented wild ginseng root has been shown to have anti-inflammatory and anti-oxidative properties[2021]. Traditionally, it's been used to fight fatigue and erectile dysfunction[2022]. Taking 200 mg of ginseng a day improved mental health and mood of 30 people, but the effects reduced to normal after 8 weeks[2023]. In another study, the 200 mg dose was also more effective in promoting mental performance and fatigue during a test compared to a 400 mg dose[2024]. So, about ½ tsp of ginseng a day is more than enough.

- **Ginger.** Ginger helps to lower inflammation, treat infectious agents, and protect against environmental stressors such as smoke and chemicals[2025,2026]. It contains gingerol, which is a powerful anti-inflammatory substance[2027]. Consuming 2 grams per day has been shown to reduce muscle pain[2028]. Amongst 247 people with osteoarthritis in the knee, those who took ginger extract had less pain and needed fewer medications[2029]. Women who took 1 gram of ginger powder for the first 3 days of their menstruation reduced their pain as effectively as drugs like ibuprofen[2030]. In type 2 diabetes, 2 grams of ginger powder a day lowered fasting blood glucose by 12% and reduced oxidized lipoproteins by 23%[2031]. Three grams a day can also lower cholesterol significantly[2032]. Taking 1.2 grams before a meal increased stomach emptying by 50% in healthy people[2033].

- **Turmeric.** Curcumin, the active compound of turmeric, has been shown to embody anti-inflammatory properties that can help in treating chronic pain and infections[2034]. It also helps to boost glutathione levels in the body. Curcumin and turmeric also have antibacterial, antiviral, and antifungal properties in humans[2035]. You can obtain it from just using curry or other Indian spices on your food.

- **Astragalus (*Astragalus membranaceus*).** Astragalus has been used in Chinese medicine for thousands of years. Research shows it protects against gastrointestinal inflammation, and has immune system boosting properties[2036]. In one study, a combination of astragalus, echinacea, and licorice herbal tincture stimulated immune cells within 24 hours of ingestion and remained active for at least 7 days[2037]. Test tube studies have shown that Astragalus membranaceus extract activates an immune response in macrophages[2038]. Latest research on Astragalus also shows that it is a very potent herb on activating autophagy pathways and can have therapeutic potential in autophagy dysregulation-associated diseases because of their biological positive effects[2039]. You can take about 250-500 mg as a supplement, ½ tsp as a tincture, or 1-2 tsp of dried root powder.

- **Rhodiola Rosea** – It's an adaptogenic herb that grows in mountainous regions. The root has many compounds known to reduce stress, fatigue, and anxiety[2040]. People who received 400 mg a day for 8 weeks reported much fewer symptoms of stress and chronic fatigue[2041]. Both 340 and 680 mg a day have been shown to improve depression, insomnia, and emotional stability whereas the placebo didn't[2042]. Taking 200 mg two hours before a cycling test enabled the subjects to exercise on average 24 seconds longer compared to a placebo[2043]. For men, it can also boost testosterone and increase virility.
- **MACA Root** - Another superfood from the Peruvian mountains and is the root of ginseng. It has numerous amounts of vitamins and minerals in it, such as magnesium zinc, copper, etc. Also, it promotes hormone functioning for both men and women[2044], as well as increases our energy production just like creatine does[2045]. It can either

be powdered or made into a tablet. You shouldn't take maca every day because of its potency. Optimally, you'd want to take maybe a teaspoon every other day.

- **Gingko Biloba** – Maidenhair, as it's called, is native to China. Not only is it a powerful antioxidant but it also improves blood circulation by increasing nitric oxide[2046]. There are other benefits on brain and eye health as well. Gingko contains alkylphenols that may cause nausea, allergic reactions, headaches, and rashes. Taking about 120-240 mg in several doses throughout the day seems to be enough. Any more than 600 mg is probably not a good idea.
- **Licorice Root** – Also known as sweet root, licorice is a common sweetener in candies and sweets. It's been used as medicine as well for centuries. The main active compounds are glycyrrhizin and deglycyrrhizinated licorice (DGL). Licorice has been shown to reduce the amount of toxic bacteria like H. pylori[2047], alleviate ulcers[2048], relieve bloating, constipation, promote immune system functioning[2049], and fight viral infections like SARS or influenza[2050]. It's most commonly used to treat coughs, digestive problems, and colds. However, excess glycyrrhiza can cause headaches, fatigue, hypertension, and even heart attacks[2051]. It's also not recommended for consumption during pregnancy or breastfeeding. Licorice can interact with many medications, resulting in negative side-effects, such as diuretics, anti-arrhythmia drugs, blood pressure medication, blood thinners, statins, and non-

steroidal anti-inflammatory drugs (NSAIDS)[2052]. Doses of 5-15 g/day are considered safe in the short-term.

- ***Schisandra chinensis*** **(five flavor fruit)** or just Schisandra is a fruit vine that grows purple-red berries. They contain lignans with many health-promoting effects. A 2013 study on animals found Schisandra's antioxidative benefits reduced liver damage and protected against lipid peroxidation[2053]. It's also shown to be anti-depressive in mice[2054], alleviate symptoms of menopause in women[2055], and block excessive amyloid beta proteins in Alzheimer's[2056]. Generally, it's used as 1.5-6 g/day of powdered product and about 3 g/day of the actual fruit. Excessive doses can cause heartburn, ulcers, reflux, or allergies.

- **Moringa** - *Moringa oleifera* is a plant from India rich in antioxidants and other nutrients. It contains vitamin C, beta-carotene, quercetin, and chlorogenic acid that can lower inflammation[2057]. A few human studies have shown it can lower blood sugar and lipids[2058]. Interestingly, moringa leaves and seeds may protect against arsenic toxicity in mice and rats[2059]. Administrating moringa root has reduced urinary oxalate levels and less kidney stone formation[2060]. It also inhibits lipid peroxidation and improves kidney functioning[2061]. Other benefits include reduced stress, anxiety, and regulation of thyroid hormones[2062, 2063]. Daily dosage should stay between 1/2-2 tsp/day of moringa powder or 1500-3000 mg/day.

- **Holy Basil** (*Ocimum sanctum)* or tulsi is another adaptogenic herb from India. It has antiviral, antibiotic, fungicidal, germicidal, and disinfectant properties[2064], which can be used to combat fever. A 2017 review revealed the potential efficacy and safety of holy basil in treating cardiovascular disease due to its antioxidant content[2065]. Because of the antibacterial effects, it can be used for oral health and as a mouth freshener[2066]. You can use holy basil as a tea to reduce inflammation and stress.
- **EGCG** - Epigallocatechin gallate (EGCG) is the main polyphenol in green tea with many health benefits. Green tea, in particular, is probably the healthiest drink in the world after mineral water. Too much green tea, however, may cause anxiety and heart palpitations because of the high caffeine content, which makes using green tea extracts or EGCG supplements a more convenient way to add extra polyphenols to your diet. Doses above 500 mg may become problematic.

Longevity Supplements

Here are some supplements targeted specifically towards longevity and life extension:

- **Resveratrol** – We all know the anti-aging benefits of resveratrol by now. The red wine special...Unfortunately, 1-2 glasses of red wine wouldn't give you a significant longevity boost and drinking more than that isn't optimal.

Taking resveratrol supplements with medications may cause unwanted side-effects like blood clotting and enzyme blocking. Most supplements have 250-500 mg per serving but studies say that to get the benefits of resveratrol you'd have to consume about 2000 mg a day. To get that amount, you're going to have to take a high-quality resveratrol supplement.

- **Sulforaphane** – The cruciferous special...It's a powerful antioxidant that turns on the Nrf2 pathway with many anti-cancer properties. Cooking broccoli and cabbage triples their sulforaphane content[2067]. Unfortunately, frozen veggies deactivate myrosinase, which is an enzyme that creates sulforaphane. Broccoli sprouts, in particular, contain dozens of times more vitamin K and sulforaphane. If you're not eating a lot of cruciferous or sprouts, then you can take 10 mgs of sulforaphane as a supplement.
- **Carnosine** – It's a combination of the amino acids beta-alanine and histidine with many anti-aging benefits. Carnosine is most known for protecting against free radicals and AGEs. This will keep the cells healthy and prevents aging of the skin. Naturally found in red meat and animal foods, supplementing carnosine has no side-effects. 7 ounces of beef has about 250 mg of carnosine but for optimal longevity, you'd want to aim for about 1000 mg a day.
- **Astaxanthin** – Wild salmon's flesh is slightly pink and reddish – that's astaxanthin. It's an incredibly powerful

antioxidant and mitochondrial supporter, which is why freshwater fish like trout and salmon are capable of surviving such harsh conditions. Astaxanthin supplementation is great for anti-aging and maintaining muscle functioning. Doses of 4-40 mg a day have been shown to be safe. Too much astaxanthin may cause an upset stomach.

- **Alpha Lipoic Acid (ALA)** – Lipoic Acid has anti-inflammatory and anti-oxidant effects on the brain, and other tissue. It helps with fat oxidation, blood sugar regulation, and cardiovascular function. ALA is found in every cell of your body and it can be obtained from vegetables and meat. Therapeutic dosages of ALA range from 600-1800 mg/day with doses above 1200 mg causing nausea and itching.
- **C60** – Buckminsterfullerene, or buckyballs, or C60 is quite a new and unstudied compound. It helps to eliminate superoxide dismutase, which gets created as a by-product of cellular metabolism[2068]. This fights reactive oxygen species and promotes longevity[2069]. C60 oil should be dissolved in oil with a centrifuge. Pure C60 can be toxic[2070]. Usually, as a supplement, C60 comes in either olive oil, avocado oil, or coconut oil. One teaspoon a day is probably enough for experimentation.
- **Glutathione** – The most important antioxidant in the body that's made of glutamine, glycine, and cysteine. Naturally, glutathione is found in sulfur-rich foods like beef, fish,

poultry, and vegetables. Glutathione decreases with age so it's a good idea to supplement it. Increasing vitamin C and selenium may help co-factor the production of glutathione. Milk thistle and curcumin can also increase glutathione. Glutathione supplement doses range from 50-600 mg/day.

- **Apigenin** – It's a compound found in plants and vegetables like parsley, onions, fruit, etc. with anti-carcinogenic and anti-inflammatory properties. Apigenin is neuroprotective and fights cancer. However, it can be toxic with 100 mg/kg causing liver toxicity in mice[2071]. Parsley is 45 mg/g apigenin. It can also be found in olive leaf and artichoke extracts. No need to take additional apigenin supplements.
- **Quercetin** – Flavonoids are amazing anti-aging compounds and quercetin is one of them. In fact, quercetin is the most consumed flavonoids in the human diet[2072]. It protects against free radicals and DNA damage. Quercetin is found in elderberries, red onions, garlic, cranberries, kale, hot peppers, kale, blueberries, and the skin of apples. Supplementation is generally safe but not very effective because of poor bioavailability.
- **Nicotinamide Riboside** – B vitamins play an important role in energy and nerve functioning. Nicotinamide Riboside (NR) is a form of B3 that gets easily converted into NAD+ and can activate sirtuins. It's mostly found in cow's milk, whey protein, and brewer's yeast. If you're not eating a lot of animal products or are deficient in B3, then nicotinamide supplements can work. Doses of 5000 mg/kg

haven't shown increased risk of death or anything like that. Nicotinamide Riboside can increase NAD+ as does Nicotinamide Mononucleotide (NMN).

- **DHEA** – DHEA (dehydroepiandrosterone) is a steroid hormone produced by the adrenal cortex. It acts as a precursor to testosterone and estrogen. DHEA levels drop with age and under stress. That's why supplementation enables to maintain better hormonal functioning and performance. DHEA can promote bone density, inhibit weight gain, alleviate depression, increase muscle strength, and improve immune functioning[2073,2074,2075,2076]. Some side-effects include acne, headaches, nausea, upset stomach, restlessness. It's not recommended for people with heart disease, diabetes, PCOS, or liver disease. Daily doses of 50-100 mg-s are deemed to be safe.
- **Glycine** – Glycine is quickly becoming one of my favorite supplements because of its wide range of benefits and ease of use. We've discussed in previous chapters how it helps to combat oxidative stress, support methylation, improve sleep, balance methionine, boost glutathione, heals fatty liver, and helps to detoxify heavy metals[2077]. It also lowers blood sugar and insulin, which doesn't really break autophagy either. That's why I love adding 3 grams to my morning coffee as well as 3 grams to my evening tea. Doses of up to 31 grams/day have proved to be safe.
- **Trimethylglycine** – TMG or betaine is a major methyl donor that supports methylation. Its primary role is to

reduce homocysteine and protect liver health thanks to being the metabolite of choline[2078]. Standard doses are between 2500-6000 mg/day taken in two doses. You can also take additional TMG if you're taking NR or NMN to counterbalance the loss of methyl groups.

Mitochondrial Functioning Supplements

Here are some supplements targeted towards improving mitochondrial functioning:

- **PQQ** - Pyrroloquinoline Quinone is a non-vitamin growth factor that supports mitochondrial function. This will have a compounding effect on everything else you do. Humans can make about 100-400 nanograms of PQQ a day[2079], mainly from dietary sources. Consuming 0.3 mg/kg PQQ is safe but 500-1000 mg/kg can cause death in rats[2080]. Foods high in PQQ are raw cacao, green tea, fermented foods, and organ meats. Taking about 20 mg of PQQ as a supplement is the optimal dose for an average weighing individual.
- **CoQ10** – Co-Enzyme Q10 is another mitochondrial supporter and antioxidant. It's important for energy production and tissue development. Found in fish, red meat, especially organ meats, and fermented foods. CoQ10 comes in two different forms — ubiquinol and ubiquinone. The CoQ10 in your blood consists of 90% ubiquinol and it's more absorbable. Therefore, ubiquinol CoQ10

supplements are better. Daily dosage ranges from 90-200 mg. Doses over 500 mg are also safe.

- **Pterostilbene** – It is a polyphenol that's chemically similar to resveratrol that can also act as a precursor to NAD. The benefits include improved insulin sensitivity, reduced cholesterol, increased cognition, and antioxidant capacity[2081]. High doses of pterostilbene can raise LDL cholesterol but there are no other known side-effects.

- **Digestive Enzymes** - Everything you do requires some enzymatic reaction whether that be moving around, speaking, blinking, or digesting food. Digestive enzymes help break down macromolecules into smaller energy units in order to absorb them better and use for different physiological processes. They can help to improve digestion, reduce bloating, fix brain fog, heal the gut, and decrease inflammation. Foods high in enzymes are ginger, pineapple, fruit, sauerkraut, and fresh food in general. Cooking destroys enzymes but you can use digestive enzyme supplements as well. There are even specific types of enzymes targeted towards either gluten, carbs, or fats and protein.

- **D-Ribose** – It's a simple carbohydrate molecule that's involved in energy production. D-Ribose can be found in all living cells as it's the structural basis of DNA and RNA. The other health benefits include reduced fatigue, improved heart health, better workouts, and kidney protection. Long-term D-Ribose supplementation may promote AGE production so you only want to use it for some hardcore workouts. In total, you can take about 5-10 grams of D-Ribose with pre-workout protein.

- **Cordyceps** – Cordyceps aren't actually mushrooms but a family of parasitic fungi that grow on the larvae of insects. They grow inside their victims, usually ants, and grow stems outside of the host's corpse. These 'zombie-parasites' have been shown to promote ATP production, reduce time to fatigue, increase oxygen uptake, and improve exercise performance[2082]. They're not necessarily anabolic or pro-longevity but they will improve your health and lifespan.

Sleep Supplements

Moving on with actual sleep supplements that come as a pill or powder:

- **Melatonin** – The main sleep hormone melatonin is also a powerful antioxidant. You don't want to rely on melatonin supplementation as it may hinder your natural ability to produce melatonin. However, using about 0,3-1 mg on some nights can be useful. Liquid melatonin is absorbed much better and gives a more sustained release. You can use melatonin to prevent and treat jet lag, shift work, and circadian misalignment[2083]. Melatonin supplementation seems to be more effective in people already suffering from sleep disorders[2084]. It can reduce the time needed to fall asleep and increase total sleep time[2085].

- **Tryptophan** – Tryptophan can cross the blood-brain barrier and get transformed into a precursor of serotonin 5-hydroxytryptamine (5-HT). 5-HT promotes drowsiness and fatigue, which is perfect for sleep. You can get tryptophan from poultry, dairy, nuts, seeds, and carbohydrates. If you take a tryptophan supplement, then 1000 mg is more than enough.
- **Glycine** – Glycine functions as an inhibitory neurotransmitter that antagonizes glutamate receptors. It's been shown to improve sleep quality[2086]. Glycine also lowers body temperature, which will help to fall asleep[2087]. About 5 grams of glycine powder a few times a day is enough of a dose but even 31 grams a day appears to be safe[2088].
- **Phosphatidylserine (PS)** is a phospholipid that's important for cell membranes and brain structures. Phosphatidylserine reduces cortisol production under stress and speeds up physical recovery at doses of 600 mg per day[2089].
- **Inositol (Vitamin B8) -** Inositol is a vitamin-like substance that's found in animals and plants. It helps with the regulation of neurochemicals. Your body produces a little bit of inositol but it may get depleted if you're experiencing a lot of emotional turmoil. Inositol

supplementation can boost serotonin and calm down the nervous system[2090].

- **GABA** - Called gamma-aminobutyric acid, it's the main inhibitory neurotransmitter and regulates the nerve impulses in the human body. Therefore, it is important for both physical and mental performance, as both are connected to the nervous system. Also, GABA is to an extent responsible for causing relaxation and calmness, helping to produce BDNF.
- **L-Theanine** –L-theanine has an alertness boosting effect but it's not as stimulating as coffee. It's an amino acid found in tea leaves, especially green tea. The release of caffeine from L-theanine is more subtle and long-lasting. That makes it a great addition to your morning coffee if you want to prevent the crash. It's generally safe and there isn't a lethal dose.
- **5-HTP** – 5-Hydroxytryptophan is a precursor to serotonin, which has anti-depressant and relaxing effects. In one study, a combination of GABA and 5-HTP reduced time to fall asleep, increased the duration of sleep, and improved overall quality[2091]. It can help with sleep, weight loss, and anxiety. 200-300 mg doses are safe but higher ones may have side-effects.
- **Bacopa monnieri** – Also known as Water hyssop is a nootropic herb that's used in traditional medicine for cognition and longevity. It can improve memory and

relieve stress. The standard dosage is 300 mg a day with the upper limit being around 750-1200 mg.

Herbal Compounds for Sleep

Let's carry on with various herbs and other natural ingredients that can promote sleep:

- **Chamomile** – You can use it in teas, infusions, extracts, and essential oils. Chamomile has anxiolytic effects and acts as an antidepressant[2092]. It's a great liquid for the evening that makes you relaxed, sedated, and calm. Don't drink too much though or you'll have to go to the bathroom more often.
- **Lavender** – You can use lavender essential oils, fragrance, and teas for improved sleep. Lavender aromatherapy may be as effective for insomnia as conventional medications but without the side effects[2093]. Supplementing lavender can cause nausea and stomach pain in some cases[2094].
- **Passionflower** – Research about the sleep benefits of passion flower shows that it can improve sleep quality. However, some placebo studies find no difference. It might help you to relax and calm down by drinking some tea but it might not.
- **Lemon Balm** – It has been used in many cultures as teas, essential oils, or extracts. By regulating the HPA axis and

GABA, lemon balm has been shown to lower stress and have anxiolytic effects[2095].

- **St. John Wort's** is a common herb used for depression and anxiety. It can also increase serotonin in the brain[2096].
- **Ginkgo Biloba** – It's also known as maidenhair with a long history of herbal medicine. Consuming 250 mg of ginkgo biloba 30-60 minutes before bed can reduce stress, promote relaxation, and promote sleep[2097].
- **Kava Kava** – Kava is a root native to the South Pacific islands that's traditionally consumed as a tea. Known for its sedative and anti-anxiety properties, kava has been found to be beneficial in relieving stress-induced insomnia[2098]
- **Valerian Root** - Studies have found valerian root helps improve the speed at which you fall asleep, depth of sleep (achieving deep sleep 36 percent faster) and improving overall sleep quality[2099]. A 2006 meta-analysis found that valerian root can improve sleep quality without side-effects[2100]. Studies say the effective dose is 300-900 mg.

Exercise Performance Supplements

Here are supplements that can improve muscle, strength, and exercise performance:

- **Creatine Monohydrate** - Creatine is an organic acid produced in the liver that helps to supply energy to cells all over the body, especially muscles. It enhances ATP production and allows for muscle fibers to contract faster, quicker, and makes them overall stronger. This means increased physical performance with explosive and strength-based movements and sprinting. However, it doesn't end there. Creatine has been found to improve cognitive functioning, as it's a nootropic as well, improving mental acuity and memory, especially in vegetarian diets. Naturally, it can be found mostly in red meat. It's dirt cheap and easy to consume, as only 5 grams per day will do wonders, and doing so won't make a person big nor bulky. You don't have to load with 30 grams of creatine a day or some other nonsense. Just take 3-5 grams a day, preferably with food.

- **Branched Chain Amino Acids.** L-Leucine, L-Isoleucine, and L-Valine are grouped together and called BCAAs because of their unique chemical structure. They're essential and have to be derived from diet. Supplementing will increase performance, muscle recovery and protein synthesis. There is no solid evidence to show any significant benefit to BCAAs. However, they can be very useful to take before fasted workouts to reduce muscle catabolism. It will protect against muscle catabolism and can even promote ketone body production.
- **Whey protein** - On a standard ketogenic diet, you would want to avoid protein shakes because they spike your insulin. If you're doing CKD or TKD you would benefit from having an easily digestible source of protein. Before you break your fast and begin your carb refeed, make a

quick shake to get the juices flowing. You can also use whey protein during fasted workouts with targeted intermitted fasting. This again will prevent muscle catabolism and will increase performance.

- **Phosphatidic Acid** – Phosphatidic Acid (PA) can regulate mTOR and promote muscle growth. It's a unique lipid molecule that turns on MPS in response to resistance training. PA can be found in foods some foods but in extremely low quantities. Vegetables like cabbage contain 0.5 mg of PA per gram. That's minute compared to the 250-750 mg doses in studies. More research about the effectiveness of PA supplementation in humans is needed but no long-term side-effects have been noted. Daily intake of 450 mg-s is optional.
- **Dextrose -** It's basically powdered glucose and very high on the glycemic index. You want to avoid it on SKD, but on CKD or TKD it's very useful for a post-workout shake with protein. It's dirt cheap and you'd want to take only 3-5 grams at once. Use it ONLY when doing the TKD or CKD because under other circumstances you're not doing your health a service. Dextrose is pure glucose and it's processed so it's definitely not optimal for autophagy or longevity. Most people don't need it and you may feel better without it but it's just an option to keep at the back of your head.
- **Colostrum** – It's sometimes called 'liquid gold' because of the yellowish color. Colostrum is the precursor to breast

milk and it's rich in immunity-boosting compounds and growth factors. As an anti-aging supplement, it may prevent tissue degeneration and skin aging. Not something I'd recommend taking every day but on workout days it can be used for muscle hypertrophy.

- **Collagen protein.** Collagen provides the fastest possible healthy tissue repair, bone renewal, and recovery after exercise. It can also boost mental clarity, reduce inflammation, clear your skin, promote joint integrity, reduces aging, and builds muscle. Naturally, it's found in tendons, and ligaments, that can be consumed by eating meat. As a supplement, it can be used as protein powder or as gelatin capsules.
- **HMB** - β-Hydroxy β-Methyl butyrate is a by-product of leucine, which is an amino acid that stimulates protein synthesis. It's been shown to reduce exercise-induced muscle damage and improve recovery[2101]. HMB can cause acute muscle anabolism and MPS independent of insulin[2102], thus it will maintain a semi-fasted state. You can take it with the intra-workout protein shake to minimize muscle catabolism. Use pure HMB powder instead of the ones with artificial sweeteners.
- **Beta-Alanine** – An amino acid that reduces fatigue and increases physical performance. It's the main ingredient of many pre-workout drinks and thus it can help you to push yourself during workouts, especially if you're training fasted. Generic pre-workout drinks are way too stimulating

and high in caffeine. Instead, take pure beta-alanine with your protein shake.

Like I said in the beginning, you should focus on getting your nutrition on point first before taking any supplements. And secondly, you don't need to take all the supplements, only the ones you're most deficient in.

Taking Supplements Protocol

- Before supplementing any specific vitamin, you'd be better off by first focusing on eating real food, getting your nutrients from that, taking blood tests to see your deficiencies, and then taking those supplements you need with food. Generally, a whole foods diet that includes both animal and plant ingredients will probably cover most of your needs.
    - Getting a DNA analysis can be an effective way to know what kind of genetic predispositions or shortcomings you might have. Especially in relation to methylation. Regular blood tests will also assess your nutrient status and tell you if you need to eat more of certain foods.
    - Use vitamin D based on your sun exposure. On a daily basis you can stick to 1000 IU but during the dark months aim for 3000-5000 IU. When you're sick, taking up to 10 000 IU for a few days can

boost the immune system. Don't use brands that contain vegetable oils or sunflower oil.

  - Iron and electrolyte deficiencies happen usually because of dehydration and excessive excretion of your body's salts. The most common reason is coffee and tea consumption. You can drink coffee while fasting but hot liquids and beverages may make you absorb less of the micronutrients. The tannins and caffeine in coffee and teas can lower the absorption rate of your supplements. They can also make you excrete more of the other electrolytes and minerals through urine, so you have to be careful with not taking your supplements together with these drinks. Otherwise, you're simply pissing them out and wasting your money.

- Take adaptogens around stress. Don't take them all the time as to avoid building a tolerance. They're especially great during the flu season. Adaptogens don't generally affect exercise adaptations negatively.
- Don't take antioxidants around exercise. Use things like vitamin C only when you're sick or caught something. Otherwise, you may weaken your body's ability to deal with oxidative stress by being too anti-inflammatory.
- Take supplements with food. The bioavailability of most supplements is much higher if you take them with food, especially fat-soluble vitamins like A, D, E, and K. There's no reason to take supplements in a fasted state unless

they're stimulatory or adaptogenic. Mixing adaptogens or herbs with your coffee or broth is not an issue either and won't fully break the fast.

Giving your body an overabundant supply of all the nutrients isn't a good idea either because it will have conflicting reactions and may cause some other issues. It's definitely a waste of money.

Chapter Sixteen: Creating Your Own Hormesis Routine

„If you fail to plan, you are planning to fail.“

Benjamin Franklin

It's true that the things we can control in life are minuscule compared to what we can't. The world is chaotic and operates regardless of how we'd like it to. This can be a source of great uncertainty but only if you let it become one.

The only way to protect yourself from the chaos of random events and dangerous situations is to prepare your body and mind. With higher levels of stress adaptation, you'll experience less damage from any stressor and will recover faster as well. What's more, your psychological resilience would be much greater, which improves the way you react. Most of the time, people aren't harmed as much by outside incidents but instead by how they themselves respond. The average person tends to lose their mind and self-control whenever something unexpected happens. Once it does, they start acting irrationally, make poor decisions, and experience additional trauma.

To mitigate the negative effects of chaos on your health, you need structure and order in your life. It's going to protect and guide you through disorder when the world around you is in flames. Certain systems, habits, and daily practices will augment stress adaptation as well as improve general responsiveness to things that will wipe out those who don't know what to do. This chapter will talk about how to create your own hormesis routine that makes you antifragile.

Why You Need Purposeful Routines

Studies show that people who have no greater purpose tend to suffer from depression, lack of motivation, apathy, and inaction[2103]. It also mediates the satisfaction between life satisfaction and suicide ideation[2104]. Amongst US adults older than 50, a stronger purpose in life is associated with decreased mortality[2105]. That might be raising children, working, contributing to your community, a creative project, friendships, a spouse or even taking care of pets. Having hobbies has also been shown to extend longevity and healthy life expectancy[2106].

This is written about very well by Viktor Frankl – the Jewish psychiatrist who spent 3 years in Auschwitz. There he endured countless amount of suffering and ordeals. Despite all of that, he came out of the imprisonment alive and changed the rest of his life for good. Frankl noticed that some prisoners, like himself, managed to deal with the given situation a lot better than others. He realized that the human condition enabled them to adapt to anything they came across. They were so close to hell as ever possible but in spite of that, there were a significant amount of people who did not break.

Most prisoners couldn't cope with reality. They saw their destiny as pre-determined and thus soon died while there were others who remained indomitable in the face of resistance. What's up with that? Frankl discovered that it had to do with how they interpreted their situations. Most importantly, how they thought that all their sufferings must've had some sort of a meaning. There wasn't…I mean…there is no justification to kill children, women, elders, and men of all ages – to torture them, starve them, enslave them, to humiliate them. But for Frankl and the prisoners there was. Behind all that torture was a greater meaning because they themselves created it there. You might think: *"Ignorant fools…How can you be so naïve? Obviously, the thing was wrong."* Perhaps you're right but it doesn't deny the effectiveness of such a mindset. It allowed them to accomplish nothing short of the impossible, although we're all capable of achieving the same thing. With that kind of a mentality, they could re-frame how they experienced their suffering.

Another essential aspect of it was how the prisoners envisioned their future and how they absorbed meaning from their vision in a particular present moment. Every prisoner's reaction depended on how they had thus far associated their situation, but a lot more important was the opportunity for free will. Free will which we all possess, even in torture. Eventually, after Frankl had emerged from hell alive, he continued his work as a psychiatrist and created his own method called 'logotherapy.' At the core of his theory is the belief that the primary motivating power of man is the search for meaning or a purpose and the therapist's job is to help find it. Despite Auschwitz, despite the years of horror and slavery, Frankl still lived until 92 and died in 1997. How tough do you have to be to pull off something like that? I believe this story can be very inspirational wherever you are in your life. But just remembering it won't mean a thing. You have to be able to take action and execute your purpose.

Friedrich Nietzsche said: *"He who has a why to live for can bear almost any how."* This should be the starting point for doing anything that's done for getting specific results. You first come to terms with what you want to accomplish and then reverse engineer yourself from where you want to be all the way down to where you are right now. Every easily recognizable milestone is sort of a mini-goal - a step along the way that marks your progress and lets you know you're on the right track.

If you already know what you want to accomplish, you should still know your why. It's the reason why you want to achieve a particular goal and what sense of meaning does it give you?

- *"Well, I want to start my own company."* - WHY
- *"So that I could change the world for the better."* - WHY
- *"So that there would be less suffering."* - WHY
- *"So that the people around me could be happier."* - WHY
- *"So that I myself could be happier."*

Ask yourself WHY 3-4 times and you'll be amazed to discover what your real motives are. Whatever goal you might set for yourself, it will always get traced back to happiness and fulfillment. It can be in any form, but what you seek is to be subjectively happy.

Our time and mental resources are finite. Everyone has the same amount of time but not everyone accomplishes as much with it. The reason why some people get so many things done is that they know how to use their time better and not be used by it. Everything they do is already planned and mapped out. All they have to do is execute, execute, execute. But more crucially, not only do these people have *a plan* but THE PLAN. It's not your average to-do-list but a fully optimized strategic blueprint that is constructed as to be as effective as possible.

People have negative associations with routines – they think that doing certain things over and over again somehow degrades the quality of their life. But that's a dabbler's way of thinking – don't be a dabbler. Instead of being a drifter and a dabbler who tries something new only for a while but then quits, instead of that, you should aim for mastery. It's a level of skill that's near excellence and perfection. Great people throughout the history mastered their craft – whether that be Leonardo da Vinci, Aristotle, Mozart, Napoleon Bonaparte, Stephen King, or whoever. They're not only admired by the masses but also have something none of the drifters will ever have – a greater purpose they pursue, and the more they do, the more alive they feel.

Mastery is achieved through rigorous practice and honing of one's skill. You'd think that creative people are only spontaneous and lack discipline but the opposite is true. They're actually the most disciplined and consistent people on the planet. Mozart started playing the piano since he was 4 years old. Stephen King writes like 4 hours every day. It's their routines and habits that allow them to reach such levels of greatness. The repeated actions and activities they follow every day. Rome wasn't built in a day, and the same applies to whatever goal you want to accomplish.

Goals vs Systems

Goal orientation is based on getting specific results. Get an A on an exam, lift X amount of weight, make a million dollars, win the championships, write a bestselling book, and so on.

Peter Drucker, the infamous business management philosopher, said: *"What gets measured gets managed."* That's the foundation to goals way of thinking. You set a specific outcome you want to achieve, come up with an action plan, then go out there to start following the steps. *There's the target, now go get it.* Simple, but not always easy. Here are a few random examples:

- Earn an annual income of 10 000
- Lose 10 pounds by the end of the year
- Run a mile under 4 minutes
- Get a degree in college
- Write a bestselling book

You can now dissect them into smaller chunks and actionable steps to follow *i.e.* earn 1000 every month or lose one pound a month. They're now mini-goals you'll follow specifically. But there are some flaws to simply having goals. For instance, achieving SOMETHING doesn't mean that that which was achieved was important. Are getting specific outcomes actually worthwhile, or is measuring them misleading you from what's truly paramount to your greater purpose in life? It's worth pondering about regularly.

Another quote by Peter Drucker: *"Doing the right thing is more important than doing the thing right."* Effectiveness is about doing the right things that are important for what you want to accomplish, whereas efficiency is about doing things right. You can be very efficient at mopping the floor or typing the keyboard very fast, but will it be effective is a completely different story. Unless you're planning to become a championship mopper or typer it's not that relevant.

Here's where systems way of thinking comes into play. I'll use the same examples of goals as systems instead.

- Learning and increasing your knowledge about marketing and finance, and constantly implementing them into your business is a system for getting wealthy.
- Eating healthy and consistent physical exercise is a system towards losing weight and maintaining excellent body composition.
- Studying, reading, and self-improvement every day is a system for getting smarter and developing yourself as a person.
- Practicing writing, researching, tinkering with one's skills is a system for becoming a master writer and thus creating a bestseller.

- Paying attention to world news and having back-up preparation plans for the worst-case scenarios is a system for surviving unexpected events.

The difference between goals and systems is that one focuses on specific results (DOING), whereas the other is based on your values, purpose, and who you are as a person (BEING).

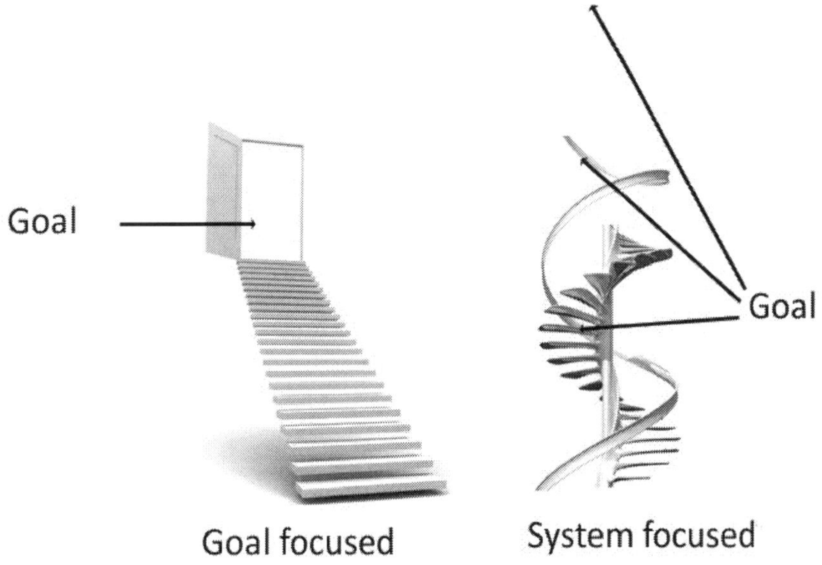

Here are a few reasons why you should focus on systems instead of goals:

#1 Goals Make You Less Happy in the Now

Wanting something and setting specific steps towards achieving them is basing your happiness on a future outcome. You're not quite there yet, but after you get the results, you'll instantly be fulfilled. At least that's what you'd expect.

The problem with this way of thinking is that you're basing your happiness on something external - the next goal, another milestone, a specific finish line you have to cross. This will make you feel as if you're currently inferior and need to get the achievement before you can start living fully in the present. Once you become accomplished, you'll experience a brief moment of fulfillment - a momentary feeling of bliss and relaxation. Then...*meh...now what?*

What's more important is to focus on the system and enjoy the process itself. The goal or the destination itself isn't worthwhile if you hate every moment of getting there. You have to walk the journey just for its own sake.

#2 Goals Can Cause Anxiety

Because of the same reason, goals can also make you more anxious and stressed out. If you have an impending list of goals, then you'll want to accomplish them. But that doesn't necessarily mean that you'll always have enough guts or willpower to do so. Most people who set goals fail to achieve them because they lack discipline, don't know what they have to do nor how to do it.

Say you set out to lose 5 pounds by the end of the month. The end-result (the future you but with 5 fewer pounds) will be constantly looming in the back of your head directing your decisions and draining your willpower. *I have to stay vigilant if I were to meet my goal.* You made some progress the first week and everything seemed to follow its right course. However, your weight loss stalled at the last week and there's no way you can meet your set goal. Now, what do you do? No magic formula will make those pounds disappear at an instant. You may still continue dieting (which is a systems-based approach) or you might give up completely, thinking to yourself that you'll never be able to get the desired results. In the worst-case scenario, you'll feel remorseful, judge yourself as being worthless and then binge on junk food. *Because why not...you already missed your goal.*

#3 Systems Are Who You Are

Goals are temporary milestones you set out to reach. Systems are permanent patterns of thinking and behavior carved into your psyche. If you were to follow a system of eating healthy and focusing on the process of becoming fit, then you would've continued to pursue your dream body despite there being a deadline.

Herein comes into importance the notion of DOING vs BEING. One is about deliberately setting out to achieve goals, the other is based on who you are as a person and what you do on a daily basis. To accomplish a goal, you have to be doing the things that allow you to do so. You can't deliberately DO a system. A system is a process of being. Both of them are effective ways of living a fulfilling life, as improving yourself entails transforming who you are currently – you have to DO in order to BE.

#4 Goals Make You Attached to the Outcome

Because goals base your happiness on a future achievement, you'll also become too attached to the outcome. *I have to accomplish the target goal weight, otherwise, I'm worthless.* You'll be constantly thinking about the final result and destination.

The more you cling onto something, the less congruent you'll become. You're trying to hold on for your life so that you wouldn't lose this thing you're after. It means that you fear becoming a lesser version of yourself if you don't have that specific outcome.

Fragile things tend to break into millions of pieces once they come into contact with another object or a force in motion. Being attached to the outcome reduces your happiness in the now and actually hinders your progress because of your analytical mind getting in your own way. An antifragile system would maintain a state of abundance and high functionality even if things aren't going as expected to the plan.

#5 Systems Can Be Updated

If you set a goal, you may fall into the trap of starting to pursue it just for its own sake. *I said that I was going to lose those 5 pounds by the end of the month and I will goddammit!* You begin to develop a self-image inside your head that corresponds with the final outcome. *I'm the guy who makes a ton of cash. I'm the goal-getter.* But because you don't want to give up your goal and seem inconsistent to yourself or others, you'll keep hammering through. Only losers quit, right?

Becoming too attached to the outcome can also blind you from what you genuinely want. Setting some goals doesn't mean that they're in alignment with your core values or purpose. Achieving SOMETHING isn't worth the effort if it doesn't contribute to your higher cause. That's the beauty of systems' way of thinking. You as a person are in constant motion and development. Your behavior, appearance, habits, even the neural network of your brain are plastic, which means that they're always changing. What you do on a daily basis creates certain patterns of thinking and acting which are based on how you adapt to the surrounding world.

Goal-oriented way of thinking limits you within a set range of parameters, whereas systems are self-constructed modes of being that are always aligned with who you are. If you decide to change some of your behavior, you'll be able to do so easily because of having maintained your freedom.

Both goals and systems, doing and being, results and purpose overlap. These two modes of thinking contribute to each other's existence and you can't be entirely without one or the other. They work in certain situations and I'm not trying to guide you away from ever setting goals. They're necessary. You're going to have to reach certain outcomes to achieve your higher purpose. To get fit and healthy, you're going to have to lose some weight. In order to be wealthy and financially independent, you have to earn some money. The difference is that systems free you from the outcome and are more consistent.

I've never been much of a goal-tracker. Don't get me wrong, I'm a goal-getter and aspire to achieve great things. But in order to do so, I'm not checking off my to-do list, which would satisfy my ego. Instead, I'm focusing on the system of becoming the greatest being of myself. Intrinsic motivation to self-actualize is more fulfilling and purposeful.

The best advice I can give you is to establish a balance between the two. You can set some goals to keep track of your progress, to know whether or not you're getting results or achieving specific outcomes. Use them as guidelines but don't get too attached to them. Focus on the systems instead - the process of being the person you're trying to be.

Goals vs systems way of thinking has been a part of my being, albeit with some slight course-adjustments, ever since I can remember. I hadn't named it before but simply followed my nature. This specific concept under this particular name I came across in Scott Adam's book *How I've Failed at Everything and Thus Succeeded.* He's the creator of the famous comic series Dilbert and he follows the same mindset.

Antifragile Standard Operating Systems

It's very important to have certain routines and structures in order to get things done. But trust me on this one: *nothing will ever go exactly according to plan."*

There's the saying: *"People underestimate what they can do in a year and overestimate what they can do in a month."* Even if you have set yourself some goals you have to give them extra time and leave room for randomness. It'd be great if everything went according to plan but that's just the nature of life. It's highly probable that the highly improbable is highly probable. And to not be swept away by these unlikely yet devastating accidents, you'd want to have some insurance.

One great way of dealing with these types of events is to create Standard Operating Procedures (SOPs). They're basically reactionary routines to put out the fire. The basic formula for SOPs is this: *IF this, THEN that.*

- IF I lose internet connection, THEN I'll read a book.
- IF I get stuck in traffic, THEN I'll call off the meeting.
- IF the gym gets closed down, THEN I'll do calisthenics in a park.
- IF I finish off with work sooner than expected, THEN I'll go for a walk.
- IF my sleep is cut short, THEN I'll have a nap in the afternoon.
- IF I start feeling dependent on caffeine, THEN I'll do a coffee detox.
- IF I get home too late in the evening, THEN I'll fast for longer instead.

What these IF/THEN statements allow you to do is create automatic standard operating procedures that give you more control over the things you can influence. You can't control the things that happen to you, but what you can change is your own response. With SOPs you react the way you planned to. Instead of following what your immediate urges and feelings tell you to, you respond in a proactive manner because more often than not your amygdala gets hijacked before the neocortex could make a more rational decision.

SOPs also give you a greater sense of agency in life - you're not going to panic when something doesn't go as expected. You can feel at ease and less stressed out in anticipation of future events. It’s going to make your daily routines and habits antifragile as well.

So, how do you create standard operating procedures?

#1 Analyze Your Systems

First, you need to go through your current systems - your daily habits, routines, behaviors. What do you do on a daily basis after all?

Most people don't have an actual understanding of what they do or why they do it. Their routines are on autopilot but they don't have a clue when nor why did they develop these ones in particular. *"They just happened,"* is the wrong mindset because more often than not it’s the result of societal conditioning, parenting, poor habits, peer pressure, and lack of knowledge of what’s optimal. If there's a much more effective way of doing something, then the rational thing would be to do it, instead of wasting time and energy on the activities that don't work nearly as well.

Look at your daily routines in chronological order - What do you do? When do you do it? Why do you do it? How do you do it? The clearer you understand yourself and your operating systems, the better you can leverage your strengths and avoid the weak spots.

#2 Identify Your Weaknesses

Secondly, you have to do exactly that - identify the weaknesses and fragilities in your routine. Bringing them to daylight helps to fix these loopholes. Ask yourself these questions:

- Where are you taking a lot of unprotected risks? Are you asking for trouble? Is it only a matter of time before this thing collapses entirely? Have you been ignoring some blatantly obvious problems for months?
- What activities don't give you as many results but have a very large potential downside? Which commitments, responsibilities, or relationships cause more stress than benefit? Are you doing something stressful just out of habit or because of past history?
- Do you even need to do a particular thing? Is there a less risky alternative? Are you settling for less when you could have something much better? Why haven't you fixed these loopholes yet?

Here, you must re-evaluate everything you do. It's very likely that you can exclude most of your activities that don't yield any significant gains but they have a chance of greater losses. You don't want to waste time and energy on activities that give you meager results. Focus on the select few instead of the dispensable many.

Cut away all the minutiae and unessential ties that are nothing but extra weight. The more mobility you have in terms of openness and freedom the quicker you can respond to chaos. Apply the

Pareto's principle and keep only the 20% most effective components of your routine while hacking away all the rest.

#3 Pierce Your Own Routine

Thirdly, now that all you have left is your best routines and greatest strengths, try to think of ways to knock them over. You know your systems better than anyone else and are aware where the greatest fragilities lie.

What situations or events would make you completely ineffective? Like, all your greatest work and most creative projects are on a hard drive, but you break it. Everything destroyed - just gone and there's no way of retrieving the data. A worse example would be forgetting the password to your Bitcoin wallet and never being able to access the funds. Or you have organized this live event but the main speaker calls and jumps out on you. Now you have a few dozen people waiting to be entertained but no one to deliver. Those kinds of weaknesses you can't exclude entirely - other people have their own interests and unexpected things are bound to happen.

To alleviate the blows, you should have solid redundancies in place, just in case the main performance driving factor stops functioning. Humans have 2 kidneys, 2 arms, 2 eyes and so should you have 2 plans. IF Plan A flops, THEN initiate Plan B.

#4 Your Mental Operating System

Fourthly, you need to change your mindset as well. Remember that perception creates reality and the way you look at things can change your subjective experience.

What are you going to do when, despite all your greatest efforts, you fail to have redundancies and you DO lose that hard drive? Will you quit your plans completely? Most people would fall into depression. Or will you use it as an opportunity and turn it into an invaluable lesson?

This is where stoic negative visualization and preconditioning comes in handy. It's going to keep the potential of chaos at the back of your head and prepares your brain for it in advance. The way you perceive the situation and how you think determines how you're going to respond.

#5 Create Your SOPs

And fifth, put your standard operating procedures in place. Because just thinking about these things won't make anything better - you have to proactively do act as well. SOPs are IF/THEN style actions that can be applied to all types of activities, starting from losing your life's greatest work and ending with running out of ink in your pen.

To make SOPs effective, you have to dissect them into 2 parts.

1. The first is the IF component. This is the chaotic event. You can't be fully certain of what's going to happen but you can try to pierce holes into your own strengths and weaknesses until you have a solid understanding of what are the biggest detriments to your results. Really dig deep and try to uncover as much of your own weak spots. Then you should also keep the worst-case scenario at the back of your head - what's really the worst thing that could happen?

2. Then you create the THEN statement. If this, then what? If not, then why not? If yes, then when? What are you going to do if this or that happens? Consider the severity of the situation, how bad is it, your circumstances, and routines and the effectiveness of your THEN statement. You should choose a more flexible response with additional redundancies that could potentially turn you antifragile and actually gain from the situation. Try to decrease the downside while maintaining high levels of responsiveness.

Don't make the mistake of having too many THEN options either because having excessive things to choose from still leaves you vulnerable and causes paralysis by analysis. Have a few extra redundancies but focus on the select few. Stick to the 80/20 rule. Well, at least 80% of the time.

Fortunately, life isn’t only one accident on top of another and we do experience plenty of positive situations as well. That’s why you should also have SOPs for the positivity - the unexpected good. Otherwise, you’ll miss these opportunities.

- What if you finish your work faster, then what? What task will you do then? Or will you use it as a time for rest and recovery? Maybe you can just spend more time with family.
- What if you reach your goals fast? Will you just let it happen and take it as it is? Or will you start analyzing what caused it to happen and try to repeat it in the future? Being

clueless about why you succeed is almost as bad as not knowing why you fail.

- What if you learn a new and very useful skill or gain some knowledge? Will you ignore it or implement it to the best of your ability in your routine?

SOPs are great for creating systems for proactive reactiveness. When the unexpected occurs, you're forced to respond, but having made it a part of your plan, you'll know how to deal with the situation a lot better. This not only reduces stress, increases your results, and gives you a greater sense of control, but also minimizes the effort of having to clean up some random mess that could've been easily prevented.

Whatever the case may be, you should keep in mind your goals, circumstances, and how you want your daily routine to look like. Use these SOPs and IF/THEN systems wisely.

Blocks for Doing Everything That Matters

All of the hormetic practices we've talked about in this book require time and commitment. You need to be doing them on a regular basis as to not lose your adaptation. The same applies to rest and recovery – you need to have some downtime to let all these benefits sink in.

For an optimal circadian rhythm and daily routine, you have to dedicate certain blocks of time to your most valuable activities, habits, goals, and values. Here's a list of the things you should include:

- **Block for Sleep** – Naturally, you want to get sufficient quality sleep every night. Otherwise, everything else will begin to slowly fall apart. For most people, that would require 6-8 hours depending on how optimized and efficient you are. I don't think it's healthy to be consistently getting less than 6 hours per night. On some days you may end up with 9-10 hours but anything beyond that is also unwanted.
- **Block for Sunlight** – You can get quite a good amount of exposure to sunlight in the morning and while going for a nature walk. However, it's probably not enough nor optimal. One study in the UK found that 13 minutes of sun exposure just 3 times a week is enough to maintain healthy vitamin D levels[2107]. That's baloney, if you ask me, especially in the UK. The problem is that over 40% of people are already vitamin D deficient and it's linked to heart disease, diabetes, cancer, and others[2108]. Wearing clothes also makes it harder to obtain vitamin D as it's absorbed through the skin. So, wearing fewer clothes would enable you to get more of it from less time. 20 minutes a day of direct exposure to the sun might be enough and you definitely don't need to over-do it. Just keep track of your blood test results.
- **Block for Work** – This is the period of time you dedicate to your most important tasks related to work, life's

purpose, personal success, mastery, productivity, etc. Cal Newport calls it Deep Work of uninterrupted focused attention where you're not checking social media or procrastinating. Optimally, you want to aim for 2-4 hours a day at a minimum but that may not be always possible. I personally have two 4-hour blocks of deep work a day.

- **Block for Time-Restricted Eating** – Food intake has a significant impact on the circadian rhythms and a smaller eating window has been shown to have unique metabolic benefits beyond just weight loss. You should definitely eat your calories only when it's light outside and not at night. In general, the minimum window to aim for is 14-16 hours fasted and 8-10 hours fed. If you want to gain more of the benefits of fasting and autophagy, then you'd probably have to restrict yourself more and aim for a 20/4 or 22/2 schedule. Refer to the chapter about intermittent fasting.
- **Block for Exercise** – It's not a successful day if you didn't move around your body or got a sweat on in some other way. Generally, 20-45 minutes of exercise a day is sufficient but working out 60-90 minutes can improve your results. It's not the length of the workouts but the effectiveness and quality that matters. From a circadian perspective, it's better to train later in the day around 2-5 PM because that's when your nervous system's been warmed up and is ready to go but any other time is better than nothing.

- **Block for Learning** – You should go to bed smarter than you woke up. That's one of the most important tasks you need to accomplish every day. Reading, learning new skills, going through courses, listening to audiobooks, are amazing ways to increase your knowledge, improve your skillset, and get better at whatever you're trying to do. You should definitely learn about the most relevant topics to you such as health, psychology, wealth accumulation, society, etc. In total, 1-2 hours spent on actual learning a day should be the minimum amount.
- **Block for Nature Bathing** – I think it's incredibly important to spend at least an hour a day in natural environments. Forest bathing or Shinrin-Yoku as it's called in Japan has many health benefits such as reduced cortisol, less anxiety, improved blood flow, better cardiovascular health, more mindfulness, circadian alignment, and higher sleep quality. I like to go for an hour-long walk in nature at noon to take a break from work and get my daily steps in. It's incredibly soothing and enjoyable.
- **Block for Rest and Reflection** – *„If you don't know where you're going any road will lead you there"* said Lewis Carroll the author of Alice in Wonderland. The famous Stoic philosopher Seneca also said: *"If one does not know to which port one is sailing, no wind is favorable."* It means you should be very clear about what your goals are and what activities you need to do to achieve them. That's why spending time in self-reflection, journaling, meditation, or just thinking will make things much easier,

reduces stress, increases your effectiveness, and improves your happiness. In total, this would take about 30-60 minutes of your day.

- **Block for Relationships** – *„No man is an island"* said the 17th century English poet John Donne. Humans can rarely survive and be happy without other people. That's why it's important to nurture and develop close connections with your loved ones as well as make new friends. Depending on your lifestyle and personality, you can spend anywhere between 2-4 hours with family. But again, it's the quality of those hours that matters. An hour of deep intimate conversation with your spouse is much more valuable than 2 hours of talking with each other while scrolling through social media or watching TV.
- **Block for Play** – Play is how children learn and develop their communication skills. It's something grown-ups tend to neglect and forget about. The purpose of play isn't to act stupid or procrastinate but to just have fun. Stress and anxiety plague a lot of people and part of the reason has to do with not enough play-time. That's why, as a hard-charging go-getter and self-disciplined person, I take my play-time very seriously and actually schedule it into my evenings. It's a way to de-stress, wind down, reflect, recover, and prepare for the next day's events. Play can be anything from watching movies, playing video games, listening to music to exercise, socializing, dancing, walking in nature, or cooking. Just do what you enjoy and makes you happy. Too much play at the wrong time and at the

expense of your health or work, however, can make you a procrastinator and not achieve much with your goals. That's why it's important to have only about 2-3 hours of actual downtime with no real purpose other than the activity itself.

Some of these blocks may overlap and cover each others' basis. If you just love working out, then it can be a form of exercise as well as meditation, play, relationship building, and fun. Likewise, you may be working together with your family and actually want to take a break from each other in the evening.

How many hours you spend on a given block depends on your particular circumstance and what your goals are. If you want to be a professional athlete or get ripped for summer, then, of course, you have to work out more often. Someone with the aspiration of building a business and earning a lot of money would also have to work longer hours and sacrifice time from other things.

Weekly Hormesis Stress Adaptation Plan

Your routines shouldn't be looked at only within the 24 hours because they actually span over the course of months and years even. There's only a certain amount of work, exercise, and play you can fit within a given day. However, the cycle of stress adaptation is running all the time and your body adapts to specifically the signals it receives on a consistent basis.

To balance performance and recovery, you need to be doing different hormetic activities on different days. This allows all the positive shifts in your physiology to sink in and also avoids over-training. What's more, you'll be building all of your body's systems in unison and prevent imbalances.

We all have distinct goals and thus have to spend different amount of time pursuing them. An athlete would naturally have to work out more often and longer because they want to improve their sports performance whereas a housewide just wants to stay healthy with the minimum effective dose. Additionally, a biohacker who's already adapted to high amounts of stress may want to stack all of these activities discussed in this book because they're an over-achiever. You know who you are.

Whatever the case, here's the minimum effective dose as well as the upper threshold for all things hormesis:

- **Resistance Training** – At minimum 2 times a week full-body workouts with either weights or calisthenics. Kettlebells and resistance bands are also great tools. For optimal muscle growth, 3-4 workouts per body part per week. If this is your main form of exercise and you just want to improve your body composition (build muscle, burn fat), you can train every other day. It's easier to maintain muscle and strength than it is to build it, which is why if you're in maintenance mode you could get away with just 2 workouts a week. Time per workout can be 30-90 minutes, depending on how long you rest and what's the intensity.

- **Aerobic Cardio** – At minimum once a week of running, cycling, rowing, or some sport for 45-60 minutes below 65% of your VO2 max. It's supposed to be an easy-going yet invigorating pace where you're able to maintain nasal breathing. If you're an endurance athlete or are preparing for a marathon then you'd need to train more often. Getting at least 10 000 steps a day is also a great milestone for daily conditioning. I personally try to aim for 12 000 on most days but even up to 15 000 steps on days where I don't workout.
- **HIIT Cardio** – Doing something anaerobic and near maximum effort is also great for both cardiovascular fitness as well as mitochondrial functioning. However, you shouldn't over-do it because the body will adapt, thus making the exercise less effective. The minimal effective dose for HIIT or Tabata cardio would be just once a week. If you really push yourself hard for 5-20 minutes with intervals, then that's all you need for maintenance. You can also dissect it into several but shorter sessions. Instead of going on a full-on 20-minute massacre, you can have 5-minute workouts 3-4 times per week. I like to do either hill sprints, plyometrics, burpees, or kettlebell swings at the end of my easier resistance training workouts.
- **Mobility and Yoga** – Staying mobile and flexible is also important for longevity. It'll also prevent injuries by increasing your range of motion. Some easier mobility exercises should be done daily like staying in a deep squat for 5-10 minutes, hanging from something for a few

minutes, or doing hip circles. Taking a yoga class or doing it at home once a week is great for keeping things in check. Regular passive stretching on rest can alleviate muscle soreness but just being flexible isn't the goal. In fact, being too flexible can leave you vulnerable to injuries because the body is too soft or nimble. You need a certain amount of tightness and rigidity to bear weights and stay strong. That's why dynamic mobility applied during an exercise is much more important and preferable. You may need to consult with a personal trainer or chiropractor to assess your muscle imbalances and shortcomings.

- **Cold Exposure** – Some form of shivering or chills every day is a great way to promote brown fat, burn a few extra calories, and trigger hormesis. Starting the day with a cold shower is like a cup of coffee on steroids and I always feel amazing afterward. It can also mitigate a bad night's sleep. At least the wakefulness part. However, I don't recommend exposing yourself to the cold after a workout or too late in the evening. When it comes to ice baths and plunges, then I try to get one in at least once a week. If you don't have a lake or a deep freezer, then just taking a longer cold shower also works.
- **Heat Exposure** – Based on research, taking the sauna more than 4 times a week gives the highest improvement in mortality and cardiovascular health. However, you have to know your current condition and tolerance to heat. Ideally, you would want to hit the sauna 3-4 times a week but even once is fine. Hell, once a month is still better than nothing.

Taking the sauna every day, however, might be too much because your body will adapt to it and the benefits slow down. That's why I like to rotate between the traditional sauna and the infrared sauna to give my body a different kind of stimulus.

- **Ketogenic Dieting** – Some form of carbohydrate restriction is a prime form of hormesis with many benefits, namely metabolic flexibility, and keto-adaptation. As said before, you don't need to be on the ketogenic diet all the time to achieve keto adaptation but you do need to go through it initially. You should also structure your carb intake based on your training load, insulin sensitivity, body composition, and metabolic profile. The ultimate goal, however, is to reach metabolic flexibility and eat some carbs every once in a while. On the cyclical ketogenic diet, you'd eat keto for 5-6 days and on the 7th day consume carbohydrates with less fat.
- **Intermittent Fasting** – Daily time restricted eating is safe and should be considered an intrinsic part of an optimal circadian rhythm. Meaning, you can do it every day because it's what our bodies thrive under. If you're doing the 16/8 method, then you don't even need to change your routine either because it's not that short nor too long of a fast. In that case, it would be still beneficial to extend it to 20 or 23 hours once a week for increased autophagy. If you're already eating once a day or doing 20/4, then it might be a good idea to add an extra meal once a week as a way to give your body a break. However, you should

primarily pay attention to your results and progress. Don't change something that isn't broken and only change if you hit a plateau.

  - When it comes to extended fasts beyond 24 hours, then they can be implemented based on your preference. Some people like to have a 48-hour fast once a week or month whereas others opt-in for a 3-day fast once a quarter. Most people don't need to fast longer than that and I think it can be more damaging than good. I personally throw in either a 2 or 3 day fast whenever I'm traveling. On average, I end up with 4-5 of these longer fasts per year and about a handful 48s.

- **Xenohormetic Compounds** – Small amounts of plant hormetics can strengthen your immune system and improve gut diversity. You should figure out the ones you react the best to and the ones that tend to cause you problems. To stay within reason, use things like curry, ginger, cayenne pepper, and adaptogens as seasoning. Instead of taking sulforaphane or resveratrol supplements, try getting them from whole foods and vegetables to avoid overdosing. I eat more vegetables on days where I don't exercise or am fasting for longer as to not make the antioxidants blunt the anabolic signal from resistance training. If you react negatively to compounds like oxalates or lectins, then it's best to limit their consumption but most people don't need to do it. Just prepare them properly and don't eat them in large quantities. It's also recommended to eat some unconventional food that you normally don't eat at least once a week to condition your palate and improve tolerance to different ingredients.

- **Red Light Therapy** – You can do red light therapy every day for 10-15 minutes. It does emanate small amounts of EMF and radiation so don't place it too close to your body. Instead, stay about 30 cm or 1 inch away and let it shine directly on your entire body. I like to use it in the afternoon but any time of the day is fine.
- **Stoic Practice of Hardship** – For one day of the month you should do something uncomfortable or challenging. It's supposed to condition your brain to tolerate adversity, especially your own inner feeling of resistance. You could take an ice bath but more often than not people start enjoying it after a while. That's why I like to either go for a longer fast or just stay as minimalistic with my enjoyment as possible. For example, I'll work longer hours than normal, not consume any entertainment, no coffee or tea, eat a very boring diet or sleep on the floor. It may sound masochistic or crazy but I'm not doing because I like it. I'm doing it because it strengthens my mind to endure any kind of discomfort both physical and mental. It also lowers my hedonic adaptation and actually makes me happier after I return to my normal habits.
- **Minor Sleep Deprivation** – Generally, it's not recommended to deliberately deprive yourself of sleep, especially if done chronically. However, sleeping slightly shorter on some days can have its benefit and hormetic

effect. First of all, you'll know what it feels like and will be able to still function sub-optimally. Always needing to get your 8 hours of sleep, depending on caffeine, and having to follow your routine all the time is fragile. You're dependent on the things that keep you optimal whether that be some supplements or your mood. If you were to be antifragile, you wouldn't need these things and would be able to perform at your peak regardless. I'm not scheduling sleep deprivation because it'll eventually happen anyway. When it does I'll just follow some of the guidelines outlined in chapter thirteen and take it as an opportunity to adapt.

- **Caffeine Consumption** – Coffee is great and I consume it on a regular basis. However, as with everything, it can eventually stop working. What's worse, some people build up such a high tolerance to caffeine that they need to drink several cups a day to feel the effect. That's why it's better to cycle off caffeine every once in a while to reset your sensitivity and avoid addiction. Signs of caffeine dependence are feeling exhausted in the morning, needing coffee to wake up properly, energy dips throughout the day, digestion problems, dragging your feet, headaches, jitters, and anxiety. You can either drink coffee only every other day or once a month go an entire week without caffeine. The same principle can be applied to the various supplements you might be taking.

It's also important to keep in mind that sometimes you should practice antifragility in your routines. Meaning, you act the opposite of what you're used to as to condition your response against unpredictable scenarios and whatnot. Here's a sample weekly plan for stress adaptation you can follow:

- **Monday – Strength and Muscle Growth**
    - Intermittent fasting 16/8
    - Red light therapy 10 minutes
    - Resistance training with weights or calisthenics for 45-90 minutes
    - Short sauna or bath for 10-15 minutes
    - Eat a high protein moderate carb paleo type diet
    - Sleep for 8-9 hours
- **Tuesday – Cardio and Recovery**
    - Intermittent fasting 16/8
    - Red light therapy 10 minutes
    - Cardio workout for 45-60 minutes
    - Longer sauna or bath for 2x10 minutes
    - Cold shower or plunge for 5 minutes
    - Eat a low carb ketogenic diet
    - Sleep for 8 hours
- **Wednesday – Strength and Conditioning**
    - Intermittent fasting 16/8
    - Resistance training with weights or calisthenics 45-60 minutes
    - Tabata session for 5 minutes
    - Eat a high carb low fat diet with high protein
    - Sleep for 8-9 hours

- **Thursday – Active Recovery Day**
    - Intermittent Fasting 20/4
    - Red light therapy 10 minutes
    - Go for a longer hike or nature walking 45-60 minutes
    - Cold shower or plunge for 5 minutes
    - Eat a low carb ketogenic diet
    - Sleep for 8 hours
- **Friday – HIIT Conditioning**
    - Intermittent fasting 20/4
    - HIIT intervals for 10-15 minutes
    - Cold shower or plunge for 1-2 minutes
    - Eat a low carb ketogenic diet
    - Sleep for 7-8 hours
- **Saturday – Strength and Muscle Growth**
    - Intermittent Fasting 16/8
    - Resistance training with weights or calisthenics 45-90 minutes
    - Eat a higher protein moderate carb paleo diet
    - Sleep for 8 hours
- **Sunday – Rest and Recovery**
    - Intermittent Fasting one meal a day
    - Go for a longer walk or hike for 45-60 minutes
    - Sauna for 20 minutes
    - Cold shower or plunge for 1-2 minutes
    - Eat a low carb ketogenic diet
    - Sleep for 7-8 hours

This is just a sample week for improving both physical strength as well as stress adaptation. A lot of these activities also promote recovery so with enough conditioning they should become easier. It's best to do at least something every day because there's always an opportunity to train any particular system whether that be cardio, heat acclimatization, cold tolerance, or just muscle function.

Conclusion

"Yesterday I was clever, so I wanted to change the world.
Today I am wise, so I am changing myself."
— Rumi

We've now reached the end of this book. Hopefully, you've got some valuable information and applicable tips from it. I surely had a lot of fun researching and writing. A lot of the times these kinds of creative projects are both educational and eye-opening in many ways. I'm forced to dig deeper into the science and try to explain it so that the readers could understand. Most importantly, I'm actively experimenting and trying things out to know how the knowledge holds true in a real-world setting.

During the writing of this book, I practiced everything that's written here wholeheartedly because I believe in it and it's proven to be effective for me. I feel immune to most stressors the average person tends to struggle with such as weight gain, emotional turmoil, inflammation, aging, brain fog, sleeping problems, bloating, hypertension, or cravings. Not that I'm superior or better than others. I actually have mediocre genetics and heritage. It's just that with the right information and consistency you can override your limiting predispositions whatever they may be and feel truly awesome every single day.

This book teaches you how to strengthen your body with different beneficial stressors but it also augments your psychological resilience against stress in general. That's probably the underappreciated aspect that will increase your confidence, brings peace of mind, prevents unnecessary anxiety, and gives more freedom from the bondage of fear or anger. You'll also realize that usually, the wiser and smarter decision is to just come to terms with the fact that the world isn't always safe and that you're going to experience countless obstacles. I'm fine with that because it lies outside of my influence. I can control only my own thoughts and actions.

Because there's so much practical information in this book you might not know where to start. To prevent you from becoming paralyzed by analysis I'm going to give you a final hierarchy of the most important stress adaptation modalities and which ones to focus on first.

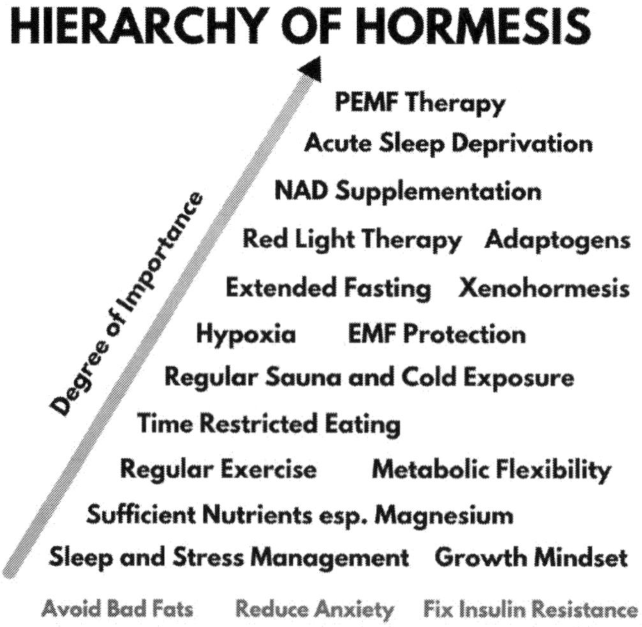

1. **Pre-Requisites for Stress Adaptation** – Before you even try to do more advanced strategies, you have to deal with the things that will undermine all attempts of inducing hormesis. They are insulin resistance, metabolic syndrome, and consumption of bad inflammatory fats. If you're full of rancid fats then mobilizing them in large quantities can make you sicker. Nutrition-related problems can be fixed with the following steps of the hierarchy. Likewise, mental turmoil, negative self-talk, and anxiety will also inhibit your efforts because it creates a vicious loop of chronic stress.
2. **Sleep and Stress Management** – At the bottom of the hierarchy, there's sleep and recovery because that's the time your body is actually getting stronger. Without enough sleep or even sub-optimal sleep, everything else becomes less effective because you go through the most powerful antioxidant and repair processes while sleeping.
3. **Growth Mindset** – It's also important to have a growth mindset because if you don't believe that you can get better or healthier then you won't. You'll just start to self-sabotage and put in lackluster effort. This includes the stoic approach to life where you're aware that bad things will happen but you're not freaking out about it because you're confident in your ability to respond appropriately.
4. **Sufficient Nutrients** – You need to obtain the essential nutrients for all the body's biochemical reactions to work properly. For a full list, refer to chapter fifteen. The most important one for stress management is probably

magnesium because it gets depleted by stress and almost everyone is deficient.

5. **Regular Exercise and Metabolic Flexibility** – Obviously, food has a profound effect on our overall health. For the sake of stress adaptation, you ought to be metabolically flexible, characterized by the ability to fast effortlessly and switch between carbs and fat for fuel. Regular exercise is also one of the most impactful things for your vitality and longevity. It's arguably more potent than any drug or supplement.
6. **Time Restricted Eating** – Periods of intermittent fasting and confined eating increase your metabolic flexibility and reduce basically every metabolic disease. It mimics the effects of calorie restriction and is a stronger stimulus for autophagy. Doing IF regularly prevents the accumulation of cellular debris and slows down aging. It also improves your ability to tolerate nutritional stress.
7. **Regular Sauna and Cold Exposure** – Taking the sauna at least once a week is amazing and doing it more often even better. It's not as important as exercise but the metabolic and cardiovascular effects are very similar. Cold baths and showers are another way to increase fat burning and adapt to harsh temperatures.
8. **Hypoxia and EMF Protection** – Breathing is the gateway between your nervous system and the outside world. It can alter your state immediately. That's why you should teach yourself proper nasal breathing. Some hypoxic training is also effective for improving tissue oxygenation and

exercise performance but not if done chronically. I would take EMF protection in moderation because avoiding it all the time will leave you too EMF sensitive. The most cost-effective thing to do is daily grounding and sleeping in a low EMF environment.

9. **Extended Fasting and Xenohormesis** – Fasting for several days is a huge stressor to the body that can certainly have negative side-effects. That's why you have to know how to prepare for it, what to do afterward, and why would you want to do it. If you're already optimized and healthy then there isn't a lot for you to gain from fasting any longer than 3 days. I like to do a 48 hour fast once a month and a 3 day fast a few times a year. In terms of xenohormesis, eating various plant compounds will improve metabolic flexibility and prevents food intolerances. Just don't macro dose polyphenolic supplements or eat something if it gives you an autoimmune flare-up.
10. **Red Light Therapy and Adaptogens** – Taking adaptogenic herbs can help to lower stress and adapt to it better. However, it might also make someone feel worse if their body isn't properly equipped. That's why it's more important to work on the previous strategies first. Red light therapy is also an amazing tool but it doesn't replace actual sunlight.
11. **NAD Supplementation** – Supplemental NAD seems to work best in people who are severely deficient of it or their mitochondria are dysfunctional. It can be life-changing for

the elderly or someone suffering from chronic fatigue. If you're already optimized, taking NAD on days you experience circadian mismatches or higher inflammation is enough.

12. **Acute Sleep Deprivation** – Having a shorter sleep on some nights can actually be beneficial because of increasing BDNF and mild autophagy. It's certainly not good for you if done chronically but you can at least stay antifragile whenever you do have a bad night's sleep. If you do sleep badly, then avoid processed carbs and sugar because your body is slightly insulin resistant.
13. **PEMF Therapy and Other Gadgets** – For some people, PEMF therapy can be life-changing because it helps them to relieve their pain. In that case, PEMF therapy could be more important than sauna or exercise because they couldn't even walk otherwise. However, for the majority of people who are already doing everything else right, these kinds of gadgets will only give them an additional 5% in their performance. Not that they're bad or that you should avoid them. It's just more cost-effective to focus on the lower half of the hierarchy.

The effects of hormesis aren't permanent and set in stone. They occur as a result of consistent implementation and following a hormetic lifestyle that balances both exertion and recovery. That's why you have to get your skin in the game and walk the walk. At the same time, be mindful of your level of adaptation and progress according to your conditioning.

About the Author

Siim Land is a bestselling author, content creator, podcaster, high-performance coach and a self-empowered being.

He creates content about Body Mind Empowerment which is human life enhancement through optimizing your physiological potential and becoming the greatest version of yourself.

The phrase *"Stay Empowered"* refers to showing up to the greatest version of yourself and earning your laurels every day.

Stay Empowered

Siim

Contact me at my blog: http://siimland.com/contact

Facebook Page: https://www.facebook.com/thesiimland/

YouTube Channel: http://www.youtube.com/c/SiimLand/

Instagram Page: https://www.instagram.com/siimland/

Twitter Page: https://twitter.com/iamsiimland

Podcast Page: https://siimland.com/body-mind-empowerment-podcast-with-siim-land/

Performance Store: http://siimland.com/store/

Books and Products: http://siimland.com/books-and-products/

Audiobooks: http://siimland.com/audiobooks/

Blog Page: http://siimland.com/blog/

More Books From the Author

You can check out all the printed and audio versions of these books at http://www.siimland.com/books-and-products/ and http://www.siimland.com/audiobooks/

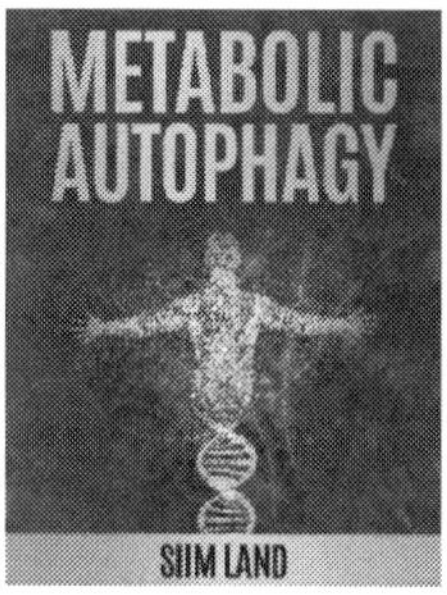

Metabolic Autophagy: Practice Intermittent Fasting and Resistance Training
to Build Muscle and Promote Longevity

Metabolic Autophagy Cookbook: Eat Foods That Boost Autophagy,
Balance mTOR for Longevity, and Build Muscle

Metabolic Autophagy Master Class Video Course

Total Sleep Optimization Video Course

Metabolic Autophagy Program + Meal Plan and Workout Routine

Keto // IF Fasting

Keto Fit Program

Keto Fasting

Keto Bodybuilding

Keto Cycle the Cyclical Ketogenic Diet Book

Simple Keto the Easiest Ketogenic Diet Book

Target Keto the Targeted Ketogenic Diet Book

Intermittent Fasting and Feasting

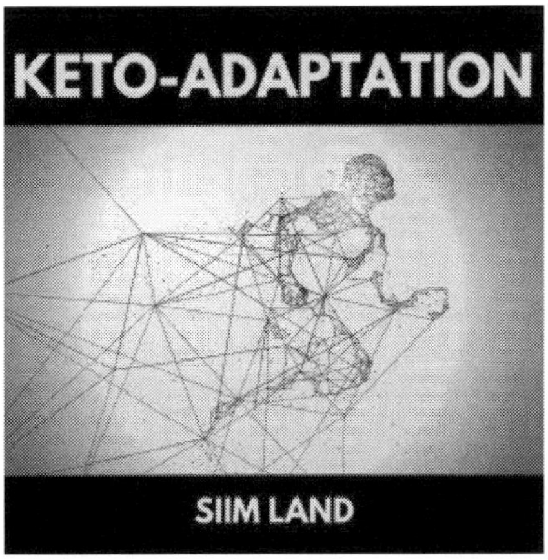

Keto Adaptation Manual

References

Here are all the studies, articles, and other references used in this book.

---

[1] Bernard (1878) 'Leçons sur les phénomènes de la vie communs aux animaux et aux vegetaux,' Paris, Bailliere JB, editor.

[2] Starling (1923) 'THE WISDOM OF THE BODY: The Harveian Oration, Delivered Before The Royal College of Physicians of London on St. Luke's Day, 1923', *British Medical Journal*, October 20;2(3277):685-90.

[3] Cannon (1963) 'Wisdom of the Body', New York, W. W. Norton & Company; Rev. and Enl. Ed edition.

[4] Ulrich-Lai et al (2017). 'Neural Regulation of Endocrine and Autonomic Stress Responses'. *Nature Reviews Neuroscience*. 10 (6): 397–409.

[5] Malenka et al (2009). 'Chapter 10: Neural and Neuroendocrine Control of the Internal Milieu'. In Sydor A, Brown RY (ed.). *Molecular Neuropharmacology: A Foundation for Clinical Neuroscience* (2nd ed.). New York: McGraw-Hill Medical. pp. 246, 248–259.

[6] Willis (2004). 'The Autonomic Nervous System and its central control'. In Berne, Robert M. (ed.). *Physiology* (5. ed.). St. Louis, Mo.: Mosby. ISBN 0323022251.

[7] Cannon, W. (1915). 'Bodily changes in pain, hunger, fear, and rage'. New York: Appleton-Century-Crofts. p. 211.

[8] Gershon, M. (1998). *The Second Brain*. New York: HarperCollins. pp. 2–7. ISBN 0-06-018252-0.

[9] Bonneaud et al. 'Assessing the cost of mounting an immune response'. Am Nat. 2003;161(3):367-379. doi:10.1086/346134

[10] Pedersen et al (1988), 'Modulation of Natural Killer Cell Activity in Peripheral Blood by Physical Exercise', Scandinavian Journal of Immunology, 27(6). Pages 673-678.

[11] Moret and Schmid-Hempel (2000) 'Survival for immunity: the price of immune system activation for bumblebee workers'. Science. 2000;290(5494):1166-1168. doi:10.1126/science.290.5494.1166

[12] Hamrick et al (2002) 'Being popular can be healthy or unhealthy: stress, social network diversity, and incidence of upper respiratory infection'. Health Psychol. 2002;21(3):294-298.

[13] Råberg et al (1998) 'On the adaptive significance of stress-induced immunosuppression'. Proc Biol Sci. 1998;265(1406):1637-1641. doi:10.1098/rspb.1998.0482

[14] Segestrom (2007) 'Stress, Energy, and Immunity: An Ecological View', Curr Dir Psychol Sci. 2007; 16(6): 326–330.

[15] Elia M (1992) 'Organ and tissue contribution to metabolic rate'. In: Kinney JM, Tucker HN, editors. Energy metabolism: Tissue determinants and cellular corollaries. New York: Raven Press; 1992. pp. 61–79.

[16] Buttgereit et al (2000) 'Bioenergetics of immune functions: fundamental and therapeutic aspects', Immunol Today. 2000;21(4):192-199. doi:10.1016/s0167-5699(00)01593-0

[17] Larson SJ, and Dunn AJ. (2001) 'Behavioral effects of cytokines', Brain Behav Immun. 2001;15(4):371-387. doi:10.1006/brbi.2001.0643

[18] Schneiderman et al (2005). "STRESS AND HEALTH: Psychological, Behavioral, and Biological Determinants". *Annual Review of Clinical Psychology*. 1: 607–628.

[19] Stojanovich and Marisavljevich (2008). 'Stress as a trigger of autoimmune disease',

Autoimmunity Reviews, 7(3), 209–213. doi:10.1016/j.autrev.2007.11.007

[20] Cohen et al (2007). "Psychological stress and disease". *JAMA*. 298 (14): 1685–7. doi:10.1001/jama.298.14.1685

[21] Herman et al (2016). 'Regulation of the Hypothalamic-Pituitary-Adrenocortical Stress Response', Comprehensive Physiology, 603–621. doi:10.1002/cphy.c150015

[22] Ahola, K., Sirén, I., Kivimäki, M., Ripatti, S., Aromaa, A., Lönnqvist, J., & Hovatta, I. (2012). Work-Related Exhaustion and Telomere Length: A Population-Based Study. PLoS ONE, 7(7), e40186. doi:10.1371/journal.pone.0040186

[23] Aydinonat, D., Penn, D. J., Smith, S., Moodley, Y., Hoelzl, F., Knauer, F., & Schwarzenberger, F. (2014). Social Isolation Shortens Telomeres in African Grey Parrots (Psittacus erithacus erithacus). PLoS ONE, 9(4), e93839. doi:10.1371/journal.pone.0093839

[24] https://www.eurekalert.org/pub_releases/2013-11/icl-lum111813.php

[25] Ornish, D., Lin, J., Chan, J. M., Epel, E., Kemp, C., Weidner, G., ... Blackburn, E. H. (2013). Effect of comprehensive lifestyle changes on telomerase activity and telomere length in men with biopsy-proven low-risk prostate cancer: 5-year follow-up of a descriptive pilot study. The Lancet Oncology, 14(11), 1112–1120. doi:10.1016/s1470-2045(13)70366-8

[26] Viner (1999). "Putting Stress in Life: Hans Selye and the Making of Stress Theory". *Social Studies of Science*. 29 (3): 391–410.

[27] Selye (1936) 'A Syndrome produced by Diverse Nocuous Agents', Nature volume 138, page32.

[28] Selye H. (1956). The Stress of Life. New York: McGraw-Hill Book Co.

[29] Humphrey and James (2005). *Anthology of Stress Revisited: Selected Works Of James H. Humphrey*. Foreword by Paul J. Rosch. Nova Science Publishers. p. viii. ISBN 9781594546402.

[30] Gozhenko et al (2009) *Pathology: Medical student's library*. Radom University. p. 272

[31] Chaitow, L. (2006). *Local Adaptation Syndromes Wholistic Solutions Depend on Contextual Thinking*. Sourced from http://www.massagetoday.com/mpacms/mt/article.php?id=13363

[32] Selye (1938) 'EXPERIMENTAL EVIDENCE SUPPORTING THE CONCEPTION OF "ADAPTATION ENERGY"', American Journal of Physiology, Volume 123, Issue 3, August 1938, Pages 758-765.

[33] Selye (1938) 'Adaptation Energy', Nature volume 141, page 926.

[34] Gorban et al (2016) 'Evolution of adaptation mechanisms: Adaptation energy, stress, and oscillating death', Journal of Theoretical Biology, Volume 405, 21 September 2016, Pages 127-139.

[35] Schkade and Schultz (2003) 'Occupational adaptation in perspectives'. In: Kramer, Paula, Hinojosa, Jim, Royeen, Charlotte Brasic (Eds.), Perspectives in Human Occupation: Participation in Life. Lippincott Williams & Wilkins, Baltimore, MD, pp. 181–221 (Chapter 7).

[36] Goldstone (1952) 'THE GENERAL PRACTITIONER AND THE GENERAL ADAPTATION SYNDROME', S.A. MEDICAL JOURNAL, February 2nd, p 88-92.

[37] Jeronimus et al (2014) 'Mutual Reinforcement Between Neuroticism and Life Experiences: A Five-Wave, 16-Year Study to Test Reciprocal Causation'. *Journal of Personality and Social Psychology*. 107 (4): 751–64.

[38] Jeronimus et al (2013). 'Negative and positive life events are associated with small but lasting change in neuroticism'. *Psychological Medicine*. 43 (11): 2403–15.

[39] Schneiderman et al (2005). "Stress and health: psychological, behavioral, and biological determinants". *Annual Review of Clinical Psychology*. 1: 607–628.

[40] Schetter et al (2011), "Resilience in the Context of Chronic Stress and Health in Adults". *Social and Personality Psychology Compass*. 5 (9): 634–52.

[41] Selye (1975). "Confusion and controversy in the stress field". *Journal of Human Stress*. 1 (2): 37–44.

[42] Selye, H. (1975). "Implications of Stress Concept". *New York State Journal of Medicine*. 75: 2139–2145.

[43] de Kloet et al (2005). "Stress and the brain: from adaptation to disease". *Nature Reviews Neuroscience*. 6 (6): 463–475.

[44] McEwen and Stellar (1993). "Stress and the individual. Mechanisms leading to disease". *Archives of Internal Medicine*. 153 (18): 2093–101.

[45] Ogden (2004). *Health Psychology: A textbook, 3rd edition*. Open University Press - McGraw-Hill Education. p. 259.

[46] Brame and Singer (2010). 'Stressing the obvious? An allostatic look at critical illness'. Critical Care Medicine, 38, S600–S607. doi:10.1097/ccm.0b013e3181f23e92

[47] McEwen and Wingfield (2003). "The concept of allostasis in biology and biomedicine". *Hormones and Behavior*. 43 (1): 2–15.

[48] Schenk, H.M.; et al. (2017). "Associations of Positive Affect and Negative Affect With Allostatic Load: A Lifelines Cohort Study". *Psychosomatic Medicine*. 80 (2): 160–166.

[49] Sterling, P.; Eyer, J. (1988). "Allostasis: A new paradigm to explain arousal pathology". In Fisher, S.; Reason, J. T. (eds.). *Handbook of life stress, cognition, and health*. Chicester, NY: Wiley.

[50] Sterling, P (12 April 2012). "Allostasis: a model of predictive regulation". Physiology & Behavior. 106 (1): 5–15.

[51] Sterling, Peter (2004). "Chapter 1. Principles of Allostasis". In Schulkin, Jay (ed.). *Allostasis, homeostasis, and the costs of physiological adaptation*. New York, NY: Cambridge University Press.

[52] Peters, A., McEwen, B. S., & Friston, K. (2017). 'Uncertainty and stress: Why it causes diseases and how it is mastered by the brain'. Progress in Neurobiology, 156, 164–188. doi:10.1016/j.pneurobio.2017.05.004

[53] Edes, Ashley; Crews, Douglas (2017). "Allostatic load and biological anthropology". *American Journal of Physical Anthropology*. 162: 44–70.

[54] Danese, A., & McEwen, B. S. (2012). 'Adverse childhood experiences, allostasis, allostatic load, and age-related disease'. Physiology & Behavior, 106(1), 29–39. doi:10.1016/j.physbeh.2011.08.019

[55] Kristenson, M., Eriksen, H. ., Sluiter, J. ., Starke, D., & Ursin, H. (2004). 'Psychobiological mechanisms of socioeconomic differences in health'. Social Science & Medicine, 58(8), 1511–1522. doi:10.1016/s0277-9536(03)00353-8

[56] Juster, R.-P., McEwen, B. S., & Lupien, S. J. (2010). 'Allostatic load biomarkers of chronic stress and impact on health and cognition'. Neuroscience & Biobehavioral Reviews, 35(1), 2–16. doi:10.1016/j.neubiorev.2009.10.002

[57] Carpenter et al (2010). 'Association between plasma IL-6 response to acute stress and early-life adversity in healthy adults'. Neuropsychopharmacology. 2010;35(13):2617-2623. doi:10.1038/npp.2010.159

[58] Carlsson et al (2014). 'Psychological stress in children may alter the immune response'. J Immunol. 2014;192(5):2071-2081. doi:10.4049/jimmunol.1301713

[59] Copeland et al (2014) 'Childhood bullying involvement predicts low-grade systemic inflammation into adulthood'. Proc Natl Acad Sci U S A. 2014;111(21):7570-7575. doi:10.1073/pnas.1323641111

[60] Ugazio, G., Koch, R.R., Recknagel, R.O., (1972). 'Mechanism of protection against carbon tetrachloride by prior carbon tetrachloride administration'. Exp. Mol. Pathol. 16, 281–285

[61] Jeggo, P., Defais, M., Samson, L., Schendel, P., (1977). 'An adaptive response of E. coli to low levels of alkylating agent: comparison with previously characterized DNA repair pathways'. Mol. Gen. Genet. 157, 1–9.

[62] Murry, C.E., Jennings, R.B., Reimer, K.A., (1986). 'Preconditioning with ischemia: a delay of

lethal cell injury in ischemic myocardium'. Circulation 74, 1124–1136.

[63] Majer, A. D., Fasanello, V. J., Tindle, K., Frenz, B. J., Ziur, A. D., Fischer, C. P., Haussmann, M. F. (2019). 'Is there an oxidative cost of acute stress? Characterization, implication of glucocorticoids and modulation by prior stress experience'. Proceedings of the Royal Society B: Biological Sciences, 286(1915), 20191698. doi:10.1098/rspb.2019.1698

[64] Koolhaas, J. M., Bartolomucci, A., Buwalda, B., de Boer, S. F., Flügge, G., Korte, S. M., Fuchs, E. (2011). 'Stress revisited: A critical evaluation of the stress concept'. Neuroscience & Biobehavioral Reviews, 35(5), 1291–1301. doi:10.1016/j.neubiorev.2011.02.003

[65] Southam CM, Erhlich J (1943) 'Effects of extracts of western red-cedar heartwood on certain wood-decaying fungi in culture'. Phytopathology 33: 517–524.

[66] Calabrese, E.J., (2005). 'Paradigm lost, paradigm found: the re-emergence of hormesis as a fundamental dose response model in the toxicological sciences'. Env. Poll. 138, 379–412.

[67] Calabrese et al (2007) 'Biological Stress Response Terminology: Integrating the Concepts of Adaptive Response and Preconditioning Stress Within a Hormetic Dose-Response Framework' Toxicol Appl Pharmacol 2007 Jul 1;222(1):122-8. doi: 10.1016/j.taap.2007.02.015.

[68] Calabrese and Baldwin (2001) 'The Frequency of U-Shaped Dose Responses in the Toxicological Literature', Toxicological Sciences, Volume 62, Issue 2, August 2001, Pages 330–338.

[69] Calabrese and Baldwin (2001) 'Hormesis: U-shaped Dose Responses and Their Centrality in Toxicology', Trends Pharmacol Sci, 2001 Jun;22(6):285-91. doi: 10.1016/s0165-6147(00)01719-3.

[70] Pareto, Vilfredo; Page, Alfred N. (1971), Translation of *Manuale di economia politica ("Manual of political economy")*, A.M. Kelley, ISBN 978-0-678-00881-2

[71] Taleb (2014) 'Antifragile: Things That Gain From Disorder', 'Prologue', p 3, Random House Trade Paperbacks.

[72] Welch (1993). "How Cells Respond to Stress", Scientific American, 1993 May;268(5):56-64. doi: 10.1038/scientificamerican0593-56.

[73] Ciechanover, A. (2005). 'Proteolysis: from the lysosome to ubiquitin and the proteasome'. Nature Reviews Molecular Cell Biology, 6(1), 79–87. doi:10.1038/nrm1552

[74] Richter, Klaus; Martin Haslbeck; Johannes Buchner (22 October 2010). "The Heat Shock Response: Life on the Verge of Death". *Molecular Cell*. 40 (2): 253–266. doi:10.1016/j.molcel.2010.10.006.

[75] Lindquist, J. A., & Mertens, P. R. (2018). Cold shock proteins: from cellular mechanisms to pathophysiology and disease. Cell Communication and Signaling, 16(1). doi:10.1186/s12964-018-0274-6

[76] Jones, P. G., & Inouye, M. (1994). The cold-shock response? a hot topic. Molecular Microbiology, 11(5), 811–818. doi:10.1111/j.1365-2958.1994.tb00359.x

[77] Wolffe, A. P., Tafuri, S., Ranjan, M., & Familari, M. (1992). The Y-box factors: a family of nucleic acid binding proteins conserved from Escherichia coli to man. The New biologist, 4(4), 290–298.

[78] Phadtare S. (2004). Recent developments in bacterial cold-shock response. Current issues in molecular biology, 6(2), 125–136.

[79] Ron, D., & Walter, P. (2007). Signal integration in the endoplasmic reticulum unfolded protein response. Nature Reviews Molecular Cell Biology, 8(7), 519–529. doi:10.1038/nrm2199

[80] Hampton, R. Y. (2000). ER stress response: Getting the UPR hand on misfolded proteins. Current Biology, 10(14), R518–R521. doi:10.1016/s0960-9822(00)00583-2

[81] Cuervo A. M. (2004). Autophagy: many paths to the same end. Molecular and cellular biochemistry, 263(1-2), 55–72.

[82] Kryston, T. B., Georgiev, A. B., Pissis, P., & Georgakilas, A. G. (2011). Role of oxidative stress and DNA damage in human carcinogenesis. Mutation research, 711(1-2), 193–201.

https://doi.org/10.1016/j.mrfmmm.2010.12.016

[83] Caldecott K. W. (2008). Single-strand break repair and genetic disease. Nature reviews. Genetics, 9(8), 619–631. https://doi.org/10.1038/nrg2380

[84] Hanawalt, P. C. (2002). Subpathways of nucleotide excision repair and their regulation. Oncogene, 21(58), 8949–8956. doi:10.1038/sj.onc.1206096

[85] Fuscoe, J. C., Zimmerman, L. J., Harrington-Brock, K., & Moore, M. M. (1994). Deletion mutations in the hprt gene of T-lymphocytes as a biomarker for genomic rearrangements important in human cancers. Carcinogenesis, 15(7), 1463–1466. https://doi.org/10.1093/carcin/15.7.1463

[86] D. Trachootham, W. Lu, M. A. Ogasawara, N. R.-D. Valle, and P. Huang, "Redox regulation of cell survival," *Antioxidants & Redox Signaling*, vol. 10, no. 8, pp. 1343–1374, 2008.

[87] Black, P. H. (2002). Stress and the inflammatory response: A review of neurogenic inflammation. Brain, Behavior, and Immunity, 16(6), 622–653. doi:10.1016/s0889-1591(02)00021-1

[88] Jain S, Gautam V, Naseem S (January 2011). "Acute-phase proteins: As diagnostic tool". *Journal of Pharmacy & Bioallied Sciences*. 3 (1): 118–27

[89] Stojanovich, L., & Marisavljevich, D. (2008). Stress as a trigger of autoimmune disease. Autoimmunity Reviews, 7(3), 209–213. doi:10.1016/j.autrev.2007.11.007

[90] Song, H., Fang, F., Tomasson, G., Arnberg, F. K., Mataix-Cols, D., Fernández de la Cruz, L., … Valdimarsdóttir, U. A. (2018). Association of Stress-Related Disorders With Subsequent Autoimmune Disease. JAMA, 319(23), 2388. doi:10.1001/jama.2018.7028

[91] Gouin, J.-P., Glaser, R., Malarkey, W. B., Beversdorf, D., & Kiecolt-Glaser, J. (2012). Chronic stress, daily stressors, and circulating inflammatory markers. Health Psychology, 31(2), 264–268. doi:10.1037/a0025536

[92] Dhabhar, F. S., Malarkey, W. B., Neri, E., & McEwen, B. S. (2012). Stress-induced redistribution of immune cells--from barracks to boulevards to battlefields: a tale of three hormones--Curt Richter Award winner. Psychoneuroendocrinology, 37(9), 1345–1368. https://doi.org/10.1016/j.psyneuen.2012.05.008

[93] Reddan, M. C., Wager, T. D., & Schiller, D. (2018). Attenuating Neural Threat Expression with Imagination. Neuron, 100(4), 994–1005.e4. doi:10.1016/j.neuron.2018.10.047

[94] Wullschleger, S., Loewith, R., & Hall, M. N. (2006). TOR Signaling in Growth and Metabolism. Cell, 124(3), 471–484. doi:10.1016/j.cell.2006.01.016

[95] Qian, S.-B., Zhang, X., Sun, J., Bennink, J. R., Yewdell, J. W., & Patterson, C. (2010). mTORC1 Links Protein Quality and Quantity Control by Sensing Chaperone Availability. Journal of Biological Chemistry, 285(35), 27385–27395. doi:10.1074/jbc.m110.120295

[96] King N, Hittinger CT, Carroll SB (July 2003). "Evolution of key cell signaling and adhesion protein families predates animal origins". *Science*. 301 (5631): 361–3.

[97] Trzaskowski B, Latek D, Yuan S, Ghoshdastider U, Debinski A, Filipek S (2012). "Action of molecular switches in GPCRs—theoretical and experimental studies". *Current Medicinal Chemistry*. 19 (8): 1090–109

[98] Chadwick, W., & Maudsley, S. (2009). The Devil is in the Dose: Complexity of Receptor Systems and Responses. Hormesis, 95–108. doi:10.1007/978-1-60761-495-1_5

[99] Gilman AG (1987). "G proteins: transducers of receptor-generated signals". *Annual Review of Biochemistry*. 56 (1): 615–49.

[100] Ali, E. S., Hua, J., Wilson, C. H., Tallis, G. A., Zhou, F. H., Rychkov, G. Y., & Barritt, G. J. (2016). The glucagon-like peptide-1 analogue exendin-4 reverses impaired intracellular Ca 2+ signalling in steatotic hepatocytes. Biochimica et Biophysica Acta (BBA) - Molecular Cell Research, 1863(9), 2135–2146. doi:10.1016/j.bbamcr.2016.05.006

[101] Serezani, C. H., Ballinger, M. N., Aronoff, D. M., & Peters-Golden, M. (2008). Cyclic AMP. American Journal of Respiratory Cell and Molecular Biology, 39(2), 127–132. doi:10.1165/rcmb.2008-0091tr

[102] Spruill, Tanya M. (7 February 2017). "Chronic Psychosocial Stress and Hypertension". *Current Hypertension Reports*. 12 (1): 10–16.

[103] Aguilera, Greti (1 January 2011). "HPA axis responsiveness to stress: Implications for healthy aging". *Experimental Gerontology*. 46 (2–3): 90–95.

[104] Lu, S. (2009). Regulation of glutathione synthesis. *Molecular Aspects of Medicine* 30 (1-2): 42–59.

[105] Lu, S. (2013). Glutathione synthesis. *Biochimica et Biophysica Acta (BBA)-General Subjects* 1830 (5): 3143–3153.

[106] Moi, P., Chan, K., Asunis, I., Cao, A., & Kan, Y. W. (1994). Isolation of NF-E2-related factor 2 (Nrf2), a NF-E2-like basic leucine zipper transcriptional activator that binds to the tandem NF-E2/AP1 repeat of the beta-globin locus control region. Proceedings of the National Academy of Sciences of the United States of America, 91(21), 9926–9930. https://doi.org/10.1073/pnas.91.21.9926

[107] Deretic, V. (2006). Autophagy as an immune defense mechanism. *Current Opinion in Immunology* 18 (4): 375–382.

[108] Glantzounis, G. & Tsimoyiannis, E. & Kappas, A. & Galaris, D. (2005). Uric acid and oxidative stress. *Current Pharmaceutical Design* 11 (32): 4145–4151.

[109] Pasalic, D. & Marinkovic, N. & Feher-Turkovic, L. (2012). Uric acid as one of the important factors in multifactorial disorders–facts and controversies. *Biochemia Medica* 22 (1): 63–75.

[110] Zhang, M., & Ying, W. (2019). NAD+ deficiency is a common central pathological factor of a number of diseases and aging: mechanisms and therapeutic implications. Antioxidants & Redox Signaling 30 (6): 890–905.

[111] Bradshaw, P. (2019). Cytoplasmic and Mitochondrial NADPH-Coupled Redox Systems in the Regulation of Aging. *Nutrients* 11 (3): 504.

[112] Arrieta, M. & Bistritz, L. & Meddings, J. (2006). Alterations in intestinal permeability. *Gut* 55 (10): 1512–1520.

[113] Mu, Q. & Kirby, J. & Reilly, C. & Luo, X. (2017). Leaky Gut As a Danger Signal for Autoimmune Diseases. *Frontiers in Immunology* 8: 598.

[114] Pahwa, R. & Singh, A. & Jialal, I. (2019). *Chronic Inflammation*. In: StatPearls [Internet]. Treasure Island (FL): StatPearls Publishing. [date of reference: 6.2.2020]

[115] Elam, M. et al. (2015). A Calcium-Collagen Chelate Dietary Supplement Attenuates Bone Loss in Postmenopausal Women with Osteopenia: A Randomized Controlled Trial. *Journal of Medicinal Food* 18 (3): 324–331.

[116] Canani, R. (2011). Potential beneficial effects of butyrate in intestinal and extraintestinal diseases. *World Journal of Gastroenterology* 17 (12): 1519.

[117] Sanz, Y. (2010). Effects of a gluten-free diet on gut microbiota and immune function in healthy adult humans. *Gut Microbes* 1 (3): 135–137.

[118] Wu, H.-J. & Wu, E. (2012). The role of gut microbiota in immune homeostasis and autoimmunity. *Gut Microbes* 3 (1): 4–14.

[119] Tatar, M. et al (2003) 'The endocrine regulation of aging by insulin-like signals', Science, Vol 299(5611), p 1346-1351.

[120] Holzenberger, M. et al (2003) 'IGF-1 receptor regulates lifespan and resistance to oxidative stress in mice', Nature, Vol 421(6919), p 182-187.

[121] Clancy, DJ. et al (2001) 'Extension of life-span by loss of CHICO, a Drosophila insulin receptor substrate protein', Vol 292(5514), Vol 104-106.

[122] Nakae, J. et al (2008) 'The FoxO transcription factors and metabolic regulation', FEBS Letters, Vol 582(1), p 54-67.

[123] Peng, SL. (2008) 'Foxo in the immune system', Oncogene, Vol 27, p 2337-2344.

[124] Kanfi, Y. et al (2012) 'The sirtuin SIRT6 regulates lifespan in male mice', Nature, Vol 483, p 218-221.

[125] Mostoslavsky, R. (2006) 'Genomic instability and aging-like phenotype in the absence of mammalian SIRT6', Cell, Vol 124(2), p 315-329.

[126] Johnson and Imai (2018) 'NAD + biosynthesis, aging, and disease', F1000Res. 2018 Feb 1;7:132.

[127] Gomes et al (2013) 'Declining NAD+ Induces a Pseudohypoxic State Disrupting Nuclear-Mitochondrial Communication during Aging', Cell. 2013 Dec 19; 155(7): 1624–1638.

[128] Wang, M.C. et al (2003) 'JNK Signaling Confers Tolerance to Oxidative Stress and Extends Lifespan in Drosophila', Developmental Cell, Vol 5(5), p 811-816.

[129] Hercus, MJ. et al (2003) 'Lifespan extension of Drosophila melanogaster through hormesis by repeated mild heat stress', Biogerontology, Vol 4(3), p 149-156.

[130] Takahashi, K., & Yamanaka, S. (2006). Induction of pluripotent stem cells from mouse embryonic and adult fibroblast cultures by defined factors. cell, 126(4), 663-676.

[131] Liu et al (2008) 'Yamanaka factors critically regulate the developmental signaling network in mouse embryonic stem cells', Cell Res. 2008 Dec;18(12):1177-89. doi: 10.1038/cr.2008.309.

[132] Lapasset, L., Milhavet, O., Prieur, A., Besnard, E., Babled, A., Ait-Hamou, N., Lemaitre, J.-M. (2011). Rejuvenating senescent and centenarian human cells by reprogramming through the pluripotent state. Genes & Development, 25(21), 2248–2253. doi:10.1101/gad.173922.111

[133] Longo et al (2015) 'A Periodic Diet that Mimics Fasting Promotes Multi-System Regeneration, Enhanced Cognitive Performance, and Healthspan', CLINICAL AND TRANSLATIONAL REPORT| VOLUME 22, ISSUE 1, P86-99, JULY 07, 2015

[134] Zhang, H., Ryu, D., Wu, Y., Gariani, K., Wang, X., Luan, P., D'Amico, D., Ropelle, E. R., Lutolf, M. P., Aebersold, R., Schoonjans, K., Menzies, K. J., & Auwerx, J. (2016). NAD^+ repletion improves mitochondrial and stem cell function and enhances life span in mice. Science (New York, N.Y.), 352(6292), 1436–1443. https://doi.org/10.1126/science.aaf2693

[135] Hucklenbroich, J., Klein, R., Neumaier, B., Graf, R., Fink, G., Schroeter, M., & Rueger, M. (2014). Aromatic-turmerone induces neural stem cell proliferation in vitro and in vivo. Stem Cell Research & Therapy, 5(4), 100. doi:10.1186/scrt500

[136] Bachstetter AD, Jernberg J, Schlunk A, Vila JL, Hudson C, et al. (2010) Spirulina Promotes Stem Cell Genesis and Protects against LPS Induced Declines in Neural Stem Cell Proliferation. PLOS ONE 5(5): e10496. https://doi.org/10.1371/journal.pone.0010496

[137] Xu, K. et al (2014) 'mTOR signaling in tumorigenesis', BBA, Vol 1846(2), p 638-654.

[138] Meléndez, A. et al (2003) 'Autophagy genes are essential for dauer development and life-span extension in C. elegans', Vol 301(5638), p 1387-1391.

[139] Lamb, CA., Yoshimori, T. and Tooze SA. (2013) 'The autophagosome: origins unknown, biogenesis complex', Nature Reviews Molecular Cell Biology, Vol 14(12), p 759-774.

[140] Klass M. and Hirsh D. (1976) 'Non-ageing developmental variant of Caenorhabditis elegans', Nature, Vol 260(5551), p 523-525.

[141] Weindruch, R., & Sohal, R. S. (1997). Seminars in medicine of the Beth Israel Deaconess Medical Center. Caloric intake and aging. The New England journal of medicine, 337(14), 986–994. https://doi.org/10.1056/NEJM199710023371407

[142] Mattison et al (2012) 'Impact of caloric restriction on health and survival in rhesus monkeys from the NIA study', Nature. 2012 Sep 13;489(7415):318-21.

[143] Beck, S. and Bharadwaj, R. (1972) 'Reversed Development and Cellular Aging in an Insect', Science, Vol 178(4066), p 1210-1211.

[144] Heilbronn, LK. et al (2005) 'Glucose tolerance and skeletal muscle gene expression in response to alternate day fasting', Obesity Research, Vol 13(3), p 574-581.

[145] Nemoto, S. et al (2005) 'SIRT1 functionally interacts with the metabolic regulator and

transcriptional coactivator PGC-1{alpha}', Journal of Biological Chemistry, Vol 280(16), p 16456-60.

[146] Sack, MN. and Finkel, T. (2012) 'Mitochondrial metabolism, sirtuins, and aging', Cold Spring Harbor Perspectives in Biology, Vol 4(12).

[147]Harman, D. (1956) 'Aging: A Theory Based on Free Radical and Radiation Chemistry', Journal of Gerontology, Vol 11(3), p 298-300.

[148] Harman, D. (1972) 'The biologic clock: the mitochondria?', Journal of the American Geriatrics Society, April 20(4), p 145-147.

[149] Sies H. (1997). Oxidative stress: oxidants and antioxidants. Experimental physiology, 82(2), 291–295. https://doi.org/10.1113/expphysiol.1997.sp004024

[150] Christen Y. (2000). Oxidative stress and Alzheimer disease. The American journal of clinical nutrition, 71(2), 621S–629S. https://doi.org/10.1093/ajcn/71.2.621s

[151] Giugliano, D., Ceriello, A., & Paolisso, G. (1996). Oxidative stress and diabetic vascular complications. Diabetes care, 19(3), 257–267. https://doi.org/10.2337/diacare.19.3.257

[152] Hitchon, C. A., & El-Gabalawy, H. S. (2004). Oxidation in rheumatoid arthritis. Arthritis research & therapy, 6(6), 265–278. https://doi.org/10.1186/ar1447

[153] Aviram M. (2000). Review of human studies on oxidative damage and antioxidant protection related to cardiovascular diseases. Free radical research, 33 Suppl, S85–S97.

[154] Miguel, J. et al (1980) 'Mitochondrial role in cell aging', Experimental Gerontology, Vol 15(6), p 575-591.

[155] Schriner, SE. et al (2005) 'Extension of murine life span by overexpression of catalase targeted to mitochondria', Science, Vol 308(5730), p 1909-1911.

[156] Morrow, G., & Tanguay, R. M. (2003). Heat shock proteins and aging in Drosophila melanogaster. Seminars in Cell & Developmental Biology, 14(5), 291–299. doi:10.1016/j.semcdb.2003.09.023

[157] Wei, YH. et al (2001) 'Mitochondrial theory of aging matures--roles of mtDNA mutation and oxidative stress in human aging', Zhonghua Yi Xue Za Zhi (Taipei), Vol 64(5), p 259-270.

[158] Trifunovic, A. et al (2005) 'Somatic mtDNA mutations cause aging phenotypes without affecting reactive oxygen species production', Proceedings of the National Academy of Sciences of the United States of America, Vol 102(50), p 17993-17998.

[159] Fontana, L. et al (2013) 'Dietary Restriction, Growth Factors and Aging: from yeast to humans', Science, Vol 328(5976), p 321-326.

[160] Pérez, V. et al (2009) 'Is the Oxidative Stress Theory of Aging Dead?', Biochimica et Biophysica Acta, Vol 1790(10), p 1005-1014.

[161] Van Raamsdonk, J.M. and Hekimi, S. (2009) 'Deletion of the Mitochondrial Superoxide Dismutase sod-2 Extends Lifespan in Caenorhabditis elegans', PLOS Genetics, Vol 5(2).

[162] Bjelakovic, G. et al (2007) 'Mortality in Randomized Trials of Antioxidant Supplements for Primary and Secondary Prevention', JAMA, Vol 297(8), p 842-857.

[163] Boffetta, P. et al (2010) 'Fruit and Vegetable Intake and Overall Cancer Risk in the European Prospective Investigation Into Cancer and Nutrition (EPIC)', JNCI: Journal of the National Cancer Institute, Vol 102(8), p 529-537.

[164]Halliwell, B. (2012) 'Free radicals and antioxidants: updating a personal view', Nutrition Reviews, Vol 70(5), p 257-265.

[165]Tapia, P. (2006) 'Sublethal mitochondrial stress with an attendant stoichiometric augmentation of reactive oxygen species may precipitate many of the beneficial alterations in cellular physiology produced by caloric restriction, intermittent fasting, exercise and dietary phytonutrients: "Mitohormesis" for health and vitality', Medical Hypotheses, Vol 66(4), p 832-843.

[166] Lamb, CA., Yoshimori, T. and Tooze SA. (2013) 'The autophagosome: origins unknown,

biogenesis complex', Nature Reviews Molecular Cell Biology, Vol 14(12), p 759-774.

[167] Lamb CA, Yoshimori T, Tooze SA. (2013) 'The autophagosome: origins unknown, biogenesis complex', Nature Reviews Molecular Cell Biology. 2013;14:759–74.

[168] Apfield, J. et al (2004) 'The AMP-activated protein kinase AAK-2 links energy levels and insulin-like signals to lifespan in C. elegans', Genes & Development, Vol 18(24), p 3004-3009.

[169] Blüher, M. et al (2003) 'Extended longevity in mice lacking the insulin receptor in adipose tissue', Science, Vol 299(5606), p 572-574.

[170] Jia, K. et al (2004) 'The TOR pathway interacts with the insulin signaling pathway to regulate C. elegans larval development, metabolism and life span', Development, Vol 131(16), p 3897-3906.

[171] Morselli, E. et al (2010) 'Caloric restriction and resveratrol promote longevity through the Sirtuin-1-dependent induction of autophagy', Cell Death & Disease, Vol 1(1).

[172] de Cabo et al (2018) 'Daily Fasting Improves Health and Survival in Male Mice Independent of Diet Composition and Calories', Cell Metabolism, VOLUME 29, ISSUE 1, P221-228.E3, JANUARY 08, 2019.

[173] NIH/National Institute on Aging. "Longer daily fasting times improve health and longevity in mice: Benefits seen regardless of calorie intake, diet composition in new study." ScienceDaily. www.sciencedaily.com/releases/2018/09/180906123305.htm (accessed January 17, 2020).

[174]Carlson, A.J. and Hoelzel, F. (1946) 'Apparent Prolongation of the Life Span of Rats by Intermittent Fasting: One Figure', The Journal of Nutrition, Vol 31(3), p 363-375.

[175] Wei, M. et al (2017) 'Fasting-mimicking diet and markers/risk factors for aging, diabetes, cancer, and cardiovascular disease', Science Translational Medicine, Vol 9(377).

[176] Fontana, L., Longo, VD., and Partridge, L. (2010) 'Dietary Restriction, Growth Factors and Aging: from yeast to humans', Science. 2010 Apr 16; 328(5976): 321–326.

[177] Alirezaei, M., Kemball, C. C., Flynn, C. T., Wood, M. R., Whitton, J. L., & Kiosses, W. B. (2010). Short-term fasting induces profound neuronal autophagy. Autophagy, 6(6), 702–710. https://doi.org/10.4161/auto.6.6.12376

[178] Jamshed, H., Beyl, R. A., Della Manna, D. L., Yang, E. S., Ravussin, E., & Peterson, C. M. (2019). Early Time-Restricted Feeding Improves 24-Hour Glucose Levels and Affects Markers of the Circadian Clock, Aging, and Autophagy in Humans. Nutrients, 11(6), 1234. https://doi.org/10.3390/nu11061234

[179] Zare, A., Hajhashemi, M., Hassan, Z. M., Zarrin, S., Pourpak, Z., Moin, M., Salarilak, S., Masudi, S., & Shahabi, S. (2011). Effect of Ramadan fasting on serum heat shock protein 70 and serum lipid profile. Singapore medical journal, 52(7), 491–495.

[180] Ehrenfried, J. A., Evers, B. M., Chu, K. U., Townsend, C. M., & Thompson, J. C. (1996). Caloric Restriction Increases the Expression of Heat Shock Protein in the Gut. Annals of Surgery, 223(5), 592–599. doi:10.1097/00000658-199605000-00015

[181] Selsby, J. T., Judge, A. R., Yimlamai, T., Leeuwenburgh, C., & Dodd, S. L. (2005). Life long calorie restriction increases heat shock proteins and proteasome activity in soleus muscles of Fisher 344 rats. Experimental gerontology, 40(1-2), 37–42. https://doi.org/10.1016/j.exger.2004.08.012

[182] Bruce-Keller, A.J., Umberger, G., McFall, R. and Mattson, M.P. (1999), Food restriction reduces brain damage and improves behavioral outcome following excitotoxic and metabolic insults. Ann Neurol., 45: 8-15. doi:10.1002/1531-8249(199901)45:1<8::AID-ART4>3.0.CO;2-V

[183] Shimazu, T., Hirschey, M. D., Newman, J., He, W., Shirakawa, K., Le Moan, N., Grueter, C. A., Lim, H., Saunders, L. R., Stevens, R. D., Newgard, C. B., Farese, R. V., Jr, de Cabo, R., Ulrich, S., Akassoglou, K., & Verdin, E. (2013). Suppression of oxidative stress by β-hydroxybutyrate, an endogenous histone deacetylase inhibitor. Science (New York, N.Y.), 339(6116), 211–214. https://doi.org/10.1126/science.1227166

[184] Newman, J. C., & Verdin, E. (2014). Ketone bodies as signaling metabolites. Trends in endocrinology and metabolism: TEM, 25(1), 42–52. https://doi.org/10.1016/j.tem.2013.09.002

[185] Youm, Y. H., Nguyen, K. Y., Grant, R. W., Goldberg, E. L., Bodogai, M., Kim, D., D'Agostino, D., Planavsky, N., Lupfer, C., Kanneganti, T. D., Kang, S., Horvath, T. L., Fahmy, T. M., Crawford, P. A., Biragyn, A., Alnemri, E., & Dixit, V. D. (2015). The ketone metabolite β-hydroxybutyrate blocks NLRP3 inflammasome-mediated inflammatory disease. Nature medicine, 21(3), 263–269. https://doi.org/10.1038/nm.3804

[186] Han, X., Tai, H., Wang, X., Wang, Z., Zhou, J., Wei, X., Ding, Y., Gong, H., Mo, C., Zhang, J., Qin, J., Ma, Y., Huang, N., Xiang, R., & Xiao, H. (2016). AMPK activation protects cells from oxidative stress-induced senescence via autophagic flux restoration and intracellular NAD(+) elevation. Aging cell, 15(3), 416–427. https://doi.org/10.1111/acel.12446

[187] Pópulo, H., Lopes, J. M., & Soares, P. (2012). The mTOR signalling pathway in human cancer. International journal of molecular sciences, 13(2), 1886–1918. https://doi.org/10.3390/ijms13021886

[188] Trafton, Anne (2018) 'Fasting boosts stem cells' regenerative capacity', MIT News Office, May 3, 2018, Accessed Online: https://news.mit.edu/2018/fasting-boosts-stem-cells-regenerative-capacity-0503

[189] Houtkooper and Auwerx (2012) 'Exploring the therapeutic space around NAD+', J Cell Biol (2012) 199 (2): 205–209.

[190] Mattson, M. P., Moehl, K., Ghena, N., Schmaedick, M., & Cheng, A. (2018). Intermittent metabolic switching, neuroplasticity and brain health. Nature reviews. Neuroscience, 19(2), 63–80. https://doi.org/10.1038/nrn.2017.156

[191] Hornsby et al (2016) 'Short-term caloric restriction enhances adult hippocampal neurogenesis and remote fear memory in a Ghsr-dependent manner', Psychoneuroendocrinology. 2016 Jan; 63: 198–207. doi: 10.1016/j.psyneuen.2015.09.023

[192] Panda S. (2016). Circadian physiology of metabolism. Science (New York, N.Y.), 354(6315), 1008–1015. https://doi.org/10.1126/science.aah4967

[193] de Cabo et al (2018) 'A time to fast', Science, Vol. 362, Issue 6416, pp. 770-775.

[194] Longo, V. D., & Mattson, M. P. (2014). Fasting: Molecular Mechanisms and Clinical Applications. Cell Metabolism, 19(2), 181–192. doi:10.1016/j.cmet.2013.12.008

[195] Zhang, P., Zhang, H., Lin, J., Xiao, T., Xu, R., Fu, Y., … Jiang, H. (2020). Insulin impedes osteogenesis of BMSCs by inhibiting autophagy and promoting premature senescence via the TGF-β1 pathway. Aging, 12(3), 2084–2100. doi:10.18632/aging.102723

[196] Wu et al (2016) 'Sources and implications of NADH/NAD+ redox imbalance in diabetes and its complications', Diabetes Metab Syndr Obes. 2016; 9: 145–153.

[197] Keys A, et al (1950) 'The Biology of Human Starvation', University of Minnesota Press, Minneapolis.

[198]Kalm, L. M., & Semba, R. D. (2005). They Starved So That Others Be Better Fed: Remembering Ancel Keys and the Minnesota Experiment. The Journal of Nutrition, 135(6), 1347–1352. doi:10.1093/jn/135.6.1347

[199] Cuervo et al (2005) 'Autophagy and aging: the importance of maintaining "clean" cells', Autophagy. 2005 Oct-Dec;1(3):131-40.

[200] Yang et al (2010) 'Defective Hepatic Autophagy in Obesity Promotes ER Stress and Causes Insulin Resistance', Cell Metab. 2010 Jun 9; 11(6): 467–478.

[201] Nakai, A., Yamaguchi, O., Takeda, T., Higuchi, Y., Hikoso, S., Taniike, M., … Otsu, K. (2007). The role of autophagy in cardiomyocytes in the basal state and in response to hemodynamic stress. Nature Medicine, 13(5), 619–624. doi:10.1038/nm1574

[202] Hua, Y., Zhang, Y., Ceylan-Isik, A. F., Wold, L. E., Nunn, J. M., & Ren, J. (2011). Chronic Akt activation accentuates aging-induced cardiac hypertrophy and myocardial contractile dysfunction: role of autophagy. Basic research in cardiology, 106(6), 1173–1191. https://doi.org/10.1007/s00395-011-0222-8

[203] Liao et al (2012) 'Macrophage Autophagy Plays a Protective Role in Advanced Atherosclerosis', Cell Metabolism, Volume 15, Issue 4, 4 April 2012, Pages 545-553.

[204] Shi et al (2012) 'Activation of autophagy by inflammatory signals limits IL-1β production by targeting ubiquitinated inflammasomes for destruction', Nat Immunol. 2012 Jan 29;13(3):255-63.

[205] Henderson and Stevens (2012) 'The Role of Autophagy in Crohn's Disease', Cells. 2012 Sep; 1(3): 492–519.

[206] Deretic and Levine (2009) 'Autophagy, immunity, and microbial adaptations', Cell Host Microbe. 2009 Jun 18;5(6):527-49.

[207] Bossy et al (2008) 'Clearing the Brain's Cobwebs: The Role of Autophagy in Neuroprotection', Curr Neuropharmacol. 2008 Jun; 6(2): 97–101.

[208] Chu, Hiutung; Khosravi, Arya; Kusumawardhani, Indah P et al. (2016) Gene-microbiota interactions contribute to the pathogenesis of inflammatory bowel disease. Science 352:1116-20

[209] Khambu et al (2018) 'Autophagy in non-alcoholic fatty liver disease and alcoholic liver disease', Liver Research, Volume 2, Issue 3, September 2018, Pages 112-119.

[210] Ruckenstuhl et al (2014) 'Lifespan extension by methionine restriction requires autophagy-dependent vacuolar acidification', PLoS Genet. 2014 May 1;10(5):e1004347.

[211] Morselli et al (2010) 'Caloric restriction and resveratrol promote longevity through the Sirtuin-1-dependent induction of autophagy', Cell Death Dis. 2010 Jan; 1(1): e10.

[212] Bagherniya, M., Butler, A. E., Barreto, G. E., & Sahebkar, A. (2018). The effect of fasting or calorie restriction on autophagy induction: A review of the literature. Ageing research reviews, 47, 183–197. https://doi.org/10.1016/j.arr.2018.08.004

[213] He, C., Sumpter, R., Jr, & Levine, B. (2012). Exercise induces autophagy in peripheral tissues and in the brain. Autophagy, 8(10), 1548–1551. https://doi.org/10.4161/auto.21327

[214] Penke, B., Bogár, F., Crul, T., Sántha, M., Tóth, M. E., & Vígh, L. (2018). Heat Shock Proteins and Autophagy Pathways in Neuroprotection: from Molecular Bases to Pharmacological Interventions. International journal of molecular sciences, 19(1), 325. https://doi.org/10.3390/ijms19010325

[215] Gomes, LC and Dikic, I. (2014) 'Autophagy in antimicrobial immunity', Mol Cell, Vol 54(2), p 224-233.

[216] Nakagawa et al (2004) 'Autophagy defends cells against invading group A Streptococcus', Science. 2004 Nov 5;306(5698):1037-40.

[217] Gutierrez et al (2004) 'Autophagy is a defense mechanism inhibiting BCG and Mycobacterium tuberculosis survival in infected macrophages', Cell. 2004 Dec 17;119(6):753-66.

[218] Birmingham et al (2006) 'Autophagy controls Salmonella infection in response to damage to the Salmonella-containing vacuole', J Biol Chem. 2006 Apr 21;281(16):11374-83. Epub 2006 Feb 22.

[219] Py et al (2007) 'Autophagy limits Listeria monocytogenes intracellular growth in the early phase of primary infection', Autophagy. 2007 Mar-Apr;3(2):117-25. Epub 2007 Mar 27.

[220] Gutierrez et al (2007) 'Protective role of autophagy against Vibrio cholerae cytolysin, a pore-forming toxin from V. cholerae', Proc Natl Acad Sci U S A. 2007 Feb 6;104(6):1829-34. Epub 2007 Jan 31.

[221] Tan et al (2009) 'Induction of autophagy by anthrax lethal toxin', Biochem Biophys Res Commun. 2009 Feb 6;379(2):293-7. doi: 10.1016/j.bbrc.2008.12.048. Epub 2008 Dec 25.

[222] Terebiznik et al (2009) 'Effect of Helicobacter pylori's vacuolating cytotoxin on the autophagy pathway in gastric epithelial cells', Autophagy. 2009 Apr;5(3):370-9. Epub 2009 Apr 19.

[223] Starr T et al (2012) 'Selective subversion of autophagy complexes facilitates completion of the Brucella intracellular cycle', Cell Host & Microbe, Vol 11(1), p 33-45.

[224] Sinha et al (2008) 'Molecular basis of the regulation of Beclin 1-dependent autophagy by the gamma-herpesvirus 68 Bcl-2 homolog M11', Autophagy. 2008 Nov;4(8):989-97. Epub 2008 Nov

18.

[225] Zhou and Spector (2009) 'Human immunodeficiency virus type-1 infection inhibits autophagy', AIDS. 2008 Mar 30; 22(6): 695–699.

[226] Chaumorcel et al (2008) 'Human cytomegalovirus controls a new autophagy-dependent cellular antiviral defense mechanism', Autophagy. 2008 Jan;4(1):46-53. Epub 2007 Oct 17.

[227] Wong et al (2008) 'Autophagosome supports coxsackievirus B3 replication in host cells', J Virol. 2008 Sep;82(18):9143-53. doi: 10.1128/JVI.00641-08. Epub 2008 Jul 2.

[228] Laddha, SV. et al (2014) 'Mutational landscape of the essential autophagy gene BECN1 in human cancers', Molecular Cancer Research: MCR, Vol 12(4), p 485-490.

[229] Kenific, CM et al (2010) 'Autophagy and metastasis: another double-edged sword', Curr Opin Cell Biol, Vol 22(2), p 241-245.

[230] Gump, JM. and Thouburn, A. (2011) 'Autophagy and apoptosis: what is the connection?', Trends in Cellular Biology, Vol 21(7), p 387-392.

[231] Gump, JM. et al (2014) 'Autophagy variation within a cell population determines cell fate through selective degradation of Fap-1', Nat Cell Biol, Vol 16(1), p 47-54.

[232] Takai, N. et al (2012) 'Bufalin, a traditional oriental medicine, induces apoptosis in human cancer cells', Asian Pacific Journal of Cancer Prevention, Vol 13(1), p 399-402.

[233] Zhao, H. et al (2017) 'Blocking autophagy enhances the pro-apoptotic effect of bufalin on human gastric cancer cells through endoplasmic reticulum stress', Biology Open, Vol 6, p 1416-1422.

[234] Hu et al (2012) 'Hypoxia-Induced Autophagy Promotes Tumor Cell Survival and Adaptation to Antiangiogenic Treatment in Glioblastoma', Molecular and Cellular Pathobiology, Cancer Res; 72(7) April 1, 2012.

[235] Bellot et al (2009) 'Hypoxia-Induced Autophagy Is Mediated through Hypoxia-Inducible Factor Induction of BNIP3 and BNIP3L via Their BH3 Domains', MOLECULAR AND CELLULAR BIOLOGY, May 2009, p. 2570–2581.

[236] Pouysségur et al (2006) 'Hypoxia signalling in cancer and approaches to enforce tumour regression', Nature. 2006 May 25;441(7092):437-43.

[237] Krock et al (2011) 'Hypoxia-induced angiogenesis: good and evil', Genes Cancer. 2011 Dec;2(12):1117-33. doi: 10.1177/1947601911423654.

[238] Huang et al (2017) 'MicroRNA-21 protects against cardiac hypoxia/reoxygenation injury by inhibiting excessive autophagy in H9c2 cells via the Akt/mTOR pathway', J Cell Mol Med. 2017 Mar;21(3):467-474. doi: 10.1111/jcmm.12990. Epub 2016 Sep 29.

[239] Chen et al (2016) 'Up-regulation of miRNA-221 inhibits hypoxia/reoxygenation-induced autophagy through the DDIT4/mTORC1 and Tp53inp1/p62 pathways', Biochem Biophys Res Commun. 2016 May 20;474(1):168-174. doi: 10.1016/j.bbrc.2016.04.090. Epub 2016 Apr 20.

[240] Lum JJ et al (2005) 'Growth factor regulation of autophagy and cell survival in the absence of apoptosis', Cell, Vol 120(2), p 237-248.

[241] Takamura, A. et al (2011) 'Autophagy-deficient mice develop multiple liver tumors', Genes Dev, Vol 25(8), p 795-800.

[242] Tomiyama et al (2012). Does cellular aging relate to patterns of allostasis? An examination of basal and stress reactive HPA axis activity and telomere length. Physiology & behavior, 106(1), 40–45. https://doi.org/10.1016/j.physbeh.2011.11.016

[243] Puterman, E., Lin, J., Krauss, J., Blackburn, E. H., & Epel, E. S. (2015). Determinants of telomere attrition over 1 year in healthy older women: stress and health behaviors matter. Molecular psychiatry, 20(4), 529–535. https://doi.org/10.1038/mp.2014.70

[244] Vitlic, A., Lord, J. M., & Phillips, A. C. (2014). Stress, ageing and their influence on functional, cellular and molecular aspects of the immune system. Age (Dordrecht, Netherlands), 36(3), 9631. https://doi.org/10.1007/s11357-014-9631-6

[245] Aw et al (2007) 'Immunosenescence: emerging challenges for an ageing population', Immunology. 2007 Apr; 120(4): 435–446.

[246] Laurberg, P., Knudsen, N., Andersen, S., Carlé, A., Pedersen, I. B., & Karmisholt, J. (2012). Thyroid Function and Obesity. European Thyroid Journal, 1(3), 159–167. doi:10.1159/000342994

[247] Samuels, M. H. (2014). Psychiatric and cognitive manifestations of hypothyroidism. Current Opinion in Endocrinology & Diabetes and Obesity, 21(5), 377–383. doi:10.1097/med.0000000000000089

[248] Bennett, W. E., & Heuckeroth, R. O. (2012). Hypothyroidism Is a Rare Cause of Isolated Constipation. Journal of Pediatric Gastroenterology and Nutrition, 54(2), 285–287. doi:10.1097/mpg.0b013e318239714f

[249] Joffe, R.T., Pearce, E.N., Hennessey, J.V., Ryan, J.J. and Stern, R.A. (2013), Subclinical hypothyroidism, mood, and cognition in older adults: a review. Int J Geriatr Psychiatry, 28: 111-118. doi:10.1002/gps.3796

[250] Dons, Robert F.; Jr, Frank H. Wians (2009). *Endocrine and metabolic disorders clinical lab testing manual* (4th ed.). Boca Raton: CRC Press. p. 10. ISBN 9781420079364.

[251] Protsiv, M., Ley, C., Lankester, J., Hastie, T., & Parsonnet, J. (2020). Decreasing human body temperature in the United States since the Industrial Revolution. eLife, 9. doi:10.7554/elife.49555

[252] Landsberg, L., Young, J. B., Leonard, W. R., Linsenmeier, R. A., & Turek, F. W. (2009). Is obesity associated with lower body temperatures? Core temperature: a forgotten variable in energy balance. Metabolism, 58(6), 871–876. doi:10.1016/j.metabol.2009.02.017

[253] Abdullatif, H. D., & Ashraf, A. P. (2006). REVERSIBLE SUBCLINICAL HYPOTHYROIDISM IN THE PRESENCE OF ADRENAL INSUFFICIENCY. Endocrine Practice, 12(5), 572–575. doi:10.4158/ep.12.5.572

[254] Ongphiphadhanakul, B., Fang, S. L., Tang, K.-T., Patwardhan, N. A., & Braverman, L. E. (1994). Tumor necrosis factor-α decreases thyrotropin-induced 5′-deiodinase activity in FRTL-5 thyroid cells. European Journal of Endocrinology, 130(5), 502–507. doi:10.1530/eje.0.1300502

[255] Bartalena, L., Brogioni, S., Grasso, L., Velluzzi, F., & Martino, E. (1994). Relationship of the increased serum interleukin-6 concentration to changes of thyroid function in nonthyroidal illness. Journal of Endocrinological Investigation, 17(4), 269–274. doi:10.1007/bf03348974

[256] Corssmit, E. P., Heyligenberg, R., Endert, E., Sauerwein, H. P., & Romijn, J. A. (1995). Acute effects of interferon-alpha administration on thyroid hormone metabolism in healthy men. The Journal of Clinical Endocrinology & Metabolism, 80(11), 3140–3144. doi:10.1210/jcem.80.11.7593416

[257] Bohr et al (1904) 'Concerning a Biologically Important Relationship - The Influence of the Carbon Dioxide Content of Blood on its Oxygen Binding', Skand. Arch. Physiol. 16, 401-412 (1904) by Ulf Marquardt for CHEM-342, January 1997, Accessed Online: https://www1.udel.edu/chem/white/C342/Bohr(1904).html

[258] Stegen, K., De Bruyne, K., Rasschaert, W., Van de Woestijne, K. P., & Van den Bergh, O. (1999). Fear-relevant images as conditioned stimuli for somatic complaints, respiratory behavior, and reduced end-tidal pCO_2. Journal of Abnormal Psychology, 108(1), 143–152. doi:10.1037/0021-843x.108.1.143

[259] Lee and Levine (1999) 'Acute respiratory alkalosis associated with low minute ventilation in a patient with severe hypothyroidism', Can J Anaesth. 1999 Feb;46(2):185-9.

[260] Woods, Hubert Frank; Cohen, Robert (1976). *Clinical and biochemical aspects of lactic acidosis*. Oxford: Blackwell Scientific

[261] Vander Heiden, M. G., Cantley, L. C., & Thompson, C. B. (2009). Understanding the Warburg effect: the metabolic requirements of cell proliferation. Science (New York, N.Y.), 324(5930), 1029–1033. https://doi.org/10.1126/science.1160809

[262] Buffenstein R. (2008). Negligible senescence in the longest living rodent, the naked mole-rat: insights from a successfully aging species. Journal of comparative physiology. B, Biochemical,

systemic, and environmental physiology, 178(4), 439–445. https://doi.org/10.1007/s00360-007-0237-5

263 Csiszar, A., Labinskyy, N., Orosz, Z., Xiangmin, Z., Buffenstein, R., & Ungvari, Z. (2007). Vascular aging in the longest-living rodent, the naked mole rat. American journal of physiology. Heart and circulatory physiology, 293(2), H919–H927. https://doi.org/10.1152/ajpheart.01287.2006

264 Maina, J. N., Maloiy, G. M. O., & Makanya, A. N. (1992). Morphology and morphometry of the lungs of two East African mole rats,Tachyoryctes splendens andHeterocephalus glaber (Mammalia, Rodentia). Zoomorphology, 112(3), 167–179. doi:10.1007/bf01633107

265 Park et al (2017). Fructose-driven glycolysis supports anoxia resistance in the naked mole-rat. Science (New York, N.Y.), 356(6335), 307–311. https://doi.org/10.1126/science.aab3896

266 Park et al (2008). Selective inflammatory pain insensitivity in the African naked mole-rat (Heterocephalus glaber). PLoS biology, 6(1), e13. https://doi.org/10.1371/journal.pbio.0060013

267 Buffenstein, R. (2005). The Naked Mole-Rat: A New Long-Living Model for Human Aging Research. The Journals of Gerontology Series A: Biological Sciences and Medical Sciences, 60(11), 1369–1377. doi:10.1093/gerona/60.11.1369

268 HENDERSON, E., WEINBERG, M., & WRIGHT, W. A. (1950). Pregnenolone. The Journal of clinical endocrinology and metabolism, 10(4), 455–474. https://doi.org/10.1210/jcem-10-4-455

269 Marx et al (2011). Pregnenolone as a novel therapeutic candidate in schizophrenia: emerging preclinical and clinical evidence. Neuroscience, 191, 78–90. https://doi.org/10.1016/j.neuroscience.2011.06.076

270 Wang et al (2012). Thyroid-Stimulating Hormone Levels within the Reference Range Are Associated with Serum Lipid Profiles Independent of Thyroid Hormones. The Journal of Clinical Endocrinology & Metabolism, 97(8), 2724–2731. doi:10.1210/jc.2012-1133

271 Rizos, C. V. (2011). Effects of Thyroid Dysfunction on Lipid Profile. The Open Cardiovascular Medicine Journal, 5(1), 76–84. doi:10.2174/1874192401105010076

272 Trexler, E. T., Smith-Ryan, A. E., & Norton, L. E. (2014). Metabolic adaptation to weight loss: implications for the athlete. Journal of the International Society of Sports Nutrition, 11(1), 7. https://doi.org/10.1186/1550-2783-11-7

273 Bevilacqua, L., Ramsey, J. J., Hagopian, K., Weindruch, R., & Harper, M. E. (2004). Effects of short- and medium-term calorie restriction on muscle mitochondrial proton leak and reactive oxygen species production. American journal of physiology. Endocrinology and metabolism, 286(5), E852–E861. https://doi.org/10.1152/ajpendo.00367.2003

274 Doucet et al (2001). Evidence for the existence of adaptive thermogenesis during weight loss. British Journal of Nutrition, 85(6), 715–723. doi:10.1079/bjn2001348

275 Esterbauer et al (1999). Uncoupling protein-3 gene expression: reduced skeletal muscle mRNA in obese humans during pronounced weight loss. Diabetologia, 42(3), 302–309. doi:10.1007/s001250051155

276 Loeffelholz and Birkenfeld (2018) The Role of Non-exercise Activity Thermogenesis in Human Obesity, In: Feingold KR, Anawalt B, Boyce A, et al., editors. Endotext [Internet]. South Dartmouth (MA): MDText.com, Inc.; 2000-. Available from: https://www.ncbi.nlm.nih.gov/books/NBK279077/

277 Mäestu, J., Jürimäe, J., Valter, I., & Jürimäe, T. (2008). Increases in ghrelin and decreases in leptin without altering adiponectin during extreme weight loss in male competitive bodybuilders. Metabolism, 57(2), 221–225. doi:10.1016/j.metabol.2007.09.004

278 Rosenbaum, M., Hirsch, J., Gallagher, D. A., & Leibel, R. L. (2008). Long-term persistence of adaptive thermogenesis in subjects who have maintained a reduced body weight. The American journal of clinical nutrition, 88(4), 906–912. https://doi.org/10.1093/ajcn/88.4.906

279 Cameron, J. D., Cyr, M. J., & Doucet, E. (2010). Increased meal frequency does not promote greater weight loss in subjects who were prescribed an 8-week equi-energetic energy-restricted diet. The British journal of nutrition, 103(8), 1098–1101. https://doi.org/10.1017/S0007114509992984

[280] Shelmet et al (1988) 'Ethanol causes acute inhibition of carbohydrate, fat, and protein oxidation and insulin resistance', J Clin Invest. 1988 Apr;81(4):1137-45.

[281] Jeong, E. A., Jeon, B. T., Shin, H. J., Kim, N., Lee, D. H., Kim, H. J., Kang, S. S., Cho, G. J., Choi, W. S., & Roh, G. S. (2011). Ketogenic diet-induced peroxisome proliferator-activated receptor-γ activation decreases neuroinflammation in the mouse hippocampus after kainic acid-induced seizures. Experimental neurology, 232(2), 195–202. https://doi.org/10.1016/j.expneurol.2011.09.001

[282] Masuda, R., Monahan, J. W., & Kashiwaya, Y. (2005). D-beta-hydroxybutyrate is neuroprotective against hypoxia in serum-free hippocampal primary cultures. Journal of neuroscience research, 80(4), 501–509. https://doi.org/10.1002/jnr.20464

[283] Haces et al (2008). Antioxidant capacity contributes to protection of ketone bodies against oxidative damage induced during hypoglycemic conditions. Experimental neurology, 211(1), 85–96. https://doi.org/10.1016/j.expneurol.2007.12.029

[284] Westman et al (2008). The effect of a low-carbohydrate, ketogenic diet versus a low-glycemic index diet on glycemic control in type 2 diabetes mellitus. Nutrition & metabolism, 5, 36. https://doi.org/10.1186/1743-7075-5-36

[285] Tieu et al (2003). D-beta-hydroxybutyrate rescues mitochondrial respiration and mitigates features of Parkinson disease. The Journal of clinical investigation, 112(6), 892–901. https://doi.org/10.1172/JCI18797

[286] Milder, J. B., Liang, L. P., & Patel, M. (2010). Acute oxidative stress and systemic Nrf2 activation by the ketogenic diet. Neurobiology of disease, 40(1), 238–244. https://doi.org/10.1016/j.nbd.2010.05.030

[287] Jarrett, S. G., Milder, J. B., Liang, L. P., & Patel, M. (2008). The ketogenic diet increases mitochondrial glutathione levels. Journal of neurochemistry, 106(3), 1044–1051. https://doi.org/10.1111/j.1471-4159.2008.05460.x

[288] Martin et al (2011). Change in food cravings, food preferences, and appetite during a low-carbohydrate and low-fat diet. Obesity (Silver Spring, Md.), 19(10), 1963–1970. https://doi.org/10.1038/oby.2011.62

[289] Poff et al (2014). Ketone supplementation decreases tumor cell viability and prolongs survival of mice with metastatic cancer. International journal of cancer, 135(7), 1711–1720. https://doi.org/10.1002/ijc.28809

[290] Shimazu et al (2013). Suppression of oxidative stress by β-hydroxybutyrate, an endogenous histone deacetylase inhibitor. Science (New York, N.Y.), 339(6116), 211–214. https://doi.org/10.1126/science.1227166

[291] Elamin et al (2018). Ketogenic Diet Modulates NAD+-Dependent Enzymes and Reduces DNA Damage in Hippocampus. Frontiers in cellular neuroscience, 12, 263. https://doi.org/10.3389/fncel.2018.00263

[292] De Rooy, L., and Hawdon, J. (2002). Nutritional Factors That Affect the Postnatal Metabolic Adaptation of Full-Term Small- and Large-for-Gestational-Age Infants. PEDIATRICS, 109(3), e42–e42. doi:10.1542/peds.109.3.e42

[293] Meira et al (2019). Ketogenic Diet and Epilepsy: What We Know So Far. Frontiers in Neuroscience, 13. doi:10.3389/fnins.2019.00005

[294] Wilder R. (1921). The effect of ketonemia on the course of epilepsy. *Mayo Clin. Proc.* 2 307–308.

[295] McCarty, M. F., DiNicolantonio, J. J., & O'Keefe, J. H. (2015). Ketosis may promote brain macroautophagy by activating Sirt1 and hypoxia-inducible factor-1. Medical hypotheses, 85(5), 631–639. https://doi.org/10.1016/j.mehy.2015.08.002

[296] Finn, P. F., & Dice, J. F. (2005). Ketone Bodies Stimulate Chaperone-mediated Autophagy. Journal of Biological Chemistry, 280(27), 25864–25870. doi:10.1074/jbc.m502456200

[297] Wang et al (2018). Ketogenic diet attenuates neuronal injury via autophagy and mitochondrial

pathways in pentylenetetrazol-kindled seizures. Brain research, 1678, 106–115. https://doi.org/10.1016/j.brainres.2017.10.009

[298] Takagi et al (2016). Mammalian autophagy is essential for hepatic and renal ketogenesis during starvation. Scientific Reports, 6(1). doi:10.1038/srep18944

[299] Phinney SD et al (1983) 'The human metabolic response to chronic ketosis without caloric restriction: preservation of submaximal exercise capability with reduced carbohydrate oxidation', Metabolism. 1983 Aug;32(8):769-76.

[300] Michael H et al (2008) 'Insulin Resistance and Hyperinsulinemia', Diabetes Care Feb 2008, 31 (Supplement 2) S262-S268.

[301] Modan et al (1985). Hyperinsulinemia. A link between hypertension obesity and glucose intolerance. The Journal of clinical investigation, 75(3), 809-17.

[302] Wang, G (2014) 'Raison d'être of insulin resistance: the adjustable threshold hypothesis', Journal of the Royal Society, Vol 11(101).

[303]Unger RH et al (2012) 'Gluttony, sloth and the metabolic syndrome: a roadmap to lipotoxicity', Trends in Endocrinology and Metabolism 21 (2010) 345–352.

[304] Kraegen, EW et al (1991) 'Development of Muscle Insulin Resistance After Liver Insulin Resistance in High-Fat–Fed Rats', Diabetes Nov 1991, 40 (11) 1397-1403.

[305] Isganaitis E and Lustig R.H. (2005) 'Fast Food, Central Nervous System Insulin Resistance, and Obesity', Arteriosclerosis, Thrombosis, and Vascular Biology. 2005;25:2451–2462.

[306] Clément L et al (2002) 'Dietary trans-10,cis-12 conjugated linoleic acid induces hyperinsulinemia and fatty liver in the mouse', J Lipid Res. 2002 Sep;43(9):1400-9.

[307] Storlien LH et al (1991) 'Influence of Dietary Fat Composition on Development of Insulin Resistance in Rats: Relationship to Muscle Triglyceride and ω-3 Fatty Acids in Muscle Phospholipid', Diabetes 1991 Feb; 40(2): 280-289.

[308] Grandl et al (2018). Short-term feeding of a ketogenic diet induces more severe hepatic insulin resistance than an obesogenic high-fat diet. The Journal of physiology, 596(19), 4597–4609. https://doi.org/10.1113/JP275173

[309] Kinzig, K. P., Honors, M. A., & Hargrave, S. L. (2010). Insulin sensitivity and glucose tolerance are altered by maintenance on a ketogenic diet. Endocrinology, 151(7), 3105–3114. https://doi.org/10.1210/en.2010-0175

[310] Boden et al (2005). Effect of a low-carbohydrate diet on appetite, blood glucose levels, and insulin resistance in obese patients with type 2 diabetes. Annals of internal medicine, 142(6), 403–411. https://doi.org/10.7326/0003-4819-142-6-200503150-00006

[311] Noakes et al (2006). Comparison of isocaloric very low carbohydrate/high saturated fat and high carbohydrate/low saturated fat diets on body composition and cardiovascular risk. Nutrition & metabolism, 3, 7. https://doi.org/10.1186/1743-7075-3-7

[312] Grill, V., & Qvigstad, E. (2000). Fatty acids and insulin secretion. British Journal of Nutrition, 83(S1), S79–S84. doi:10.1017/s0007114500000994

[313] Berry, M. N., Phillips, J. W., Henly, D. C., & Clark, D. G. (1993). Effects of fatty acid oxidation on glucose utilization by isolated hepatocytes. FEBS letters, 319(1-2), 26–30. https://doi.org/10.1016/0014-5793(93)80030-x

[314] Hue, L., & Taegtmeyer, H. (2009). The Randle cycle revisited: a new head for an old hat. American journal of physiology. Endocrinology and metabolism, 297(3), E578–E591. https://doi.org/10.1152/ajpendo.00093.2009

[315] Gaesser, G. A., & Brooks, G. A. (1980). Glycogen repletion following continuous and intermittent exercise to exhaustion. Journal of applied physiology: respiratory, environmental and exercise physiology, 49(4), 722–728. https://doi.org/10.1152/jappl.1980.49.4.722

[316] Pascoe DD, et al (1993), 'Glycogen resynthesis in skeletal muscle following resistive exercise', Med Sci Sports Exerc, Vol 25(3), p 349-54.

[317] Bevilacqua et al (1990) 'Operation of Randle's cycle in patients with NIDDM', Diabetes. 1990 Mar;39(3):383-9.

[318] Delarue and Magnan (2007) 'Free fatty acids and insulin resistance', Curr Opin Clin Nutr Metab Care. 2007 Mar;10(2):142-8.

[319] Shuldiner and McLenithan (2004) 'Genes and pathophysiology of type 2 diabetes: more than just the Randle cycle all over again', J Clin Invest. 2004 Nov;114(10):1414-7.

[320] Bergman et al (2006), Why Visceral Fat is Bad: Mechanisms of the Metabolic Syndrome. Obesity, 14: 16S-19S.

[321] Jeffery et al (2015). Rapid depot-specific activation of adipocyte precursor cells at the onset of obesity. Nature Cell Biology, 17(4), 376–385. doi:10.1038/ncb3122

[322] Świderska et al (November 5th 2018). Role of PI3K/AKT Pathway in Insulin-Mediated Glucose Uptake, Blood Glucose Levels, Leszek Szablewski, IntechOpen, DOI: 10.5772/intechopen.80402. Available from: https://www.intechopen.com/books/blood-glucose-levels/role-of-pi3k-akt-pathway-in-insulin-mediated-glucose-uptake

[323] Shoelson, S. E., Lee, J., & Goldfine, A. B. (2006). Inflammation and insulin resistance. The Journal of clinical investigation, 116(7), 1793-801.

[324] Huang YJ et al (1997) 'Amelioration of insulin resistance and hypertension in a fructose-fed rat model with fish oil supplementation', Metabolism Clinical and Experimental, November 1997Volume 46, Issue 11, Pages 1252–1258.

[325] Balkau et al (2008). Physical activity and insulin sensitivity: the RISC study. Diabetes, 57(10), 2613–2618. https://doi.org/10.2337/db07-1605

[326] Lund, S., Holman, G. D., Schmitz, O., & Pedersen, O. (1995). Contraction stimulates translocation of glucose transporter GLUT4 in skeletal muscle through a mechanism distinct from that of insulin. Proceedings of the National Academy of Sciences of the United States of America, 92(13), 5817-21.

[327] Shiloah E et al (2003) 'Effect of Acute Psychotic Stress in Nondiabetic Subjects on β-Cell Function and Insulin Sensitivity', Diabetes Care 2003 May; 26(5): 1462-1467.

[328] Piroli GG et al (2007) 'Corticosterone Impairs Insulin-Stimulated Translocation of GLUT4 in the Rat Hippocampus', Neuroendocrinology 2007;85:71–80.

[329] Paul-Labrador M. et al (2006) 'Effects of a randomized controlled trial of transcendental meditation on components of the metabolic syndrome in subjects with coronary heart disease', Arch Intern Med. 2006 Jun 12;166(11):1218-24.

[330] Donga et al (2010). A Single Night of Partial Sleep Deprivation Induces Insulin Resistance in Multiple Metabolic Pathways in Healthy Subjects. The Journal of Clinical Endocrinology & Metabolism, 95(6), 2963–2968. doi:10.1210/jc.2009-2430

[331] Clément L et al (2002) 'Dietary trans-10,cis-12 conjugated linoleic acid induces hyperinsulinemia and fatty liver in the mouse', J Lipid Res. 2002 Sep;43(9):1400-9.

[332] Attvall S et al (1993) 'Smoking induces insulin resistance--a potential link with the insulin resistance syndrome', J Intern Med. 1993 Apr;233(4):327-32.

[333] Basciano H et al (2005) 'Fructose, insulin resistance, and metabolic dyslipidemia', Nutrition & Metabolism20052:5.

[334] Spaulding, S. W., Chopra, I. J., Sherwin, R. S., & Lyall, S. S. (1976). EFFECT OF CALORIC RESTRICTION AND DIETARY COMPOSITION ON SERUM T3AND REVERSE T3IN MAN. The Journal of Clinical Endocrinology & Metabolism, 42(1), 197–200. doi:10.1210/jcem-42-1-197

[335] Fery et al (1982) 'Hormonal and metabolic changes induced by an isocaloric isoproteinic ketogenic diet in healthy subjects', Diabete Metab. 1982 Dec;8(4):299-305.

[336] Paz-Filho, G., Wong, M.-L., Licinio, J., & Mastronardi, C. (2012). Leptin therapy, insulin sensitivity, and glucose homeostasis. Indian Journal of Endocrinology and Metabolism, 16(9), 549. doi:10.4103/2230-8210.105571

[337] Berglund et al (2012). Direct leptin action on POMC neurons regulates glucose homeostasis and hepatic insulin sensitivity in mice. Journal of Clinical Investigation, 122(3), 1000–1009. doi:10.1172/jci59816

[338] Hasselbalch et al (1995). Blood-brain barrier permeability of glucose and ketone bodies during short-term starvation in humans. American Journal of Physiology-Endocrinology and Metabolism, 268(6), E1161–E1166. doi:10.1152/ajpendo.1995.268.6.e1161

[339] Abdul Rahim, M. B. H. B. (2016). Gut microbial metabolome: regulation of host metabolism by SCFAs. <i>Imperial College London</i>. https://doi.org/10.25560/42223

[340] Ardawi (1998) 'Glutamine and ketone-body metabolism in the gut of streptozotocin-diabetic rats', Biochem J. 1988 Jan 15; 249(2): 565–572.

[341] Keenan, M. M., & Chi, J.-T. (2015). Alternative Fuels for Cancer Cells. The Cancer Journal, 21(2), 49–55. doi:10.1097/ppo.0000000000000104

[342] Helmholtz Association of German Research Centres. (2013, July 25). Rapamycin: Limited anti-aging effects. ScienceDaily. Retrieved May 29, 2020 from www.sciencedaily.com/releases/2013/07/130725141715.htm

[343] Volek JS. et al (2016) 'Metabolic characteristics of keto-adapted ultra-endurance runners', Metabolism, Vol 65(3), p 100-110.

[344] Campbell et al (2020). Intermittent Energy Restriction Attenuates the Loss of Fat Free Mass in Resistance Trained Individuals. A Randomized Controlled Trial. Journal of Functional Morphology and Kinesiology, 5(1), 19. doi:10.3390/jfmk5010019

[345] al-Saady et al (1989) 'High fat, low carbohydrate, enteral feeding lowers PaCO2 and reduces the period of ventilation in artificially ventilated patients', Intensive Care Med. 1989;15(5):290-5.

[346] Valdes, A. M., Walter, J., Segal, E., & Spector, T. D. (2018). Role of the gut microbiota in nutrition and health. BMJ, k2179. doi:10.1136/bmj.k2179

[347] Hawrelak and Myers (2004) 'The causes of intestinal dysbiosis: a review', Altern Med Rev. 2004 Jun;9(2):180-97.

[348] Paoli, A., Mancin, L., Bianco, A., Thomas, E., Mota, J. F., & Piccini, F. (2019). Ketogenic Diet and Microbiota: Friends or Enemies? Genes, 10(7), 534. doi:10.3390/genes10070534

[349] Richardson, H. B. (1929). THE RESPIRATORY QUOTIENT. Physiological Reviews, 9(1), 61–125. doi:10.1152/physrev.1929.9.1.61

[350] Ellis, Amy C; Hyatt, Tanya C; Gower, Barbara A; Hunter, Gary R (2017-05-02). "Respiratory Quotient Predicts Fat Mass Gain in Premenopausal Women". *Obesity (Silver Spring, Md.).* 18 (12): 2255–2259.

[351] Cronise et al (2017) Oxidative Priority, Meal Frequency, and the Energy Economy of Food and Activity: Implications for Longevity, Obesity, and Cardiometabolic Disease, Metab Syndr Relat Disord. 2017 Feb 1; 15(1): 6–17.

[352] Shelmet, J. J., Reichard, G. A., Skutches, C. L., Hoeldtke, R. D., Owen, O. E., & Boden, G. (1988). Ethanol causes acute inhibition of carbohydrate, fat, and protein oxidation and insulin resistance. The Journal of clinical investigation, 81(4), 1137–1145. https://doi.org/10.1172/JCI113428

[353] Schoenfeld, B. J., Aragon, A. A., Wilborn, C. D., Krieger, J. W., & Sonmez, G. T. (2014). Body composition changes associated with fasted versus non-fasted aerobic exercise. Journal of the International Society of Sports Nutrition, 11(1), 54. https://doi.org/10.1186/s12970-014-0054-7

[354] Bartrina J. A. (2007). Ortorexia o la obsesión por la dieta saludable [Orthorexia or when a healthy diet becomes an obsession]. Archivos latinoamericanos de nutricion, 57(4), 313–315.

[355] McGing, Brian (2009). "Mithridates VI", Encyclopaedia Iranica, p 43.

[356] Celsus, De Medicina, Book V, 23.3. (Loeb, 1935)

[357] Hojte, Jakob Munk. (2009) "The Death and Burial of Mithridates VI", in Mithridates VI and the

Pontic Kingdom, Aarhus University Press, p 121-131, Accessed Online: http://www.mithridat-eupator.ru/biblioteka/Mithridates_VI_and_the_Pontic_Kingdom_Hojte_2009.pdf

[358] Olszak, T., An, D., Zeissig, S., Vera, M. P., Richter, J., Franke, A., Glickman, J. N., Siebert, R., Baron, R. M., Kasper, D. L., & Blumberg, R. S. (2012). Microbial exposure during early life has persistent effects on natural killer T cell function. Science (New York, N.Y.), 336(6080), 489–493. https://doi.org/10.1126/science.1219328

[359] Blustein, J., & Liu, J. (2015). Time to consider the risks of caesarean delivery for long term child health. BMJ, 350(jun09 3), h2410–h2410. doi:10.1136/bmj.h2410

[360] Mattson MP (2012) Energy intake and exercise as determinants of brain health and vulnerability to injury and disease. Cell Metabolism 16: 706-722

[361] Arumugam TV, Phillips TM, Cheng A, Morrell CH, Mattson MP, Wan R (2010) Age and energy intake interact to modify cell stress pathways and stroke outcome. Ann Neurol. 67:41-52.

[362] Yang JL, Lin YT, Chuang PC, Bohr VA, Mattson MP (2014). BDNF and exercise enhance neuronal DNA repair by stimulating CREB-mediated production of apurinic/apyrimidinic endonuclease 1. Neuromolecular Med. 16:161-74

[363] Li et al (2019) 'Habitual tea drinking modulates brain efficiency: evidence from brain connectivity evaluation', Aging (Albany NY). 2019; 11:3876-3890. https://doi.org/10.18632/aging.102023

[364] van Dam, R. M., & Hu, F. B. (2005). Coffee consumption and risk of type 2 diabetes: a systematic review. JAMA, 294(1), 97–104. https://doi.org/10.1001/jama.294.1.97

[365] Maia, L., & de Mendonca, A. (2002). Does caffeine intake protect from Alzheimer's disease? European Journal of Neurology, 9(4), 377–382. doi:10.1046/j.1468-1331.2002.00421.x

[366] Santos, C., Costa, J., Santos, J., Vaz-Carneiro, A., & Lunet, N. (2010). Caffeine intake and dementia: systematic review and meta-analysis. Journal of Alzheimer's disease : JAD, 20 Suppl 1, S187–S204. https://doi.org/10.3233/JAD-2010-091387

[367] Larsson, S. C., & Wolk, A. (2007). Coffee Consumption and Risk of Liver Cancer: A Meta-Analysis. Gastroenterology, 132(5), 1740–1745. doi:10.1053/j.gastro.2007.03.044

[368] Birringer, M. (2011). Hormetics: Dietary Triggers of an Adaptive Stress Response. Pharmaceutical Research, 28(11), 2680–2694. doi:10.1007/s11095-011-0551-1

[369] Lamming, Dudley W.; Wood, Jason G.; Sinclair, David A. (2004). "Small molecules that regulate lifespan: Evidence for xenohormesis". *Molecular Microbiology*. 53 (4): 1003–9. doi:10.1111/j.1365-2958.2004.04209.x.

[370] Hooper, P. L., Hooper, P. L., Tytell, M., & Vígh, L. (2010). Xenohormesis: health benefits from an eon of plant stress response evolution. Cell Stress and Chaperones, 15(6), 761–770. doi:10.1007/s12192-010-0206-x

[371] Howitz, K.T., Bitterman, K.J., Cohen, H.Y., Lamming, D.W., Lavu, S., Wood, J.G., et al. (2003) Small molecule activators of sirtuins extend Saccharomyces cerevisiae lifespan. Nature 425: 191–196.

[372] Senger DR, Li D, Jaminet SC, Cao S (2016) Activation of the Nrf2 Cell Defense Pathway by Ancient Foods: Disease Prevention by Important Molecules and Microbes Lost from the Modern Western Diet. PLOS ONE 11(2): e0148042. https://doi.org/10.1371/journal.pone.0148042

[373] Espinosa-Diez, C., Miguel, V., Mennerich, D., Kietzmann, T., Sánchez-Pérez, P., Cadenas, S., & Lamas, S. (2015). Antioxidant responses and cellular adjustments to oxidative stress. Redox biology, 6, 183–197. https://doi.org/10.1016/j.redox.2015.07.008

[374] Dinkova-Kostova et al (2002). Direct evidence that sulfhydryl groups of Keap1 are the sensors regulating induction of phase 2 enzymes that protect against carcinogens and oxidants. Proceedings of the National Academy of Sciences of the United States of America, 99(18), 11908–11913. https://doi.org/10.1073/pnas.172398899

[375] Fourquet, S., Guerois, R., Biard, D., & Toledano, M. B. (2010). Activation of NRF2 by

nitrosative agents and H2O2 involves KEAP1 disulfide formation. The Journal of biological chemistry, 285(11), 8463–8471. https://doi.org/10.1074/jbc.M109.051714

[376] Itoh et al (1997). An Nrf2/small Maf heterodimer mediates the induction of phase II detoxifying enzyme genes through antioxidant response elements. Biochemical and biophysical research communications, 236(2), 313–322. https://doi.org/10.1006/bbrc.1997.6943

[377] Dodson et al (2015). KEAP1-NRF2 signalling and autophagy in protection against oxidative and reductive proteotoxicity. The Biochemical journal, 469(3), 347–355. https://doi.org/10.1042/BJ20150568

[378] Bartolini, D., Dallaglio, K., Torquato, P., Piroddi, M., & Galli, F. (2018). Nrf2-p62 autophagy pathway and its response to oxidative stress in hepatocellular carcinoma. Translational research : the journal of laboratory and clinical medicine, 193, 54–71. https://doi.org/10.1016/j.trsl.2017.11.007

[379] Jiang, T., Harder, B., Rojo de la Vega, M., Wong, P. K., Chapman, E., & Zhang, D. D. (2015). p62 links autophagy and Nrf2 signaling. Free radical biology & medicine, 88(Pt B), 199–204. https://doi.org/10.1016/j.freeradbiomed.2015.06.014

[380] Wu et al (2006). Upregulation of heme oxygenase-1 by Epigallocatechin-3-gallate via the phosphatidylinositol 3-kinase/Akt and ERK pathways. Life Sciences, 78(25), 2889–2897. doi:10.1016/j.lfs.2005.11.013

[381] Zhou et al (2014). Epigallocatechin-3-gallate (EGCG), a green tea polyphenol, stimulates hepatic autophagy and lipid clearance. PloS one, 9(1), e87161. https://doi.org/10.1371/journal.pone.0087161

[382] Lee et al (2016). Green Tea (-)-Epigallotocatechin-3-Gallate Induces PGC-1α Gene Expression in HepG2 Cells and 3T3-L1 Adipocytes. Preventive Nutrition and Food Science, 21(1), 62–67. doi:10.3746/pnf.2016.21.1.62

[383] Jo et al (2014). Sulforaphane induces autophagy through ERK activation in neuronal cells. FEBS letters, 588(17), 3081–3088. https://doi.org/10.1016/j.febslet.2014.06.036

[384] Bao et al (2014). Grape seed proanthocyanidin extracts ameliorate podocyte injury by activating peroxisome proliferator-activated receptor-γ coactivator 1α in low-dose streptozotocin-and high-carbohydrate/high-fat diet-induced diabetic rats. Food & Function, 5(8), 1872. doi:10.1039/c4fo00340c

[385] Wang, L., Huang, W., & Zhan, J. (2019). Grape Seed Proanthocyanidins Induce Autophagy and Modulate Survivin in HepG2 Cells and Inhibit Xenograft Tumor Growth in Vivo. Nutrients, 11(12), 2983. doi:10.3390/nu11122983

[386] Thyagarajan et al (2010). Triterpenes from Ganoderma Lucidum induce autophagy in colon cancer through the inhibition of p38 mitogen-activated kinase (p38 MAPK). Nutrition and cancer, 62(5), 630–640. https://doi.org/10.1080/01635580903532390

[387] Hung et al (2009). 6-Shogaol, an active constituent of dietary ginger, induces autophagy by inhibiting the AKT/mTOR pathway in human non-small cell lung cancer A549 cells. Journal of agricultural and food chemistry, 57(20), 9809–9816. https://doi.org/10.1021/jf902315e

[388] Schadich et al (2016). Effects of Ginger Phenylpropanoids and Quercetin on Nrf2-ARE Pathway in Human BJ Fibroblasts and HaCaT Keratinocytes. BioMed Research International, 2016, 1–6. doi:10.1155/2016/2173275

[389] Xiao et al (2013). Curcumin induces autophagy via activating the AMPK signaling pathway in lung adenocarcinoma cells. Journal of pharmacological sciences, 123(2), 102–109. https://doi.org/10.1254/jphs.13085fp

[390] Rao et al (1995). Chemoprevention of colon carcinogenesis by dietary curcumin, a naturally occurring plant phenolic compound. Cancer research, 55(2), 259–266.

[391] Tapia et al (2012) 'Curcumin Induces Nrf2 Nuclear Translocation and Prevents Glomerular Hypertension, Hyperfiltration, Oxidant Stress, and the Decrease in Antioxidant Enzymes in 5/6 Nephrectomized Rats', Oxid Med Cell Longev. 2012; 2012: 269039.

[392] Nabavi et al (2018). Regulation of autophagy by polyphenols: Paving the road for treatment of

neurodegeneration. Biotechnology advances, 36(6), 1768–1778. https://doi.org/10.1016/j.biotechadv.2017.12.001

[393] Scapagnini et al (2011). Modulation of Nrf2/ARE Pathway by Food Polyphenols: A Nutritional Neuroprotective Strategy for Cognitive and Neurodegenerative Disorders. Molecular Neurobiology, 44(2), 192–201. doi:10.1007/s12035-011-8181-5

[394] Pietrocola et al (2014). Coffee induces autophagy in vivo. Cell cycle (Georgetown, Tex.), 13(12), 1987–1994. https://doi.org/10.4161/cc.28929

[395] Fu et al (2014). Resveratrol inhibits breast cancer stem-like cells and induces autophagy via suppressing Wnt/β-catenin signaling pathway. PloS one, 9(7), e102535. https://doi.org/10.1371/journal.pone.0102535

[396] Liuzzi et al (2014). Zinc and autophagy. Biometals : an international journal on the role of metal ions in biology, biochemistry, and medicine, 27(6), 1087–1096. https://doi.org/10.1007/s10534-014-9773-0

[397] Rodríguez et al (2009). Food phenolics and lactic acid bacteria. International journal of food microbiology, 132(2-3), 79–90. https://doi.org/10.1016/j.ijfoodmicro.2009.03.025

[398] Sato et al (2016). Azuki bean (Vigna angularis) extract reduces oxidative stress and stimulates autophagy in the kidneys of streptozotocin-induced early diabetic rats. Canadian journal of physiology and pharmacology, 94(12), 1298–1303. https://doi.org/10.1139/cjpp-2015-0540

[399] Xing et al (2014). Salidroside Stimulates Mitochondrial Biogenesis and Protects against H2O2-Induced Endothelial Dysfunction. Oxidative Medicine and Cellular Longevity, 2014, 1–13. doi:10.1155/2014/904834

[400] Zhang et al (2013). Antioxidant and Nrf2 inducing activities of luteolin, a flavonoid constituent in Ixeris sonchifolia Hance, provide neuroprotective effects against ischemia-induced cellular injury. Food and Chemical Toxicology, 59, 272–280. doi:10.1016/j.fct.2013.05.058

[401] Tsai et al (2017). Docosahexaenoic acid increases the expression of oxidative stress-induced growth inhibitor 1 through the PI3K/Akt/Nrf2 signaling pathway in breast cancer cells. Food and chemical toxicology : an international journal published for the British Industrial Biological Research Association, 108(Pt A), 276–288. https://doi.org/10.1016/j.fct.2017.08.010

[402] Jung et al (2013). Docosahexaenoic acid improves vascular function via up-regulation of SIRT1 expression in endothelial cells. Biochemical and Biophysical Research Communications, 437(1), 114–119. doi:10.1016/j.bbrc.2013.06.049

[403] Hao et al (2010). Hydroxytyrosol promotes mitochondrial biogenesis and mitochondrial function in 3T3-L1 adipocytes. The Journal of Nutritional Biochemistry, 21(7), 634–644. doi:10.1016/j.jnutbio.2009.03.012

[404] Rigacci et al (2015). Oleuropein aglycone induces autophagy via the AMPK/mTOR signalling pathway: a mechanistic insight. Oncotarget, 6(34), 35344–35357. https://doi.org/10.18632/oncotarget.6119

[405] Khanfar et al (2015). Olive Oil-derived Oleocanthal as Potent Inhibitor of Mammalian Target of Rapamycin: Biological Evaluation and Molecular Modeling Studies. Phytotherapy research : PTR, 29(11), 1776–1782. https://doi.org/10.1002/ptr.5434

[406] Stefek, M., & Karasu, C. (2011). Eye Lens in Aging and Diabetes: Effect of Quercetin. Rejuvenation Research, 14(5), 525–534. doi:10.1089/rej.2011.1170

[407] Nieman (2010) 'QUERCETIN'S BIOACTIVE EFFECTS IN HUMAN ATHLETES', CURRENT TOPICS IN NUTRACEUTICAL RESEARCH Vol. 8, No. 1, pp. xxx-xxx, 2010.

[408] Rho, J. M., & Rogawski, M. A. (2007). The Ketogenic Diet: Stoking the Powerhouse of the Cell. Epilepsy Currents, 7(2), 58–60. doi:10.1111/j.1535-7511.2007.00170.x

[409] Ben-Dor et al (2005). Carotenoids activate the antioxidant response element transcription system. Molecular cancer therapeutics, 4(1), 177–186.

[410] Yaku et al (2012). The enhancement of phase 2 enzyme activities by sodium butyrate in normal

intestinal epithelial cells is associated with Nrf2 and p53. Molecular and cellular biochemistry, 370(1-2), 7–14. https://doi.org/10.1007/s11010-012-1392-x

[411] Ames, B. N., Profet, M., & Gold, L. S. (1990). Dietary pesticides (99.99% all natural). Proceedings of the National Academy of Sciences, 87(19), 7777–7781. doi:10.1073/pnas.87.19.7777

[412] Mennen et al (2005). Risks and safety of polyphenol consumption. The American Journal of Clinical Nutrition, 81(1), 326S–329S. doi:10.1093/ajcn/81.1.326s

[413] Pfister, J. (1999). Behavioral Strategies for Coping with Poisonous Plants. Bulletin, 45.

[414] Perez, 'How do Wild Herbivores Cope with Plant Toxins?', BEHAVE, Stories of Applied Animal Behavior, University of Idaho and Washington State University, Accessed Online [25.02.20] at https://www.webpages.uidaho.edu/range556/Appl_BEHAVE/projects/toxins-wildlife.htm

[415] M. E. Fowler (*1983*) PLANT POISONING IN FREE-LIVING WILD ANIMALS: A REVIEW. Journal of Wildlife Diseases: January 1983, Vol. 19, No. 1, pp. 34-43.

[416] Chu, Michael; Seltzer, Terry F. (May 20, 2010). "Myxedema Coma Induced by Ingestion of Raw Bok Choy". *New England Journal of Medicine.* 362 (20): 1945–1946. doi:10.1056/NEJMc0911005.

[417] ASTWOOD et al (1949). 1-5-Vinyl-2-thiooxazolidone, an antithyroid compound from yellow turnip and from Brassica seeds. The Journal of biological chemistry, 181(1), 121–130.

[418] Dekker, M., Verkerk, R., & Jongen, W. M. . (2000). Predictive modelling of health aspects in the food production chain: a case study on glucosinolates in cabbage. Trends in Food Science & Technology, 11(4-5), 174–181. doi:10.1016/s0924-2244(00)00062-5

[419] Samuni et al (2013). "Friend or foe? Disparate plant–animal interactions of two congeneric rodents". *Evolutionary Ecology.* 27 (6): 1069–1080. doi:10.1007/s10682-013-9655-x.

[420] Ishida, M; Hara, M; Fukino, N; Kakizaki, T; Morimitsu, Y (2014). "Glucosinolate metabolism, functionality and breeding for the improvement of Brassicaceae vegetables". *Breeding Science.* 64 (1): 48–59. doi:10.1270/jsbbs.64.48.

[421] Cornell University Department of Animal Science 'Glucosinolates (Goitrogenic Glycosides)', Cornell CALS, Accessed Online: http://poisonousplants.ansci.cornell.edu/toxicagents/glucosin.html

[422] Lynn, A., Fuller, Z., Collins, A. R., & Ratcliffe, B. (2015). Comparison of the effect of raw and blanched-frozen broccoli on DNA damage in colonocytes. Cell Biochemistry and Function, 33(5), 266–276. doi:10.1002/cbf.3106

[423] Ratcliffe B, Collins AR, Glass HJ, Hillman K. (1999) 'The effect of physical processing on the protective effect of broccoli in relation to damage to DNA in colonocytes'. In Natural Antioxidants and Anticarcinogens in Nutrition, Health and Disease. Kumpulainen J. T, Salonen JT (eds). The Royal Society of Chemistry: Cambridge, UK, 1999; p 440–442.

[424] Ratcliffe et al (2000) 'The Effect of Cooking on the Protective Effect of Broccoli Against Damage to DNA in Colonocytes', In Dietary Anticarcinogens and Antimutagens—Chemical and Biological Aspects. Johnson IT, Fenwick GR (eds). The Royal Society of Chemistry: Cambridge, UK, 2000; 161–164.

[425] Lynn et al (2006) 'Cruciferous vegetables and colo-rectal cancer', Proceedings of The Nutrition Society, 65(1):135-44, DOI: 10.1079/PNS2005486.

[426] de Figueiredo et al (2015). The antioxidant properties of organosulfur compounds (sulforaphane). Recent patents on endocrine, metabolic & immune drug discovery, 9(1), 24–39. https://doi.org/10.2174/1872214809666150505164138

[427] Kim, J. K., & Park, S. U. (2016). Current potential health benefits of sulforaphane. EXCLI journal, 15, 571–577. https://doi.org/10.17179/excli2016-485

[428] Houghton, C. A., Fassett, R. G., & Coombes, J. S. (2016). Sulforaphane and Other Nutrigenomic Nrf2 Activators: Can the Clinician's Expectation Be Matched by the Reality?. Oxidative medicine and cellular longevity, 2016, 7857186. https://doi.org/10.1155/2016/7857186

[429] Yuan, G., Sun, B., Yuan, J., & Wang, Q. (2009). Effects of different cooking methods on health-promoting compounds of broccoli. Journal of Zhejiang University SCIENCE B, 10(8), 580–588. doi:10.1631/jzus.b0920051

[430] Bahadoran, Z., Mirmiran, P., & Azizi, F. (2013). Potential efficacy of broccoli sprouts as a unique supplement for management of type 2 diabetes and its complications. Journal of medicinal food, 16(5), 375–382. https://doi.org/10.1089/jmf.2012.2559

[431] Senanayake et al (2012). The dietary phase 2 protein inducer sulforaphane can normalize the kidney epigenome and improve blood pressure in hypertensive rats. American journal of hypertension, 25(2), 229–235. https://doi.org/10.1038/ajh.2011.200

[432] Riedl, M. A., Saxon, A., & Diaz-Sanchez, D. (2009). Oral sulforaphane increases Phase II antioxidant enzymes in the human upper airway. Clinical immunology (Orlando, Fla.), 130(3), 244–251. https://doi.org/10.1016/j.clim.2008.10.007

[433] Armah et al (2015). Diet rich in high glucoraphanin broccoli reduces plasma LDL cholesterol: Evidence from randomised controlled trials. Molecular nutrition & food research, 59(5), 918–926. https://doi.org/10.1002/mnfr.201400863

[434] Kim, H. J., Barajas, B., Wang, M., & Nel, A. E. (2008). Nrf2 activation by sulforaphane restores the age-related decrease of T(H)1 immunity: role of dendritic cells. The Journal of allergy and clinical immunology, 121(5), 1255–1261.e7. https://doi.org/10.1016/j.jaci.2008.01.016

[435] Johansson, N. L., Pavia, C. S., & Chiao, J. W. (2008). Growth inhibition of a spectrum of bacterial and fungal pathogens by sulforaphane, an isothiocyanate product found in broccoli and other cruciferous vegetables. Planta medica, 74(7), 747–750. https://doi.org/10.1055/s-2008-1074520

[436] Bahadoran, Z., Mirmiran, P., & Azizi, F. (2013). Potential efficacy of broccoli sprouts as a unique supplement for management of type 2 diabetes and its complications. Journal of medicinal food, 16(5), 375–382. https://doi.org/10.1089/jmf.2012.2559

[437] Kikuchi et al (2015). Sulforaphane-rich broccoli sprout extract improves hepatic abnormalities in male subjects. World journal of gastroenterology, 21(43), 12457–12467. https://doi.org/10.3748/wjg.v21.i43.12457

[438] Kim et al (2017). Sulforaphane epigenetically enhances neuronal BDNF expression and TrkB signaling pathways. Molecular nutrition & food research, 61(2), 10.1002/mnfr.201600194. https://doi.org/10.1002/mnfr.201600194

[439] Tortorella et al (2015). Dietary Sulforaphane in Cancer Chemoprevention: The Role of Epigenetic Regulation and HDAC Inhibition. Antioxidants & redox signaling, 22(16), 1382–1424. https://doi.org/10.1089/ars.2014.6097

[440] Kensler et al (2013). Keap1-nrf2 signaling: a target for cancer prevention by sulforaphane. Topics in current chemistry, 329, 163–177. https://doi.org/10.1007/128_2012_339

[441] Yang et al (2018). Sulforaphane induces autophagy by inhibition of HDAC6-mediated PTEN activation in triple negative breast cancer cells. Life Sciences, 213, 149–157. doi:10.1016/j.lfs.2018.10.034

[442] Liu et al (2017). Sulforaphane promotes ER stress, autophagy, and cell death: implications for cataract surgery. Journal of molecular medicine (Berlin, Germany), 95(5), 553–564. https://doi.org/10.1007/s00109-016-1502-4

[443] Juge, N., Mithen, R. F., & Traka, M. (2007). Molecular basis for chemoprevention by sulforaphane: a comprehensive review. Cellular and Molecular Life Sciences, 64(9), 1105–1127. doi:10.1007/s00018-007-6484-5

[444] Sharma et al (2010). Role of Lipid Peroxidation in Cellular Responses tod,l-Sulforaphane, a Promising Cancer Chemopreventive Agent. Biochemistry, 49(14), 3191–3202. doi:10.1021/bi100104e

[445] Baier et al (2014). Off-target effects of sulforaphane include the derepression of long terminal repeats through histone acetylation events. The Journal of Nutritional Biochemistry, 25(6), 665–668. doi:10.1016/j.jnutbio.2014.02.007

[446] Socala et al (2017) 'Increased seizure susceptibility and other toxicity symptoms following acute sulforaphane treatment in mice', April 2017Toxicology and Applied Pharmacology 326.

[447] Smith, T. K. (2003). Effects of Brassica vegetable juice on the induction of apoptosis and aberrant crypt foci in rat colonic mucosal crypts in vivo. Carcinogenesis, 24(3), 491–495. doi:10.1093/carcin/24.3.491

[448] Moon, J. K., Kim, J. R., Ahn, Y. J., & Shibamoto, T. (2010). Analysis and anti-Helicobacter activity of sulforaphane and related compounds present in broccoli (Brassica oleracea L.) sprouts. Journal of agricultural and food chemistry, 58(11), 6672–6677. https://doi.org/10.1021/jf1003573

[449] Sood, S., & Nagpal, M. (2013). Role of curcumin in systemic and oral health: An overview. Journal of Natural Science, Biology and Medicine, 4(1), 3. doi:10.4103/0976-9668.107253

[450] Lal et al (1999) 'Efficacy of curcumin in the management of chronic anterior uveitis', Phytother Res. 1999 Jun;13(4):318-22.

[451] Usharani et al (2008). Effect of NCB-02, Atorvastatin and Placebo on Endothelial Function, Oxidative Stress and Inflammatory Markers in Patients with Type 2 Diabetes Mellitus. Drugs in R & D, 9(4), 243–250. doi:10.2165/00126839-200809040-00004

[452] Biswas et al (2005). Curcumin Induces Glutathione Biosynthesis and Inhibits NF-κB Activation and Interleukin-8 Release in Alveolar Epithelial Cells: Mechanism of Free Radical Scavenging Activity. Antioxidants & Redox Signaling, 7(1-2), 32–41. doi:10.1089/ars.2005.7.32

[453] McNally et al (2007) 'Curcumin induces heme oxygenase 1 through generation of reactive oxygen species, p38 activation and phosphatase inhibition', Int J Mol Med. 2007 Jan;19(1):165-72.

[454] Atsumi, T., Fujisawa, S., & Tonosaki, K. (2005). Relationship between intracellular ROS production and membrane mobility in curcumin- and tetrahydrocurcumin-treated human gingival fibroblasts and human submandibular gland carcinoma cells. Oral Diseases, 11(4), 236–242. doi:10.1111/j.1601-0825.2005.01067.x

[455] Sahebkar et al (2013). Curcuminoids Modulate Pro-Oxidant-Antioxidant Balance but not the Immune Response to Heat Shock Protein 27 and Oxidized LDL in Obese Individuals. Phytotherapy Research, 27(12), 1883–1888. doi:10.1002/ptr.4952

[456] Agarwal et al (2010) 'Detoxification and antioxidant effects of curcumin in rats experimentally exposed to mercury', Journal of Applied Toxicology, Volume 30, Issue 5, July 2010, Pages 457-468.

[457] García-Niño, W. R., & Pedraza-Chaverrí, J. (2014). Protective effect of curcumin against heavy metals-induced liver damage. Food and Chemical Toxicology, 69, 182–201. doi:10.1016/j.fct.2014.04.016

[458] Bulmus et al (2013) Protective effects of curcumin on antioxidant status, body weight gain, and reproductive parameters in male rats exposed to subchronic 2,3,7,8-tetrachlorodibenzo-*p*-dioxin, Toxicological & Environmental Chemistry, 95:6, 1019-1029, DOI: 10.1080/02772248.2013.829061

[459] Noorafshan et al (2017). Restorative effects of curcumin on sleep-deprivation induced memory impairments and structural changes of the hippocampus in a rat model. Life Sciences, 189, 63–70. doi:10.1016/j.lfs.2017.09.018

[460] Xu et al (2006). Curcumin reverses the effects of chronic stress on behavior, the HPA axis, BDNF expression and phosphorylation of CREB. Brain Research, 1122(1), 56–64. doi:10.1016/j.brainres.2006.09.009

[461] Yi et al (2020). Curcumin attenuates cognitive impairment by enhancing autophagy in chemotherapy. Neurobiology of Disease, 136, 104715. doi:10.1016/j.nbd.2019.104715

[462] Wang et al (2017). Curcumin protects neuronal cells against status-epilepticus-induced hippocampal damage through induction of autophagy and inhibition of necroptosis. Canadian Journal of Physiology and Pharmacology, 95(5), 501–509. doi:10.1139/cjpp-2016-0154

[463] Kim et al (2012). Curcumin-induced autophagy contributes to the decreased survival of oral cancer cells. Archives of Oral Biology, 57(8), 1018–1025. doi:10.1016/j.archoralbio.2012.04.005

[464] Kawamori et al (1999) 'Chemopreventive effect of curcumin, a naturally occurring anti-inflammatory agent, during the promotion/progression stages of colon cancer', Cancer Res. 1999 Feb 1;59(3):597-601.

[465] Goodpasture, C. E., & Arrighi, F. E. (1976). Effects of food seasonings on the cell cycle and chromosome morphology of mammalian cells in vitro with special reference to turmeric. Food and Cosmetics Toxicology, 14(1), 9–14. doi:10.1016/s0015-6264(76)80356-2

[466] Fang, J., Lu, J., & Holmgren, A. (2005). Thioredoxin Reductase Is Irreversibly Modified by Curcumin. Journal of Biological Chemistry, 280(26), 25284–25290. doi:10.1074/jbc.m414645200

[467] Ahsan, H., & Hadi, S. M. (1998). Strand scission in DNA induced by curcumin in the presence of Cu(II). Cancer Letters, 124(1), 23–30. doi:10.1016/s0304-3835(97)00442-4

[468] Sakano, K., & Kawanishi, S. (2002). Metal-mediated DNA damage induced by curcumin in the presence of human cytochrome P450 isozymes. Archives of Biochemistry and Biophysics, 405(2), 223–230. doi:10.1016/s0003-9861(02)00302-8

[469] Dance-Barnes et al (2009). Lung tumor promotion by curcumin. Carcinogenesis, 30(6), 1016–1023. doi:10.1093/carcin/bgp082

[470] National Toxicology Program (1993). NTP Toxicology and Carcinogenesis Studies of Turmeric Oleoresin (CAS No. 8024-37-1) (Major Component 79%-85% Curcumin, CAS No. 458-37-7) in F344/N Rats and B6C3F1 Mice (Feed Studies). National Toxicology Program technical report series, 427, 1–275.

[471] Sharma, R. A. (2004). Phase I Clinical Trial of Oral Curcumin: Biomarkers of Systemic Activity and Compliance. Clinical Cancer Research, 10(20), 6847–6854. doi:10.1158/1078-0432.ccr-04-0744

[472] Sharma, R. A., Gescher, A. J., & Steward, W. P. (2005). Curcumin: The story so far. European Journal of Cancer, 41(13), 1955–1968. doi:10.1016/j.ejca.2005.05.009

[473] Tayyem et al (2006). Curcumin Content of Turmeric and Curry Powders. Nutrition and Cancer, 55(2), 126–131. doi:10.1207/s15327914nc5502_2

[474] Jiao et al (2009). Curcumin, a cancer chemopreventive and chemotherapeutic agent, is a biologically active iron chelator. Blood, 113(2), 462–469. doi:10.1182/blood-2008-05-155952

[475] Hunnicutt, J., He, K., & Xun, P. (2014). Dietary Iron Intake and Body Iron Stores Are Associated with Risk of Coronary Heart Disease in a Meta-Analysis of Prospective Cohort Studies. The Journal of Nutrition, 144(3), 359–366. doi:10.3945/jn.113.185124

[476] Shoba et al (1998). Influence of Piperine on the Pharmacokinetics of Curcumin in Animals and Human Volunteers. Planta Medica, 64(04), 353–356. doi:10.1055/s-2006-957450

[477] Galeotti et al (2008). Flavonoids from carnation (Dianthus caryophyllus) and their antifungal activity. Phytochemistry Letters, 1(1), 44–48. doi:10.1016/j.phytol.2007.10.001

[478] Spencer JP (2008). "Flavonoids: modulators of brain function?". *British Journal of Nutrition.* 99 (E-S1): ES60–77. doi:10.1017/S0007114508965776.

[479] Haytowitz et al 'Sources of Flavonoids in the U.S. Diet Using USDA's Updated Database on the Flavonoid Content of Selected Foods.', USDA's Database on the Flavonoid Content, Accessed Online 25.02.20 at https://www.ars.usda.gov/ARSUserFiles/80400525/Articles/AICR06_flav.pdf

[480] Izzi et al (2012). "The effects of dietary flavonoids on the regulation of redox inflammatory networks". *Frontiers in Bioscience.* 17 (7): 2396–2418. doi:10.2741/4061.

[481] Gomes et al (2012). "Trihydroxyflavones with antioxidant and anti-inflammatory efficacy". *BioFactors.* 38 (5): 378–386. doi:10.1002/biof.1033.

[482] Williams RJ, Spencer JP, Rice-Evans C (2004). "Flavonoids: antioxidants or signalling molecules?". *Free Radical Biology & Medicine.* 36 (7): 838–49. doi:10.1016/j.freeradbiomed.2004.01.001.

[483] Cazarolli et al (2008). "Flavonoids: Prospective Drug Candidates". *Mini-Reviews in Medicinal Chemistry.* 8 (13): 1429–1440. doi:10.2174/138955708786369564.

[484] Yamamoto Y, Gaynor RB (2001). "Therapeutic potential of inhibition of the NF-κB pathway in the treatment of inflammation and cancer". *Journal of Clinical Investigation*. 107 (2): 135–42. doi:10.1172/JCI11914.

[485] Cushnie TP, Lamb AJ (2011). "Recent advances in understanding the antibacterial properties of flavonoids". *International Journal of Antimicrobial Agents*. 38 (2): 99–107. doi:10.1016/j.ijantimicag.2011.02.014.

[486] Cushnie TP, Lamb AJ (2005). "Antimicrobial activity of flavonoids" (PDF). *International Journal of Antimicrobial Agents*. 26 (5): 343–356. doi:10.1016/j.ijantimicag.2005.09.002.

[487] Friedman M (2007). "Overview of antibacterial, antitoxin, antiviral, and antifungal activities of tea flavonoids and teas". *Molecular Nutrition & Food Research*. 51 (1): 116–134. doi:10.1002/mnfr.200600173.

[488] de Sousa et al (2007). "Phosphoprotein levels, MAPK activities and NFkappaB expression are affected by fisetin". *J Enzyme Inhib Med Chem*. 22 (4): 439–444. doi:10.1080/14756360601162063.

[489] Schuier M, Sies H, Illek B, Fischer H (2005). "Cocoa-related flavonoids inhibit CFTR-mediated chloride transport across T84 human colon epithelia". *J. Nutr*. 135 (10): 2320–5. doi:10.1093/jn/135.10.2320.

[490] EFSA Panel on Dietetic Products, Nutrition and Allergies (NDA) (2010). "Scientific Opinion on the substantiation of health claims related to various food(s)/food constituent(s) and protection of cells from premature aging, antioxidant activity, antioxidant content and antioxidant properties, and protection of DNA, proteins and lipids from oxidative damage pursuant to Article 13(1) of Regulation (EC) No 1924/20061". *EFSA Journal*. 8 (2): 1489. doi:10.2903/j.efsa.2010.1489.

[491] González, C. A., Sala, N., & Rokkas, T. (2013). Gastric Cancer: Epidemiologic Aspects. Helicobacter, 18, 34–38. doi:10.1111/hel.12082

[492] Woo HD, Kim J (2013). "Dietary flavonoid intake and smoking-related cancer risk: a meta-analysis". *PLoS ONE*. 8 (9): e75604.

[493] Siasos et al (2013). "Flavonoids in atherosclerosis: An overview of their mechanisms of action". *Current Medicinal Chemistry*. 20 (21): 2641–2660. doi:10.2174/0929867311320210003.

[494] Van Dam, R. M., Naidoo, N., & Landberg, R. (2013). Dietary flavonoids and the development of type 2 diabetes and cardiovascular diseases. Current Opinion in Lipidology, 24(1), 25–33. doi:10.1097/mol.0b013e32835bcdff

[495] Wang, X., Ouyang, Y. Y., Liu, J., & Zhao, G. (2013). Flavonoid intake and risk of CVD: a systematic review and meta-analysis of prospective cohort studies. British Journal of Nutrition, 111(1), 1–11. doi:10.1017/s000711451300278x

[496] Miksicek (1993) 'Commonly occurring plant flavonoids have estrogenic activity', Molecular Pharmacology July 1993, 44 (1) 37-43.

[497] Messina, M. (2016). Soy and Health Update: Evaluation of the Clinical and Epidemiologic Literature. Nutrients, 8(12), 754. doi:10.3390/nu8120754

[498] Bar-El Dadon, S., & Reifen, R. (2010). Soy as an Endocrine Disruptor: Cause for Caution? Journal of Pediatric Endocrinology and Metabolism, 23(9). doi:10.1515/jpem.2010.138

[499] Chavarro et al (2008). Soy food and isoflavone intake in relation to semen quality parameters among men from an infertility clinic. Human Reproduction, 23(11), 2584–2590. doi:10.1093/humrep/den243

[500] Habito, R. C., Montalto, J., Leslie, E., & Ball, M. J. (2000). Effects of replacing meat with soyabean in the diet on sex hormone concentrations in healthy adult males. British Journal of Nutrition, 84(4), 557–563. doi:10.1017/s0007114500001872

[501] Frémont L. (2000). Biological effects of resveratrol. Life sciences, 66(8), 663–673. https://doi.org/10.1016/s0024-3205(99)00410-5

[502] Smoliga, J. M., Baur, J. A., & Hausenblas, H. A. (2011). Resveratrol and health--a

comprehensive review of human clinical trials. Molecular nutrition & food research, 55(8), 1129–1141. https://doi.org/10.1002/mnfr.201100143

[503] Rege, S. D., Geetha, T., Griffin, G. D., Broderick, T. L., & Babu, J. R. (2014). Neuroprotective effects of resveratrol in Alzheimer disease pathology. Frontiers in aging neuroscience, 6, 218. https://doi.org/10.3389/fnagi.2014.00218

[504] Baur J. A. (2010). Resveratrol, sirtuins, and the promise of a DR mimetic. Mechanisms of ageing and development, 131(4), 261–269. https://doi.org/10.1016/j.mad.2010.02.007

[505] Howitz et al (2003). Small molecule activators of sirtuins extend Saccharomyces cerevisiae lifespan. Nature, 425(6954), 191–196. https://doi.org/10.1038/nature01960

[506] Xiong et al (2015). Activation of miR-34a/SIRT1/p53 signaling contributes to cochlear hair cell apoptosis: implications for age-related hearing loss. Neurobiology of aging, 36(4), 1692–1701. https://doi.org/10.1016/j.neurobiolaging.2014.12.034

[507] Campagna, M., & Rivas, C. (2010). Antiviral activity of resveratrol. Biochemical Society Transactions, 38(1), 50–53. doi:10.1042/bst0380050

[508] Timmers et al (2011). Caloric restriction-like effects of 30 days of resveratrol supplementation on energy metabolism and metabolic profile in obese humans. Cell metabolism, 14(5), 612–622. https://doi.org/10.1016/j.cmet.2011.10.002

[509] Xu et al (2014). Novel role of resveratrol: suppression of high-mobility group protein box 1 nucleocytoplasmic translocation by the upregulation of sirtuin 1 in sepsis-induced liver injury. Shock (Augusta, Ga.), 42(5), 440–447. https://doi.org/10.1097/SHK.0000000000000225

[510] dos Santos et al (2006). Resveratrol increases glutamate uptake and glutamine synthetase activity in C6 glioma cells. Archives of biochemistry and biophysics, 453(2), 161–167. https://doi.org/10.1016/j.abb.2006.06.025

[511] Soleas, G. J., Diamandis, E. P., & Goldberg, D. M. (1997). Resveratrol: A molecule whose time has come? And gone? Clinical Biochemistry, 30(2), 91–113. doi:10.1016/s0009-9120(96)00155-5

[512] Chiva-Blanch et al (2012). Dealcoholized red wine decreases systolic and diastolic blood pressure and increases plasma nitric oxide: short communication. Circulation research, 111(8), 1065–1068. https://doi.org/10.1161/CIRCRESAHA.112.275636

[513] Chen et al (2016). Resveratrol Attenuates Trimethylamine-N-Oxide (TMAO)-Induced Atherosclerosis by Regulating TMAO Synthesis and Bile Acid Metabolism via Remodeling of the Gut Microbiota. mBio, 7(2). doi:10.1128/mbio.02210-15

[514] Vang et al (2011). What is new for an old molecule? Systematic review and recommendations on the use of resveratrol. PloS one, 6(6), e19881. https://doi.org/10.1371/journal.pone.0019881

[515] Kjær et al (2017). No Beneficial Effects of Resveratrol on the Metabolic Syndrome: A Randomized Placebo-Controlled Clinical Trial. The Journal of Clinical Endocrinology & Metabolism, 102(5), 1642–1651. doi:10.1210/jc.2016-2160

[516] Kjaer et al (2015). Resveratrol reduces the levels of circulating androgen precursors but has no effect on, testosterone, dihydrotestosterone, PSA levels or prostate volume. A 4-month randomised trial in middle-aged men. The Prostate, 75(12), 1255–1263. doi:10.1002/pros.23006

[517] Heebøll et al (2016). Placebo-controlled, randomised clinical trial: high-dose resveratrol treatment for non-alcoholic fatty liver disease. Scandinavian Journal of Gastroenterology, 51(4), 456–464. doi:10.3109/00365521.2015.1107620

[518] Ahmad, A., Syed, F. A., Singh, S., & Hadi, S. M. (2005). Prooxidant activity of resveratrol in the presence of copper ions: Mutagenicity in plasmid DNA. Toxicology Letters, 159(1), 1–12. doi:10.1016/j.toxlet.2005.04.001

[519] Gadacha et al (2009). Resveratrol opposite effects on rat tissue lipoperoxidation: pro-oxidant during day-time and antioxidant at night. Redox Report, 14(4), 154–158. doi:10.1179/135100009x466131

[520] Cunningham, E. (2010). Are there Foods that Should Be Avoided if a Patient Is Sensitive to

Salicylates? Journal of the American Dietetic Association, 110(6), 976. doi:10.1016/j.jada.2010.04.020

521 Duthie, G. G., & Wood, A. D. (2011). Natural salicylates: foods, functions and disease prevention. Food & Function, 2(9), 515. doi:10.1039/c1fo10128e

522 Sharma, J. N., & Mohammed, L. A. (2006). The role of leukotrienes in the pathophysiology of inflammatory disorders: Is there a case for revisiting leukotrienes as therapeutic targets? Inflammopharmacology, 14(1-2), 10–16. doi:10.1007/s10787-006-1496-6

523 Baenkler (2008) 'Salicylate Intolerance, Pathophysiology, Clinical Spectrum, Diagnosis and Treatment', Dtsch Arztebl Int. 2008 Feb; 105(8): 137–142.

524 Sommer et al (2016). A novel treatment adjunct for aspirin exacerbated respiratory disease: the low-salicylate diet: a multicenter randomized control crossover trial. International Forum of Allergy & Rhinology, 6(4), 385–391. doi:10.1002/alr.21678

525 Noonan, S. C., & Savage, G. P. (1999). Oxalate content of foods and its effect on humans. Asia Pacific journal of clinical nutrition, 8(1), 64–74.

526 Holmes, R. P., Goodman, H. O., & Assimos, D. G. (2001). Contribution of dietary oxalate to urinary oxalate excretion. Kidney International, 59(1), 270–276. doi:10.1046/j.1523-1755.2001.00488.x

527 Heaney, R. P., & Weaver, C. M. (1989). Oxalate: effect on calcium absorbability. The American Journal of Clinical Nutrition, 50(4), 830–832. doi:10.1093/ajcn/50.4.830

528 Kelsay, J. L., & Prather, E. S. (1983). Mineral balances of human subjects consuming spinach in a low-fiber diet and in a diet containing fruits and vegetables. The American Journal of Clinical Nutrition, 38(1), 12–19. doi:10.1093/ajcn/38.1.12

529 Luck et al (2013). Human serum transferrin: is there a link among autism, high oxalate levels, and iron deficiency anemia?. Biochemistry, 52(46), 8333–8341. https://doi.org/10.1021/bi401190m

530 Makkapati, S., D'Agati, V. D., & Balsam, L. (2018). "Green Smoothie Cleanse" Causing Acute Oxalate Nephropathy. American Journal of Kidney Diseases, 71(2), 281–286. doi:10.1053/j.ajkd.2017.08.002

531 Worcester, E. M., & Coe, F. L. (2010). Calcium Kidney Stones. New England Journal of Medicine, 363(10), 954–963. doi:10.1056/nejmcp1001011

532 DEATH FROM RHUBARB LEAVES DUE TO OXALIC ACID POISONING. (1919). Journal of the American Medical Association, 73(8), 627. doi:10.1001/jama.1919.02610340059028

533 FARRE, M. (1989). FATAL OXALIC ACID POISONING FROM SORREL SOUP. The Lancet, 334(8678-8679), 1524. doi:10.1016/s0140-6736(89)92967-x

534 Gul, Z., & Monga, M. (2014). Medical and Dietary Therapy for Kidney Stone Prevention. Korean Journal of Urology, 55(12), 775. doi:10.4111/kju.2014.55.12.775

535 Harper and Mansell (1991) 'Treatment of enteric hyperoxaluria', Postgrad Med J (1991) 67, 219 - 222.

536 Siener, R., Ebert, D., Nicolay, C., & Hesse, A. (2003). Dietary risk factors for hyperoxaluria in calcium oxalate stone formers. Kidney International, 63(3), 1037–1043. doi:10.1046/j.1523-1755.2003.00807.x

537 Kaufman et al (2008). Oxalobacter formigenesMay Reduce the Risk of Calcium Oxalate Kidney Stones. Journal of the American Society of Nephrology, 19(6), 1197–1203. doi:10.1681/asn.2007101058

538 Mittal, R. D., Kumar, R., Bid, H. K., & Mittal, B. (2005). Effect of Antibiotics on Oxalobacter formigenes Colonization of Human Gastrointestinal Tract. Journal of Endourology, 19(1), 102–106. doi:10.1089/end.2005.19.102

539 Van Damme et al (1998) 'Handbook of plant lectins: properties and biomedical applications'. London: Wiley; 1998. pp. 31–50.

[540] PUSZTAI, A., GREER, F., & GRANT, G. (1989). Specific uptake of dietary lectins into the systemic circulation of rats. Biochemical Society Transactions, 17(3), 481–482. doi:10.1042/bst0170481

[541] Wang et al (1998). Identification of intact peanut lectin in peripheral venous blood. The Lancet, 352(9143), 1831–1832. doi:10.1016/s0140-6736(05)79894-9

[542] Goldstein IJ, Poretz RD. Isolation and chemical properties of lectins. In: Liener IE, Sharon N, Goldstein IJ, editors. The lectins. Orlando: Academic Press; 1986.

[543] Rodhouse, J. C., Haugh, C. A., Roberts, D., & Gilbert, R. J. (1990). Red kidney bean poisoning in the UK: an analysis of 50 suspected incidents between 1976 and 1989. Epidemiology and Infection, 105(3), 485–491. doi:10.1017/s095026880004810x

[544] McGee, Harold (2004). *On food and cooking: the science and lore of the kitchen.* New York: Scribner. p. 714.

[545] Reed, Jess D. (1 May 1995). "Nutritional toxicology of tannins and related polyphenols in forage legumes". *Journal of Animal Science.* 73 (5): 1516–1528. doi:10.2527/1995.7351516x.

[546] Griffiths, D. W. (1986). The Inhibition of Digestive Enzymes by Polyphenolic Compounds. Nutritional and Toxicological Significance of Enzyme Inhibitors in Foods, 509–516. doi:10.1007/978-1-4757-0022-0_29

[547] Sarwar Gilani, G., Wu Xiao, C., & Cockell, K. A. (2012). Impact of Antinutritional Factors in Food Proteins on the Digestibility of Protein and the Bioavailability of Amino Acids and on Protein Quality. British Journal of Nutrition, 108(S2), S315–S332. doi:10.1017/s0007114512002371

[548] Griffiths et al (2016). Food Antioxidants and Their Anti-Inflammatory Properties: A Potential Role in Cardiovascular Diseases and Cancer Prevention. Diseases, 4(4), 28. doi:10.3390/diseases4030028

[549] Peluso et al (2018). Effects of High Consumption of Vegetables on Clinical, Immunological, and Antioxidant Markers in Subjects at Risk of Cardiovascular Diseases. Oxidative Medicine and Cellular Longevity, 2018, 1–9. doi:10.1155/2018/5417165

[550] Duthie et al (2017). Effect of increasing fruit and vegetable intake by dietary intervention on nutritional biomarkers and attitudes to dietary change: a randomised trial. European Journal of Nutrition, 57(5), 1855–1872. doi:10.1007/s00394-017-1469-0

[551] Loft, S., & Poulsen, H. E. (1996). Cancer risk and oxidative DNA damage in man. Journal of Molecular Medicine, 74(6), 297–312. doi:10.1007/bf00207507

[552] Vivekananthan, D. P., Penn, M. S., Sapp, S. K., Hsu, A., & Topol, E. J. (2003). Use of antioxidant vitamins for the prevention of cardiovascular disease: meta-analysis of randomised trials. The Lancet, 361(9374), 2017–2023. doi:10.1016/s0140-6736(03)13637-9

[553] Howarth, N. C., Saltzman, E., & Roberts, S. B. (2001). Dietary fiber and weight regulation. Nutrition reviews, 59(5), 129–139. https://doi.org/10.1111/j.1753-4887.2001.tb07001.x

[554] Duncan, K. H., Bacon, J. A., & Weinsier, R. L. (1983). The effects of high and low energy density diets on satiety, energy intake, and eating time of obese and nonobese subjects. The American journal of clinical nutrition, 37(5), 763–767. https://doi.org/10.1093/ajcn/37.5.763 .

[555] Gostic, K. et al. (2019). Childhood immune imprinting to influenza A shapes birth year-specific risk during seasonal H1N1 and H3N2 epidemics. *PLOS Pathogens* 15 (12): e1008109.

[556] Mina, M. et al. (2019). Measles virus infection diminishes preexisting antibodies that offer protection from other pathogens. *Science* 366 (6465): 599–606.

[557] Dhabhar, F. S., Malarkey, W. B., Neri, E., & McEwen, B. S. (2012). Stress-induced redistribution of immune cells--from barracks to boulevards to battlefields: a tale of three hormones--Curt Richter Award winner. Psychoneuroendocrinology, 37(9), 1345–1368. https://doi.org/10.1016/j.psyneuen.2012.05.008

[558] Rohleder N. (2012). Acute and chronic stress induced changes in sensitivity of peripheral inflammatory pathways to the signals of multiple stress systems --2011 Curt Richter Award Winner.

Psychoneuroendocrinology, 37(3), 307–316. https://doi.org/10.1016/j.psyneuen.2011.12.015

[559] Steptoe, A., Hamer, M., & Chida, Y. (2007). The effects of acute psychological stress on circulating inflammatory factors in humans: a review and meta-analysis. Brain, behavior, and immunity, 21(7), 901–912. https://doi.org/10.1016/j.bbi.2007.03.011

[560] Segerstrom, S. C., & Miller, G. E. (2004). Psychological stress and the human immune system: a meta-analytic study of 30 years of inquiry. Psychological bulletin, 130(4), 601–630. https://doi.org/10.1037/0033-2909.130.4.601

[561] Gouin, J. P., Glaser, R., Malarkey, W. B., Beversdorf, D., & Kiecolt-Glaser, J. (2012). Chronic stress, daily stressors, and circulating inflammatory markers. Health psychology : official journal of the Division of Health Psychology, American Psychological Association, 31(2), 264–268. https://doi.org/10.1037/a0025536

[562] Ershler W. B. (1993). Interleukin-6: a cytokine for gerontologists. Journal of the American Geriatrics Society, 41(2), 176–181. https://doi.org/10.1111/j.1532-5415.1993.tb02054.x

[563] de Brouwer et al (2013). Immune responses to stress after stress management training in patients with rheumatoid arthritis. Arthritis research & therapy, 15(6), R200. https://doi.org/10.1186/ar4390

[564] Karagkouni, A., Alevizos, M., & Theoharides, T. C. (2013). Effect of stress on brain inflammation and multiple sclerosis. Autoimmunity reviews, 12(10), 947–953. https://doi.org/10.1016/j.autrev.2013.02.006

[565] Kennedy, P. J., Cryan, J. F., Quigley, E. M., Dinan, T. G., & Clarke, G. (2014). A sustained hypothalamic-pituitary-adrenal axis response to acute psychosocial stress in irritable bowel syndrome. Psychological medicine, 44(14), 3123–3134. https://doi.org/10.1017/S003329171400052X

[566] Pawelec, G., Akbar, A., Caruso, C., Solana, R., Grubeck-Loebenstein, B., & Wikby, A. (2005). Human immunosenescence: is it infectious?. Immunological reviews, 205, 257–268. https://doi.org/10.1111/j.0105-2896.2005.00271.x

[567] Pall, M. & Levine, S. (2015). Nrf2, a master regulator of detoxification and also antioxidant, anti-inflammatory and other cytoprotective mechanisms, is raised by health promoting factors. *Sheng Li Xue Bao* 67 (1): 1–18.

[568] Antunes et al (2016) 'Arterial thickness and immunometabolism: the mediating role of chronic exercise', Curr Cardiol Rev, 12 (2016), pp. 47-51.

[569] Estaki et al (2016). Cardiorespiratory fitness as a predictor of intestinal microbial diversity and distinct metagenomic functions. Microbiome, 4(1). doi:10.1186/s40168-016-0189-7

[570] Nieman (1999) 'Exercise, Infection, and Immunity: Practical Applications', In Military Strategies for Sustainment of Nutrition and Immune Function in the Field. Washington (DC): National Academies Press (US); 1999. 17, Exercise, Infection, and Immunity: Practical Applications. Available from: https://www.ncbi.nlm.nih.gov/books/NBK230961/

[571] Millet, G. Y., Martin, V., & Temesi, J. (2018). The role of the nervous system in neuromuscular fatigue induced by ultra-endurance exercise. Applied Physiology, Nutrition, and Metabolism, 43(11), 1151–1157. doi:10.1139/apnm-2018-0161

[572] Shinkai et al (1997) 'Aging, exercise, training, and the immune system', Exerc Immunol Rev, 3 (1997), pp. 68-95.

[573] Grande, A. J., Reid, H., Thomas, E. E., Nunan, D., & Foster, C. (2016). Exercise prior to influenza vaccination for limiting influenza incidence and its related complications in adults. Cochrane Database of Systematic Reviews. doi:10.1002/14651858.cd011857.pub2

[574] Simpson, R. J., Kunz, H., Agha, N., & Graff, R. (2015). Exercise and the Regulation of Immune Functions. Molecular and Cellular Regulation of Adaptation to Exercise, 355–380. doi:10.1016/bs.pmbts.2015.08.001

[575] NIEMAN, D. C., HENSON, D. A., AUSTIN, M. D., & BROWN, V. A. (2005). Immune Response to a 30-Minute Walk. Medicine & Science in Sports & Exercise, 37(1), 57–62. doi:10.1249/01.mss.0000149808.38194.21

[576] Karstoft et al (2016). 'Exercise and type 2 diabetes: focus on metabolism and inflammation', Immunology & Cell Biology, 94(2). doi:10.1111/imcb.2016.94.issue-2

[577] Pedersen, B. K. (2017). Anti-inflammatory effects of exercise: role in diabetes and cardiovascular disease. European Journal of Clinical Investigation, 47(8), 600–611. doi:10.1111/eci.12781

[578] Mackinnon, L. T., Chick, T. W., van As, A., & Tomasi, T. B. (1987). The Effect of Exercise on Secretory and Natural Immunity. Advances in Experimental Medicine and Biology, 869–876. doi:10.1007/978-1-4684-5344-7_102

[579] Cardillo et al (2017). Synthetic Lethality Exploitation by an Anti–Trop-2-SN-38 Antibody–Drug Conjugate, IMMU-132, Plus PARP Inhibitors inBRCA1/2–wild-type Triple-Negative Breast Cancer. Clinical Cancer Research, 23(13), 3405–3415. doi:10.1158/1078-0432.ccr-16-2401

[580] Pedersen et al (1988), 'Modulation of Natural Killer Cell Activity in Peripheral Blood by Physical Exercise', Scandinavian Journal of Immunology, 27(6). doi:10.1111/sji.1988.27.issue-6

[581] Tvede et al (1989) 'Effect of Physical Exercise on Blood Mononuclear Cell Subpopulations and in Vitro Proliferative Responses', Scandinavian Journal of Immunology, Volume 29, Issue 3, March 1989, Pages 383-389.

[582] Nieman et al (2018). Metabolic recovery from heavy exertion following banana compared to sugar beverage or water only ingestion: A randomized, crossover trial. PLOS ONE, 13(3), e0194843. doi:10.1371/journal.pone.0194843

[583] Ahmed et al (2014) The protective effects of a polyphenol-enriched protein powder on exercise-induced susceptibility to virus infection. Phytother. Res. 28: 1829–36

[584] Barrett, B. et al. (2012). Meditation or exercise for preventing acute respiratory infection: a randomized controlled trial. *Annals of Family Medicine* 10 (4): 337–346.

[585] Weidner, T. & Cranston, T. & Schurr, T. & Kaminsky, L. (1998). The effect of exercise training on the severity and duration of a viral upper respiratory illness. *Medicine & Science in Sports & Exercise* 30 (11): 1578–1583.

[586] Cannon, J., & Kluger, M. (1983). Endogenous pyrogen activity in human plasma after exercise. Science, 220(4597), 617–619. doi:10.1126/science.6836306

[587] Fitzgerald, L. (1991). Overtraining increases the susceptibility to infection. *International Journal of Sports Medicine* 12 (Suppl 1): S5–S8.

[588] Sears, M. E., Kerr, K. J., & Bray, R. I. (2012). Arsenic, Cadmium, Lead, and Mercury in Sweat: A Systematic Review. Journal of Environmental and Public Health, 2012, 1–10. doi:10.1155/2012/184745

[589] Hussain, J., & Cohen, M. (2018). Clinical Effects of Regular Dry Sauna Bathing: A Systematic Review. *Evidence-Based Complementary and Alternative Medicine* 2018: 1857413.

[590] Kunutsor, S. & Laukkanen, T. & Laukkanen, J. (2017). Sauna bathing reduces the risk of respiratory diseases: a long-term prospective cohort study. *European Journal of Epidemiology* 32 (12): 1107–1111.

[591] Xue, J. et al. (2016). Short-Term Heat Shock Affects Host-Virus Interaction in Mice Infected with Highly Pathogenic Avian Influenza Virus H5N1. *Frontiers in Microbiology* 7: 924.

[592] Hirayama, E., Atagi, H., Hiraki, A., & Kim, J. (2004). Heat shock protein 70 is related to thermal inhibition of nuclear export of the influenza virus ribonucleoprotein complex. Journal of virology, 78(3), 1263–1270. https://doi.org/10.1128/jvi.78.3.1263-1270.2004

[593] Conti, C., De Marco, A., Mastromarino, P., Tomao, P., & Santoro, M. G. (1999). Antiviral effect of hyperthermic treatment in rhinovirus infection. Antimicrobial agents and chemotherapy, 43(4), 822–829.

[594] Pujhari et al (2019). Heat shock protein 70 (Hsp70) mediates Zika virus entry, replication, and egress from host cells. Emerging microbes & infections, 8(1), 8–16. https://doi.org/10.1080/22221751.2018.1557988

[595] Ryan, M., & Levy, M. M. (2003). Clinical review: Fever in intensive care unit patients. Crit Care 7, 221. doi:10.1186/cc1879.

[596] Castellani J. & M Brenner, I. & Rhind, S. (2002). Cold exposure: human immune responses and intracellular cytokine expression. *Medicine & Science in Sports & Exercise* 34 (12): 2013–2020. Review.

[597] Janský, L. et al. (1996). Immune system of cold-exposed and cold-adapted humans. *European Journal of Applied Physiology and Occupational Physiology* 72 (5-6): 445–450.

[598] Siems, W. G., van Kuijk, F. J., Maass, R., & Brenke, R. (1994). Uric acid and glutathione levels during short-term whole body cold exposure. Free radical biology & medicine, 16(3), 299–305. https://doi.org/10.1016/0891-5849(94)90030-2

[599] Anton, S. et al. (2018). Flipping the Metabolic Switch: Understanding and Applying the Health Benefits of Fasting. *Obesity (Silver Spring, Md.)* 26 (2): 254–268.

[600] Wang, A. et al. (2016). Opposing effects of fasting metabolism on tissue tolerance in bacterial and viral inflammation. *Cell* 166 (6): 1512–1525.

[601] Cheng et al. (2014). Prolonged Fasting Reduces IGF-1/PKA to Promote Hematopoietic-Stem-Cell-Based Regeneration and Reverse Immunosuppression. Cell Stem Cell, 14(6), 810–823. doi:10.1016/j.stem.2014.04.014

[602] Fontana, L., Partridge, L., & Longo, V. D. (2010). Extending Healthy Life Span--From Yeast to Humans. Science, 328(5976), 321–326. doi:10.1126/science.1172539

[603] Milder, J. & Patel, M. (2012). Modulation of oxidative stress and mitochondrial function by the ketogenic diet. *Epilepsy Research* 100 (3): 295–303.

[604] Brownlow, M. & Jung, S. & Moore, R. & Bechmann, N. & Jankord, R. (2017). Nutritional Ketosis Affects Metabolism and Behavior in Sprague-Dawley Rats in Both Control and Chronic Stress Environments. *Frontiers in Molecular Neuroscience* 10: 129.

[605] Youm et al (2015). The ketone metabolite β-hydroxybutyrate blocks NLRP3 inflammasome-mediated inflammatory disease. Nature medicine, 21(3), 263–269. https://doi.org/10.1038/nm.3804

[606] Tate, M. D., & Mansell, A. (2018). An update on the NLRP3 inflammasome and influenza: the road to redemption or perdition?. Current opinion in immunology, 54, 80–85. https://doi.org/10.1016/j.coi.2018.06.005

[607] Masino, S. A., & Ruskin, D. N. (2013). Ketogenic Diets and Pain. Journal of Child Neurology, 28(8), 993–1001. doi:10.1177/0883073813487595

[608] Moser (2011) 'The Role of AMPK in Viral Infection', Publicly Accessible Penn Dissertations. 975. https://repository.upenn.edu/edissertations/975

[609] Xie et al (2015). Activation of AMPK restricts coxsackievirus B3 replication by inhibiting lipid accumulation. Journal of molecular and cellular cardiology, 85, 155–167. https://doi.org/10.1016/j.yjmcc.2015.05.021

[610] Mankouri et al (2010). Enhanced hepatitis C virus genome replication and lipid accumulation mediated by inhibition of AMP-activated protein kinase. Proceedings of the National Academy of Sciences, 107(25), 11549–11554. doi:10.1073/pnas.0912426107

[611] Wang et al (2020). AMPK and Akt/mTOR signalling pathways participate in glucose-mediated regulation of hepatitis B virus replication and cellular autophagy. Cellular microbiology, 22(2), e13131. https://doi.org/10.1111/cmi.13131

[612] Geerlings, S. E., & Hoepelman, A. I. . (1999). Immune dysfunction in patients with diabetes mellitus (DM). FEMS Immunology & Medical Microbiology, 26(3-4), 259–265. doi:10.1111/j.1574-695x.1999.tb01397.x

[613] Ringsdorf, W. Jr. & Cheraskin, E. & Ramsay, R. Jr. (1976). Sucrose, neutrophilic phagocytosis and resistance to disease. *Dental Survey* 52 (12): 46–48.

[614] Sanchez, A. et al. (1973). Role of sugars in human neutrophilic phagocytosis. *The American Journal of Clinical Nutrition* 26 (11): 1180–1184.

[615] Goldberg, E. et al. (2019). Ketogenic diet activates protective γδ T cell responses against influenza virus infection. *Science Immunology* 4 (41): eaav2026.

[616] Katona, P. & Katona-Apte, J. (2008). The Interaction between Nutrition and Infection. Clinical Infectious Diseases 46 (10): 1582–1588.

[617] Bailey, R. L., West, K. P., Jr, & Black, R. E. (2015). The epidemiology of global micronutrient deficiencies. Annals of nutrition & metabolism, 66 Suppl 2, 22–33. https://doi.org/10.1159/000371618

[618] Nair, R. & Maseeh, A. (2012). Vitamin D: The "sunshine" vitamin. *Journal of Pharmacology & Pharmacotherapeutics* 3 (2): 118–126.

[619] Scheiermann, C. & Kunisaki, Y. & Frenette, P. (2013). Circadian control of the immune system. Nature reviews. *Immunology* 13 (3): 190–198.

[620] Labrecque, N. & Cermakian, N. (2015). Circadian Clocks in the Immune System. *Journal of Biological Rhythms* 30 (4): 277–290. Review.

[621] Segerstrom, S. & Miller, G. (2004). Psychological stress and the human immune system: a meta-analytic study of 30 years of inquiry. *Psychological Bulletin* 130 (4): 601–630.

[622] Dhabhar, F. (2014). Effects of stress on immune function: the good, the bad, and the beautiful. *Immunologic Research* 58 (2-3): 193–210.

[623] Baričević, I. & Nedić, O. & Anna Nikolić, J. & Nedeljković, J. (2004). The insulin-like growth factor system in the circulation of patients with viral infections. *Clinical Chemistry and Laboratory Medicine (CCLM)* 42 (10): 1127–1131.

[624] Segerstrom S. C. (2005). Optimism and immunity: do positive thoughts always lead to positive effects?. Brain, behavior, and immunity, 19(3), 195–200. https://doi.org/10.1016/j.bbi.2004.08.003

[625] Nes, L. S., & Segerstrom, S. C. (2006). Dispositional optimism and coping: a meta-analytic review. Personality and social psychology review : an official journal of the Society for Personality and Social Psychology, Inc, 10(3), 235–251. https://doi.org/10.1207/s15327957pspr1003_3

[626] Reiter, R. & Tan, D. & Fuentes-Broto, L. (2010). Melatonin: a multitasking molecule. *Progress in Brain Research* 181: 127–151.

[627] Boga, J. et al. (2019). Therapeutic potential of melatonin related to its role as an autophagy regulator: A review. *Journal of Pineal Research* 66 (1): e12534.

[628] Klein (1993) 'Stress and infections', J Fla Med Assoc. 1993 Jun;80(6):409-11.

[629] Ibarra-Coronado et al (2015) 'The Bidirectional Relationship between Sleep and Immunity against Infections', Journal of Immunology Research, Volume 2015 |Article ID 678164 | 14 pages | https://doi.org/10.1155/2015/678164

[630] Cohen et al (2009). Sleep Habits and Susceptibility to the Common Cold. Archives of Internal Medicine, 169(1), 62. doi:10.1001/archinternmed.2008.505

[631] Irwin (1994). Partial sleep deprivation reduces natural killer cell activity in humans. Psychosomatic Medicine, 56(6), 493–498. doi:10.1097/00006842-199411000-00004

[632] Prather et al (2012) 'Sleep and Antibody Response to Hepatitis B Vaccination', Sleep. 2012 Aug 1; 35(8): 1063–1069.

[633] Lange, T., Perras, B., Fehm, H. L., & Born, J. (2003). Sleep Enhances the Human Antibody Response to Hepatitis A Vaccination. Psychosomatic Medicine, 65(5), 831–835. doi:10.1097/01.psy.0000091382.61178.f1

[634] Besedovsky, L., Lange, T., & Born, J. (2011). Sleep and immune function. Pflügers Archiv - European Journal of Physiology, 463(1), 121–137. doi:10.1007/s00424-011-1044-0

[635] Imeri, L., & Opp, M. R. (2009). How (and why) the immune system makes us sleep. Nature Reviews Neuroscience, 10(3), 199–210. doi:10.1038/nrn2576

[636] Carrillo-Vico (2013). Melatonin: buffering the immune system. *International Journal of Molecular Sciences* 14 (4): 8638–8683.

[637] Asif, N. & Iqbal, R. & Nazir, C. (2017). Human immune system during sleep. *American Journal of Clinical and Experimental Immunology* 6 (6): 92–96.

[638] Srinivasan, V. et al. (2005). Melatonin, immune function and aging. *Immunity & Ageing* 2: 17.

[639] Katona, P. & Katona-Apte, J. (2008). The Interaction between Nutrition and Infection. *Clinical Infectious Diseases* 46 (10): 1582–1588.

[640] Bailey, R. & West Jr, K. & Black, R. (2015). The epidemiology of global micronutrient deficiencies. *Annals of Nutrition and Metabolism* 66 (Suppl. 2): 22–33.

[641] Zhou, W. & Zuo, X. & Li, J. & Yu, Z. (2016). Effects of nutrition intervention on the nutritional status and outcomes of pediatric patients with pneumonia. *Minerva Pediatrica* 68 (1): 5-10.

[642] Aranow, C. (2011). Vitamin D and the immune system. *Journal of Investigative Medicine* 59 (6): 881–886.

[643] Illescas-Montes, R., Melguizo-Rodríguez, L., Ruiz, C., & Costela-Ruiz, V. J. (2019). Vitamin D and autoimmune diseases. Life sciences, 233, 116744. https://doi.org/10.1016/j.lfs.2019.116744

[644] Ginde, A. & Mansbach, J. & Camargo, C. Jr. (2009). Association between serum 25-hydroxyvitamin D level and upper respiratory tract infection in the Third National Health and Nutrition Examination Survey. Archives of Internal Medicine 169 (4): 384–390.

[645] Urashima, M. et al. (2010). Randomized trial of vitamin D supplementation to prevent seasonal influenza A in schoolchildren. *The American Journal of Clinical Nutrition* 91 (5): 1255–1260.

[646] Bergman, P. & Lindh, Å. & Björkhem-Bergman, L. & Lindh, J. (2013). Vitamin D and respiratory tract infections: a systematic review and meta-analysis of randomized controlled trials. *PloS one* 8 (6): e65835.

[647] Martineau, A. et al. (2017). Vitamin D supplementation to prevent acute respiratory tract infections: systematic review and meta-analysis of individual participant data. *BMJ* 356: i6583.

[648] Hathcock, J. N., Shao, A., Vieth, R., & Heaney, R. (2007). Risk assessment for vitamin D. The American Journal of Clinical Nutrition, 85(1), 6–18. doi:10.1093/ajcn/85.1.6

[649] Ginde, A. A., Mansbach, J. M., & Camargo, C. A., Jr (2009). Association between serum 25-hydroxyvitamin D level and upper respiratory tract infection in the Third National Health and Nutrition Examination Survey. Archives of internal medicine, 169(4), 384–390. https://doi.org/10.1001/archinternmed.2008.560

[650] Martineau et al (2017). Vitamin D supplementation to prevent acute respiratory tract infections: systematic review and meta-analysis of individual participant data. BMJ (Clinical research ed.), 356, i6583. https://doi.org/10.1136/bmj.i6583

[651] Camargo et al (2019). Effect of Monthly High-Dose Vitamin D Supplementation on Acute Respiratory Infections in Older Adults: A Randomized Controlled Trial. Clinical Infectious Diseases. doi:10.1093/cid/ciz801

[652] Raharusun, P., Priambada, S., Budiarti, C., Agung, E., & Budi, C. (2020). Patterns of COVID-19 Mortality and Vitamin D: An Indonesian Study. SSRN Electronic Journal. doi:10.2139/ssrn.3585561

[653] Alipio, M. (2020). Vitamin D Supplementation Could Possibly Improve Clinical Outcomes of Patients Infected with Coronavirus-2019 (COVID-2019). SSRN Electronic Journal. doi:10.2139/ssrn.3571484

[654] Liu, X., Baylin, A., & Levy, P. D. (2018). Vitamin D deficiency and insufficiency among US adults: prevalence, predictors and clinical implications. The British journal of nutrition, 119(8), 928–936. https://doi.org/10.1017/S0007114518000491

[655] Watkins, J. (2020). Preventing a covid-19 pandemic. BMJ, m810. doi:10.1136/bmj.m810

[656] Pérez-Barrios et al (2016). Prevalence of hypercalcemia related to hypervitaminosis D in clinical practice. Clinical Nutrition, 35(6), 1354–1358. doi:10.1016/j.clnu.2016.02.017

[657] Shea, M. K., & Holden, R. M. (2012). Vitamin K Status and Vascular Calcification: Evidence

from Observational and Clinical Studies. Advances in Nutrition, 3(2), 158–165. doi:10.3945/an.111.001644

[658] Shea et al (2009). Vitamin K supplementation and progression of coronary artery calcium in older men and women. The American Journal of Clinical Nutrition, 89(6), 1799–1807. doi:10.3945/ajcn.2008.27338

[659] Tinggi, U. (2008). Selenium: its role as antioxidant in human health. *Environmental Health and Preventive Medicine* 13 (2): 102–108.

[660] Schwarz, K. (1996). Oxidative stress during viral infection: A review. *Free Radical Biology and Medicine* 21 (5): 641–649.

[661] Guillin, O. & Vindry, C. & Ohlmann, T. & Chavatte, L. (2019). Selenium, selenoproteins and viral infection. Nutrients 11 (9): 2101.

[662] Moya, M. et al. (2013). Potentially-toxic and essential elements profile of AH1N1 patients in Mexico City. *Scientific Reports* 3: 1284.

[663] Sauve, A. (2008). NAD+ and vitamin B3: from metabolism to therapies. *Journal of Pharmacology and Experimental Therapeutics* 324 (3): 883-893.

[664] Murray, M. (2003). Nicotinamide: an oral antimicrobial agent with activity against both Mycobacterium tuberculosis and human immunodeficiency virus. *Clinical Infectious Diseases* 36 (4): 453-460.

[665] Drouin, G. & Godin, J.-R. & Page, B. (2011). The Genetics of Vitamin C Loss in Vertebrates. *Current Genomics* 12 (5): 371–378.

[666] Lenton, K. et al. (2003). Vitamin C augments lymphocyte glutathione in subjects with ascorbate deficiency. *The American Journal of Clinical Nutrition* 77 (1): 189–195.

[667] Hemilä, H., & Chalker, E. (2013). Vitamin C for preventing and treating the common cold. *Cochrane Database of Systematic Reviews* 31 (1): CD000980.

[668] Gorton, H. & Jarvis, K. (1999). The effectiveness of vitamin C in preventing and relieving the symptoms of virus-induced respiratory infections. *Journal of Manipulative and Physiological Therapeutics* 22 (8): 530–533.

[669] Higdon, J. et al. (2018). *Vitamin C.* Linus Pauling Micronutrient Information Center. <https://lpi.oregonstate.edu/mic/vitamins/vitamin-C#common-cold-treatment> [date of reference: 17.02.2020]

[670] Cai, Y. et al. (2015). A new mechanism of vitamin C effects on A/FM/1/47 (H1N1) virus-induced pneumonia in restraint-stressed mice. *BioMed Research International* 2015: 675149.

[671] Hemilä, H. & Chalker, E. (2020). Vitamin C may reduce the duration of mechanical ventilation in critically ill patients: a meta-regression analysis. *Journal of Intensive Care* 8 (1): 15.

[672] Ernster, L. & Dallner, G. (1995). Biochemical, physiological and medical aspects of ubiquinone function. *Biochimica et Biophysica Acta (BBA)-Molecular Basis of Disease* 1271 (1): 195–204.

[673] Lee, B.-J. & Huang, Y.-C. & Chen, S.-J. & Lin, P.-T. (2012). Coenzyme Q10 supplementation reduces oxidative stress and increases antioxidant enzyme activity in patients with coronary artery disease. *Nutrition* 28 (3): 250–255.

[674] McCall, K. & Huang, C. & Fierke, C. (2000). Function and mechanism of zinc metalloenzymes. *The Journal of Nutrition* 130 (5): 1437S–1446S.

[675] Berg, J. (1990). Zinc fingers and other metal-binding domains. *Receptor* 29: 31.

[676] Barnett, J. & Hamer, D. & Meydani, S. (2010). Low zinc status: a new risk factor for pneumonia in the elderly? *Nutrition Reviews* 68 (1): 30–37.

[677] Plum, L. & Rink, L. & Haase, H. (2010). The essential toxin: impact of zinc on human health. *International Journal of Environmental Research and Public Health* 7 (4): 1342–1365.

[678] Hemilä, H. & Chalker, E. (2017). Zinc for preventing and treating the common cold. *Cochrane Database of Systematic* Reviews 2017 (9): CD012808.

[679] Allan, G. & Arroll, B. (2014). Prevention and treatment of the common cold: making sense of the evidence. *CMAJ* 186 (3): 190–199.

[680] Te Velthuis et al (2010). Zn2+ Inhibits Coronavirus and Arterivirus RNA Polymerase Activity In Vitro and Zinc Ionophores Block the Replication of These Viruses in Cell Culture. PLoS Pathogens, 6(11), e1001176. doi:10.1371/journal.ppat.1001176

[681] Reiss, C. & Komatsu, T. (1998). Does nitric oxide play a critical role in viral infections? *Journal of Virology* 72 (6): 4547-4551.

[682] Åkerström, S. et al. (2005). Nitric oxide inhibits the replication cycle of severe acute respiratory syndrome coronavirus. *Journal of Virology* 79 (3): 1966-1969.

[683] Sanchez, M. & Ochoa, A. & Foster, T. (2016). Development and evaluation of a host-targeted antiviral that abrogates herpes simplex virus replication through modulation of arginine-associated metabolic pathways. *Antiviral Research* 132: 13–25.

[684] Ng, W, & Tate, M. & Brooks, A. & Reading, P. (2012). Soluble host defense lectins in innate immunity to influenza virus. *Journal of Biomedicine & Biotechnology* 2012: 732191.

[685] Casals et al (2018). The Role of Collectins and Galectins in Lung Innate Immune Defense. *Frontiers in Immunology* 9: 1998.

[686] Holmskov, U. (2000). Collectins and collectin receptors in innate immunity. *APMIS Supplementum* 100: 1–59.

[687] Traboulsi, H. et al. (2015). The Flavonoid Isoliquiritigenin Reduces Lung Inflammation and Mouse Morbidity during Influenza Virus Infection. *Antimicrobial Agents and Chemotherapy* 59 (10): 6317–6327.

[688] Cinatl, J. et al. (2003). Glycyrrhizin, an active component of liquorice roots, and replication of SARS-associated coronavirus. *The Lancet* 361 (9374): 2045–2046.

[689] Roxas, M. & Jurenka, J. (2007). Colds and influenza: a review of diagnosis and conventional, botanical, and nutritional considerations. *Alternative Medicine Reviews* 12 (1): 25–48.

[690] Berlutti, F. et al. (2011). Antiviral properties of lactoferrin--a natural immunity molecule. *Molecules (Basel, Switzerland)* 16 (8): 6992–7018.

[691] Wakabayashi, H. et al. (2014). Lactoferrin for prevention of common viral infections. *Journal of Infection and Chemotherapy* 20 (11): 666–671.

[692] Scala, M. et al. (2017). Lactoferrin-derived Peptides Active towards Influenza: Identification of Three Potent Tetrapeptide Inhibitors. *Scientific Reports* 7 (1): 10593.

[693] Tasala, T. et al. (2018). Concentration-dependent Activation of Inflammatory/Anti-inflammatory Functions of Macrophages by Hydrolyzed Whey Protein. *Anticancer Research* 38 (7): 4299–4304.

[694] Cruzat et al (2018). Glutamine: Metabolism and Immune Function Supplementation and Clinical Translation. *Nutrients* 10 (11): 1564.

[695] Calder, P. & Yaqoob, P. (1999). Glutamine and the immune system. Amino Acids 17 (3): 227–241. Review.

[696] Chang, W-K. & Yang, K. & Shaio, M.-F. (1999). Effect of Glutamine on Th1 and Th2 Cytokine Responses of Human Peripheral Blood Mononuclear Cells. *Clinical Immunology* 93 (3): 294–301.

[697] Zhou, Q. et al. (2019). Randomised placebo-controlled trial of dietary glutamine supplements for postinfectious irritable bowel syndrome. *Gut* 68 (6): 996–1002.

[698] Rao, R., & Samak, G. (2012). Role of Glutamine in Protection of Intestinal Epithelial Tight Junctions. *Journal of Epithelial Biology & Pharmacology* 5 (Suppl 1-M7): 47–54.

[699] Ozdal, T. et al. (2016). The Reciprocal Interactions between Polyphenols and Gut Microbiota and Effects on Bioaccessibility. *Nutrients* 8 (2): 78.

[700] Zakay-Rones, Z. & Thom, E. & Wollan, T. & Wadstein, J. (2004). Randomized Study of the Efficacy and Safety of Oral Elderberry Extract in the Treatment of Influenza A and B Virus Infections. *Journal of International Medical Research* 32 (2): 132–140.

[701] Ulbricht, C. et al. (2014). An Evidence-Based Systematic Review of Elderberry and Elderflower (Sambucus nigra) by the Natural Standard Research Collaboration. *Journal of Dietary Supplements* 11 (1): 80–120.

[702] Hosseini, B. et al. (2018). Effects of fruit and vegetable consumption on inflammatory biomarkers and immune cell populations: a systematic literature review and meta-analysis. *The American Journal of Clinical Nutrition* 108 (1): 136–155.

[703] Gibson, A. et al. (2012). Effect of fruit and vegetable consumption on immune function in older people: a randomized controlled trial. *The American Journal of Clinical Nutrition* 96 (6): 1429–1436.

[704] Yan, F. & Polk, D. (2011). Probiotics and immune health. *Current Opinion in Gastroenterology* 27 (6): 496–501.

[705] Rezac, S. & Kok, C. & Heermann, M. & Hutkins, R. (2018). Fermented Foods as a Dietary Source of Live Organisms. *Frontiers in Microbiology* 9: 1785.

[706] Olivares, M. et al. (2008). Dietary deprivation of fermented foods causes a fall in innate immune response. Lactic acid bacteria can counteract the immunological effect of this deprivation. *Journal of Dairy Research* 73 (4): 492–498.

[707] Corthésy, B. (2013). Multi-faceted functions of secretory IgA at mucosal surfaces. *Frontiers in immunology* 4: 185.

[708] Lehtoranta, L. & Pitkäranta, A. & Korpela, R. (2014). Probiotics in respiratory virus infections. *European Journal of Clinical Microbiology and Infectious Diseases* 33 (8): 1289–1302. Review.

[709] Pregliasco, F. et al. (2008). A new chance of preventing winter diseases by the administration of synbiotic formulations. *Journal of Clinical Gastroenterology* 42 (Suppl 3 Pt 2): S224–S33.

[710] Lefevre, M. et al. (2015). Probiotic strain Bacillus subtilis CU1 stimulates immune system of elderly during common infectious disease period: a randomized, double-blind placebo-controlled study. *Immunity and Ageing* 12 (1): 24.

[711] de Vos, P. et al. (2017). Lactobacillus plantarum Strains Can Enhance Human Mucosal and Systemic Immunity and Prevent Non-steroidal Anti-inflammatory Drug Induced Reduction in T Regulatory Cells. *Frontiers in immunology* 8: 1000.

[712] Grimble, R. (2006). The effects of sulfur amino acid intake on immune function in humans. *The Journal of Nutrition* 136 (6 Suppl): 1660S–1665S.

[713] Doleman, J. et al. (2017). The contribution of alliaceous and cruciferous vegetables to dietary sulphur intake. *Food chemistry* 234: 38–45.

[714] Fashner, J. & Ericson, K. & Werner, S. (2012). Treatment of the common cold in children and adults. *American Family Physician* 86 (2): 153–159. Review.

[715] Kiyohara, H. et al. (2011). Patchouli alcohol: in vitro direct anti-influenza virus sesquiterpene in Pogostemon cablin Benth. *Journal of Natural Medicines* 66 (1): 55–61.

[716] Li, Y.-C. et al. (2012). Oral administration of patchouli alcohol isolated from Pogostemonis Herba augments protection against influenza viral infection in mice. *International Immunopharmacology* 12 (1): 294–301.

[717] Swamy, M. & Akhtar, M. & Sinniah, U. (2016). Antimicrobial Properties of Plant Essential Oils against Human Pathogens and Their Mode of Action: An Updated Review. *Evidence-based Complementary and Alternative Medicine* 2016: 3012462.

[718] Somerville, V. & Braakhuis, A. & Hopkins, W. (2016). Effect of Flavonoids on Upper Respiratory Tract Infections and Immune Function: A Systematic Review and Meta-Analysis. *Advances in nutrition (Bethesda, Md.)* 7 (3): 488–497.

[719] Somerville, V. & Moore, R. & Braakhuis, A. (2019). The Effect of Olive Leaf Extract on Upper

Respiratory Illness in High School Athletes: A Randomised Control Trial. *Nutrients* 11 (2): 358.

[720] Bianchini, F. & Vainio, H. (2001). Allium vegetables and organosulfur compounds: do they help prevent cancer? *Environmental Health Perspectives* 109 (9): 893–902.

[721] Iciek, M. & Kwiecień, I. & Włodek, L. (2009). Biological properties of garlic and garlic-derived organosulfur compounds. *Environmental and Molecular Mutagenesis* 50 (3): 247–265.

[722] Guo, N. et al. (1993). Demonstration of the anti-viral activity of garlic extract against human cytomegalovirus in vitro. *Chinese Medicine Journal (Engl)* 106 (2): 93–96.

[723] Harris, J. & Cottrell, S. & Plummer, S. & Lloyd, D. (2001). Antimicrobial properties of Allium sativum (garlic). *Applied Microbiology and Biotechnology* 57 (3): 282–286.

[724] Kyo, E. & Uda, N. & Kasuga, S. & Itakura, Y. (2001). Immunomodulatory effects of aged garlic extract. *The Journal of Nutrition* 131 (3s): 1075s–1079s.

[725] Tran, G-B. & Pham, T-V. & Trinh, N-N. (2018). *Black Garlic and Its Therapeutic Benefits*. In: Medicinal Plants - Use in Prevention and Treatment of Diseases. intechopen.85042.

[726] Kim, M. et al. (2014). Aged black garlic exerts anti-inflammatory effects by decreasing no and proinflammatory cytokine production with less cytoxicity in LPS-stimulated raw 264.7 macrophages and LPS-induced septicemia mice. *Journal of Medicinal Food* 17 (10): 1057–1063.

[727] Rodriguez-Garcia, I. et al. (2016). Oregano Essential Oil as an Antimicrobial and Antioxidant Additive in Food Products. *Critical Reviews in Food Science and Nutrition* 56 (10): 1717–1727

[728] Liu, Q. et al. (2017). Antibacterial and Antifungal Activities of Spices. *International Journal of Molecular Aciences* 18 (6): 1283.

[729] Mokhtar, M. et al. (2017). Antimicrobial activity of selected polyphenols and capsaicinoids identified in pepper (Capsicum annuum L.) and their possible mode of interaction. *Current Microbiology* 74 (11): 1253–1260.

[730] Marini, E. & Magi, G. & Mingoia, M. & Pugnaloni, A. & Facinelli, B. (2015). Antimicrobial and anti-virulence activity of capsaicin against erythromycin-resistant, cell-invasive group a streptococci. *Frontiers in Microbiology* 6: 1281.

[731] Forester, S. & Lambert, J. (2011). The role of antioxidant versus pro-oxidant effects of green tea polyphenols in cancer prevention. *Molecular Nutrition & Food Research* 55 (6): 844–854.

[732] Chen, D. et al. (2008). Tea polyphenols, their biological effects and potential molecular targets. *Histology and Histopathology* 23 (4): 487–496.

[733] Reygaert, W. (2014). The antimicrobial possibilities of green tea. *Frontiers in Microbiology* 5: 434.

[734] Maddocks, S. & Jenkins, R. (2013). Honey: a sweet solution to the growing problem of antimicrobial resistance? *Future Microbiology* 8 (11): 1419–1429. Review.

[735] Mandal, M. & Mandal, S. (2011). Honey: its medicinal property and antibacterial activity. *Asian Pacific Journal of Tropical Biomedicine* 1 (2): 154.

[736] Komosinska-Vassev et al (2015). Bee pollen: chemical composition and therapeutic application. *Evidence-based Complementary and Alternative Medicine* 2015: 297425.

[737] Paul, I. M. (2007). Effect of Honey, Dextromethorphan, and No Treatment on Nocturnal Cough and Sleep Quality for Coughing Children and Their Parents. Archives of Pediatrics & Adolescent Medicine, 161(12), 1140.

[738] Tanzi, M. & Gabay, M. (2002). Association between honey consumption and infant botulism. *Pharmacotherapy* 22 (11): 1479–1483. Review.

[739] Nomicos, E. (2007). Myrrh: medical marvel or myth of the magi? *Holistic Nursing Practice* 21 (6): 308–323.

[740] de Rapper et al (2012). The additive and synergistic antimicrobial effects of select frankincense and myrrh oils--a combination from the pharaonic pharmacopoeia. *Letters in Applied Microbiology* 54 (4): 352–358.

[741] Bello, S. et al. (2014). Tobacco smoking increases the risk for death from pneumococcal pneumonia. *Chest* 146 (4): 1029–1037.

[742] Arcavi, L. & Benowitz, N. (2004). Cigarette smoking and infection. *Archives of Internal Medicine* 164 (20): 2206–2216. Review.

[743] Cai, G. (2020). Tobacco-Use Disparity in Gene Expression of ACE2, the Receptor of 2019-nCov. *Preprints* 2020: 2020020051.

[744] Zhang et al (2008). Alcohol abuse, immunosuppression, and pulmonary infection. *Current Drug Abuse Reviews* 1 (1): 56–67. Review.

[745] Calabrese, E. & Baldwin, L. (2001). U-shaped dose-responses in biology, toxicology, and public health. *Annual Reviews in Public Health* 22: 15–33. Review.

[746] Ng, C. et al. (2014). Heated vegetable oils and cardiovascular disease risk factors. *Vascular Pharmacology* 61 (1): 1–9.

[747] Maszewska, M. et al. (2018). Oxidative Stability of Selected Edible Oils. *Molecules (Basel, Switzerland)* 23 (7): 1746.

[748] Yamagishi, S. et al. (2012). Role of advanced glycation end products (AGEs) and oxidative stress in vascular complications in diabetes. *Biochimica et Biophysica Acta* 1820 (5): 663–671.

[749] de Punder, K. & Pruimboom, L. (2013). The dietary intake of wheat and other cereal grains and their role in inflammation. *Nutrients* 5 (3): 771–787.

[750] Nionelli, L. & Rizzello, C. (2016). Sourdough-Based Biotechnologies for the Production of Gluten-Free Foods. *Foods (Basel, Switzerland)* 5 (3): 65.

[751] Greco, L. et al. (2011). Safety for patients with celiac disease of baked goods made of wheat flour hydrolyzed during food processing. *Clinical Gastroenterology and Hepatology* 9 (1): 24–29.

[752] Roszkowska et al (2019). Non-Celiac Gluten Sensitivity: A Review. *Medicina (Kaunas, Lithuania)* 55 (6): 222.

[753] Ringsdorf, W. Jr. & Cheraskin, E. & Ramsay, R. Jr. (1976). Sucrose, neutrophilic phagocytosis and resistance to disease. *Dental Survey* 52 (12): 46–48.

[754] Sanchez, A. et al. (1973). Role of sugars in human neutrophilic phagocytosis. *The American Journal of Clinical Nutrition* 26 (11): 1180–1184.

[755] Spreadbury, I. (2012). Comparison with ancestral diets suggests dense acellular carbohydrates promote an inflammatory microbiota, and may be the primary dietary cause of leptin resistance and obesity. *Diabetes Metabolic Syndrome and Obesity: Targets and Therapy* 5: 175–189.

[756] Buyken A. et al. (2010). Carbohydrate nutrition and inflammatory disease mortality in older adults. *The American Journal of Clinical Nutrition* 92 (3): 634–643.

[757] Dickinson et al (2008). High-glycemic index carbohydrate increases nuclear factor-kappaB activation in mononuclear cells of young, lean healthy subjects. *The American Journal of Clinical Nutrition* 87 (5): 1188–1193.

[758] Alagawany, M. et al. (2019). Omega-3 and Omega-6 Fatty Acids in Poultry Nutrition: Effect on Production Performance and Health. *Animals: an open access journal from MDPI* 9 (8): 573.

[759] USDA. (2015). Epidemiologic and Other Analyses of HPAI-Affected Poultry Flocks: June 15, 2015 Report. <https://www.aphis.usda.gov/animal_health/animal_dis_spec/poultry/downloads/Epidemiologic-Analysis-June-15-2015.pdf> [date of reference: 11.02.2020]

[760] Joosen, A. et al. (2009). Effect of processed and red meat on endogenous nitrosation and DNA damage. *Carcinogenesis* 30 (8): 1402–1407.

[761] Song, P. & Wu, L. & Guan, W. (2015). Dietary Nitrates, Nitrites, and Nitrosamines Intake and the Risk of Gastric Cancer: A Meta-Analysis. *Nutrients* 7 (12): 9872–9895.

[762] Estevez et al (2019). Emerging Marine Biotoxins in Seafood from European Coasts: Incidence and Analytical Challenges. *Foods (Basel, Switzerland)* 8 (5): 149.

[763] Foran, J. et al. (2005). Quantitative analysis of the benefits and risks of consuming farmed and wild salmon. *The Journal of Nutrition* 135 (11): 2639–2643.

[764] Hu, X. et al. (2019). Environmental Cadmium Enhances Lung Injury by Respiratory Syncytial Virus Infection. *The American Journal of Pathology* 189(8): 1513–1525.

[765] Chavance, M. & Herbeth, B. & Lemoine, A. & Zhu, B. (1993). Does multivitamin supplementation prevent infections in healthy elderly subjects? A controlled trial. *International Journal of Vitamin & Nutrition Research* 63 (1): 11–16.

[766] El-Kadiki, A. & Sutton, A. (2005). Role of multivitamins and mineral supplements in preventing infections in elderly people: systematic review and meta-analysis of randomised controlled trials. *BMJ* 330 (7496): 871.

[767] Barringer et al (2003). Effect of a multivitamin and mineral supplement on infection and quality of life. A randomized, double-blind, placebo-controlled trial. *Annals of Internal Medicine* 138 (5): 365–371.

[768] Higdon, J. et al. (2015). *Vitamin E.* Linus Pauling Micronutrient Center. <https://lpi.oregonstate.edu/mic/vitamins/vitamin-E> [date of reference: 6.2.202]

[769] Wu, Q. et al. (2015). Vitamin E intake and the lung cancer risk among female nonsmokers: a report from the Shanghai Women's Health Study. *International Journal of Cancer* 136 (3): 610–617.

[770] Barrett, B. (2003). Medicinal properties of Echinacea: A critical review. *Phytomedicine* 10 (1): 66–86.

[771] Karsch-Völk, M. et al. (2014). Echinacea for preventing and treating the common cold. *The Cochrane Database of Systematic Reviews* 2 (2): CD000530.

[772] Byleveld, P. & Pang, G. & Clancy, R. & Roberts, D. (1999). Fish Oil Feeding Delays Influenza Virus Clearance and Impairs Production of Interferon-γ and Virus-Specific Immunoglobulin A in the Lungs of Mice. *The Journal of Nutrition* 129 (2): 328–335.

[773] Schwerbrock, N. & Karlsson, E. & Shi, Q. & Sheridan, P. & Beck, M. (2009). Fish Oil-Fed Mice Have Impaired Resistance to Influenza Infection. *The Journal of Nutrition* 139 (8): 1588–1594.

[774] Gutiérrez, S. & Svahn, S., & Johansson, M. (2019). Effects of Omega-3 Fatty Acids on Immune Cells. *International Journal of Molecular Sciences* 20 (20): 5028.

[775] Vernacchio, L. & Kelly, J. & Kaufman, D. & Mitchell, A. (2008). Pseudoephedrine Use Among US Children, 1999-2006: Results From the Slone Survey. *Pediatrics* 122(6): 1299–1304.

[776] Hutton, N. et al. (1991). Effectiveness of an antihistamine-decongestant combination for young children with the common cold: A randomized, controlled clinical trial. *The Journal of Pediatrics* 118 (1): 125–130.

[777] Paul, I. (2004). Effect of Dextromethorphan, Diphenhydramine, and Placebo on Nocturnal Cough and Sleep Quality for Coughing Children and Their Parents. *Pediatrics* 114 (1): e85–e90.

[778] Kenia, P., Houghton, T. & Beardsmore, C. (2008). Does inhaling menthol affect nasal patency or cough? *Pediatric Pulmonology* 43 (6): 532–537.

[779] Prymula, R. et al. (2009). Effect of prophylactic paracetamol administration at time of vaccination on febrile reactions and antibody responses in children: two open-label, randomised controlled trials. *The Lancet* 374 (9698): 1339–1350.

[780] Voiriot, G. et al. (2019). Risks Related to the Use of Non-Steroidal Anti-Inflammatory Drugs in Community-Acquired Pneumonia in Adult and Pediatric Patients. *Journal of Clinical Medicine* 8 (6): 786.

[781] Earn, D. & Andrews, P. & Bolker, B. (2014). Population-level effects of suppressing fever. *Proceedings. Biological Sciences* 281 (1778): 20132570.

[782] Phadtare S. (2004). Recent developments in bacterial cold-shock response. Current issues in molecular biology, 6(2), 125–136.

[783] Lindquist, J. A., & Mertens, P. R. (2018). Cold shock proteins: from cellular mechanisms to

pathophysiology and disease. Cell communication and signaling : CCS, 16(1), 63. https://doi.org/10.1186/s12964-018-0274-6

784 Jones, P. G., & Inouye, M. (1994). The cold-shock response--a hot topic. Molecular microbiology, 11(5), 811–818. https://doi.org/10.1111/j.1365-2958.1994.tb00359.x

785 Wolffe, A. P., Tafuri, S., Ranjan, M., & Familari, M. (1992). The Y-box factors: a family of nucleic acid binding proteins conserved from Escherichia coli to man. The New biologist, 4(4), 290–298.

786 Lage, H., Surowiak, P., & Holm, P. S. (2008). YB-1 als potenzielles Ziel für die Tumortherapie [YB-1 as a potential target in cancer therapy]. Der Pathologe, 29 Suppl 2, 187–190. https://doi.org/10.1007/s00292-008-1030-2

787 Anderson, E. C., & Catnaigh, P. Ó. (2015). Regulation of the expression and activity of Unr in mammalian cells. Biochemical Society transactions, 43(6), 1241–1246. https://doi.org/10.1042/BST20150165

788 Sund-Levander, M., Forsberg, C., & Wahren, L. K. (2002). Normal oral, rectal, tympanic and axillary body temperature in adult men and women: a systematic literature review. Scandinavian Journal of Caring Sciences, 16(2), 122–128. doi:10.1046/j.1471-6712.2002.00069.x

789 Van Marken Lichtenbelt, W. D., & Daanen, H. A. M. (2003). Cold-induced metabolism. Current Opinion in Clinical Nutrition and Metabolic Care, 6(4), 469–475. doi:10.1097/01.mco.0000078992.96795.5f

790 Van Marken Lichtenbelt et al (2002). Individual variation in body temperature and energy expenditure in response to mild cold. American Journal of Physiology-Endocrinology and Metabolism, 282(5), E1077–E1083. doi:10.1152/ajpendo.00020.2001

791 Makinen, T. M. (2010). Different types of cold adaptation in humans. Frontiers in Bioscience, S2(3), 1047–1067. doi:10.2741/s117

792 Wijers et al (2008). Human Skeletal Muscle Mitochondrial Uncoupling Is Associated with Cold Induced Adaptive Thermogenesis. PLoS ONE, 3(3), e1777. doi:10.1371/journal.pone.0001777

793 Haman, F. (2010). Metabolic requirements of shivering humans. Frontiers in Bioscience, S2(3), 1155–1168. doi:10.2741/s124

794 Weber, J.-M., & Haman, F. (2005). Fuel selection in shivering humans. Acta Physiologica Scandinavica, 184(4), 319–329. doi:10.1111/j.1365-201x.2005.01465.x

795 Jedema et al (2008). Chronic cold exposure increases RGS7 expression and decreases alpha(2)-autoreceptor-mediated inhibition of noradrenergic locus coeruleus neurons. The European journal of neuroscience, 27(9), 2433–2443. https://doi.org/10.1111/j.1460-9568.2008.06208.x

796 Rymaszewska, J., Ramsey, D., & Chładzińska-Kiejna, S. (2008). Whole-body cryotherapy as adjunct treatment of depressive and anxiety disorders. Archivum Immunologiae et Therapiae Experimentalis, 56(1), 63–68. doi:10.1007/s00005-008-0006-5

797 Hirvonen et al (2006). Effectiveness of different cryotherapies on pain and disease activity in active rheumatoid arthritis. A randomised single blinded controlled trial. Clinical and experimental rheumatology, 24(3), 295–301.

798 Hinkka, H., Väättänen, S., Ala-Peijari, S., & Nummi, T. (2016). Effects of cold mist shower on patients with inflammatory arthritis: a crossover controlled clinical trial. Scandinavian Journal of Rheumatology, 46(3), 206–209. doi:10.1080/03009742.2016.1199733

799 Aihara T, Tsuruta F (2016) Cold Shock as a Possible Remedy for Neurodegenerative Disease. Int J Neurol Neurother 3:053. 10.23937/2378-3001/3/4/1053

800 Pfeiffer, J. R., McAvoy, B. L., Fecteau, R. E., Deleault, K. M., & Brooks, S. A. (2011). CARHSP1 is required for effective tumor necrosis factor alpha mRNA stabilization and localizes to processing bodies and exosomes. Molecular and cellular biology, 31(2), 277–286. https://doi.org/10.1128/MCB.00775-10

801 Jacobs et al (2013). "Cooling for newborns with hypoxic ischaemic encephalopathy". *The*

Cochrane Database of Systematic Reviews. 1 (1): CD003311. doi:10.1002/14651858.CD003311.pub3.

[802] Imbeault, P., Dépault, I., & Haman, F. (2009). Cold exposure increases adiponectin levels in men. Metabolism: clinical and experimental, 58(4), 552–559. https://doi.org/10.1016/j.metabol.2008.11.017

[803] Schrauwen, P., & van Marken Lichtenbelt, W. D. (2016). Combatting type 2 diabetes by turning up the heat. Diabetologia, 59(11), 2269–2279. doi:10.1007/s00125-016-4068-3

[804] Hanssen et al (2015). Short-term cold acclimation improves insulin sensitivity in patients with type 2 diabetes mellitus. Nature Medicine, 21(8), 863–865. doi:10.1038/nm.3891

[805] Fan et al (2006). Nuclease sensitive element binding protein 1 gene disruption results in early embryonic lethality. Journal of cellular biochemistry, 99(1), 140–145. https://doi.org/10.1002/jcb.20911

[806] Elatmani et al (2011). The RNA-binding protein Unr prevents mouse embryonic stem cells differentiation toward the primitive endoderm lineage. Stem cells (Dayton, Ohio), 29(10), 1504–1516. https://doi.org/10.1002/stem.712

[807] Van Marken Lichtenbelt et al (2009). Cold-Activated Brown Adipose Tissue in Healthy Men. New England Journal of Medicine, 360(15), 1500–1508. doi:10.1056/nejmoa0808718

[808] Saito, M. (2013). Brown Adipose Tissue as a Regulator of Energy Expenditure and Body Fat in Humans. Diabetes & Metabolism Journal, 37(1), 22. doi:10.4093/dmj.2013.37.1.22

[809] Nie et al (2015) 'Cold exposure stimulates lipid metabolism, induces inflammatory response in the adipose tissue of mice and promotes the osteogenic differentiation of BMMSCs via the p38 MAPK pathway in vitro', Int J Clin Exp Pathol. 2015; 8(9): 10875–10886.

[810] Dempersmier et al (2015). Cold-Inducible Zfp516 Activates UCP1 Transcription to Promote Browning of White Fat and Development of Brown Fat. Molecular Cell, 57(2), 235–246. doi:10.1016/j.molcel.2014.12.005

[811] Hanssen et al (2015). Short-term Cold Acclimation Recruits Brown Adipose Tissue in Obese Humans. Diabetes, 65(5), 1179–1189. doi:10.2337/db15-1372

[812] Acosta et al (2018). Physiological responses to acute cold exposure in young lean men. PLOS ONE, 13(5), e0196543. doi:10.1371/journal.pone.0196543

[813] Brand, MD. (2000) 'Uncoupling to survive? The role of mitochondrial inefficiency in ageing', Experimental Gerontology, Vol 35(6-7), p 811-820.

[814] Keipert, S. et al (2013) 'Skeletal muscle uncoupling-induced longevity in mice is linked to increased substrate metabolism and induction of the endogenous antioxidant defense system', The American Physiological Society, Vol 304(5), p E495-E506.

[815] Obradovich et al (2017) 'Nighttime temperature and human sleep loss in a changing climate', Science Advances, 26 May 2017: Vol. 3, no. 5, e1601555, DOI: 10.1126/sciadv.1601555.

[816] Buhr, E. D., Yoo, S.-H., & Takahashi, J. S. (2010). Temperature as a Universal Resetting Cue for Mammalian Circadian Oscillators. Science, 330(6002), 379–385. doi:10.1126/science.1195262

[817] Siems et al (1994). Uric acid and glutathione levels during short-term whole body cold exposure. Free radical biology & medicine, 16(3), 299–305. https://doi.org/10.1016/0891-5849(94)90030-2

[818] Huttunen, P., Kokko, L., & Ylijukuri, V. (2004). Winter swimming improves general well-being. International Journal of Circumpolar Health, 63(2), 140–144. doi:10.3402/ijch.v63i2.17700

[819] Søberg (2018) 'The Impact of Winter Swimming on Brown Adipose Tissue Recruitment and Metabolic Health in Middel-aged Obese Pre-diabetic Subjects.', Center for Inflammation and Metabolism/ Center for Physical Activity Research, Copenhagen, Denmark, 2100

[820] Lubkowska et al (2019). The Effects of Swimming Training in Cold Water on Antioxidant Enzyme Activity and Lipid Peroxidation in Erythrocytes of Male and Female Aged Rats. International Journal of Environmental Research and Public Health, 16(4), 647. doi:10.3390/ijerph16040647

[821] Janský et al (1996). Immune system of cold-exposed and cold-adapted humans. European Journal of Applied Physiology and Occupational Physiology, 72-72(5-6), 445–450. doi:10.1007/bf00242274

[822] Castellani et al (2002) 'Cold exposure: human immune responses and intracellular cytokine expression', Medicine & Science in Sports & Exercise: December 2002 - Volume 34 - Issue 12 - p 2013-2020.

[823] Buijze et al (2016) 'The Effect of Cold Showering on Health and Work: A Randomized Controlled Trial', PLoS One. 2016; 11(9): e0161749.

[824] Srámek et al (2000). Human physiological responses to immersion into water of different temperatures. European journal of applied physiology, 81(5), 436–442. https://doi.org/10.1007/s004210050065

[825] Goldstein, J., Pollitt, N. S., & Inouye, M. (1990). Major cold shock protein of Escherichia coli. Proceedings of the National Academy of Sciences, 87(1), 283–287. doi:10.1073/pnas.87.1.283

[826] Obokata, J., Ohme, M., & Hayashida, N. (1991). Nucleotide sequence of a cDNA clone encoding a putative glycine-rich protein of 19.7 kDa in Nicotiana sylvestris. Plant Molecular Biology, 17(4), 953–955. doi:10.1007/bf00037080

[827] Tafuri, S. R., & Wolffe, A. P. (1990). Xenopus Y-box transcription factors: molecular cloning, functional analysis and developmental regulation. Proceedings of the National Academy of Sciences, 87(22), 9028–9032. doi:10.1073/pnas.87.22.9028

[828] Miquel, J., Lundgren, P. R., Bensch, K. G., & Atlan, H. (1976). Effects of temperature on the life span, vitality and fine structure of Drosophila melanogaster. Mechanisms of Ageing and Development, 5, 347–370. doi:10.1016/0047-6374(76)90034-8

[829] Van Voorhies, W. A., & Ward, S. (1999). Genetic and environmental conditions that increase longevity in Caenorhabditis elegans decrease metabolic rate. Proceedings of the National Academy of Sciences of the United States of America, 96(20), 11399–11403. https://doi.org/10.1073/pnas.96.20.11399

[830] Kelly et al (2013). Effect of temperature on the rate of ageing: an experimental study of the blowfly Calliphora stygia. PloS one, 8(9), e73781. https://doi.org/10.1371/journal.pone.0073781

[831] Gunay, F., Alten, B., & Ozsoy, E. D. (2010). Estimating reaction norms for predictive population parameters, age specific mortality, and mean longevity in temperature-dependent cohorts of Culex quinquefasciatus Say (Diptera: Culicidae). Journal of vector ecology : journal of the Society for Vector Ecology, 35(2), 354–362. https://doi.org/10.1111/j.1948-7134.2010.00094.x

[832] Hofmann, S., Cherkasova, V., Bankhead, P., Bukau, B., & Stoecklin, G. (2012). Translation suppression promotes stress granule formation and cell survival in response to cold shock. Molecular biology of the cell, 23(19), 3786–3800. https://doi.org/10.1091/mbc.E12-04-0296

[833] Protter, D., & Parker, R. (2016). Principles and Properties of Stress Granules. Trends in cell biology, 26(9), 668–679. https://doi.org/10.1016/j.tcb.2016.05.004

[834] Nivon, M., Richet, E., Codogno, P., Arrigo, A. P., & Kretz-Remy, C. (2009). Autophagy activation by NFkappaB is essential for cell survival after heat shock. Autophagy, 5(6), 766–783. https://doi.org/10.4161/auto.8788

[835] Martins, R., Lithgow, G. J., & Link, W. (2016). Long live FOXO: unraveling the role of FOXO proteins in aging and longevity. Aging cell, 15(2), 196–207. https://doi.org/10.1111/acel.12427

[836] Imae, M., Fu, Z., Yoshida, A., Noguchi, T., & Kato, H. (2003). Nutritional and hormonal factors control the gene expression of FoxOs, the mammalian homologues of DAF-16. Journal of molecular endocrinology, 30(2), 253–262. https://doi.org/10.1677/jme.0.03002536

[837] Palacios et al (2009). Diet and exercise signals regulate SIRT3 and activate AMPK and PGC-1alpha in skeletal muscle. Aging, 1(9), 771–783. https://doi.org/10.18632/aging.100075

[838] Ropelle et al (2009). Acute exercise modulates the Foxo1/PGC-1α pathway in the liver of diet-induced obesity rats. The Journal of Physiology, 587(9), 2069–2076. doi:10.1113/jphysiol.2008.164202

[839] Donovan, M. R., & Marr, M. T., 2nd (2016). dFOXO Activates Large and Small Heat Shock Protein Genes in Response to Oxidative Stress to Maintain Proteostasis in Drosophila. The Journal of biological chemistry, 291(36), 19042–19050. https://doi.org/10.1074/jbc.M116.723049

[840] Polesello, C., & Le Bourg, E. (2017). A mild cold stress that increases resistance to heat lowers FOXO translocation in Drosophila melanogaster. Biogerontology, 18(5), 791–801. https://doi.org/10.1007/s10522-017-9722-8

[841] Zhou et al (2012). FOXO3 induces FOXO1-dependent autophagy by activating the AKT1 signaling pathway. Autophagy, 8(12), 1712–1723. https://doi.org/10.4161/auto.21830

[842] Sengupta, A., Molkentin, J. D., & Yutzey, K. E. (2009). FoxO transcription factors promote autophagy in cardiomyocytes. The Journal of biological chemistry, 284(41), 28319–28331. https://doi.org/10.1074/jbc.M109.024406

[843] Jang et al (2018) 'AMPK contributes to autophagosome maturation and lysosomal fusion', Scientific Reports volume 8, Article number: 12637 (2018).

[844] Kim, J., Kundu, M., Viollet, B., & Guan, K.-L. (2011). AMPK and mTOR regulate autophagy through direct phosphorylation of Ulk1. Nature Cell Biology, 13(2), 132–141. doi:10.1038/ncb2152

[845] Kunutsor et al (2018) Longitudinal associations of sauna bathing with inflammation and oxidative stress: the KIHD prospective cohort study. Annals of Medicine 50:5, pages 437-442.

[846] Neutelings, T., Lambert, C. A., Nusgens, B. V., & Colige, A. C. (2013). Effects of mild cold shock (25°C) followed by warming up at 37°C on the cellular stress response. PloS one, 8(7), e69687. https://doi.org/10.1371/journal.pone.0069687

[847] Tanida, I., Minematsu-Ikeguchi, N., Ueno, T., & Kominami, E. (2005). Lysosomal turnover, but not a cellular level, of endogenous LC3 is a marker for autophagy. Autophagy, 1(2), 84–91. https://doi.org/10.4161/auto.1.2.1697

[848] Zhou, J., Ng, S. B., & Chng, W. J. (2013). LIN28/LIN28B: an emerging oncogenic driver in cancer stem cells. The international journal of biochemistry & cell biology, 45(5), 973–978. https://doi.org/10.1016/j.biocel.2013.02.006

[849] Lu et al (2017). YB-1 expression promotes pancreatic cancer metastasis that is inhibited by microRNA-216a. Experimental cell research, 359(2), 319–326. https://doi.org/10.1016/j.yexcr.2017.07.039

[850] Schniepp et al (2002) 'The effects of cold-water immersion on power output and heart rate in elite cyclists', J Strength Cond Res. 2002 Nov;16(4):561-6.

[851] Bleakley et al (2012). Cold-water immersion (cryotherapy) for preventing and treating muscle soreness after exercise. Cochrane Database of Systematic Reviews. doi:10.1002/14651858.cd008262.pub2

[852] Hohenauer et al (2015). The Effect of Post-Exercise Cryotherapy on Recovery Characteristics: A Systematic Review and Meta-Analysis. PLOS ONE, 10(9), e0139028. doi:10.1371/journal.pone.0139028

[853] Roberts et al (2015). Post-exercise cold water immersion attenuates acute anabolic signalling and long-term adaptations in muscle to strength training. The Journal of Physiology, 593(18), 4285–4301. doi:10.1113/jp270570

[854] Cannon, B., & Nedergaard, J. (2010). Nonshivering thermogenesis and its adequate measurement in metabolic studies. Journal of Experimental Biology, 214(2), 242–253. doi:10.1242/jeb.050989

[855] Pedersen, B. K., & Hoffman-Goetz, L. (2000). Exercise and the Immune System: Regulation, Integration, and Adaptation. Physiological Reviews, 80(3), 1055–1081. doi:10.1152/physrev.2000.80.3.1055

[856] National Institutes of Health (2008) 'Flu Virus Fortified In Colder Weather', NIH Research Matters, March 10, 2008, Accessed Online: https://www.nih.gov/news-events/nih-research-matters/flu-virus-fortified-colder-weather

[857] Lowen, A. C., & Steel, J. (2014). Roles of Humidity and Temperature in Shaping Influenza

Seasonality. Journal of Virology, 88(14), 7692–7695. doi:10.1128/jvi.03544-13

[858] Jacobs, S. E., Lamson, D. M., St. George, K., & Walsh, T. J. (2013). Human Rhinoviruses. Clinical Microbiology Reviews, 26(1), 135–162. doi:10.1128/cmr.00077-12

[859] Nakamura et al (1988). Optimal Temperature for Synthesis of DNA, RNA, and Protein by Human Testis in Vitro. Archives of Andrology, 20(1), 41–44. doi:10.3109/01485018808987051

[860] Jain, S., Bruot, B. C., & Stevenson, J. R. (1996). Cold swim stress leads to enhanced splenocyte responsiveness to concanavalin a, decreased serum testosterone, and increased serum corticosterone, glucose, and protein. Life Sciences, 59(3), 209–218. doi:10.1016/0024-3205(96)00286-x

[861] Sakamoto et al (1991). Effects of Physical Exercise and Cold Stimulation on Serum Testosterone Level in men. Nippon Eiseigaku Zasshi (Japanese Journal of Hygiene), 46(2), 635–638. doi:10.1265/jjh.46.635

[862] Vale et al (2017). Chronic non-freezing cold injury results in neuropathic pain due to a sensory neuropathy. Brain, 140(10), 2557–2569. doi:10.1093/brain/awx215

[863] Hannuksela, M. L., & Ellahham, S. (2001). Benefits and risks of sauna bathing. The American journal of medicine, 110(2), 118–126. https://doi.org/10.1016/s0002-9343(00)00671-9

[864] CDC, NCHS. Underlying Cause of Death 1999-2018 on CDC WONDER Online Database, released 2015. Accessed Online: https://wonder.cdc.gov/ucd-icd10.html

[865] Laukkanen, T., Khan, H., Zaccardi, F., & Laukkanen, J. A. (2015). Association Between Sauna Bathing and Fatal Cardiovascular and All-Cause Mortality Events. JAMA Internal Medicine, 175(4), 542. doi:10.1001/jamainternmed.2014.8187

[866] The JAMA Network Journals. (2015, February 23). Sauna use associated with reduced risk of cardiac, all-cause mortality. ScienceDaily. Retrieved May 29, 2020 from www.sciencedaily.com/releases/2015/02/150223122602.htm

[867] Kokura et al. (2007) Whole body hyperthermia improves obesity-induced insulin resistance in diabetic mice. International journal of hyperthermia : the of icial journal of European Society for Hyperthermic Oncology, North American Hyperthermia Group 23, 259-265, doi:10.1080/02656730601176824

[868] Ricardo et al (2011) Heat acclimation responses of an ultra-endurance running group preparing for hot desert-based competition. European Journal of Sport Science, 1-11

[869] Scoon et al (2007) Effect of post-exercise sauna bathing on the endurance performance of competitive male runners. Journal of science and medicine in sport / Sports Medicine Australia 10, 259-262, doi:10.1016/j.jsams.2006.06.009

[870] King et al (1985) Muscle metabolism during exercise in the heat in unacclimatized and acclimatized humans. J Appl Physiol 59, 1350-1354.

[871] Michael N. Sawka, C. B. W., Kent B. Pandolf. (2011) Thermoregulatory Responses to Acute Exercise-Heat Stress and Heat Acclimation. Handbook of Physiology, Environmental Physiology.

[872] Scoon et al (2007) Effect of post-exercise sauna bathing on the endurance performance of competitive male runners. Journal of science and medicine in sport / Sports Medicine Australia 10, 259-262, doi:10.1016/j.jsams.2006.06.009

[873] Selsby, J. T. et al. (2007) Intermittent hyperthermia enhances skeletal muscle regrowth and attenuates oxidative damage following reloading. J Appl Physiol (1985) 102, 1702-1707, doi:10.1152/japplphysiol.00722.2006

[874] Khamwong, P., Paungmali, A., Pirunsan, U., & Joseph, L. (2015). Prophylactic Effects of Sauna on Delayed-Onset Muscle Soreness of the Wrist Extensors. Asian journal of sports medicine, 6(2), e25549. https://doi.org/10.5812/asjsm.6(2)2015.25549

[875] Velloso, C. P. (2008) Regulation of muscle mass by growth hormone and IGF-I. British journal of pharmacology 154, 557-568, doi:10.1038/bjp.2008.153

[876] Leppaluoto, J. et al. (1986) Endocrine effects of repeated sauna bathing. Acta physiologica Scandinavica 128, 467-470, doi:10.1111/j.1748-1716.1986.tb08000.x

[877] Kukkonen-Harjula, K. et al (1989) Haemodynamic and hormonal responses to heat exposure in a Finnish sauna bath. European journal of applied physiology and occupational physiology 58, 543-550

[878] Leppaluoto, J. et al. (1986) Endocrine effects of repeated sauna bathing. Acta physiologica Scandinavica 128, 467-470, doi:10.1111/j.1748-1716.1986.tb08000.x

[879] Nikesitch, N., & Ling, S. C. W. (2015). Molecular mechanisms in multiple myeloma drug resistance. Journal of Clinical Pathology, 69(2), 97–101. doi:10.1136/jclinpath-2015-203414

[880] Hartmann (1958) Asiatic flu in 1957; sauna baths as prophylactic measure. *Hippokrates* 1958;29:153-4.

[881] Kunutsor et al (2017) Sauna bathing reduces the risk of respiratory diseases: a long-term prospective cohort study. *Eur J Epidemiol* 2017;32:1107-11.

[882] Kunutsor (2017) Frequent sauna bathing may reduce the risk of pneumonia in middle-aged Caucasian men: The KIHD prospective cohort study. *Respir Med* 2017;132:161-3.

[883] Laurent H. (2009) Control of typhus fever in Finland during World War II. *Vesalius*;15:71-9.

[884] Ernst E, Pecho E, Wirz P, et al. Regular sauna bathing and the incidence of common colds. *Ann Med* 1990;22:225-7.

[885] WHO 'First data on stability and resistance of SARS coronavirus compiled by members of WHO laboratory network', Emergencies preparedness, response, Accessed Online 16th March at https://www.who.int/csr/sars/survival_2003_05_04/en/

[886] Chan-Yeung, M. (2003). Outbreak of severe acute respiratory syndrome in Hong Kong Special Administrative Region: case report. BMJ, 326(7394), 850–852. doi:10.1136/bmj.326.7394.850

[887] Laukkanen, T., Kunutsor, S., Kauhanen, J., & Laukkanen, J. A. (2016). Sauna bathing is inversely associated with dementia and Alzheimer's disease in middle-aged Finnish men. Age and Ageing, 46(2), 245–249. doi:10.1093/ageing/afw212

[888] Santoro, M. G. (2000). Heat shock factors and the control of the stress response. Biochemical Pharmacology, 59(1), 55–63. doi:10.1016/s0006-2952(99)00299-3

[889] Selsby, J. T. et al. (2007) Intermittent hyperthermia enhances skeletal muscle regrowth and attenuates oxidative damage following reloading. J Appl Physiol (1985) 102, 1702-1707, doi:10.1152/japplphysiol.00722.2006

[890] Naito, H. et al. (2000) Heat stress attenuates skeletal muscle atrophy in hindlimb-unweighted rats. J Appl Physiol 88, 359-363

[891] Guo, S., Wharton, W., Moseley, P., & Shi, H. (2007). Heat shock protein 70 regulates cellular redox status by modulating glutathione-related enzyme activities. Cell Stress & Chaperones, 12(3), 245. doi:10.1379/csc-265.1

[892] Deffit, S. N., & Blum, J. S. (2015). A central role for HSC70 in regulating antigen trafficking and MHC class II presentation. Molecular Immunology, 68(2), 85–88. doi:10.1016/j.molimm.2015.04.007

[893] Fan et al(2005). Novel cardioprotective role of a small heat-shock protein, Hsp20, against ischemia/reperfusion injury. Circulation, 111(14), 1792–1799. https://doi.org/10.1161/01.CIR.0000160851.41872.C6

[894] McLemore et al (2005). Role of the Small Heat Shock Proteins in Regulating Vascular Smooth Muscle Tone. Journal of the American College of Surgeons, 201(1), 30–36. doi:10.1016/j.jamcollsurg.2005.03.017

[895] Khazaeli et al (1997) Heat-induced longevity extension in Drosophila. I. Heat treatment, mortality, and thermotolerance. The journals of gerontology. Series A, Biological sciences and medical sciences 52, B48-52

[896] Lithgow, et al (1995) Thermotolerance and extended life-span conferred by single-gene mutations and induced by thermal stress. Proceedings of the National Academy of Sciences of the United States of America 92, 7540-7544

[897] Tatar, M., Khazaeli, A. A., & Curtsinger, J. W. (1997). Chaperoning extended life. Nature, 390(6655), 30–30. doi:10.1038/36237

[898] Kumsta, C., Chang, J. T., Schmalz, J., & Hansen, M. (2017). Hormetic heat stress and HSF-1 induce autophagy to improve survival and proteostasis in C. elegans. Nature Communications, 8(1). doi:10.1038/ncomms14337

[899] Yang, Y., Xing, D., Zhou, F., & Chen, Q. (2010). Mitochondrial autophagy protects against heat shock-induced apoptosis through reducing cytosolic cytochrome c release and downstream caspase-3 activation. Biochemical and biophysical research communications, 395(2), 190–195. https://doi.org/10.1016/j.bbrc.2010.03.155

[900] Imai Y, et al. (1998) Acute myocardial infarction induced by alternating exposure to heat in a sauna and rapid cooling in cold water. Cardiology. 1998;90(4):299-301.

[901] Papp A. (2002) Sauna-related burns: a review of 154 cases treated in Kuopio University Hospital Burn Center 1994-2000. Burns. 2002 Feb;28(1):57-9.

[902] Guddat, S. S., & Tsokos, M. (2007). Verbrennung auf dem Saunaofen [Death on a sauna stove]. Archiv fur Kriminologie, 220(3-4), 115–119.

[903] "Finalist dies at World Sauna event in Finland". *BBC News Online*. 2010-08-08.

[904] Brown-Woodman et al (1984). The effect of a single sauna exposure on spermatozoa. Archives of andrology, 12(1), 9–15. https://doi.org/10.3109/01485018409161141

[905] Crinnion WJ. (2011) Sauna as a valuable clinical tool for cardiovascular, autoimmune, toxicant-induced and other chronic health problems. Altern Med Rev. Sep;16(3):215-25.

[906] Mitchell JE, et al. (1991) Sauna abuse as a clinical feature of bulimia nervosa. Psychosomatics. 1991 Fall;32(4):417-9.

[907] Axelrod YK, Diringer MN. (2006) Temperature management in acute neurologic disorders. *Crit Care Clin* 2006;22:767-85; abstract x.

[908] Sund-Levander, M., Forsberg, C., & Wahren, L. K. (2002). Normal oral, rectal, tympanic and axillary body temperature in adult men and women: a systematic literature review. Scandinavian Journal of Caring Sciences, 16(2), 122–128. doi:10.1046/j.1471-6712.2002.00069.x

[909] Jokinen E, Va¨lima¨ki I, Marniemi J, et al. (1991) Children in sauna: hormonal adjustments to intensive short thermal stress. Acta Physiol Scand. 1991;142:437– 442.

[910] Jokinen E. (1989) Children's physiological adjustment to heat stress during Finnish sauna bath as studied in a climatic chamber. University of Turku, Finland, Thesis.

[911] Jokinen E, Va¨lima¨ki I, Antila K, et al. (1990) Children in sauna: cardiovascular adjustment. Pediatrics. 1990;86:282–288.

[912] Heckmann, J. G., Rauch, C., Seidler, S., Dutsch, M. & Kasper, B. (2005) Sauna stroke syndrome. Journal of stroke and cerebrovascular diseases : the of icial journal of National Stroke Association 14, 138-139, doi:10.1016/j.jstrokecerebrovasdis.2005.01.006

[913] Benouis et al (2008). Peut-on boire les eaux de bains maures << Hammam >> ? : cas des bains de la ville de Sidi-Bel-Abbès [Is Turkish bath water potable?: The baths of Sidi-Bel-Abbes]. Sante (Montrouge, France), 18(2), 63–66. https://doi.org/10.1684/san.2008.0112

[914] Metzger WJ, et al. (1976) Sauna-takers disease. Hypersensitivity pneumonitis due to contaminated water in a home sauna. JAMA. 1976 Nov 8;236(19):2209-11.

[915] Puranen, M., Syrjänen, K., & Syrjänen, S. (1996). Transmission of Genital Human Papillomavirus Infections is Unlikely through the Floor and Seats of Humid Dwellings in Countries of High-Level Hygiene. Scandinavian Journal of Infectious Diseases, 28(3), 243–246. doi:10.3109/00365549609027165

[916] Harkins, M. S. (2009). Exercise Regulates Heat Shock Proteins and Nitric Oxide. Exercise and Sport Sciences Reviews, 37(2), 73–77. doi:10.1097/jes.0b013e31819c2e7a

[917] Archer, A. E., Von Schulze, A. T., & Geiger, P. C. (2017). Exercise, heat shock proteins and insulin resistance. Philosophical Transactions of the Royal Society B: Biological Sciences, 373(1738), 20160529. doi:10.1098/rstb.2016.0529

[918] Thompson, H. S., Scordilis, S. P., Clarkson, P. M., & Lohrer, W. A. (2001). A single bout of eccentric exercise increases HSP27 and HSC/HSP70 in human skeletal muscle. Acta Physiologica Scandinavica, 171(2), 187–193. doi:10.1046/j.1365-201x.2001.00795.x

[919] Zare et al (2011). Effect of Ramadan fasting on serum heat shock protein 70 and serum lipid profile. Singapore medical journal, 52(7), 491–495.

[920] Schliess, F., Richter, L., vom Dahl, S., & Haussinger, D. (2006). Cell hydration and mTOR-dependent signalling. Acta Physiologica, 187(1-2), 223–229. doi:10.1111/j.1748-1716.2006.01547.x

[921] Simon et al (1995). Heat shock protein 70 overexpression affects the response to ultraviolet light in murine fibroblasts. Evidence for increased cell viability and suppression of cytokine release. The Journal of clinical investigation, 95(3), 926–933. https://doi.org/10.1172/JCI117800

[922] Matsuda et al (2010) 'Prevention of UVB radiation-induced epidermal damage by expression of heat shock protein 70', J Biol Chem. 2010 Feb 19;285(8):5848-58. doi: 10.1074/jbc.M109.063453. Epub 2009 Dec 14.

[923] Janssens et al (2013). Acute Effects of Capsaicin on Energy Expenditure and Fat Oxidation in Negative Energy Balance. PLoS ONE, 8(7), e67786. doi:10.1371/journal.pone.0067786

[924] Dauncey, M. J. (1981). Influence of mild cold on 24 h energy expenditure, resting metabolism and diet-induced thermogenesis. British Journal of Nutrition, 45(2), 257–267. doi:10.1079/bjn19810102

[925] Lykken, David; Tellegen, Auke (1996). "Happiness Is a Stochastic Phenomenon" (PDF). *Psychological Science*. 7 (3): 186–189.

[926] Urban Dictionary (2005) 'Top Definition: stoic', by ski-ski December 26, 2005, Accessed Online: https://www.urbandictionary.com/define.php?term=stoic

[927] John Sellars. *Stoicism (Ancient Philosophies) University of California Press; First edition (July 19, 2006)*, p. 32.

[928] Segerstrom S. C. (2005). Optimism and immunity: do positive thoughts always lead to positive effects?. Brain, behavior, and immunity, 19(3), 195–200. https://doi.org/10.1016/j.bbi.2004.08.003

[929] Nes, L. S., & Segerstrom, S. C. (2006). Dispositional optimism and coping: a meta-analytic review. Personality and social psychology review : an official journal of the Society for Personality and Social Psychology, Inc, 10(3), 235–251. https://doi.org/10.1207/s15327957pspr1003_3

[930] Herodian; Bekker, Immanuel, 1785-1871, ed (1855) Herodiani ab excessu divi Marci libri octo, Lipsiae, sumptibus et typis B.G. Teubneri, i.2.4, tr. Echols.

[931] Marcus Aurelius, *Meditations*, Book II.I

[932] Anand, P (2016). *Happiness Explained.* Oxford University Press.

[933] Robert M. Sapolsky. (1998) *Why Zebras Don't Get Ulcers: An Updated Guide To Stress, Stress Related Diseases, and Coping.* 2nd Rev Ed.

[934] McEwen and Stellar (1993) 'Stress and the individual. Mechanisms leading to disease', Arch Intern Med. 1993 Sep 27;153(18):2093-101.

[935] Arnsten, A. F. T. (2015). Stress weakens prefrontal networks: molecular insults to higher cognition. Nature Neuroscience, 18(10), 1376–1385. doi:10.1038/nn.4087

[936] Goleman D. (1995) *Emotional Intelligence: Why It Can Matter More than IQ.* Bantam Books.

[937] Freedman, Joshua. (2009) "Hijacking of the Amygdala" Accessed Online: https://web.archive.org/web/20091122194535/http:/www.inspirations-unlimited.net/images/Hijack.pdf

[938] Pall ML (2007) *Explaining Unexplained Illnesses: Disease Paradigm for Chronic Fatigue Syndrome, Multiple Chemical Sensitivity, Fibromyalgia, Post-Traumatic Stress Disorder, and Gulf War Syndrome* Harrington Park Press

[939] Maes M, Twisk FN, Kubera M, Ringel K (2012) Evidence for inflammation and activation of cell-mediated immunity in Myalgic Encephalomyelitis/Chronic Fatigue Syndrome (ME/CFS): increased interleukin-1, tumor necrosis factor-α, PMN-elastase, lysozyme and neopterin *Journal of Affective Disorders* 136(3):933-9

[940] Van Houdenhove B Van Den Eede F and Luyten P (2009) Does hypothalamic-pituitary-adrenal axis hypofunction in chronic fatigue syndrome reflect a 'crash' in the stress system? *Medical Hypotheses* 72(6):701-5

[941] Jason LA Sorenson M Porter N Belkairous N (2011) An Etiological Model for Myalgic Encephalomyelitis/Chronic Fatigue Syndrome *Neuroscience & Medicine* 2:14-27

[942] Robertson et al (2015). "Resilience training in the workplace from 2003 to 2014: A systematic review". *Journal of Occupational and Organizational Psychology*. 88 (3): 533–562.

[943] Livingston R.B. (1966). "Brain mechanisms in conditioning and learning". *Neurosciences Research Program Bulletin*. 4 (3): 349–354.

[944] Ortiz et al (2014). Hippocampal brain-derived neurotrophic factor mediates recovery from chronic stress-induced spatial reference memory deficits. The European journal of neuroscience, 40(9), 3351–3362. https://doi.org/10.1111/ejn.12703

[945] Giese, et al (2013). The Interplay of Stress and Sleep Impacts BDNF Level. PLoS ONE, 8(10), e76050. doi:10.1371/journal.pone.0076050

[946] Molendijk et al (2012). Serum BDNF concentrations show strong seasonal variation and correlations with the amount of ambient sunlight. PloS one, 7(11), e48046. https://doi.org/10.1371/journal.pone.0048046

[947] Li et al (2017). Acupuncture Improved Neurological Recovery after Traumatic Brain Injury by Activating BDNF/TrkB Pathway. Evidence-based complementary and alternative medicine : eCAM, 2017, 8460145. https://doi.org/10.1155/2017/8460145

[948] Angelucci et al (2007). Investigating the neurobiology of music: brain-derived neurotrophic factor modulation in the hippocampus of young adult mice. Behavioural pharmacology, 18(5-6), 491–496. https://doi.org/10.1097/FBP.0b013e3282d28f50

[949] Szuhany, K. L., Bugatti, M., & Otto, M. W. (2015). A meta-analytic review of the effects of exercise on brain-derived neurotrophic factor. Journal of Psychiatric Research, 60, 56–64. doi:10.1016/j.jpsychires.2014.10.003

[950] Xu et al (2006). Curcumin reverses the effects of chronic stress on behavior, the HPA axis, BDNF expression and phosphorylation of CREB. Brain research, 1122(1), 56–64. https://doi.org/10.1016/j.brainres.2006.09.009

[951] Katz and Meiri (2006) 'Brain-Derived Neurotrophic Factor Is Critically Involved in Thermal-Experience-Dependent Developmental Plasticity', Journal of Neuroscience 12 April 2006, 26 (15) 3899-3907; DOI: https://doi.org/10.1523/JNEUROSCI.0371-06.2006

[952] Kalat, J. W. (2013). Biological Psychology. Cengage Learning, p. 383

[953] McGlone, F., Wessberg, J., & Olausson, H. (2014). Discriminative and Affective Touch: Sensing and Feeling. Neuron, 82(4), 737–755. doi:10.1016/j.neuron.2014.05.001

[954] Grewen et al (2003) 'Warm partner contact is related to lower cardiovascular reactivity', Behav Med. 2003 Fall;29(3):123-30.

[955] Hawkins, David (2013) 'Letting Go: The Pathway to Surrender', Veritas; 1st edition.

[956] Steptoe, A., & Molloy, G. J. (2007). Personality and heart disease. Heart, 93(7), 783–784. doi:10.1136/hrt.2006.109355

[957] Stults-Kolehmainen and Sinha (2013) 'The Effects of Stress on Physical Activity and Exercise',

Sports Medicine, Volume 44, pages 81–121(2014).

[958] Seligman (1972) 'Learned Helplessness', Annual Review of Medicine, 23(1). doi:10.1146/med.1972.23.issue-1

[959] Abramson, L. Y., Seligman, M. E. P., & Teasdale, J. D. (1978). Learned helplessness in humans: Critique and reformulation. *Journal of Abnormal Psychology, 87,* 49-74. doi:10.1037/0021-843X.87.1.49

[960] Seligman, M. E. P. (1975). *Helplessness: On Depression, Development, and Death.* San Francisco: W. H. Freeman.

[961] Maier, Steven F.; Seligman, Martin E. P. (July 2016). "Learned helplessness at fifty: Insights from neuroscience". *Psychological Review.* 123 (4): 349–367. doi:10.1037/rev0000033.

[962] Sullivan, D.R.; Liu, X; Corwin, D.S. (2012). "Learned Helplessness Among Families and Surrogate Decision-makers of Patients Admitted to Medical, Surgical and Trauma Intensive Care Units". *Chest.* 142 (6): 1440–1446.

[963] Henry, P.C. (2005). "Life stress, explanatory style, hopelessness, and occupational stress". *International Journal of Stress Management.* 12 (3): 241–56.

[964] Bennett, K.K.; Elliott, M. (2005). "Pessimistic explanatory style and Cardiac Health: What is the relation and the mechanism that links them?". *Basic and Applied Social Psychology.* 27 (3): 239–48.

[965] Donovan, W.L.; Leavitt, L.A.; Walsh, R.O. (1990). "Maternal self-efficacy: Illusory control and its effect on susceptibility to learned helplessness". *Child Development.* 61 (5): 1638–47.

[966] Raufelder, D., Regner, N., & Wood, M. A. (2018). Test anxiety and learned helplessness is moderated by student perceptions of teacher motivational support. *Educational Psychology, 38,* 54-74.

[967] Ramirez, E.; Maldonado, A.; Martos, R. (1992). "Attribution modulate immunization against learned helplessness in humans". *Journal of Personality and Social Psychology.* 62: 139–46.

[968] Rodin, J. (1986). "Aging and health: Effects of the sense of control". *Science.* 233 (4770): 1271–6.

[969] Cox et al (2012). "Stereotypes, Prejudice, and Depression The Integrated Perspective" (PDF). *Original Paper.* 7 (5): 427–449.

[970] Brown et al (2016). Daily poverty-related stress and coping: Associations with child learned helplessness. *Family Relations, 65,* 591-602. doi:10.1111/fare.12217

[971] Strigo et al (2008). Association of major depressive disorder with altered functional brain response during anticipation and processing of heat pain. *Archives of General Psychiatry, 65,* 1275-1284. doi:10.1001/archpsyc.65.11.1275

[972] Hammack et al (2012). "Overlapping neurobiology of learned helplessness and conditioned defeat: Implications for PTSD and mood disorders". *Original Paper.* University of Vermont. 62 (2): 565–575. doi:10.1016/j.neuropharm.2011.02.024.

[973] Amat et al (2005). Medial prefrontal cortex determines how stressor controllability affects behavior and dorsal raphe nucleus. Nature Neuroscience, 8(3), 365–371. doi:10.1038/nn1399

[974] Maier, S. F., & Seligman, M. E. P. (2016). Learned helplessness at fifty: Insights from neuroscience. *Psychological Review, 123,* 349-367. doi:10.1037/rev0000033

[975] Mayo Clinic. (2017). Transcranial magnetic stimulation. *Mayo Clinic – Tests & Procedures.* Retrieved from https://www.mayoclinic.org/tests-procedures/transcranial-magnetic-stimulation/about/pac-20384625

[976] Greenwood, Benjamin; Fleshner, Monika (2008). "Exercise, Learned Helplessness, and the Stress-Resistant Brain" (PDF). *Original Paper.* University of Colorado-Boulder and Department of Integrative Physiology. 10 (2): 81–98. doi:10.1007/s12017-008-8029-y.

[977] Altmaier, E.M.; Happ, D.A. (1985). "Coping skills training's immunization effects against learned helplessness". *Journal of Social and Clinical Psychology.* 3 (2): 181–9.

[978] Thornton, J.W.; Powell, G.D. (1974). "Immunization to and alleviation of learned helplessness in man". *American Journal of Psychology*. 87 (3): 351–67.

[979] Dweck, C. (1975). The role of expectations and attributions in the alleviation of learned helplessness. *Journal of Personality and Social Psychology, 31,* 674-685. doi:10.1037/h0077149

[980] Fisher A. L. (2004). Of worms and women: sarcopenia and its role in disability and mortality. Journal of the American Geriatrics Society, 52(7), 1185–1190. https://doi.org/10.1111/j.1532-5415.2004.52320.x

[981] Goodpaster, BH. (2006) 'The loss of skeletal muscle strength, mass, and quality in older adults: the health, aging and body composition study', The Journals of Gerontology. Series A, Biological Sciences and Medical Sciences, Vol 61(10), p 1059-1064.

[982] Metter et al (2002). Skeletal muscle strength as a predictor of all-cause mortality in healthy men. The journals of gerontology. Series A, Biological sciences and medical sciences, 57(10), B359–B365. https://doi.org/10.1093/gerona/57.10.b359

[983] Strasser, B. et al (2018) 'Role of Dietary Protein and Muscular Fitness on Longevity and Aging', Aging and Disease, Vol 9(1), p 119-132.

[984] Hasten, DL. et al (2000) 'Resistance exercise acutely increases MHC and mixed muscle protein synthesis rates in 78-84 and 23-32 yr olds', Americal Journal of Physiology. Endocrinology and Metabolism, Vol 278(4), E620-626.

[985] West, DW. et al (2011) 'Rapid aminoacidemia enhances myofibrillar protein synthesis and anabolic intramuscular signaling responses after resistance exercise', Americal Journal of Clinical Nutrition, Vol 94(3), p 795-803.

[986] Kraschnewski, JL. et al (2016) 'Is strength training associated with mortality benefits? A 15year cohort study of US older adults', Prev Med, Vol 87, p 121-127.

[987] American Heart Association (2018) 'American Heart Association Recommendations for Physical Activity in Adults and Kids', Accessed: https://www.heart.org/en/healthy-living/fitness/fitness-basics/aha-recs-for-physical-activity-in-adults#.WdIG-xO0NBwˇ

[988] Volaklis, KA. et al (2015) 'Muscular strength as a strong predictor of mortality: A narrative review', European Journal of Internal Medicine, Vol 26(5), p 303-310.

[989]Clark, J. (2015) 'Diet, exercise or diet with exercise: comparing the effectiveness of treatment options for weight loss and changes in fitness for adults (16-65 years old) who are overfat, or obese; systematic review and meta-analysis', Journal of Diabetes & Metabolic Disorders, Vol 14(31).

[990] Pedersen, BK. (2011) 'Muscles and their myokines', The Journal of Experimental Biology, Vol 214(Pt2), p 337-346.

[991] Kent-Braun, JA et al (1985) 'Skeletal muscle contractile and noncontractile components in young and older women and men', Journal of Applied Physiology, Vol 88(2), p 662-668.

[992] Srikanthan, P. and Karlamangla, AS. (2011) 'Relative Muscle Mass Is Inversely Associated with Insulin Resistance and Prediabetes. Findings from The Third National Health and Nutrition Examination Survey', The Journal of Clinical Endocrinology & Metabolism, Volume 96, Issue 9, 1 September 2011, Pages 2898–2903.

[993] Owusu-Ansah, E. et al (2013) 'Muscle Mitohormesis Promotes Longevity via Systemic Repression of Insulin Signaling' Cell, Vol 155(3), p 699-712.

[994] Weichhart, T. (2012) 'Mammalian target of rapamycin: a signaling kinase for every aspect of cellular life', Methods in Molecular Biology, Vol 821, p 1-14.

[995] Wullschleger, S. et al (2006) 'TOR Signaling in Growth and Metabolism', Cell, Vol 124(3), p 471-484.

[996]Kim, DH. et al (2002) 'mTOR Interacts with Raptor to Form a Nutrient-Sensitive Complex that Signals to the Cell Growth Machinery', Cell, Vol 110(2), p 163-175.

[997] Bond, P. (2016) 'Regulation of mTORC1 by growth factors, energy status, amino acids and mechanical stimuli at a glance', JISSN, Vol 13, p 8.

[998] Adegoke, OA. et al (2012) 'mTORC1 and the regulation of skeletal muscle anabolism and mass', Applied Physiology, Nutrition, and Metabolism, Vol 37(3), p 395-406.

[999]Sarbassov, DD. et al (2004) 'Rictor, a Novel Binding Partner of mTOR, Defines a Rapamycin-Insensitive and Raptor-Independent Pathway that Regulates the Cytoskeleton', Current Biology, Vol 14(14), p 1296-1302.

[1000] Yin, Y. et al (2015) 'mTORC2 promotes type I insulin-like growth factor receptor and insulin receptor activation through the tyrosine kinase activity of mTOR', Cell Research, Vol 26, p 46-65.

[1001] Yoon, MS. and Choi, CS. (2016) 'The role of amino acid-induced mammalian target of rapamycin complex 1(mTORC1) signaling in insulin resistance', Exp Mol Med, Vol 48(1), e201.

[1002] Shimobayashi, M. and Hall, MN. (2014) 'Making new contacts: the mTOR network in metabolism and signalling crosstalk', Nat Rev Mol Cell Biol, Vol 15(3), p 155-162.

[1003] Hassay, N. and Sonenberg, N. (2004) 'Upstream and downstream of mTOR', Genes & Dev, Vol 18, p 1926-1945.

[1004] Tokunaga, C. et al (2004) 'mTOR integrates amino acid- and energy-sensing pathways', BBRC, Vol 313(2), p 443-446.

[1005] Rabinowitz, JD. and White, E. (2010) 'Autophagy and metabolism', Science, Vol 330(6009), p 1344-8.

[1006] Rodriguez, J. et al (2014) 'Myostatin and the skeletal muscle atrophy and hypertrophy signaling pathways', Cell Mol Life Sci, Vol 71(22), p 4361-71.

[1007] Trendelenburg, AU. et al (2009) 'Myostatin reduces Akt/TORC1/p70S6K signaling, inhibiting myoblast differentiation and myotube size', Am J Physiol Cell Physiol, Vol 296(6), p C1258-70.

[1008] Sancak, Y. et al (2008) 'The Rag GTPases bind raptor and mediate amino acid signaling to mTORC1', Science, Vol 320(5882), p 1496-501.

[1009] Ebato, C. et al (2008) 'Autophagy is important in islet homeostasis and compensatory increase of beta cell mass in response to high-fat diet', Cell Metabolism, Vol 8(4), p 325-32.

[1010] Wolfson, RL. et al (2016) 'Sestrin2 is a leucine sensor for the mTORC1 pathway', Science, Vol 351(6268), p 43-8.

[1011] Witard et al (2016) 'Protein Considerations for Optimising Skeletal Muscle Mass in Healthy Young and Older Adults', Nutrients. 2016 Apr; 8(4): 181.

[1012] Albert, FJ. et al (2015) 'USEFULNESS OF B-HYDROXY-B-METHYLBUTYRATE (HMB) SUPPLEMENTATION IN DIFFERENT SPORTS: AN UPDATE AND PRACTICAL IMPLICATIONS', Nutricion Hospitalaria, Vol 32(1), p 20-33.

[1013] Jacobs, BL. et al (2013) 'Eccentric contractions increase the phosphorylation of tuberous sclerosis complex-2 (TSC2) and alter the targeting of TSC2 and the mechanistic target of rapamycin to the lysosome', J Physiol, Vol 591(18), p 4611-20.

[1014] Calixto et al (2014) 'Acute effects of movement velocity on blood lactate and growth hormone responses after eccentric bench press exercise in resistance-trained men', Biology of Sport, 16 Oct 2014, 31(4):289-294, DOI: 10.5604/20831862.1127287

[1015] Joy, JM. et al (2014) 'Phosphatidic acid enhances mTOR signaling and resistance exercise induced hypertrophy', Nutr Metab (Lond), Vol 11, p 29.

[1016] Yoon, MS et al (2011) 'Phosphatidic acid activates mammalian target of rapamycin complex 1 (mTORC1) kinase by displacing FK506 binding protein 38 (FKBP38) and exerting an allosteric effect', J Biol Chem, Vol 286(34), p 29568-74.

[1017] Tanaka, T. et al (2012) 'Quantification of phosphatidic acid in foodstuffs using a thin-layer-chromatography-imaging technique', J Agric Food Chem, Vol 60(16), p 4156-61.

[1018] Ogasawara, R. et al (2013) 'Ursolic acid stimulates mTORC1 signaling after resistance exercise in rat skeletal muscle', Am J Physiol Endocrinol Metab, Vol 305(6), E760-5.

[1019] Deldicque, L. et al (2008) 'Effects of resistance exercise with and without creatine supplementation on gene expression and cell signaling in human skeletal muscle', J Appl Physiol 1985, Vol 104(2), p 371-8.

[1020] Jorquera, G. et al (2013) 'Testosterone signals through mTOR and androgen receptor to induce muscle hypertrophy', Med Sci Sports Exerc, Vol 45(9), p 1712-20.

[1021] Altamirano, F. et al (2009) 'Testosterone induces cardiomyocyte hypertrophy through mammalian target of rapamycin complex 1 pathway', J Endocrinol, Vol 202(2), p 299-307.

[1022] Xu, K. et al (2014) 'mTOR signaling in tumorigenesis', BBA, Vol 1846(2), p 638-654.

[1023] Zoncu, R. et al (2010) 'mTOR: from growth signal integration to cancer, diabetes and ageing', Molecular Cell Biology, Vol 12, p 21-35.

[1024] Chano, T. et al (2007) 'RB1CC1 insufficiency causes neuronal atrophy through mTOR signaling alteration and involved in the pathology of Alzheimer's diseases', Vol 1168(7), p 97-105.

[1025] Selkoe, DJ. (2008) 'Soluble oligomers of the amyloid β-protein impair synaptic plasticity and behavior', Behavioural Brain Research, Vol 192(1), p 106-113.

[1026] Powell, J. D., Pollizzi, K. N., Heikamp, E. B., & Horton, M. R. (2012). Regulation of Immune Responses by mTOR. Annual Review of Immunology, 30(1), 39–68. doi:10.1146/annurev-immunol-020711-075024

[1027] Araki et al (2011) 'TOR in the immune system', Curr Opin Cell Biol. 2011 Dec; 23(6): 707–715.

[1028] Vierck, J. (2000). SATELLITE CELL REGULATION FOLLOWING MYOTRAUMA CAUSED BY RESISTANCE EXERCISE. Cell Biology International, 24(5), 263–272. doi:10.1006/cbir.2000.0499

[1029] Paul, A. C., & Rosenthal, N. (2002). Different modes of hypertrophy in skeletal muscle fibers. The Journal of Cell Biology, 156(4), 751–760. doi:10.1083/jcb.200105147

[1030] Vierck et al (2000) Satellite cell regulation following myotrauma caused by resistance exercise. Cell Biol Int 24: 263–272.

[1031] Bodine et al (2001). Akt/mTOR pathway is a crucial regulator of skeletal muscle hypertrophy and can prevent muscle atrophy in vivo. Nature Cell Biology, 3(11), 1014–1019. doi:10.1038/ncb1101-1014

[1032] Toigo, M and Boutellier, U. (2006) New fundamental resistance exercise determinants of molecular and cellular muscle adaptations. Eur J Appl Physiol 97: 643–663.

[1033] Kramer, H. F., & Goodyear, L. J. (2007). Exercise, MAPK, and NF-κB signaling in skeletal muscle. Journal of Applied Physiology, 103(1), 388–395. doi:10.1152/japplphysiol.00085.2007

[1034] Roux, P. P., & Blenis, J. (2004). ERK and p38 MAPK-Activated Protein Kinases: a Family of Protein Kinases with Diverse Biological Functions. Microbiology and Molecular Biology Reviews, 68(2), 320–344. doi:10.1128/mmbr.68.2.320-344.2004

[1035] Hameed (2004) The effect of recombinant human growth hormone and resistance training on IGF-I mRNA expression in the muscles of elderly men. J Physiol 555: 231–240.

[1036] Świderska et al (2020). Role of PI3K/AKT Pathway in Insulin-Mediated Glucose Uptake. Blood Glucose Levels. doi:10.5772/intechopen.80402

[1037] Loebel, CC and Kraemer, WJ. (1998) A brief review: Testosterone and resistance exercise in men. J Strength Cond Res 12: 57–63.

[1038] Buresh, R., Berg, K., & French, J. (2009). The Effect of Resistive Exercise Rest Interval on Hormonal Response, Strength, and Hypertrophy With Training. Journal of Strength and Conditioning Research, 23(1), 62–71. doi:10.1519/jsc.0b013e318185f14a

[1039] Kvorning et al (2006) Suppression of endogenous testosterone production attenuates the response to strength training: A randomized, placebo-controlled, and blinded intervention study. Am J Physiol: Endocrinol Metab 291: E1325– E1332.

[1040] Kadi et al (2000) The expression of androgen receptors in human neck and limb muscles: Effects of training and self-administration of androgenic-anabolic steroids. Histochem Cell Biol

113: 25–29.

[1041] Hartman ML et al (1993) 'Normal control of growth hormone secretion.' Horm Res 40: 37–47.

[1042] Mccall et al (1999). Acute and Chronic Hormonal Responses to Resistance Training Designed to Promote Muscle Hypertrophy. Canadian Journal of Applied Physiology, 24(1), 96–107. doi:10.1139/h99-009

[1043] Millar et al (1997) Mammary protein synthesis is acutely regulated by the cellular hydration state. Biochem Biophys Res Comm 230: 351–355.

[1044] Grant et al (2000). Regulation of protein synthesis in lactating rat mammary tissue by cell volume. Biochimica et Biophysica Acta (BBA) - General Subjects, 1475(1), 39–46. doi:10.1016/s0304-4165(00)00045-3

[1045] Sjogaard, G., Adams, R. P., & Saltin, B. (1985). Water and ion shifts in skeletal muscle of humans with intense dynamic knee extension. American Journal of Physiology-Regulatory, Integrative and Comparative Physiology, 248(2), R190–R196. doi:10.1152/ajpregu.1985.248.2.r190

[1046] Takarada et al (2000). Effects of resistance exercise combined with moderate vascular occlusion on muscular function in humans. Journal of Applied Physiology, 88(6), 2097–2106. doi:10.1152/jappl.2000.88.6.2097

[1047] Takarada et al (2000) Applications of vascular occlusion diminish disuse atrophy of knee extensor muscles. Med Sci Sport Exerc 32: 2035–2039.

[1048] Vandenburgh, HH. (1987) Motion into mass: How does tension stimulate muscle growth? Med Sci Sport Exerc 19(5 Suppl.): S142–S149.

[1049] Suzuki, YJ and Ford, GD. (1999) Redox regulation of signal transduction in cardiac and smooth muscle. J Mol and Cell Cardiol 31: 345–353.

[1050] Carey Smith, R., & Rutherford, O. M. (1995). The role of metabolites in strength training. European Journal of Applied Physiology and Occupational Physiology, 71(4), 332–336. doi:10.1007/bf00240413

[1051] Schoenfeld (2010) 'THE MECHANISMS OF MUSCLE HYPERTROPHY AND THEIR APPLICATION TO RESISTANCE TRAINING', Journal of Strength and Conditioning Research, 24(10)/2857–2872.

[1052] Jones, D. A., & Rutherford, O. M. (1987). Human muscle strength training: the effects of three different regimens and the nature of the resultant changes. The Journal of Physiology, 391(1), 1–11. doi:10.1113/jphysiol.1987.sp016721

[1053] Goldberg, AL, Etlinger, JD, Goldspink, DF, and Jablecki, C. (1975) 'Mechanism of work-induced hypertrophy of skeletal muscle.' Med Sci Sport Exerc 7: 185–198, 1975.

[1054] Hill and Goldspink (2003) 'Expression and splicing of the insulin-like growth factor gene in rodent muscle is associated with muscle satellite (stem) cell activation following local tissue damage', J Physiol. 2003 Jun 1; 549(Pt 2): 409–418.

[1055] Folland et al (2002) 'Fatigue is not a necessary stimulus for strength gains during resistance training', British Journal of Sports Medicine 36(5):370-3; discussion 374.

[1056] Shinohara, M., Kouzaki, M., Yoshihisa, T., & Fukunaga, T. (1997). Efficacy of tourniquet ischemia for strength training with low resistance. European Journal of Applied Physiology and Occupational Physiology, 77(1-2), 189–191. doi:10.1007/s004210050319

[1057] Tesch, P. A., Colliander, E. B., & Kaiser, P. (1986). Muscle metabolism during intense, heavy-resistance exercise. European Journal of Applied Physiology and Occupational Physiology, 55(4), 362–366. doi:10.1007/bf00422734

[1058] Schoenfeld, B. J. (2010). The Mechanisms of Muscle Hypertrophy and Their Application to Resistance Training. Journal of Strength and Conditioning Research, 24(10), 2857–2872. doi:10.1519/jsc.0b013e3181e840f3

[1059] Thissen, JP et al (1994) 'Nutritional regulation of the insulin-like growth factors', Endocr Rev, Vol 15(1), p 80-101.

[1060] Guo, W. et al (2011) 'Sirt1 overexpression in neurons promotes neurite outgrowth and cell survival through inhibition of the mTOR signaling', J Neurosci Res, Vol 89(11), p 1723-36.

[1061] Ghosh HS, McBurney M, Robbins PD. (2010) 'SIRT1 negatively regulates the mammalian target of rapamycin', PLoS One, 2010 Feb 15, Vol 5(2), e9199.

[1062] Drummond et al (2009) Rapamycin administration in humans blocks the contraction-induced increase in skeletal muscle protein synthesis. J Physiol. 2009 Apr 1;587(Pt 7):1535-46.

[1063] Shimizu et al (2011) Crosstalk between glucocorticoid receptor and nutritional sensor mTOR in skeletal muscle. Cell Metab. 2011 Feb 2;13(2):170-82.

[1064] Muller et al (2001) Effects of fasting and pegvisomant on the GH-releasing hormone and GH-releasing peptide-6 stimulated growth hormone secretion. Clin Endocrinol (Oxf). 2001 Oct;55(4):461-7.

[1065] Anthony et al (2007). Feeding meals containing soy or whey protein after exercise stimulates protein synthesis and translation initiation in the skeletal muscle of male rats. The Journal of nutrition, 137(2), 357–362. https://doi.org/10.1093/jn/137.2.357

[1066] Moore et al (2008). Ingested protein dose response of muscle and albumin protein synthesis after resistance exercise in young men. The American Journal of Clinical Nutrition, 89(1), 161–168. doi:10.3945/ajcn.2008.26401

[1067] Burd, NA. et al (2011) 'Enhanced Amino Acid Sensitivity of Myofibrillar Protein Synthesis Persists for up to 24 h after Resistance Exercise in Young Men', The Journal of Nutrition, Volume 141, Issue 4, 1 April 2011, Pages 568–573.

[1068] Arnal, MA. et al (2000) 'Protein feeding pattern does not affect protein retention in young women', J Nutr, Vol 130(7), p 1700-4.

[1069] Soeters, MR et al (2009) 'Intermittent fasting does not affect whole-body glucose, lipid, or protein metabolism', Am J Clin Nutr, Vol 90(5), p 1244-51.

[1070] Varady, KA. (2011) 'Intermittent versus daily calorie restriction: which diet regimen is more effective for weight loss?', Obes Rev, Vol 12(7), e593-601.

[1071] Stote KS et al (2007) 'A controlled trial of reduced meal frequency without caloric restriction in healthy, normal-weight, middle-aged adults', Am J Clin Nutr, Vol 85(4), p 981-8.

[1072] Keogh, JB et al (2014) 'Effects of intermittent compared to continuous energy restriction on short-term weight loss and long-term weight loss maintenance', Clin Obes, Vol 4(3), p 150-6.

[1073] Young-Kim et al (2018) 'Update on maximal anabolic response to dietary protein', Clinical Nutrition, Volume 37, Issue 2, April 2018, Pages 411-418.

[1074] Lee, B. C., Kaya, A., & Gladyshev, V. N. (2016). Methionine restriction and life-span control. Annals of the New York Academy of Sciences, 1363, 116–124. https://doi.org/10.1111/nyas.12973

[1075] Finkelstein J. D. (1990). Methionine metabolism in mammals. The Journal of nutritional biochemistry, 1(5), 228–237. https://doi.org/10.1016/0955-2863(90)90070-2

[1076] Kapahi et al (2010). With TOR, less is more: a key role for the conserved nutrient-sensing TOR pathway in aging. Cell metabolism, 11(6), 453–465. https://doi.org/10.1016/j.cmet.2010.05.001

[1077] Levine et al (2014). Low protein intake is associated with a major reduction in IGF-1, cancer, and overall mortality in the 65 and younger but not older population. Cell metabolism, 19(3), 407–417. https://doi.org/10.1016/j.cmet.2014.02.006

[1078] Svensson et al (2012). Both low and high serum IGF-I levels associate with cancer mortality in older men. The Journal of clinical endocrinology and metabolism, 97(12), 4623–4630. https://doi.org/10.1210/jc.2012-2329

[1079] Burgers et al (2011). Meta-analysis and dose-response metaregression: circulating insulin-like growth factor I (IGF-I) and mortality. The Journal of clinical endocrinology and metabolism, 96(9), 2912–2920. https://doi.org/10.1210/jc.2011-1377

[1080] Cappola et al (2001). Association of IGF-I levels with muscle strength and mobility in older women. The Journal of clinical endocrinology and metabolism, 86(9), 4139–4146. https://doi.org/10.1210/jcem.86.9.7868

1081 Puche, J. E., & Castilla-Cortázar, I. (2012). Human conditions of insulin-like growth factor-I (IGF-I) deficiency. Journal of translational medicine, 10, 224. https://doi.org/10.1186/1479-5876-10-224

1082 Speakman, J. R., Mitchell, S. E., & Mazidi, M. (2016). Calories or protein? The effect of dietary restriction on lifespan in rodents is explained by calories alone. Experimental Gerontology, 86, 28–38. doi:10.1016/j.exger.2016.03.011.

1083 Pugh et al (1999) Dietary intervention at middle age: caloric restriction but not dehydroepiandrosterone sulfate increases lifespan and lifetime cancer incidence in mice. Cancer Res. 1999 Apr 1;59(7):1642-8. PubMed PMID: 10197641.

1084 Davis et al (1983) Differential effects of dietary caloric and protein restriction in the aging rat. Exp Gerontol. 1983;18(6):427-35. PubMed PMID: 6673988.

1085 Lee et al (2008). Lifespan and reproduction in Drosophila: New insights from nutritional geometry. Proceedings of the National Academy of Sciences of the United States of America, 105(7), 2498–2503. https://doi.org/10.1073/pnas.0710787105

1086 Ruckenstuhl et al (2014). Lifespan extension by methionine restriction requires autophagy-dependent vacuolar acidification. PLoS genetics, 10(5), e1004347. https://doi.org/10.1371/journal.pgen.1004347

1087 Brind et al (2011) 'Dietary glycine supplementation mimics lifespan extension by dietary methionine restriction in Fisher 344 rats', Biochemistry/Molecular Biology, Vol. 25, No. 1_supplement, April 2011.

1088 Jiao, J., & Demontis, F. (2017). Skeletal muscle autophagy and its role in sarcopenia and organismal aging. Current opinion in pharmacology, 34, 1–6. https://doi.org/10.1016/j.coph.2017.03.009

1089 Dalle et al (2017) 'The Role of Inflammation in Age-Related Sarcopenia', Front. Physiol., 12 December 2017 | https://doi.org/10.3389/fphys.2017.01045

1090 Ju et al (2016). Autophagy plays a role in skeletal muscle mitochondrial biogenesis in an endurance exercise-trained condition. The journal of physiological sciences : JPS, 66(5), 417–430. https://doi.org/10.1007/s12576-016-0440-9

1091 Park et al (2019). Sarcopenia targeting with autophagy mechanism by exercise. BMB reports, 52(1), 64–69. https://doi.org/10.5483/BMBRep.2019.52.1.292

1092 Alway et al (2017). Mitochondria Initiate and Regulate Sarcopenia. Exercise and sport sciences reviews, 45(2), 58–69. https://doi.org/10.1249/JES.0000000000000101

1093 Jamart et al (2012). Modulation of autophagy and ubiquitin-proteasome pathways during ultra-endurance running. Journal of applied physiology (Bethesda, Md. : 1985), 112(9), 1529–1537. https://doi.org/10.1152/japplphysiol.00952.2011

1094 Sanchez et al (2012) 'The role of AMP-activated protein kinase in the coordination of skeletal', Am J Physiol Cell Physiol 303: C475–C485, 2012.

1095 Marzetti, E., Lees, H. A., Wohlgemuth, S. E., & Leeuwenburgh, C. (2009). Sarcopenia of aging: underlying cellular mechanisms and protection by caloric restriction. BioFactors (Oxford, England), 35(1), 28–35. https://doi.org/10.1002/biof.5

1096 Feng et al (2011). Mitochondrial dynamic remodeling in strenuous exercise-induced muscle and mitochondrial dysfunction: Regulatory effects of hydroxytyrosol. Free Radical Biology and Medicine, 50(10), 1437–1446. doi:10.1016/j.freeradbiomed.2011.03.001

1097 Fan et al (2016). Autophagy as a Potential Target for Sarcopenia. Journal of cellular physiology, 231(7), 1450–1459. https://doi.org/10.1002/jcp.25260

1098 Chen et al (2018) 'Association between high-fasting insulin levels and metabolic syndrome in non-diabetic middle-aged and elderly populations: a community-based study in Taiwan', BMJ Open

2018;8:e016554. doi: 10.1136/bmjopen-2017-016554

[1099] Dyck, P. J. (1989). Hypoxic neuropathy Does hypoxia play a role in diabetic neuropathy?: The 1988 Robert Wartenberg Lecture. Neurology, 39(1), 111–111. doi:10.1212/wnl.39.1.111

[1100] Hill J. A. (2011). Autophagy in cardiac plasticity and disease. Pediatric cardiology, 32(3), 282–289. https://doi.org/10.1007/s00246-010-9883-6

[1101] Nakai et al (2007). The role of autophagy in cardiomyocytes in the basal state and in response to hemodynamic stress. Nature medicine, 13(5), 619–624. https://doi.org/10.1038/nm1574

[1102] Singh, R., & Cuervo, A. M. (2012). Lipophagy: Connecting Autophagy and Lipid Metabolism. International Journal of Cell Biology, 2012, 1–12. doi:10.1155/2012/282041

[1103] Ding, W. X., & Yin, X. M. (2012). Mitophagy: mechanisms, pathophysiological roles, and analysis. Biological chemistry, 393(7), 547–564. https://doi.org/10.1515/hsz-2012-0119

[1104] Dalton et al (2003). Plasma C-reactive protein levels in severe diabetic ketoacidosis. Annals of clinical and laboratory science, 33(4), 435–442.

[1105] Hoffman et al (2012). Autophagy in the brains of young patients with poorly controlled T1DM and fatal diabetic ketoacidosis. Experimental and molecular pathology, 93(2), 273–280. https://doi.org/10.1016/j.yexmp.2011.10.007

[1106] Burgers et al (2011). Meta-analysis and dose-response metaregression: circulating insulin-like growth factor I (IGF-I) and mortality. The Journal of clinical endocrinology and metabolism, 96(9), 2912–2920. https://doi.org/10.1210/jc.2011-1377

[1107] Liu et al (2009). Hepatic autophagy is suppressed in the presence of insulin resistance and hyperinsulinemia: inhibition of FoxO1-dependent expression of key autophagy genes by insulin. The Journal of biological chemistry, 284(45), 31484–31492. https://doi.org/10.1074/jbc.M109.033936

[1108] Yang et al (2010). Defective hepatic autophagy in obesity promotes ER stress and causes insulin resistance. Cell metabolism, 11(6), 467–478. https://doi.org/10.1016/j.cmet.2010.04.005

[1109] Nakai et al (2007) The role of autophagy in cardiomyocytes in the basal state and in response to hemodynamic stress. *Nat Med* 2007; 13: 619–624.

[1110] Hua et al (2011) Chronic Akt activation accentuates aging-induced cardiac hypertrophy and myocardial contractile dysfunction: Role of autophagy. *Basic Res Cardiol* 2011;106: 1173–1191.

[1111] Varma et al Insulin-induced glycogen accumulation is associated with increased glycogen-autophagy in, Proceedings of the Australian Physiological Society, Accessed Online: http://aups.org.au/Proceedings/44/103P/103P.pdf

[1112] Perrotta, I. (2013). The use of electron microscopy for the detection of autophagy in human atherosclerosis. Micron, 50, 7–13. doi:10.1016/j.micron.2013.03.007

[1113] Liao et al (2012). Macrophage Autophagy Plays a Protective Role in Advanced Atherosclerosis. Cell Metabolism, 15(4), 545–553. doi:10.1016/j.cmet.2012.01.022

[1114] LaRocca et al (2012), Translational evidence that impaired autophagy contributes to arterial ageing. The Journal of Physiology, 590: 3305-3316. doi:10.1113/jphysiol.2012.229690

[1115] Razani et al (2012). Autophagy Links Inflammasomes to Atherosclerotic Progression. Cell Metabolism, 15(4), 534–544. doi:10.1016/j.cmet.2012.02.011

[1116] Shi et al (2012). Activation of autophagy by inflammatory signals limits IL-1β production by targeting ubiquitinated inflammasomes for destruction. Nature immunology, 13(3), 255–263. https://doi.org/10.1038/ni.2215

[1117] Yoshizaki et al (2012). Autophagy regulates inflammation in adipocytes. Biochemical and biophysical research communications, 417(1), 352–357. https://doi.org/10.1016/j.bbrc.2011.11.114

[1118] Henderson, P., & Stevens, C. (2012). The role of autophagy in Crohn's disease. Cells, 1(3), 492–519. https://doi.org/10.3390/cells1030492

[1119] Luciani et al (2010). Defective CFTR induces aggresome formation and lung inflammation in cystic fibrosis through ROS-mediated autophagy inhibition. Nature cell biology, 12(9), 863–875. https://doi.org/10.1038/ncb2090

[1120] Deretic, V., & Levine, B. (2009). Autophagy, immunity, and microbial adaptations. Cell host & microbe, 5(6), 527–549. https://doi.org/10.1016/j.chom.2009.05.016

[1121] Cheng et al (2014). Prolonged Fasting Reduces IGF-1/PKA to Promote Hematopoietic-Stem-Cell-Based Regeneration and Reverse Immunosuppression. Cell Stem Cell, 14(6), 810–823. doi:10.1016/j.stem.2014.04.014

[1122] Komatsu et al (2006). Loss of autophagy in the central nervous system causes neurodegeneration in mice. Nature, 441(7095), 880–884. https://doi.org/10.1038/nature04723

[1123] Bossy, B., Perkins, G., & Bossy-Wetzel, E. (2008). Clearing the brain's cobwebs: the role of autophagy in neuroprotection. Current neuropharmacology, 6(2), 97–101. https://doi.org/10.2174/157015908784533897

[1124] Li Q (2017) The role of autophagy in Alzheimer's disease. J Syst Integr Neurosci 3: DOI: 10.15761/JSIN.1000172

[1125] Huett, A., & Xavier, R. J. (2010). Autophagy at the gut interface: mucosal responses to stress and the consequences for inflammatory bowel diseases. Inflammatory bowel diseases, 16(1), 152–174. https://doi.org/10.1002/ibd.20991

[1126] Hiutung (2017) 'Microbiome-induced autophagy as a novel therapy for inflammatory bowel disease', National Institute of Diabetes and Digestive and Kidney Diseases (NIDDK), Accessed Online: https://grantome.com/grant/NIH/F32-DK100109-01A1

[1127] Zhao et al (2003). Lysosomal enzymes promote mitochondrial oxidant production, cytochrome c release and apoptosis. European journal of biochemistry, 270(18), 3778–3786. https://doi.org/10.1046/j.1432-1033.2003.03765.x

[1128] Stroikin et al (2004). Inhibition of autophagy with 3-methyladenine results in impaired turnover of lysosomes and accumulation of lipofuscin-like material. European journal of cell biology, 83(10), 583–590. https://doi.org/10.1078/0171-9335-00433

[1129] Kim et al (2018). Autophagy in Human Skin Fibroblasts: Impact of Age. International journal of molecular sciences, 19(8), 2254. https://doi.org/10.3390/ijms19082254

[1130] Monfrecola et al (2016), Mechanistic target of rapamycin (mTOR) expression is increased in acne patients' skin. Exp Dermatol, 25: 153-155. doi:10.1111/exd.12885

[1131] Khambu et al (2018). Autophagy in non-alcoholic fatty liver disease and alcoholic liver disease. Liver Research, 2(3), 112–119. doi:10.1016/j.livres.2018.09.004

[1132] Newman, AB. et al (2006) 'Strength, but not muscle mass, is associated with mortality in the health, aging and body composition study cohort', The Journals of Gerontology. Series A, Biological Sciences and Medical Sciences, Vol 61(1), p 72-77.

[1133] Goodpaster, BH. et al (1985) 'Attenuation of skeletal muscle and strength in the elderly: The Health ABC Study', Journal of Applied Physiology (1985), Vol 90(6), p 2157-2165.

[1134] Sayer, AA. and Kirkwood, T. (2015) 'Grip strength and mortality: a biomarker of ageing?', The Lancet, Vol 386(9990), p 226-227.

[1135] Leong, DP. et al (2015) 'Prognostic value of grip strength: findings from the Prospective Urban Rural Epidemiology (PURE) study', Lancet, Vol 386(9990), p 266-273.

[1136] Kim, Y. et al (2017) 'Independent and joint associations of grip strength and adiposity with all-cause and cardiovascular disease mortality in 403,199 adults: the UK Biobank study', American Journal of Clinical Nutrition, Vol 106(3), p 773-782.

[1137] Bouchard, DR. et al (2011) 'Association between muscle mass, leg strength, and fat mass with physical function in older adults: influence of age and sex', Journal of Aging and Health, Vol 23(2), p 313-328.

[1138] Bautmans, I. et al (2009) 'Sarcopenia and functional decline: pathophysiology, prevention and therapy', Vol 64(4), p 303-316.

[1139] Zebis, MK. et al (2011) 'Implementation of neck/shoulder exercises for pain relief among industrial workers: a randomized controlled trial', BMC Musculoskeletal Disorders, Vol 12, p 205.

[1140] Pereira, LM. et al (2012) 'Comparing the Pilates method with no exercise or lumbar stabilization for pain and functionality in patients with chronic low back pain: systematic review and meta-analysis', Clinical Rehabilitation, Vol 26(1), p 10-20.

[1141] Caelle, MC. and Fernandez, ML. (2010) 'Effects of resistance training on the inflammatory response', Nutrition Research and Practice, Vol 4(4), p 259-269.

[1142] Loenneke et al (2015) Exercise with Blood Flow Restriction: An Updated Evidence-Based Approach for Enhanced Muscular Development. Sports Medicine, 45 (3). pp. 313-325.

[1143] Loenneke, J.P., Slattery, K.M. and Dascombe, B.J. (2015) Exercise with Blood Flow Restriction: An Updated Evidence-Based Approach for Enhanced Muscular Development. Sports Medicine, 45 (3). pp. 313-325.

[1144] Garber et al (2011). Quantity and Quality of Exercise for Developing and Maintaining Cardiorespiratory, Musculoskeletal, and Neuromotor Fitness in Apparently Healthy Adults. Medicine & Science in Sports & Exercise, 43(7), 1334–1359. doi:10.1249/mss.0b013e318213fefb

[1145] Slysz et al (2016) 'The efficacy of blood flow restricted exercise: A systematic review & meta-analysis', Journal of Science and Medicine in Sports, Volume 19, Issue 8, Pages 669–675. DOI: https://doi.org/10.1016/j.jsams.2015.09.005

[1146] Schoenfeld BJ et al (2016) 'Effects of Resistance Training Frequency on Measures of Muscle Hypertrophy: A Systematic Review and Meta-Analysis', Sports Med, Vol 46(11), p 1689-1697.

[1147] Schoenfeld BJ et al (2015) 'Influence of Resistance Training Frequency on Muscular Adaptations in Well-Trained Men', J Strength Cond Res, Vol 29(7), p 1821-9.

[1148]Rafael, Z. et al (2018) 'High Resistance-Training Frequency Enhances Muscle Thickness in Resistance-Trained Men', The Journal of Strength & Conditioning.

[1149] Cadore, EL. et al (2008) 'Hormonal responses to resistance exercise in long-term trained and untrained middle-aged men', J Strength Cond Res, Vol 22(5), p 1617-24.

[1150] Ochi, E. et al (2018) 'Higher Training Frequency Is Important for Gaining Muscular Strength Under Volume-Matched Training', Frontiers in Physiology 02 July 2018.

[1151] Hartman, MJ. et al (2007) 'Comparisons between twice-daily and once-daily training sessions in male weight lifters', Int J Sports Physiol Perform, Vol 2(2), p 159-69.

[1152] Studenski, S. (2011). Gait Speed and Survival in Older Adults. JAMA, 305(1), 50. doi:10.1001/jama.2010.1923

[1153] Abellan Van Kan et al (2009). Gait speed at usual pace as a predictor of adverse outcomes in community-dwelling older people an International Academy on Nutrition and Aging (IANA) Task Force. The Journal of Nutrition, Health & Aging, 13(10), 881–889. doi:10.1007/s12603-009-0246-z

[1154] Fleg, JL. et al (2005) 'Accelerated longitudinal decline of aerobic capacity in healthy older adults', Circulation, Vol 112(5), p 674-682.

[1155] Tabata, I. et al (1996) 'Effects of moderate-intensity endurance and high-intensity intermittent training on anaerobic capacity and VO2max', Medicine and Science in Sports and Exercise, Vol 28(10), p 1327-1330.

[1156] Häkkinen, K. et al (2003) 'Neuromuscular adaptations during concurrent strength and endurance training versus strength training', Eur J Appl Physiol, Vol 89(1), p 42-52.

[1157] Robinson, M. et al (2017) 'Enhanced Protein Translation Underlies Improved Metabolic and Physical Adaptations to Different Exercise Training Modes in Young and Old Humans', Clinical and Translation Report, Vol 25(3), p 581-592.

[1158] Ropelle, ER. et al (2009) 'Acute exercise modulates the Foxol/PGC-1α pathway', Journal of Physiology, p 2069-2076.

[1159] Cantó C. et al (2010) 'Interdependence of AMPK and SIRT1 for metabolic adaptation to fasting and exercise in skeletal muscle', Cell Metabolism, Vol 11(3), p 213-219.

[1160] Costford, SR. et al (2010) 'Skeletal muscle NAMPT is induced by exercise in humans', American Journal of Physiology. Endocrinology and Metabolism, Vol 298(1), p E117-26.

[1161] Dale et al (2014). Unexpected Benefits of Intermittent Hypoxia: Enhanced Respiratory and Nonrespiratory Motor Function. Physiology, 29(1), 39–48. doi:10.1152/physiol.00012.2013

[1162] Welch, H.G. (1987). Effects of hypoxia and hyperoxia on human performance. Exercise and Sports Science Review 15: 191-220.

[1163] Burtscher M. (2013). Effects of living at higher altitudes on mortality: a narrative review. Aging and disease, 5(4), 274–280. https://doi.org/10.14336/AD.2014.0500274

[1164] Fang, Y., Tan, J., & Zhang, Q. (2015). Signaling pathways and mechanisms of hypoxia-induced autophagy in the animal cells. Cell Biology International, 39(8), 891–898. doi:10.1002/cbin.10463

[1165] Daskalaki et al (2018) 'Hypoxia and Selective Autophagy in Cancer Development and Therapy', Front. Cell Dev. Biol., 10 September 2018 | https://doi.org/10.3389/fcell.2018.00104

[1166] Hoppeler and Vogt (2001) 'Muscle tissue adaptations to hypoxia.', J Exp Biol. 2001 Sep;204(Pt 18):3133-9.

[1167] Connolly et al (1989). Tumor vascular permeability factor stimulates endothelial cell growth and angiogenesis. Journal of Clinical Investigation, 84(5), 1470–1478. doi:10.1172/jci114322

[1168] Yuan et al (2008). Induction of HIF-1α expression by intermittent hypoxia: Involvement of NADPH oxidase, Ca2+signaling, prolyl hydroxylases, and mTOR. Journal of Cellular Physiology, 217(3), 674–685. doi:10.1002/jcp.21537

[1169] Fulda et al (2010). "Cellular Stress Responses: Cell Survival and Cell Death". *International Journal of Cell Biology*. 2010: 214074. doi:10.1155/2010/214074.

[1170] Winkelmayer et al (2012). Altitude and the risk of cardiovascular events in incident US dialysis patients. Nephrology Dialysis Transplantation, 27(6), 2411–2417. doi:10.1093/ndt/gfr681

[1171] Shrestha et al (2012) 'Blood pressure in inhabitants of high altitude of Western Nepal.', JNMA J Nepal Med Assoc. 2012 Oct-Dec;52(188):154-8.

[1172] Manukhina, E. B., Downey, H. F., & Mallet, R. T. (2006). Role of Nitric Oxide in Cardiovascular Adaptation to Intermittent Hypoxia. Experimental Biology and Medicine, 231(4), 343–365. doi:10.1177/153537020623100401

[1173] Longmore et al (2006). *Mini Oxford Handbook of Clinical Medicine*. Oxford University Press. p. 874. ISBN 978-0198570714.

[1174] Zhu et al (2010). Intermittent Hypoxia Promotes Hippocampal Neurogenesis and Produces Antidepressant-Like Effects in Adult Rats. Journal of Neuroscience, 30(38), 12653–12663. doi:10.1523/jneurosci.6414-09.2010

[1175] Xie et al (2010). Brain-derived neurotrophic factor rescues and prevents chronic intermittent hypoxia-induced impairment of hippocampal long-term synaptic plasticity. Neurobiology of Disease, 40(1), 155–162. doi:10.1016/j.nbd.2010.05.020

[1176] Cai et al (2013). Hypoxia-inducible factor 1 is required for remote ischemic preconditioning of the heart. Proceedings of the National Academy of Sciences, 110(43), 17462–17467. doi:10.1073/pnas.1317158110

[1177] Lo et al (2013). Sleeping altitude and sudden cardiac death. American Heart Journal, 166(1), 71–75. doi:10.1016/j.ahj.2013.04.003

[1178] Michiels (2004) 'Physiological and Pathological Responses to Hypoxia', Am J Pathol. 2004 Jun; 164(6): 1875–1882. doi: 10.1016/S0002-9440(10)63747-9

[1179] Bellot et al (2009). Hypoxia-Induced Autophagy Is Mediated through Hypoxia-Inducible Factor Induction of BNIP3 and BNIP3L via Their BH3 Domains. Molecular and Cellular Biology, 29(10), 2570–2581. doi:10.1128/mcb.00166-09

[1180] Pouysségur et al (2006). Hypoxia signalling in cancer and approaches to enforce tumour regression. Nature, 441(7092), 437–443. doi:10.1038/nature04871

[1181] Schofield, C. J., & Ratcliffe, P. J. (2004). Oxygen sensing by HIF hydroxylases. Nature Reviews Molecular Cell Biology, 5(5), 343–354. doi:10.1038/nrm1366

[1182] Krock, B. L., Skuli, N., & Simon, M. C. (2011). Hypoxia-Induced Angiogenesis: Good and Evil. Genes & Cancer, 2(12), 1117–1133. doi:10.1177/1947601911423654

[1183] Hu et al (2012). Hypoxia-Induced Autophagy Promotes Tumor Cell Survival and Adaptation to Antiangiogenic Treatment in Glioblastoma. Cancer Research, 72(7), 1773–1783. doi:10.1158/0008-5472.can-11-3831

[1184] Woorons et al (2014). Swimmers can train in hypoxia at sea level through voluntary hypoventilation. Respiratory Physiology & Neurobiology, 190, 33–39. doi:10.1016/j.resp.2013.08.022

[1185] Weenink et al (2020). Perioperative Hyperoxyphobia: Justified or Not? Benefits and Harms of Hyperoxia during Surgery. Journal of Clinical Medicine, 9(3), 642. doi:10.3390/jcm9030642

[1186] Gottlieb et al (2009) 'Hypoxia, Not the Frequency of Sleep Apnea, Induces Acute Hemodynamic Stress in Patients with Chronic Heart Failure', J Am Coll Cardiol. 2009 Oct 27; 54(18): 1706–1712.

[1187] Valeo, T. (2011). Sleep Apnea-Induced Hypoxia Implicates as a Factor in Cognitive Decline. Neurology Today, 11(17), 1. doi:10.1097/01.nt.0000406009.37645.94

[1188] NHLBI, 'Sleep Apnea', Accessed Online: https://www.nhlbi.nih.gov/health-topics/sleep-apnea

[1189] Rao A, ed. (2012). *Principles and Practice of Pedodontics* (3rd ed.). New Delhi: Jaypee Brothers Medical Pub. pp. 169, 170.

[1190] Khansari et al (1990). "Effects of stress on the immune system". *Immunology Today*. 11 (5): 170–175

[1191] Peake et al (2015). Modulating exercise-induced hormesis: Does less equal more?. Journal of applied physiology (Bethesda, Md. : 1985), 119(3), 172–189. https://doi.org/10.1152/japplphysiol.01055.2014

[1192] Cook et al (2017). Blackcurrant Alters Physiological Responses and Femoral Artery Diameter during Sustained Isometric Contraction. Nutrients, 9(6), 556. doi:10.3390/nu9060556

[1193] Balsalobre-Fernández et al (2018). The effects of beetroot juice supplementation on exercise economy, rating of perceived exertion and running mechanics in elite distance runners: A double-blinded, randomized study. PLOS ONE, 13(7), e0200517. doi:10.1371/journal.pone.0200517

[1194] Trexler et al (2014). Effects of pomegranate extract on blood flow and running time to exhaustion. Applied physiology, nutrition, and metabolism, 39(9), 1038–1042. https://doi.org/10.1139/apnm-2014-0137

[1195] Merry, T. L., & Ristow, M. (2016). Do antioxidant supplements interfere with skeletal muscle adaptation to exercise training? The Journal of Physiology, 594(18), 5135–5147. doi:10.1113/jp270654

[1196] Paulsen et al (2014). Vitamin C and E supplementation alters protein signalling after a strength training session, but not muscle growth during 10 weeks of training. The Journal of Physiology, 592(24), 5391–5408. doi:10.1113/jphysiol.2014.279950

[1197] Clifford et al (2019). The effects of vitamin C and E on exercise-induced physiological adaptations: a systematic review and Meta-analysis of randomized controlled trials. Critical Reviews in Food Science and Nutrition, 1–11. doi:10.1080/10408398.2019.1703642

[1198] Massaro et al (2019). Effect of Cocoa Products and Its Polyphenolic Constituents on Exercise Performance and Exercise-Induced Muscle Damage and Inflammation: A Review of Clinical Trials. Nutrients, 11(7), 1471. https://doi.org/10.3390/nu11071471

[1199] Bell et al (2016). The Effects of Montmorency Tart Cherry Concentrate Supplementation on Recovery Following Prolonged, Intermittent Exercise. Nutrients, 8(7), 441. doi:10.3390/nu8070441

[1200] Braakhuis, A. J. (2012). Effect of Vitamin C Supplements on Physical Performance. Current

Sports Medicine Reports, 11(4), 180–184. doi:10.1249/jsr.0b013e31825e19cd

[1201] Braakhuis, A. J., & Hopkins, W. G. (2015). Impact of Dietary Antioxidants on Sport Performance: A Review. Sports Medicine, 45(7), 939–955. doi:10.1007/s40279-015-0323-x

[1202] Djordjevic et al (2012) 'Effect of astaxanthin supplementation on muscle damage and oxidative stress markers in elite young soccer players', J Sports Med Phys Fitness. 2012 Aug;52(4):382-92.

[1203] Sousa, M., Teixeira, V. H., & Soares, J. (2013). Dietary strategies to recover from exercise-induced muscle damage. International Journal of Food Sciences and Nutrition, 65(2), 151–163. doi:10.3109/09637486.2013.849662

[1204] McKenna et al (2006). N-acetylcysteine attenuates the decline in muscle Na+,K+-pump activity and delays fatigue during prolonged exercise in humans. The Journal of Physiology, 576(1), 279–288. doi:10.1113/jphysiol.2006.115352

[1205] Koechlin et al (2004). Does Oxidative Stress Alter Quadriceps Endurance in Chronic Obstructive Pulmonary Disease? American Journal of Respiratory and Critical Care Medicine, 169(9), 1022–1027. doi:10.1164/rccm.200310-1465oc

[1206] Matuszczak et al (2005) 'Effects of N-acetylcysteine on glutathione oxidation and fatigue during handgrip exercise', Muscle Nerve. 2005 Nov;32(5):633-8.

[1207] SLATTERY et al (2014). Effect of N-acetylcysteine on Cycling Performance after Intensified Training. Medicine & Science in Sports & Exercise, 46(6), 1114–1123. doi:10.1249/mss.0000000000000222

[1208] Braakhuis, A. J., & Hopkins, W. G. (2015). Impact of Dietary Antioxidants on Sport Performance: A Review. Sports Medicine, 45(7), 939–955. doi:10.1007/s40279-015-0323-x

[1209] Zhadin (2001) 'Review of russian literature on biological action of DC and low-frequency AC magnetic fields', Bioelectromagnetics. 2001 Jan;22(1):27-45.

[1210] International Agency for Research on Cancer (2011) 'IARC CLASSIFIES RADIOFREQUENCY ELECTROMAGNETIC FIELDS AS POSSIBLY CARCINOGENIC TO HUMANS', World Health Organization, Press Release No 208, Accessed Online: https://www.iarc.fr/wp-content/uploads/2018/07/pr208_E.pdf

[1211] INTERPHONE Study Group (2010). Brain tumour risk in relation to mobile telephone use: results of the INTERPHONE international case-control study. International journal of epidemiology, 39(3), 675–694. https://doi.org/10.1093/ije/dyq079

[1212] Grellier, J., Ravazzani, P., & Cardis, E. (2014). Potential health impacts of residential exposures to extremely low frequency magnetic fields in Europe. Environment International, 62, 55–63. doi:10.1016/j.envint.2013.09.017

[1213] Huss et al (2018). Occupational extremely low frequency magnetic fields (ELF-MF) exposure and hematolymphopoietic cancers – Swiss National Cohort analysis and updated meta-analysis. Environmental Research, 164, 467–474. doi:10.1016/j.envres.2018.03.022

[1214] Sun et al (2016) 'Extremely Low Frequency Electromagnetic Fields Facilitate Vesicle Endocytosis by Increasing Presynaptic Calcium Channel Expression at a Central Synapse', Scientific Reports volume 6, Article number: 21774 (2016)

[1215] Li et al (2011) 'Maternal Exposure to Magnetic Fields During Pregnancy in Relation to the Risk of Asthma in Offspring', Arch Pediatr Adolesc Med. 2011;165(10):945-950. doi:10.1001/archpediatrics.2011.135

[1216] Adams et al (2014). Effect of mobile telephones on sperm quality: A systematic review and meta-analysis. Environment International, 70, 106–112. doi:10.1016/j.envint.2014.04.015

[1217] Shafirstein and Moros (2011) 'Modelling millimetre wave propagation and absorption in a high resolution skin model: The effect of sweat glands', Physics in Medicine and Biology 56(5):1329-39. DOI: 10.1088/0031-9155/56/5/007

[1218] Prost et al (1994) Experimental studies on the influence of millimeter radiation on light transmission through the lens, Klin Oczna. 1994 Aug-Sep;96(8-9):257-9.

[1219] Potekhina et al (1992) The effect of low-intensity millimeter-range electromagnetic radiation on the cardiovascular system of the white rat, Fiziol Zh SSSR Im I M Sechenova. 1992 Jan;78(1):35-41.

[1220] Kolomytseva et al (2002) Suppression of nonspecific resistance of the body under the effect of extremely high frequency electromagnetic radiation of low intensity, Biofizika. 2002 Jan-Feb;47(1):71-7.

[1221] Soghomonyan, D., Trchounian, K., & Trchounian, A. (2016). Millimeter waves or extremely high frequency electromagnetic fields in the environment: what are their effects on bacteria? Applied Microbiology and Biotechnology, 100(11), 4761–4771. doi:10.1007/s00253-016-7538-0

[1222] EMFscientist.org (2015) 'International Appeal: Scientists call for Protection from Non-ionizing Electromagnetic Field Exposure', To: United Nations

[1223] Gerd et al (2017) 'Scientists warn of potential serious health effects of 5G', Accessed Online: https://ehtrust.org/wp-content/uploads/Scientist-5G-appeal-2017.pdf

[1224] Johansson et al (2014) 'Health risk from wireless? The debate is over', Accessed Online: https://electromagnetichealth.org/electromagnetic-health-blog/article-by-professor-olle-johansson-health-risk-from-wireless-the-debate-is-over/

[1225] Pall, M. L. (2015). Scientific evidence contradicts findings and assumptions of Canadian Safety Panel 6: microwaves act through voltage-gated calcium channel activation to induce biological impacts at non-thermal levels, supporting a paradigm shift for microwave/lower frequency electromagnetic field action. Reviews on Environmental Health, 30(2). doi:10.1515/reveh-2015-0001

[1226] Pall, M. L. (2013). Electromagnetic fields actviaactivation of voltage-gated calcium channels to produce beneficial or adverse effects. Journal of Cellular and Molecular Medicine, 17(8), 958–965. doi:10.1111/jcmm.12088

[1227] Pall (2016) 'Electromagnetic Fields Act Similarly in Plants as in Animals: Probable Activation of Calcium Channels via Their Voltage Sensor', Current Chemical Biology, Volume 10 , Issue 1 , 2016, DOI : 10.2174/2212796810666160419160433

[1228] Beckman, J. S., & Koppenol, W. H. (1996). Nitric oxide, superoxide, and peroxynitrite: the good, the bad, and ugly. American Journal of Physiology-Cell Physiology, 271(5), C1424–C1437. doi:10.1152/ajpcell.1996.271.5.c1424

[1229] Thomson, L. (2015). 3-Nitrotyrosine Modified Proteins in Atherosclerosis. Disease Markers, 2015, 1–8. doi:10.1155/2015/708282

[1230] Pacher, P., Beckman, J. S., & Liaudet, L. (2007). Nitric Oxide and Peroxynitrite in Health and Disease. Physiological Reviews, 87(1), 315–424. doi:10.1152/physrev.00029.2006

[1231] Van der Loo et al (2000). Enhanced Peroxynitrite Formation Is Associated with Vascular Aging. The Journal of Experimental Medicine, 192(12), 1731–1744. doi:10.1084/jem.192.12.1731

[1232] Stewart et al (2006). Ionizing Radiation Accelerates the Development of Atherosclerotic Lesions in ApoE−/− Mice and Predisposes to an Inflammatory Plaque Phenotype Prone to Hemorrhage. The American Journal of Pathology, 168(2), 649–658. doi:10.2353/ajpath.2006.050409

[1233] ICNIRP (2009) 'GUIDELINES ON LIMITS OF EXPOSURE TO STATIC MAGNETIC FIELDS', HEALTH PHYSICS 96(4):504-514; 2009.

[1234] Peoples et al (2019). Mitochondrial dysfunction and oxidative stress in heart disease. Experimental & Molecular Medicine, 51(12). doi:10.1038/s12276-019-0355-7

[1235] Dhalla et al (2000). Role of oxidative stress in cardiovascular diseases. Journal of Hypertension, 18(6), 655–673. doi:10.1097/00004872-200018060-00002

[1236] Yingxin et al (2012) 'Roles of oxidative stress in synchrotron radiation X-ray-induced testicular damage of rodents', Int J Physiol Pathophysiol Pharmacol. 2012; 4(2): 108–114.

[1237] Fessel, J. P., & Oldham, W. M. (2018). Pyridine Dinucleotides from Molecules to Man. Antioxidants & Redox Signaling, 28(3), 180–212. doi:10.1089/ars.2017.7120

[1238] Kim et al (2017). Long-term exposure to 835 MHz RF-EMF induces hyperactivity, autophagy and demyelination in the cortical neurons of mice. Scientific Reports, 7(1). doi:10.1038/srep41129

[1239] Li et al (2018). The Protective Effect of Autophagy on DNA Damage in Mouse Spermatocyte-Derived Cells Exposed to 1800 MHz Radiofrequency Electromagnetic Fields. Cellular Physiology and Biochemistry, 48(1), 29–41. doi:10.1159/000491660

[1240] LaRocca et al (2012) 'Translational evidence that impaired autophagy contributes to arterial ageing', The Journal of Physiology, 590: 3305-3316. doi:10.1113/jphysiol.2012.229690

[1241] Razani et al (2012). Autophagy Links Inflammasomes to Atherosclerotic Progression. Cell Metabolism, 15(4), 534–544. doi:10.1016/j.cmet.2012.02.011

[1242] Liao et al (2012). Macrophage Autophagy Plays a Protective Role in Advanced Atherosclerosis. Cell Metabolism, 15(4), 545–553. doi:10.1016/j.cmet.2012.01.022

[1243] Liu et al (2008). Changes in autophagy after traumatic brain injury. Journal of cerebral blood flow and metabolism : official journal of the International Society of Cerebral Blood Flow and Metabolism, 28(4), 674–683. https://doi.org/10.1038/sj.jcbfm.9600587

[1244] White, H., & Venkatesh, B. (2011). Clinical review: ketones and brain injury. Critical care (London, England), 15(2), 219. https://doi.org/10.1186/cc10020

[1245] Lamark, T., & Johansen, T. (2012). Aggrephagy: Selective Disposal of Protein Aggregates by Macroautophagy. International Journal of Cell Biology, 2012, 1–21. doi:10.1155/2012/736905

[1246] Huber et al (2003) 'Radio frequency electromagnetic field exposure in humans: Estimation of SAR distribution in the brain, effects on sleep and heart rate', BioElectroMagnetics, Volume 24, Issue 4, Pages 262-276.

[1247] Kim et al (2020). Effects of Recess Depth Under the Gate Area on the Vth-Shift for Fabricating Normally-Off Field Effect Transistors on AlGaN/GaN Heterostructures. Journal of Nanoscience and Nanotechnology, 20(7), 4170–4175. doi:10.1166/jnn.2020.17783

[1248] Henz et al (2018) 'Mobile Phone Chips Reduce Increases in EEG Brain Activity Induced by Mobile Phone-Emitted Electromagnetic Fields', Front Neurosci. 2018; 12: 190.

[1249] Feinendegen (2014) 'Evidence for beneficial low level radiation effects and radiation hormesis', The British Journal of Radiology 2005 78:925, 3-7.

[1250] Wolff, S. (1998). The adaptive response in radiobiology: evolving insights and implications. Environmental Health Perspectives, 106(suppl 1), 277–283. doi:10.1289/ehp.98106s1277

[1251] Muller, H. J. (1927). "Artificial Transmutation of the Gene" (PDF). *Science*. 66 (1699): 84–87.

[1252] Crow, J. F.; Abrahamson, S. (1997). "Seventy Years Ago: Mutation Becomes Experimental". *Genetics*. 147 (4): 1491–1496.

[1253] "Hermann J. Muller - Nobel Lecture". *Nobel Prize*. 12 December 1946.

[1254] Hosoi, Y., & Sakamoto, K. (1993). Suppressive effect of low dose total body irradiation on lung metastasis: dose dependency and effective period. Radiotherapy and oncology : journal of the European Society for Therapeutic Radiology and Oncology, 26(2), 177–179. https://doi.org/10.1016/0167-8140(93)90101-d

[1255] Luckey TD. (1980) Hormesis with ionizing radiation. Boca Raton, Fla: CRC Press.

[1256] Cohen B. L. (1995). Test of the linear-no threshold theory of radiation carcinogenesis for inhaled radon decay products. Health physics, 68(2), 157–174. https://doi.org/10.1097/00004032-199502000-00002

[1257] Calabrese, E. J. (2011). Muller's Nobel Prize Lecture: When Ideology Prevailed Over Science. Toxicological Sciences, 126(1), 1–4. doi:10.1093/toxsci/kfr338

[1258] E. J. Calabrese, (2011) Muller's Nobel lecture on dose–response for ionizing radiation: ideology or science? *Archives of Toxicology* 2011, 85(12), 1495-1498.

[1259] Kasperson, Roger E.; Stallen, Pieter Jan M. (1991). *Communicating Risks to the Public: International Perspectives*. Berlin: Springer Science and Media. pp. 160–2.

[1260] Perucchi, M; Domenighetti, G (1990). "The Chernobyl accident and induced abortions: Only one-way information". *Scandinavian Journal of Work, Environment & Health.* 16 (6): 443–4. doi:10.5271/sjweh.1761.

[1261] Dolk, H.; Nichols, R. (1999). "Evaluation of the impact of Chernobyl on the prevalence of congenital anomalies in 16 regions of Europe. EUROCAT Working Group". *International Journal of Epidemiology.* 28 (5): 941–8. doi:10.1093/ije/28.5.941.

[1262] Castronovo, Frank P. (1999). "Teratogen update: Radiation and chernobyl". *Teratology.* 60 (2): 100–6. doi:10.1002/(sici)1096-9926(199908)60:2<100::aid-tera14>3.3.co;2-8.

[1263] Ozasa et al (2012). Studies of the Mortality of Atomic Bomb Survivors, Report 14, 1950–2003: An Overview of Cancer and Noncancer Diseases. Radiation Research, 177(3), 229–243. doi:10.1667/rr2629.1

[1264] Russo et al (2011) 'Cellular adaptive response to chronic radiation exposure in interventional cardiologists', European Heart Journal, Volume 33, Issue 3, February 2012, Pages 408–414, https://doi.org/10.1093/eurheartj/ehr263

[1265] Mortazavi 'High Background Radiation Areas of Ramsar, Iran', Accessed Online: http://www.angelfire.com/mo/radioadaptive/ramsar.html

[1266] Sohrabi, Mehdi; Babapouran, Mozhgan (2005), "New public dose assessment from internal and external exposures in low- and elevated-level natural radiation areas of Ramsar, Iran", *International Congress Series*, 1276: 169–174, doi:10.1016/j.ics.2004.11.102

[1267] Mosavi-Jarrahi et al (2005), "Mortality and morbidity from cancer in the population exposed to high level of natural radiation area in Ramsar, Iran", *International Congress Series*, 1276: 106–109, doi:10.1016/j.ics.2004.11.109

[1268] Zakeri et al (2011), "Chromosome aberrations in peripheral blood lymphocytes of individuals living in high background radiation areas of Ramsar, Iran", *Radiation and Environmental Biophysics*, 50 (4): 571–578, doi:10.1007/s00411-011-0381-x

[1269] Tabarraie et al (2008), "Impact of High Natural Background Radiation on Woman's Primary Infertility", *Research Journal of Biological Sciences*, 3 (5): 534–536

[1270] National Research Council (2006) 'Health Risks from Exposure to Low Levels of Ionizing Radiation. BEIR VII Phase 2 (2006) doi:10.17226/11340

[1271] NCRP report. Ncrppublications.org.

[1272] Tubiana et al (2005) 'Dose–effect relationships and the estimation of the carcinogenic effects of low doses of ionizing radiation', Joint Report n° 2, Académie Nationale de Médecine, Institut de France—Académie des Sciences, Accessed Online: http://www.academiemedecine.fr/actualites/rapports.asp Edition Nucleon (Paris 2005) ISBN 2-84332-018-6

[1273] Aurengo; et al. (2005). "Dose-effect relationships and estimation of the carcinogenic effects of low doses of ionizing radiation" (PDF). Académie des Sciences & Académie nationale de Médecine.

[1274] Health Physics Society (2019) Radiation Risk in Perspective, Position Statement of the Health Physics Society, PS010-4, Accessed Online: https://hps.org/documents/radiationrisk.pdf

[1275] The American Nuclear Society (2001) Health Effects of Low-Level Radiation. Position Statement 41, Accessed Online: https://ans.org/pi/ps/docs/ps41.pdf

[1276] Tubiana, M., Feinendegen, L. E., Yang, C., & Kaminski, J. M. (2009). The linear no-threshold relationship is inconsistent with radiation biologic and experimental data. Radiology, 251(1), 13–22. https://doi.org/10.1148/radiol.2511080671

[1277] Calabrese E. J. (2004). Hormesis: a revolution in toxicology, risk assessment and medicine. EMBO reports, 5 Spec No(Suppl 1), S37–S40. https://doi.org/10.1038/sj.embor.7400222

[1278] Cuttler and Pollycove (2003) 'Can Cancer Be Treated with Low Doses of Radiation?', Journal of American Physicians and Surgeons. 2003;8:108–11.

[1279] Farooque et al (2011). Low-dose radiation therapy of cancer: role of immune enhancement. Expert review of anticancer therapy, 11(5), 791–802. https://doi.org/10.1586/era.10.217

[1280] Doss M. (2013). The importance of adaptive response in cancer prevention and therapy. Medical physics, 40(3), 030401. https://doi.org/10.1118/1.4773027

[1281] Sakamoto K. (2004). Radiobiological basis for cancer therapy by total or half-body irradiation. Nonlinearity in biology, toxicology, medicine, 2(4), 293–316. https://doi.org/10.1080/15401420490900254

[1282] Tubiana et al (2011). A new method of assessing the dose-carcinogenic effect relationship in patients exposed to ionizing radiation. A concise presentation of preliminary data. Health physics, 100(3), 296–299. https://doi.org/10.1097/hp.0b013e31820a1b35

[1283] Loprinzi (2018) 'Short-Term Fasting Before Chemotherapy in Treating Patients With Cancer', National Cancer Institute (NCI), Rochester, Minnesota, United States, 55905

[1284] Marziali (2012) 'Fasting weakens cancer in mice', USC News, Accessed Online: https://news.usc.edu/29428/fasting-weakens-cancer-in-mice/

[1285] Carroll et al (2014). Developments in low level light therapy (LLLT) for dentistry. Dental materials : official publication of the Academy of Dental Materials, 30(5), 465–475. https://doi.org/10.1016/j.dental.2014.02.006

[1286] Mitchell, U. H., & Mack, G. L. (2013). Low-level laser treatment with near-infrared light increases venous nitric oxide levels acutely: a single-blind, randomized clinical trial of efficacy. American journal of physical medicine & rehabilitation, 92(2), 151–156. https://doi.org/10.1097/PHM.0b013e318269d70a

[1287] NASA (2011) 'NASA Light Technology Successfully Reduces Cancer Patients Painful Side Effects from Radiation and Chemotherapy', Accessed Online: https://www.nasa.gov/topics/nasalife/features/heals.html

[1288] Hayworth et al (2010). In vivo low-level light therapy increases cytochrome oxidase in skeletal muscle. Photochemistry and photobiology, 86(3), 673–680. https://doi.org/10.1111/j.1751-1097.2010.00732.x

[1289] Hu, D., Zhu, S., & Potas, J. R. (2016). Red LED photobiomodulation reduces pain hypersensitivity and improves sensorimotor function following mild T10 hemicontusion spinal cord injury. Journal of neuroinflammation, 13(1), 200. https://doi.org/10.1186/s12974-016-0679-3

[1290] Ranjbar, R., & Takhtfooladi, M. A. (2016). The effects of photobiomodulation therapy on Staphylococcus aureus infected surgical wounds in diabetic rats. A microbiological, histopathological, and biomechanical study. Acta cirurgica brasileira, 31(8), 498–504. https://doi.org/10.1590/S0102-865020160080000001

[1291] Al-Watban F. A. (2009). Laser therapy converts diabetic wound healing to normal healing. Photomedicine and laser surgery, 27(1), 127–135. https://doi.org/10.1089/pho.2008.2406

[1292] Antonialli et al (2014). Phototherapy in skeletal muscle performance and recovery after exercise: effect of combination of super-pulsed laser and light-emitting diodes. Lasers in medical science, 29(6), 1967–1976. https://doi.org/10.1007/s10103-014-1611-7

[1293] Al Rashoud et al (2014). Efficacy of low-level laser therapy applied at acupuncture points in knee osteoarthritis: a randomised double-blind comparative trial. Physiotherapy, 100(3), 242–248. https://doi.org/10.1016/j.physio.2013.09.007

[1294] Antonialli et al (2014). Phototherapy in skeletal muscle performance and recovery after exercise: effect of combination of super-pulsed laser and light-emitting diodes. Lasers in medical science, 29(6), 1967–1976. https://doi.org/10.1007/s10103-014-1611-7

[1295] Greco et al (1989). Increase in RNA and protein synthesis by mitochondria irradiated with helium-neon laser. Biochemical and biophysical research communications, 163(3), 1428–1434. https://doi.org/10.1016/0006-291x(89)91138-8

[1296] Karu, T. I., & Kolyakov, S. F. (2005). Exact action spectra for cellular responses relevant to phototherapy. Photomedicine and laser surgery, 23(4), 355–361.

https://doi.org/10.1089/pho.2005.23.355

[1297] Chung et al (2012). The nuts and bolts of low-level laser (light) therapy. Annals of biomedical engineering, 40(2), 516–533. https://doi.org/10.1007/s10439-011-0454-7

[1298] Alves et al (2014). Effects of low-level laser therapy on skeletal muscle repair: a systematic review. American journal of physical medicine & rehabilitation, 93(12), 1073–1085. https://doi.org/10.1097/PHM.0000000000000158

[1299] Hashmi et al (2010). Role of low-level laser therapy in neurorehabilitation. PM & R : the journal of injury, function, and rehabilitation, 2(12 Suppl 2), S292–S305. https://doi.org/10.1016/j.pmrj.2010.10.013

[1300] Höfling et al (2012). Assessment of the effects of low-level laser therapy on the thyroid vascularization of patients with autoimmune hypothyroidism by color Doppler ultrasound. ISRN endocrinology, 2012, 126720. https://doi.org/10.5402/2012/126720

[1301] Aziz-Jalali et al (2012). Comparison of Red and Infrared Low-level Laser Therapy in the Treatment of Acne Vulgaris. Indian journal of dermatology, 57(2), 128–130. https://doi.org/10.4103/0019-5154.94283

[1302] Wunsch, A., & Matuschka, K. (2014). A controlled trial to determine the efficacy of red and near-infrared light treatment in patient satisfaction, reduction of fine lines, wrinkles, skin roughness, and intradermal collagen density increase. Photomedicine and laser surgery, 32(2), 93–100. https://doi.org/10.1089/pho.2013.3616

[1303] Nam et al (2017). The Efficacy and Safety of 660 nm and 411 to 777 nm Light-Emitting Devices for Treating Wrinkles. Dermatologic surgery : official publication for American Society for Dermatologic Surgery [et al.], 43(3), 371–380. https://doi.org/10.1097/DSS.0000000000000981

[1304] Kharkwal et al (2011). Photodynamic therapy for infections: clinical applications. Lasers in surgery and medicine, 43(7), 755–767. https://doi.org/10.1002/lsm.21080

[1305] König et al (2000). Red light kills bacteria via photodynamic action. Cellular and molecular biology (Noisy-le-Grand, France), 46(7), 1297–1303.

[1306] Ferraresi et al (2012). Low-level laser (light) therapy (LLLT) on muscle tissue: performance, fatigue and repair benefited by the power of light. Photonics & lasers in medicine, 1(4), 267–286. https://doi.org/10.1515/plm-2012-0032

[1307] Lanzafame et al (2013). The growth of human scalp hair mediated by visible red light laser and LED sources in males. Lasers in surgery and medicine, 45(8), 487–495. https://doi.org/10.1002/lsm.22173

[1308] Friedman, S., & Schnoor, P. (2017). Novel Approach to Treating Androgenetic Alopecia in Females With Photobiomodulation (Low-Level Laser Therapy). Dermatologic surgery : official publication for American Society for Dermatologic Surgery [et al.], 43(6), 856–867. https://doi.org/10.1097/DSS.0000000000001114

[1309] College of Health Sciences (2015) 'A novel approach to using light to treat macular degeneration', University of Wisconsin-Milwaukee, Accessed Online: https://uwm.edu/healthsciences/news/a-novel-approach-to-using-light-to-treat-macular-degeneration/

[1310] Naeser et al (2011). Improved cognitive function after transcranial, light-emitting diode treatments in chronic, traumatic brain injury: two case reports. Photomedicine and laser surgery, 29(5), 351–358. https://doi.org/10.1089/pho.2010.2814

[1311] Salahaldin et al (2012). Low-level laser therapy in patients with complaints of tinnitus: a clinical study. ISRN otolaryngology, 2012, 132060. https://doi.org/10.5402/2012/132060

[1312] Choi et al (2016) 'Amber Light (590 nm) Induces the Breakdown of Lipid Droplets through Autophagy-Related Lysosomal Degradation in Differentiated Adipocytes', Scientific Reports volume 6, Article number: 28476 https://doi.org/10.1038/srep28476

[1313] Fettig (2017) 'A Retrospective Analysis of the Effects of Red Light LED Therapy on Body Contouring', Vevazz, LLC, Dallas, Texas, United States, 75204

[1314] Denda, M., & Fuziwara, S. (2008). Visible radiation affects epidermal permeability barrier recovery: selective effects of red and blue light. The Journal of investigative dermatology, 128(5), 1335–1336. https://doi.org/10.1038/sj.jid.5701168

[1315] Kim et al (2018). Autophagy in Human Skin Fibroblasts: Impact of Age. International journal of molecular sciences, 19(8), 2254. https://doi.org/10.3390/ijms19082254

[1316] Sil, P., Wong, S. W., & Martinez, J. (2018). More Than Skin Deep: Autophagy Is Vital for Skin Barrier Function. Frontiers in immunology, 9, 1376. https://doi.org/10.3389/fimmu.2018.01376

[1317] Tashiro et al (2014). Age-related disruption of autophagy in dermal fibroblasts modulates extracellular matrix components. Biochemical and biophysical research communications, 443(1), 167–172. https://doi.org/10.1016/j.bbrc.2013.11.066

[1318] Zhao et al (2012). Red light and the sleep quality and endurance performance of Chinese female basketball players. Journal of athletic training, 47(6), 673–678. https://doi.org/10.4085/1062-6050-47.6.08

[1319] Obayashi et al (2014). Independent associations of exposure to evening light and nocturnal urinary melatonin excretion with diabetes in the elderly. Chronobiology international, 31(3), 394–400. https://doi.org/10.3109/07420528.2013.864299

[1320] Obayashi et al (2013). Exposure to light at night, nocturnal urinary melatonin excretion, and obesity/dyslipidemia in the elderly: a cross-sectional analysis of the HEIJO-KYO study. The Journal of clinical endocrinology and metabolism, 98(1), 337–344. https://doi.org/10.1210/jc.2012-2874

[1321] Hood, D. A., Zak, R., & Pette, D. (1989). Chronic stimulation of rat skeletal muscle induces coordinate increases in mitochondrial and nuclear mRNAs of cytochrome-c-oxidase subunits. European journal of biochemistry, 179(2), 275–280. https://doi.org/10.1111/j.1432-1033.1989.tb14551.x

[1322] Sutbeyaz, S., Sezer, N., & Koseoglu, B. (2005).The effect of pulsed electromagnetic fields in the treatment of cervical osteoarthritis: a randomized, double-blind, sham-controlled trial. Rheumatology International, 26(4), 320-324.doi:10.1007/s00296-005-0600-3

[1323] Ganesan et al (2009) 'Low frequency pulsed electromagnetic field - A viable alternative therapy for arthritis', Indian Journal of Experimental Biology, Vol. 47, December 2009, pp 939-948.

[1324] Thomas et al (2007) 'A randomized, double-blind, placebo-controlled clinical trial using a low-frequency magnetic field in the treatment of musculoskeletal chronic pain', Pain Res Manag. 2007 Winter; 12(4): 249–258.

[1325] Huang et al (2008) 'Clinical update of pulsed electromagnetic fields on osteoporosis.' Chinese Medical Journal [01 Oct 2008, 121(20):2095-2099].

[1326] Cohen et al (2016) 'Repetitive deep transcranial magnetic stimulation for motor symptoms in Parkinson's disease: A feasibility study', Clinical Neurology and Neurosurgery, 24 Nov 2015, 140:73-78, DOI: 10.1016/j.clineuro.2015.11.017

[1327] Lee et al (2015) 'Treatment of Surgical Site Infection in Posterior Lumbar Interbody Fusion', Asian Spine J. 2015 Dec; 9(6): 841–848.

[1328] Iuchi et al (2016) 'Molecular hydrogen regulates gene expression by modifying the free radical chain reaction-dependent generation of oxidized phospholipid mediators', Sci Rep. 2016; 6: 18971.

[1329] Alkhatib and Al Zailaey (2015). MEDICAL AND ENVIRONMENTAL APPLICATIONS OF ACTIVATED CHARCOAL: REVIEW ARTICLE. European Scientific Journal, ESJ, 11(3). Retrieved from http://eujournal.org/index.php/esj/article/view/4987

[1330] Yokota et al (2015). Protective effect of molecular hydrogen against oxidative stress caused by peroxynitrite derived from nitric oxide in rat retina. Clinical & Experimental Ophthalmology, 43(6), 568–577. doi:10.1111/ceo.12525

[1331] Chan et al (1995). Effects of three dietary phytochemicals from tea, rosemary and turmeric on inflammation-induced nitrite production. Cancer Letters, 96(1), 23–29. doi:10.1016/0304-3835(95)03913-h

[1332] Landauer et al (2003) 'Genistein treatment protects mice from ionizing radiation injury', November 2003Journal of Applied Toxicology 23(6):379-85, DOI: 10.1002/jat.904

[1333] ZHOU, Y., & MI, M.-T. (2005). Genistein Stimulates Hematopoiesis and Increases Survival in Irradiated Mice. Journal of Radiation Research, 46(4), 425–433. doi:10.1269/jrr.46.425

[1334] Choudhary et al (1998). Modulation of radioresponse of glyoxalase system by curcumin. Journal of Ethnopharmacology, 64(1), 1–7. doi:10.1016/s0378-8741(98)00064-6

[1335] Srinivasan et al (2006). Protective effect of curcumin on γ-radiation induced DNA damage and lipid peroxidation in cultured human lymphocytes. Mutation Research/Genetic Toxicology and Environmental Mutagenesis, 611(1-2), 96–103. doi:10.1016/j.mrgentox.2006.07.002

[1336] Inano, H., & Onoda, M. (2002). Radioprotective action of curcumin extracted from Curcuma longa LINN: inhibitory effect on formation of urinary 8-hydroxy-2′-deoxyguanosine, tumorigenesis, but not mortality, induced by γ-ray irradiation. International Journal of Radiation Oncology*Biology*Physics, 53(3), 735–743. doi:10.1016/s0360-3016(02)02794-3

[1337] Lee et al (2010) 'Radioprotective effect of American ginseng on human lymphocytes at 90 minutes postirradiation: a study of 40 cases', Journal of Alternative and Complementary Medicine (New York, N.Y.), 01 May 2010, 16(5):561-567, DOI: 10.1089/acm.2009.0590

[1338] Singh SP, Abraham SK, Kesavan PC. (1995) In vivo radioprotection with garlic extract. Mutat Res. Dec;345(3-4):147-53.

[1339] Singh SP, Abraham SK, Kesavan PC. (1996) Radioprotection of mice following garlic pretreatment. Br J Cancer Suppl. Jul;27:S102-4.

[1340] Alaoui-Youssefi A, Lamproglou I, Drieu K, Emerit I. (1999) Anticlastogenic effects of Ginkgo biloba extract (EGb 761) and some of its constituents in irradiated rats. Mutat Res. Sep 15;445(1):99-104.

[1341] Emerit I, Arutyunyan R, Oganesian N, et al. (1995) Radiation-induced clastogenic factors: anticlastogenic effect of Ginkgo biloba extract. Free Radic Biol Med. Jun;18(6):985-91.

[1342] Emerit I, Oganesian N, Sarkisian T, et al. (1995) Clastogenic factors in the plasma of Chernobyl accident recovery workers: anticlastogenic effect of Ginkgo biloba extract. Radiat Res. 1995 Nov;144(2):198-205.

[1343] Chen B, Zhou XC. (2003) Protective effect of natural dietary antioxidants on space radiation-induced damages. Space Med Med Eng (Beijing). 2003;16 Suppl:514-8.

[1344] Yong et al (2009) High dietary antioxidant intakes are associated with decreased chromosome translocation frequency in airline pilots. Am J Clin Nutr. 2009 Nov;90(5):1402-10.

[1345] Ben-Amotz A, Yatziv S, Sela M, et al. (1998) Effect of natural beta-carotene supplementation in children exposed to radiation from the Chernobyl accident. Radiat Environ Biophys. 1998 Oct;37(3):187-93.

[1346] BURNS, F. J., CHEN, S., XU, G., WU, F., & TANG, M.-S. (2002). The Action of a Dietary Retinoid on Gene Expression and Cancer Induction in Electron-irradiated Rat Skin. Journal of Radiation Research, 43(S), S229–S232. doi:10.1269/jrr.43.s229

[1347] Schoenfeld et al (2017). O 2 ·− and H 2 O 2 -Mediated Disruption of Fe Metabolism Causes the Differential Susceptibility of NSCLC and GBM Cancer Cells to Pharmacological Ascorbate. Cancer Cell, 31(4), 487–500.e8. doi:10.1016/j.ccell.2017.02.018

[1348] Creagan et al (1979). Failure of High-Dose Vitamin C (Ascorbic Acid) Therapy to Benefit Patients with Advanced Cancer. New England Journal of Medicine, 301(13), 687–690. doi:10.1056/nejm197909273011303

[1349] Yanagisawa A. Effect of Vitamin C and anti-oxidative nutrition on radiation-induced gene expression in Fukushima nuclear plant workers. Free download of full presentation at http://www.doctoryourself.com/Radiation_VitC.pptx.pdf

[1350] Narra et al (1993) 'Vitamin C as a Radioprotector Against Iodine-131 In Vivo', J Nucl Med April 1, 1993 vol. 34, no. 4 637-640.

[1351] Prasad et al (2003) Alpha-tocopheryl succinate, the most effective form of vitamin E for adjuvant cancer treatment: a review. J Am Coll Nutr. 2003 Apr;22(2):108-17.

[1352] Singh (2006) Induction of cytokines by radioprotective tocopherol analogs. Exp Mol Pathol. 2006 Aug;81(1):55-61.

[1353] Selig et al (1993) Radioprotective effect of N-acetylcysteine on granulocyte/macrophage colony-forming cells of human bone marrow. J Cancer Res Clin Oncol. 1993;119(6):346-9.

[1354] Hans von Euler-Chelpin (1930) Fermentation of Sugars and Fermentative Enzymes Nobel Lecture NobelPrize.org. Nobel Media AB 2020. Accessed Online: <https://www.nobelprize.org/prizes/chemistry/1929/euler-chelpin/lecture/>

[1355] Cantó, C., Menzies, K. J., & Auwerx, J. (2015). NAD+ Metabolism and the Control of Energy Homeostasis: A Balancing Act between Mitochondria and the Nucleus. Cell Metabolism, 22(1), 31–53. doi:10.1016/j.cmet.2015.05.023

[1356] Goody, M. F., & Henry, C. A. (2018). A need for NAD+ in muscle development, homeostasis, and aging. Skeletal muscle, 8(1), 9. https://doi.org/10.1186/s13395-018-0154-1

[1357] Ryu et al (2016). NAD+ repletion improves muscle function in muscular dystrophy and counters global PARylation. Science Translational Medicine, 8(361), 361ra139–361ra139. doi:10.1126/scitranslmed.aaf5504

[1358] Gomes et al (2013). Declining NAD+ Induces a Pseudohypoxic State Disrupting Nuclear-Mitochondrial Communication during Aging. Cell, 155(7), 1624–1638. doi:10.1016/j.cell.2013.11.037

[1359] Hou et al (2018). NAD+ supplementation normalizes key Alzheimer's features and DNA damage responses in a new AD mouse model with introduced DNA repair deficiency. Proceedings of the National Academy of Sciences of the United States of America, 115(8), E1876–E1885. https://doi.org/10.1073/pnas.1718819115

[1360] Sheline et al (2000). Zinc-Induced Cortical Neuronal Death: Contribution of Energy Failure Attributable to Loss of NAD+and Inhibition of Glycolysis. The Journal of Neuroscience, 20(9), 3139–3146. doi:10.1523/jneurosci.20-09-03139.2000

[1361] Yang et al (2007). Nutrient-Sensitive Mitochondrial NAD+ Levels Dictate Cell Survival. Cell, 130(6), 1095–1107. doi:10.1016/j.cell.2007.07.035

[1362] Trafton (2018) 'Study suggests method for boosting growth of blood vessels and muscle', MIT News Office, March 22, 2018, Accessed Online: https://news.mit.edu/2018/study-suggests-method-boost-growth-blood-vessels-muscle-0322

[1363] Fang et al (2016). NAD + Replenishment Improves Lifespan and Healthspan in Ataxia Telangiectasia Models via Mitophagy and DNA Repair. Cell Metabolism, 24(4), 566–581. doi:10.1016/j.cmet.2016.09.004

[1364] Li et al (2017) 'A conserved NAD+ binding pocket that regulates protein-protein interactions during aging', Science 24 Mar 2017: Vol. 355, Issue 6331, pp. 1312-1317, DOI: 10.1126/science.aad8242

[1365] Johnson, S., & Imai, S. I. (2018). NAD + biosynthesis, aging, and disease. F1000Research, 7, 132. https://doi.org/10.12688/f1000research.12120.1

[1366] Massudi et al (2012). Age-associated changes in oxidative stress and NAD+ metabolism in human tissue. PloS one, 7(7), e42357. https://doi.org/10.1371/journal.pone.0042357

[1367] Yang, Y., & Sauve, A. A. (2016). NAD + metabolism: Bioenergetics, signaling and manipulation for therapy. Biochimica et Biophysica Acta (BBA) - Proteins and Proteomics, 1864(12), 1787–1800. doi:10.1016/j.bbapap.2016.06.014

[1368] Gomes et al (2013). Declining NAD(+) induces a pseudohypoxic state disrupting nuclear-mitochondrial communication during aging. Cell, 155(7), 1624–1638. https://doi.org/10.1016/j.cell.2013.11.037

[1369] Gong et al (2013). Nicotinamide riboside restores cognition through an upregulation of proliferator-activated receptor-γ coactivator 1α regulated β-secretase 1 degradation and mitochondrial gene expression in Alzheimer's mouse models. Neurobiology of aging, 34(6), 1581–1588. https://doi.org/10.1016/j.neurobiolaging.2012.12.005

[1370] Schultz, M. B., & Sinclair, D. A. (2016). Why NAD(+) Declines during Aging: It's Destroyed. Cell metabolism, 23(6), 965–966. https://doi.org/10.1016/j.cmet.2016.05.022

[1371] Tirumurugaan et al (2008). Regulation of the cd38 promoter in human airway smooth muscle cells by TNF-alpha and dexamethasone. Respiratory research, 9(1), 26. https://doi.org/10.1186/1465-9921-9-26

[1372] Camacho-Pereira et al (2016). CD38 Dictates Age-Related NAD Decline and Mitochondrial Dysfunction through an SIRT3-Dependent Mechanism. Cell metabolism, 23(6), 1127–1139. https://doi.org/10.1016/j.cmet.2016.05.006

[1373] Zhou et al (2016) 'Hepatic NAD+ deficiency as a therapeutic target for non-alcoholic fatty liver disease in ageing', British Journal of Pharmacology, 173(15). doi:10.1111/bph.v173.15

[1374] Billingham, L. K., & Chandel, N. S. (2019). NAD-biosynthetic pathways regulate innate immunity. Nature Immunology, 20(4), 380–382. doi:10.1038/s41590-019-0353-x

[1375] Bruzzone et al (2009). Catastrophic NAD+ Depletion in Activated T Lymphocytes through Nampt Inhibition Reduces Demyelination and Disability in EAE. PLoS ONE, 4(11), e7897. doi:10.1371/journal.pone.0007897

[1376] Tullius et al (2014). NAD+ protects against EAE by regulating CD4+ T-cell differentiation. Nature Communications, 5(1). doi:10.1038/ncomms6101

[1377] Elkhal et al (2016). NAD+ regulates Treg cell fate and promotes allograft survival via a systemic IL-10 production that is CD4+ CD25+ Foxp3+ T cells independent. Scientific Reports, 6(1). doi:10.1038/srep22325

[1378] Bhat et al (2016). Imaging the NADH:NAD+ Homeostasis for Understanding the Metabolic Response of Mycobacterium to Physiologically Relevant Stresses. Frontiers in Cellular and Infection Microbiology, 6. doi:10.3389/fcimb.2016.00145

[1379] Heer et al (2020). Coronavirus and PARP expression dysregulate the NAD Metabolome: a potentially actionable component of innate immunity. doi:10.1101/2020.04.17.047480

[1380] Cameron et al (2019). Inflammatory macrophage dependence on NAD+ salvage is a consequence of reactive oxygen species–mediated DNA damage. Nature Immunology, 20(4), 420–432. doi:10.1038/s41590-019-0336-y

[1381] Moffett, J. R., & Namboodiri, M. A. (2003). Tryptophan and the immune response. Immunology & Cell Biology, 81(4), 247–265. doi:10.1046/j.1440-1711.2003.t01-1-01177.x

[1382] Caito, S. W., & Aschner, M. (2016). NAD+Supplementation Attenuates Methylmercury Dopaminergic and Mitochondrial Toxicity inCaenorhabditis Elegans. Toxicological Sciences, 151(1), 139–149. doi:10.1093/toxsci/kfw030

[1383] Hu, X. et al. (2019). Environmental Cadmium Enhances Lung Injury by Respiratory Syncytial Virus Infection. *The American Journal of Pathology* 189(8): 1513–1525.

[1384] Grahnert et al (2010). Review: NAD +: A modulator of immune functions. Innate Immunity, 17(2), 212–233. doi:10.1177/1753425910361989

[1385] Boshoff et al (2008). Biosynthesis and Recycling of Nicotinamide Cofactors inMycobacterium tuberculosis. Journal of Biological Chemistry, 283(28), 19329–19341. doi:10.1074/jbc.m800694200

[1386] Mesquita et al (2015). Exploring NAD+ metabolism in host–pathogen interactions. Cellular and Molecular Life Sciences, 73(6), 1225–1236. doi:10.1007/s00018-015-2119-4

[1387] Beste et al (2007). GSMN-TB: a web-based genome-scale network model of Mycobacterium tuberculosis metabolism. Genome Biology, 8(5), R89. doi:10.1186/gb-2007-8-5-r89

[1388] Vilchèze et al (2010). NAD+auxotrophy is bacteriocidal for the tubercle bacilli. Molecular Microbiology, 76(2), 365–377. doi:10.1111/j.1365-2958.2010.07099.x

[1389] Yang et al (2017). Pterostilbene, a Methoxylated Resveratrol Derivative, Efficiently Eradicates Planktonic, Biofilm, and Intracellular MRSA by Topical Application. Frontiers in Microbiology, 8. doi:10.3389/fmicb.2017.01103

[1390] Murray, M. F. (2003). Nicotinamide: An Oral Antimicrobial Agent with Activity against BothMycobacterium tuberculosisand Human Immunodeficiency Virus. Clinical Infectious Diseases, 36(4), 453–460. doi:10.1086/367544

[1391] Hwang, E. S., & Song, S. B. (2017). Nicotinamide is an inhibitor of SIRT1 in vitro, but can be a stimulator in cells. Cellular and Molecular Life Sciences, 74(18), 3347–3362. doi:10.1007/s00018-017-2527-8

[1392] Wolff et al (2015). A Redox Regulatory System Critical for Mycobacterial Survival in Macrophages and Biofilm Development. PLOS Pathogens, 11(4), e1004839. doi:10.1371/journal.ppat.1004839

[1393] Rozwarski, D. A. (1998). Modification of the NADH of the Isoniazid Target (InhA) from Mycobacterium tuberculosis. Science, 279(5347), 98–102. doi:10.1126/science.279.5347.98

[1394] Kouhpayeh et al (2020). The Molecular Story of COVID-19; NAD+ Depletion Addresses All Questions in this Infection. doi:10.20944/preprints202003.0346.v1

[1395] Stein, L. R., & Imai, S. (2012). The dynamic regulation of NAD metabolism in mitochondria. Trends in Endocrinology & Metabolism, 23(9), 420–428. doi:10.1016/j.tem.2012.06.005

[1396] Imai, S., & Guarente, L. (2016). It takes two to tango: NAD+ and sirtuins in aging/longevity control. Npj Aging and Mechanisms of Disease, 2(1). doi:10.1038/npjamd.2016.17

[1397] Imai, S., & Yoshino, J. (2013). The importance of NAMPT/NAD/SIRT1 in the systemic regulation of metabolism and ageing. Diabetes, obesity & metabolism, 15 Suppl 3(0 3), 26–33. https://doi.org/10.1111/dom.12171

[1398] Nakahata et al (2008). The NAD+-Dependent Deacetylase SIRT1 Modulates CLOCK-Mediated Chromatin Remodeling and Circadian Control. Cell, 134(2), 329–340. doi:10.1016/j.cell.2008.07.002

[1399] Nakahata, Y., & Bessho, Y. (2016). The Circadian NAD+Metabolism: Impact on Chromatin Remodeling and Aging. BioMed Research International, 2016, 1–7. doi:10.1155/2016/3208429

[1400] Nakahata et al (2009). Circadian Control of the NAD+ Salvage Pathway by CLOCK-SIRT1. Science, 324(5927), 654–657. doi:10.1126/science.1170803

[1401] Valentinuzzi et al (1997) 'Effects of aging on the circadian rhythm of wheel-running activity in C57BL/6 mice', American Journal of Physiology-Regulatory, Integrative and Comparative Physiology 1997 273:6, R1957-R1964

[1402] Yamazaki et al (2002). Effects of aging on central and peripheral mammalian clocks. Proceedings of the National Academy of Sciences, 99(16), 10801–10806. doi:10.1073/pnas.152318499

[1403] Kondratov, R. V. (2006). Early aging and age-related pathologies in mice deficient in BMAL1, the core componentof the circadian clock. Genes & Development, 20(14), 1868–1873. doi:10.1101/gad.1432206

[1404] Ohkubo, R., & Chen, D. (2017). Aging: rewiring the circadian clock. Nature Structural & Molecular Biology, 24(9), 687–688. doi:10.1038/nsmb.3461

[1405] Hatori et al (2012) 'Time-Restricted Feeding without Reducing Caloric Intake Prevents Metabolic Diseases in Mice Fed a High-Fat Diet', Cell Metabolism, VOLUME 15, ISSUE 6, P848-860, JUNE 06, 2012, DOI:https://doi.org/10.1016/j.cmet.2012.04.019

[1406] Kanfi, Y. et al (2012) 'The sirtuin SIRT6 regulates lifespan in male mice', Nature, Vol 483, p 218-221.

[1407] Mostoslavsky, R. (2006) 'Genomic instability and aging-like phenotype in the absence of mammalian SIRT6', Cell, Vol 124(2), p 315-329.

[1408] Sedelnikova, OA. et al (2004), 'Senescing human cells and ageing mice accumulate DNA lesions with unrepairable double-strand breaks', Nat Cell Biology, Vol 6(2), p 168-170.

[1409] Chung, S. et al (2010) 'Regulation of SIRT1 in cellular functions: role of polyphenols', Archives of Biochemistry and Biophysics, Vol 501(1), p 79-90.

[1410] Wang, RH. et al (2008) 'Impaired DNA damage response, genome instability, and tumorigenesis in SIRT1 mutant mice', Cancer Cell, Vol 14(4), p 312-323.

[1411] Lee, H.I. et al (2008) 'A role for the NAD-dependent deacetylase Sirt1 in the regulation of autophagy', PNAS, Vol 105(9), p 3374-3379.

[1412] Zhang et al (2016). The potential regulatory roles of NAD(+) and its metabolism in autophagy. Metabolism: clinical and experimental, 65(4), 454–462. https://doi.org/10.1016/j.metabol.2015.11.010

[1413] Cea et al (2012). Targeting NAD+ salvage pathway induces autophagy in multiple myeloma cells via mTORC1 and extracellular signal-regulated kinase (ERK1/2) inhibition. Blood, 120(17), 3519–3529. https://doi.org/10.1182/blood-2012-03-416776

[1414] Bogan, K. L., & Brenner, C. (2008). Nicotinic Acid, Nicotinamide, and Nicotinamide Riboside: A Molecular Evaluation of NAD+Precursor Vitamins in Human Nutrition. Annual Review of Nutrition, 28(1), 115–130. doi:10.1146/annurev.nutr.28.061807.155443

[1415] Gross, C. J., & Henderson, L. M. (1983). Digestion and Absorption of NAD by the Small Intestine of the Rat. The Journal of Nutrition, 113(2), 412–420. doi:10.1093/jn/113.2.412

[1416] Reece et al (2011), '*Campbell biology* (9th ed.)', Boston: Benjamin Cummings. pp. 164–181.

[1417] Maalouf et al (2007). Ketones inhibit mitochondrial production of reactive oxygen species production following glutamate excitotoxicity by increasing NADH oxidation. Neuroscience, 145(1), 256–264. doi:10.1016/j.neuroscience.2006.11.065

[1418] Fulco et al (2008). Glucose Restriction Inhibits Skeletal Myoblast Differentiation by Activating SIRT1 through AMPK-Mediated Regulation of Nampt. Developmental Cell, 14(5), 661–673. doi:10.1016/j.devcel.2008.02.004

[1419] Caton et al (2011). FRUCTOSE INDUCES GLUCONEOGENESIS AND LIPOGENESIS THROUGH A SIRT1-DEPENDENT MECHANISM. Journal of Endocrinology. doi:10.1530/joe-10-0190

[1420] Hayashida et al. (2010) Fasting promotes the expression of SIRT1, an NAD+-dependent protein deacetylase, via activation of PPARα in mice. Mol Cell Biochem 339, 285–292 (2010). https://doi.org/10.1007/s11010-010-0391-z

[1421] Chen et al (2008). Tissue-specific regulation of SIRT1 by caloric restriction. Genes & Development, 22(13), 1753–1757. doi:10.1101/gad.1650608

[1422] Williams et al (2009). Oxaloacetate supplementation increases lifespan inCaenorhabditis elegansthrough an AMPK/FOXO-dependent pathway. Aging Cell, 8(6), 765–768. doi:10.1111/j.1474-9726.2009.00527.x

[1423] Costford et al (2010). Skeletal muscle NAMPT is induced by exercise in humans. American Journal of Physiology-Endocrinology and Metabolism, 298(1), E117–E126. doi:10.1152/ajpendo.00318.2009

[1424] Raynes (2013) "SIRT1 Regulation of the Heat Shock Response in an HSF1-Dependent Manner and the Impact of Caloric Restriction" (2013). Graduate Theses and Dissertations. University of South Florida. https://scholarcommons.usf.edu/etd/4567

[1425] Mills et al (2016) 'Long-Term Administration of Nicotinamide Mononucleotide Mitigates Age-Associated Physiological Decline in Mice', Cell Metabolism, VOLUME 24, ISSUE 6, P795-806, DECEMBER 13, 2016, DOI:https://doi.org/10.1016/j.cmet.2016.09.013

[1426] Xin et al (2018) 'Nutritional Ketosis Increases NAD+/NADH Ratio in Healthy Human Brain: An in Vivo Study by 31P-MRS', Front. Nutr., 12 July 2018 | https://doi.org/10.3389/fnut.2018.00062

[1427] Courchesne-Loyer et al (2017) Inverse relationship between brain glucose and ketone metabolism in adults during short-term moderate dietary ketosis: a dual tracer quantitative positron emission tomography study. *J Cereb Blood Flow Metab.* 37:2485–93. doi: 10.1177/0271678X16669366

[1428] Newman, J. C., & Verdin, E. (2014). β-hydroxybutyrate: Much more than a metabolite. Diabetes Research and Clinical Practice, 106(2), 173–181. doi:10.1016/j.diabres.2014.08.009

[1429] Shimazu et al (2013). Suppression of oxidative stress by β-hydroxybutyrate, an endogenous histone deacetylase inhibitor. Science (New York, N.Y.), 339(6116), 211–214. https://doi.org/10.1126/science.1227166

[1430] Newman, J. C., & Verdin, E. (2014). Ketone bodies as signaling metabolites. Trends in endocrinology and metabolism: TEM, 25(1), 42–52. https://doi.org/10.1016/j.tem.2013.09.002

[1431] Youm et al (2015). The ketone metabolite β-hydroxybutyrate blocks NLRP3 inflammasome-mediated inflammatory disease. Nature medicine, 21(3), 263–269. https://doi.org/10.1038/nm.3804

[1432] Wu, J., Jin, Z., Zheng, H., & Yan, L. J. (2016). Sources and implications of NADH/NAD(+) redox imbalance in diabetes and its complications. Diabetes, metabolic syndrome and obesity : targets and therapy, 9, 145–153. https://doi.org/10.2147/DMSO.S106087

[1433] Han et al (2016). AMPK activation protects cells from oxidative stress-induced senescence via autophagic flux restoration and intracellular NAD(+) elevation. Aging cell, 15(3), 416–427. https://doi.org/10.1111/acel.12446

[1434] Sahlin K. (1983). NADH and NADPH in human skeletal muscle at rest and during ischaemia. Clinical physiology (Oxford, England), 3(5), 477–485. https://doi.org/10.1111/j.1475-097x.1983.tb00856.x

[1435] Henriksson, J., Katz, A., & Sahlin, K. (1986). Redox state changes in human skeletal muscle after isometric contraction. The Journal of physiology, 380, 441–451. https://doi.org/10.1113/jphysiol.1986.sp016296

[1436] Graham et al (1978). NAD in muscle of man at rest and during exercise. Pflugers Archiv : European journal of physiology, 376(1), 35–39. https://doi.org/10.1007/BF00585245

[1437] Close et al (2016). New strategies in sport nutrition to increase exercise performance. Free Radical Biology and Medicine, 98, 144–158. doi:10.1016/j.freeradbiomed.2016.01.016

[1438] Mayevsky, A., & Rogatsky, G. G. (2007). Mitochondrial function in vivo evaluated by NADH fluorescence: from animal models to human studies. American journal of physiology. Cell physiology, 292(2), C615–C640. https://doi.org/10.1152/ajpcell.00249.2006

[1439] Godfraind-de Becker A. (1972). Heat production and fluorescence changes of toad sartorius muscle during aerobic recovery after a short tetanus. The Journal of physiology, 223(3), 719–734. https://doi.org/10.1113/jphysiol.1972.sp009871

[1440] Jöbsis, F. F., & Stainsby, W. N. (1968). Oxidation of NADH during contractions of circulated mammalian skeletal muscle. Respiration physiology, 4(3), 292–300. https://doi.org/10.1016/0034-5687(68)90035-2

[1441] Cantó et al (2010). Interdependence of AMPK and SIRT1 for metabolic adaptation to fasting and exercise in skeletal muscle. Cell metabolism, 11(3), 213–219. https://doi.org/10.1016/j.cmet.2010.02.006

[1442] Koltai et al (2010). Exercise alters SIRT1, SIRT6, NAD and NAMPT levels in skeletal muscle of aged rats. Mechanisms of ageing and development, 131(1), 21–28. https://doi.org/10.1016/j.mad.2009.11.002

[1443] Shibata et al (1994) 'Effects of Exercise on the Metabolism of NAD in Rats', Bioscience Biotechnology and Biochemistry 58(10):1763-1766, DOI: 10.1271/bbb.58.1763

[1444] Graham et al (1978). NAD in muscle of man at rest and during exercise. Pflugers Archiv : European journal of physiology, 376(1), 35–39. https://doi.org/10.1007/BF00585245

[1445] Sahlin K. (1985). NADH in human skeletal muscle during short-term intense exercise. Pflugers Archiv : European journal of physiology, 403(2), 193–196. https://doi.org/10.1007/BF00584099

[1446] Sahlin, K., Gorski, J., & Edström, L. (1990). Influence of ATP turnover and metabolite changes on IMP formation and glycolysis in rat skeletal muscle. The American journal of physiology, 259(3 Pt 1), C409–C412. https://doi.org/10.1152/ajpcell.1990.259.3.C409

[1447] Katz, A., & Sahlin, K. (1987). Effect of decreased oxygen availability on NADH and lactate contents in human skeletal muscle during exercise. Acta physiologica Scandinavica, 131(1), 119–127. https://doi.org/10.1111/j.1748-1716.1987.tb08213.x

[1448] White, A. T., & Schenk, S. (2012). NAD(+)/NADH and skeletal muscle mitochondrial adaptations to exercise. American journal of physiology. Endocrinology and metabolism, 303(3), E308–E321. https://doi.org/10.1152/ajpendo.00054.2012

[1449] Stipp (2015) 'Beyond Resveratrol: The Anti-Aging NAD Fad', Scientific American, Accessed Online: https://blogs.scientificamerican.com/guest-blog/beyond-resveratrol-the-anti-aging-nad-fad/

[1450] Trammell et al (2016). Nicotinamide riboside is uniquely and orally bioavailable in mice and humans. Nature communications, 7, 12948. https://doi.org/10.1038/ncomms12948

[1451] Conze, D., Crespo-Barreto, J., & Kruger, C. (2016). Safety assessment of nicotinamide riboside, a form of vitamin B3. Human & Experimental Toxicology, 35(11), 1149–1160. https://doi.org/10.1177/0960327115626254

[1452] Trammell et al (2016). Nicotinamide riboside is uniquely and orally bioavailable in mice and humans. Nature communications, 7, 12948. https://doi.org/10.1038/ncomms12948

[1453] Cantó et al (2012). The NAD(+) precursor nicotinamide riboside enhances oxidative metabolism and protects against high-fat diet-induced obesity. Cell metabolism, 15(6), 838–847. https://doi.org/10.1016/j.cmet.2012.04.022

[1454] Kourtzidis et al (2016). The NAD+ precursor nicotinamide riboside decreases exercise performance in rats. Journal of the International Society of Sports Nutrition, 13(1). doi:10.1186/s12970-016-0143-x

[1455] Ristow M. (2014). Unraveling the truth about antioxidants: mitohormesis explains ROS-induced health benefits. Nature medicine, 20(7), 709–711. https://doi.org/10.1038/nm.3624

[1456] Frederick et al (2015). Increasing NAD synthesis in muscle via nicotinamide phosphoribosyltransferase is not sufficient to promote oxidative metabolism. The Journal of biological chemistry, 290(3), 1546–1558. https://doi.org/10.1074/jbc.M114.579565

[1457] Hou et al (2018). NAD+ supplementation normalizes key Alzheimer's features and DNA damage responses in a new AD mouse model with introduced DNA repair deficiency. Proceedings of the National Academy of Sciences of the United States of America, 115(8), E1876–E1885. https://doi.org/10.1073/pnas.1718819115

[1458] Grozio et al (2019). Slc12a8 is a nicotinamide mononucleotide transporter. Nature Metabolism, 1(1), 47–57. doi:10.1038/s42255-018-0009-4

[1459] Mills et al (2016). Long-Term Administration of Nicotinamide Mononucleotide Mitigates Age-Associated Physiological Decline in Mice. Cell metabolism, 24(6), 795–806. https://doi.org/10.1016/j.cmet.2016.09.013

[1460] Gomes et al (2013). Declining NAD(+) induces a pseudohypoxic state disrupting nuclear-mitochondrial communication during aging. Cell, 155(7), 1624–1638. https://doi.org/10.1016/j.cell.2013.11.037

[1461] Bogan and Brenner (2008) 'Nicotinic Acid, Nicotinamide, and Nicotinamide Riboside: A Molecular Evaluation of NAD+ Precursor Vitamins in Human Nutrition', Annual Review of Nutrition, Vol. 28:115-130, https://doi.org/10.1146/annurev.nutr.28.061807.155443

[1462] Ratajczak et al (2016) 'NRK1 controls nicotinamide mononucleotide and nicotinamide riboside metabolism in mammalian cells', Nature Communications volume 7, Article number: 13103 (2016)

[1463] Grozio et al (2019). Slc12a8 is a nicotinamide mononucleotide transporter. Nature Metabolism, 1(1), 47–57. doi:10.1038/s42255-018-0009-4

[1464] Trammell et al (2016). Nicotinamide riboside is uniquely and orally bioavailable in mice and humans. Nature communications, 7, 12948. https://doi.org/10.1038/ncomms12948

[1465] FDA (2015) 'Expert Panel Statement for Niagen™ (Nicotinamide Riboside Chloride)', Accessed Online: https://www.fda.gov/files/food/published/GRAS-Notice-000635--Nicotinamide-riboside-chloride.pdf

[1466] Conze, D., Brenner, C., & Kruger, C. L. (2019). Safety and Metabolism of Long-term Administration of NIAGEN (Nicotinamide Riboside Chloride) in a Randomized, Double-Blind, Placebo-controlled Clinical Trial of Healthy Overweight Adults. Scientific reports, 9(1), 9772. https://doi.org/10.1038/s41598-019-46120-z

[1467] Spaans et al (2015). NADPH-generating systems in bacteria and archaea. Frontiers in microbiology, 6, 742. https://doi.org/10.3389/fmicb.2015.00742

[1468] Spaans SK, Weusthuis RA, van der Oost J, Kengen SW (2015). "NADPH-generating systems in bacteria and archaea". *Frontiers in Microbiology*. 6: 742.

[1469] Rush et al (1985). Organic hydroperoxide-induced lipid peroxidation and cell death in isolated hepatocytes. Toxicology and applied pharmacology, 78(3), 473–483. https://doi.org/10.1016/0041-008x(85)90255-8

[1470] Pompella et al (2003). The changing faces of glutathione, a cellular protagonist. Biochemical Pharmacology, 66(8), 1499–1503. doi:10.1016/s0006-2952(03)00504-5

[1471] Flagg et al (1994). Dietary glutathione intake and the risk of oral and pharyngeal cancer. American journal of epidemiology, 139(5), 453–465. https://doi.org/10.1093/oxfordjournals.aje.a117028

[1472] Jones et al (1992). Glutathione in foods listed in the National Cancer Institute's Health Habits and History Food Frequency Questionnaire. Nutrition and cancer, 17(1), 57–75. https://doi.org/10.1080/01635589209514173

[1473] Bianchini, F., & Vainio, H. (2001). Allium vegetables and organosulfur compounds: do they help prevent cancer?. Environmental health perspectives, 109(9), 893–902. https://doi.org/10.1289/ehp.01109893

[1474] Sparnins et al (1988). Effects of organosulfur compounds from garlic and onions on benzo[a]pyrene-induced neoplasia and glutathione S-transferase activity in the mouse. Carcinogenesis, 9(1), 131–134. https://doi.org/10.1093/carcin/9.1.131

[1475] Hultberg, B., Andersson, A., & Isaksson, A. (2002). Lipoic acid increases glutathione production and enhances the effect of mercury in human cell lines. Toxicology, 175(1-3), 103–110. https://doi.org/10.1016/s0300-483x(02)00060-4

[1476] Paschalis et al (2018). N-acetylcysteine supplementation increases exercise performance and reduces oxidative stress only in individuals with low levels of glutathione. Free radical biology & medicine, 115, 288–297. https://doi.org/10.1016/j.freeradbiomed.2017.12.007

[1477] Sonthalia et al (2018). Glutathione for skin lightening: a regnant myth or evidence-based verity?. Dermatology practical & conceptual, 8(1), 15–21. https://doi.org/10.5826/dpc.0801a04

[1478] Brown, D. I., & Griendling, K. K. (2009). Nox proteins in signal transduction. Free radical biology & medicine, 47(9), 1239–1253. https://doi.org/10.1016/j.freeradbiomed.2009.07.023

[1479] Meijles, D. N., & Pagano, P. J. (2016). Nox and Inflammation in the Vascular Adventitia. Hypertension, 67(1), 14–19. doi:10.1161/hypertensionaha.115.03622

[1480] Ogawa et al (October 2008). "The association of elevated reactive oxygen species levels from neutrophils with low-grade inflammation in the elderly". *Immunity & Ageing*. 5: 13.

[1481] Pizzorno J. (2014). Glutathione!. Integrative medicine (Encinitas, Calif.), 13(1), 8–12.

[1482] McCarty et al (2019) 'Activated Glycine Receptors May Decrease Endosomal NADPH Oxidase Activity by Opposing ClC-3-Mediated Efflux of Chloride from Endosomes', Medical Hypotheses 123, DOI: 10.1016/j.mehy.2019.01.012

[1483] Brind et al (2011) 'Dietary glycine supplementation mimics lifespan extension by dietary methionine restriction in Fisher 344 rats', The FASEB Journal 2011 25:1_supplement, 528.2-528.2

[1484] Gannon et al (2002) 'The metabolic response to ingested glycine', The American Journal of

Clinical Nutrition, Volume 76, Issue 6, December 2002, Pages 1302–1307, https://doi.org/10.1093/ajcn/76.6.1302

[1485] Richardson B. (2003). DNA methylation and autoimmune disease. Clinical immunology (Orlando, Fla.), 109(1), 72–79. https://doi.org/10.1016/s1521-6616(03)00206-7

[1486] Li et al (2017). m6A mRNA methylation controls T cell homeostasis by targeting the IL-7/STAT5/SOCS pathways. Nature, 548(7667), 338–342. doi:10.1038/nature23450

[1487] Morales-Nebreda et al (2019). DNA methylation as a transcriptional regulator of the immune system. Translational Research, 204, 1–18. doi:10.1016/j.trsl.2018.08.001

[1488] Richardson B. (2003). DNA methylation and autoimmune disease. Clinical immunology (Orlando, Fla.), 109(1), 72–79. https://doi.org/10.1016/s1521-6616(03)00206-7

[1489] Townsend et al (2003) 'The importance of glutathione in human disease', Biomedicine & Pharmacotherapy 57 (2003) 145–155.

[1490] Yeh et al (2007). Chronic alcoholism alters systemic and pulmonary glutathione redox status. American journal of respiratory and critical care medicine, 176(3), 270–276. https://doi.org/10.1164/rccm.200611-1722OC

[1491] Mathangi et al (2012). Effect of REM sleep deprivation on the antioxidant status in the brain of Wistar rats. Annals of neurosciences, 19(4), 161–164. https://doi.org/10.5214/ans.0972.7531.190405

[1492] Pruitt et al (2006). Inhibition of SIRT1 reactivates silenced cancer genes without loss of promoter DNA hypermethylation. PLoS genetics, 2(3), e40. https://doi.org/10.1371/journal.pgen.0020040

[1493] Lozoya OA, Martinez-Reyes I, Wang T, Grenet D, Bushel P, et al. (2018) Mitochondrial nicotinamide adenine dinucleotide reduced (NADH) oxidation links the tricarboxylic acid (TCA) cycle with methionine metabolism and nuclear DNA methylation. PLOS Biology 16(4): e2005707. https://doi.org/10.1371/journal.pbio.2005707

[1494] Figueiredo et al (2009). Folic Acid and Risk of Prostate Cancer: Results From a Randomized Clinical Trial. JNCI Journal of the National Cancer Institute, 101(6), 432–435. doi:10.1093/jnci/djp019

[1495] Schultz, M. B., & Sinclair, D. A. (2016). Why NAD(+) Declines during Aging: It's Destroyed. Cell metabolism, 23(6), 965–966. https://doi.org/10.1016/j.cmet.2016.05.022

[1496] Obeid R. (2013). The metabolic burden of methyl donor deficiency with focus on the betaine homocysteine methyltransferase pathway. Nutrients, 5(9), 3481–3495. https://doi.org/10.3390/nu5093481

[1497] Shorter et al (2015). Consequences of dietary methyl donor supplements: Is more always better? Progress in Biophysics and Molecular Biology, 118(1-2), 14–20. doi:10.1016/j.pbiomolbio.2015.03.007

[1498] Oudi et al (2010). Homocysteine and markers of inflammation in acute coronary syndrome. Experimental and clinical cardiology, 15(2), e25–e28.

[1499] McRae M. P. (2013). Betaine supplementation decreases plasma homocysteine in healthy adult participants: a meta-analysis. Journal of chiropractic medicine, 12(1), 20–25. https://doi.org/10.1016/j.jcm.2012.11.001

[1500] Sun et al (2017). Comparison of the effects of nicotinic acid and nicotinamide degradation on plasma betaine and choline levels. Clinical nutrition (Edinburgh, Scotland), 36(4), 1136–1142. https://doi.org/10.1016/j.clnu.2016.07.016

[1501] Dollerup et al (2018). A randomized placebo-controlled clinical trial of nicotinamide riboside in obese men: safety, insulin-sensitivity, and lipid-mobilizing effects. The American Journal of Clinical Nutrition, 108(2), 343–353. doi:10.1093/ajcn/nqy132

[1502] Fischer, D. J., Knight, L. L., & Vestal, R. E. (1991). Fulminant hepatic failure following low-dose sustained-release niacin therapy in hospital. The Western journal of medicine, 155(4), 410–

412.

[1503] Martens et al (2018) 'Chronic nicotinamide riboside supplementation is well-tolerated and elevates NAD+ in healthy middle-aged and older adults', Nature Communications volume 9, Article number: 1286 (2018)

[1504] Selhub et al (2000). Relationship between plasma homocysteine and vitamin status in the Framingham study population. Impact of folic acid fortification. Public health reviews, 28(1-4), 117–145.

[1505] Ganguly, P., & Alam, S. F. (2015). Role of homocysteine in the development of cardiovascular disease. Nutrition Journal, 14(1). doi:10.1186/1475-2891-14-6

[1506] Alberts B, Johnson A, Lewis J, et al. (2002). *Molecular Biology of the Cell* (4th ed.). New York: Garland Science.

[1507] Harvard T.H. Chan, The Nutrition Source, 'Omega-3 Fatty Acids: An Essential Contribution', Accessed Online: https://www.hsph.harvard.edu/nutritionsource/what-should-you-eat/fats-and-cholesterol/types-of-fat/omega-3-fats/

[1508] Gutiérrez, S., Svahn, S. L., & Johansson, M. E. (2019). Effects of Omega-3 Fatty Acids on Immune Cells. International Journal of Molecular Sciences, 20(20), 5028. doi:10.3390/ijms20205028

[1509]The Source for Objective Science-based DHA/EPA Omega-3 Information, DIFFERENTIATION OF ALA (PLANT SOURCES) FROM DHA + EPA (MARINE SOURCES) AS DIETARY OMEGA-3 FATTY ACIDS FOR HUMAN HEALTH, Accessed Online: http://www.dhaomega3.org/Overview/Differentiation-of-ALA-plant-sources-from-DHA-+-EPA-marine-sources-as-Dietary-Omega-3-Fatty-Acids-for-Human-Health

[1510] Garg, A (1998) 'High-monounsaturated-fat diets for patients with diabetes mellitus: a meta-analysis', Am J Clin Nutr. 1998 Mar;67(3 Suppl):577S-582S.

[1511] Finucane, OM et al (2015) 'Monounsaturated fatty acid-enriched high-fat diets impede adipose NLRP3 inflammasome-mediated IL-1β secretion and insulin resistance despite obesity', Diabetes. 2015 Jun;64(6):2116-28.

[1512]Kris-Etherton et al (2000). Polyunsaturated fatty acids in the food chain in the United States. The American Journal of Clinical Nutrition, 71(1), 179S–188S. doi:10.1093/ajcn/71.1.179s

[1513] Leaf A and Weber PC (1988) 'Cardiovascular effects of n-3 fatty acids', AGRIS, Accessed: http://agris.fao.org/agris-search/search.do?recordID=US8845581.

[1514] Russo GL (2009) 'Dietary n-6 and n-3 polyunsaturated fatty acids: from biochemistry to clinical implications in cardiovascular prevention', Biochem Pharmacol. 2009 Mar 15;77(6):937-46.

[1515] Gago-Dominguez, M., Jiang, X., & Castelao, J. E. (2007). Lipid peroxidation, oxidative stress genes and dietary factors in breast cancer protection: a hypothesis. Breast cancer research : BCR, 9(1), 201. https://doi.org/10.1186/bcr1628

[1516] Bowe, W. P., & Logan, A. C. (2010). Clinical implications of lipid peroxidation in acne vulgaris: old wine in new bottles. Lipids in Health and Disease, 9(1), 141. doi:10.1186/1476-511x-9-141

[1517] Matsushita et al (2015) 'T cell lipid peroxidation induces ferroptosis and prevents immunity to infection', J Exp Med. 2015 Apr 6; 212(4): 555–568. doi: 10.1084/jem.20140857

[1518] Mandal and Chatterjee (1980) 'Ultraviolet- and Sunlight-Induced Lipid Peroxidation in Liposomal Membrane', Radiation Research, Vol. 83, No. 2 (Aug., 1980), pp. 290-302.

[1519] Altavilla et al (2001). Inhibition of Lipid Peroxidation Restores Impaired Vascular Endothelial Growth Factor Expression and Stimulates Wound Healing and Angiogenesis in the Genetically Diabetic Mouse. Diabetes, 50(3), 667–674. doi:10.2337/diabetes.50.3.667

[1520] *Nutrition Week* Mar 22, 1991 21:12:2-3

[1521] Barnard N et al (2014) 'Saturated and trans fats and dementia: a systematic review', Neurobiology of Aging, Volume 35, Supplement 2, September 2014, Pages S65-S73.

[1522] Pase CS et al (2013) 'Influence of perinatal trans fat on behavioral responses and brain oxidative status of adolescent rats acutely exposed to stress', Neuroscience. 2013 Sep 5;247:242-52.

[1523] Trevisan, M. (2001). Correlates of Markers of Oxidative Status in the General Population. American Journal of Epidemiology, 154(4), 348–356. doi:10.1093/aje/154.4.348

[1524] Kromhout et al (1985). The inverse relation between fish consumption and 20-year mortality from coronary heart disease. The New England journal of medicine, 312(19), 1205–1209. https://doi.org/10.1056/NEJM198505093121901

[1525] Oomen et al (2000). Fish consumption and coronary heart disease mortality in Finland, Italy, and The Netherlands. American journal of epidemiology, 151(10), 999–1006. https://doi.org/10.1093/oxfordjournals.aje.a010144

[1526] Keli, S. O., Feskens, E. J., & Kromhout, D. (1994). Fish consumption and risk of stroke. The Zutphen Study. Stroke, 25(2), 328–332. https://doi.org/10.1161/01.str.25.2.328

[1527] Hooper et al (2006). Risks and benefits of omega 3 fats for mortality, cardiovascular disease, and cancer: systematic review. BMJ (Clinical research ed.), 332(7544), 752–760. https://doi.org/10.1136/bmj.38755.366331.2F

[1528] Norwegian Scientific Committee for Food Safety (2011) 'Description of the processes in the value chain and risk assessment of decomposition substances and oxidation products in fish oils', Opinion of Steering Committee of the Norwegian Scientific Committee for Food Safety, 08-504-4-final, Accessed Online: https://web.archive.org/web/20160909213119/http://english.vkm.no/dav/0fd42c8b08.pdf

[1529] Consumerlab.com (2018) 'Fish Oil and Omega-3 and -7 Supplements Review (Including Krill, Algae, Calamari, and Sea Buckthorn)', Product Reviews, Accessed Online: https://www.consumerlab.com/reviews/fish_oil_supplements_review/omega3/

[1530] Ulven et al (2010). Metabolic Effects of Krill Oil are Essentially Similar to Those of Fish Oil but at Lower Dose of EPA and DHA, in Healthy Volunteers. Lipids, 46(1), 37–46. doi:10.1007/s11745-010-3490-4

[1531] Martin et al (2010). Intestinal digestion of fish oils and ω-3 concentrates under in vitro conditions. European Journal of Lipid Science and Technology, 112(12), 1315–1322. doi:10.1002/ejlt.201000329

[1532] Lu et al (2006). Preventive effects of Spirulina platensis on skeletal muscle damage under exercise-induced oxidative stress. European journal of applied physiology, 98(2), 220–226. https://doi.org/10.1007/s00421-006-0263-0

[1533] Queiroz et al (2003). Protective effects of Chlorella vulgaris in lead-exposed mice infected with Listeria monocytogenes. International immunopharmacology, 3(6), 889–900. https://doi.org/10.1016/S1567-5769(03)00082-1

[1534] Fernández-Friera et al (2017). Normal LDL-Cholesterol Levels Are Associated With Subclinical Atherosclerosis in the Absence of Risk Factors. Journal of the American College of Cardiology, 70(24), 2979–2991. doi:10.1016/j.jacc.2017.10.024

[1535] University of California - Los Angeles. (2009, January 13). Most Heart Attack Patients' Cholesterol Levels Did Not Indicate Cardiac Risk. ScienceDaily. Retrieved March 31, 2020 from www.sciencedaily.com/releases/2009/01/090112130653.htm

[1536] Nago N et al (2011) 'Low cholesterol is associated with mortality from stroke, heart disease, and cancer: the Jichi Medical School Cohort Study', J Epidemiol. 2011;21(1):67-74.

[1537] Bae JM et al (2012) 'Low Cholesterol is Associated with Mortality from Cardiovascular Diseases: A Dynamic Cohort Study in Korean Adults', J Korean Med Sci. 2012 Jan; 27(1): 58–63.

[1538] Parthasarathy S et al (2012) 'Oxidized Low-Density Lipoprotein', Methods Mol Biol. 2010; 610: 403–417.

[1539] Vanderlaan PA et al (2009) 'VLDL best predicts aortic root atherosclerosis in LDL receptor deficient mice', J Lipid Res. 2009 Mar; 50(3): 376–385.

[1540] Münzel et al (2008). "Pathophysiology, diagnosis and prognostic implications of endothelial dysfunction". *Annals of Medicine*. 40 (3): 180–96.

[1541] Maruhashi, T; Kihara, Y; Higashi, Y (2018). "Assessment of endothelium-independent vasodilation: From methodology to clinical perspectives". *Journal of Hypertension*. 36 (7): 1460–1467.

[1542] Briasoulis et al (Apr 2012). "Endothelial dysfunction and atherosclerosis: focus on novel therapeutic approaches". *Recent Pat Cardiovasc Drug Discov*. 7 (1): 21–32.

[1543]Lecerf, JM and de Lorgeril, M (2011) 'Dietary cholesterol: from physiology to cardiovascular risk', British Journal of Nutrition (2011), 106, 6–14.

[1544] Calderon et al (1999). "Stress, stress reduction and hypercholesterolemia in African Americans: a review". *Ethnicity & Disease*. 9 (3): 451–462.

[1545] Kanner J. (2007). Dietary advanced lipid oxidation endproducts are risk factors to human health. Molecular nutrition & food research, 51(9), 1094–1101. https://doi.org/10.1002/mnfr.200600303

[1546] Aviram, M., & Eias, K. (1993). Dietary Olive Oil Reduces Low-Density Lipoprotein Uptake by Macrophages and Decreases the Susceptibility of the Lipoprotein to Undergo Lipid Peroxidation. Annals of Nutrition and Metabolism, 37(2), 75–84. doi:10.1159/000177753

[1547] Huang et al (2002) 'Effects of vitamin C and vitamin E on in vivo lipid peroxidation: results of a randomized controlled trial', The American Journal of Clinical Nutrition, Volume 76, Issue 3, September 2002, Pages 549–555, https://doi.org/10.1093/ajcn/76.3.549

[1548] Miller III et al (1998) 'Effect of Dietary Patterns on Measures of Lipid Peroxidation Results From a Randomized Clinical Trial', (Circulation. 1998;98:2390-2395.

[1549] Kanter et al (1993). Effects of an antioxidant vitamin mixture on lipid peroxidation at rest and postexercise. Journal of applied physiology (Bethesda, Md. : 1985), 74(2), 965–969. https://doi.org/10.1152/jappl.1993.74.2.965

[1550] Jaarin, K., & Kamisah, Y. (2012). Repeatedly Heated Vegetable Oils and Lipid Peroxidation. Lipid Peroxidation. doi:10.5772/46076

[1551] Chisté et al (2014) 'Carotenoids inhibit lipid peroxidation and hemoglobin oxidation, but not the depletion of glutathione induced by ROS in human erythrocytes', Life Sciences, Volume 99, Issues 1–2, 18 March 2014, Pages 52-60.

[1552] Zhang et al (1991). Carotenoids enhance gap junctional communication and inhibit lipid peroxidation in C3H/10T1/2 cells: relationship to their cancer chemopreventive action. Carcinogenesis, 12(11), 2109–2114. https://doi.org/10.1093/carcin/12.11.2109

[1553] McNulty et al (2007). Differential effects of carotenoids on lipid peroxidation due to membrane interactions: X-ray diffraction analysis. Biochimica et Biophysica Acta (BBA) - Biomembranes, 1768(1), 167–174. doi:10.1016/j.bbamem.2006.09.010

[1554] Lu et al (2006). Preventive effects of Spirulina platensis on skeletal muscle damage under exercise-induced oxidative stress. European journal of applied physiology, 98(2), 220–226. https://doi.org/10.1007/s00421-006-0263-0

[1555] Onuegbo et al (2018) 'Consumption of Soymilk Reduces Lipid Peroxidation But May Lower Micronutrient Status in Apparently Healthy Individuals' Journal of Medicinal FoodVol. 21, No. 5, https://doi.org/10.1089/jmf.2017.0094

[1556] Jabbari et al (2005). Comparison between swallowing and chewing of garlic on levels of serum lipids, cyclosporine, creatinine and lipid peroxidation in Renal Transplant Recipients, Lipids in Health and Disease, 4(1), 11. doi:10.1186/1476-511x-4-11

[1557] Reddy, A. C., & Lokesh, B. R. (1994). Effect of dietary turmeric (Curcuma longa) on iron-induced lipid peroxidation in the rat liver. Food and chemical toxicology : an international journal published for the British Industrial Biological Research Association, 32(3), 279–283. https://doi.org/10.1016/0278-6915(94)90201-1

[1558] Ji Z. (2010). Targeting DNA damage and repair by curcumin. Breast cancer : basic and clinical research, 4, 1–3.

[1559] Wang et al (1999) 'C60 and Water-Soluble Fullerene Derivatives as Antioxidants Against Radical-Initiated Lipid Peroxidation', J. Med. Chem. 1999, 42, 22, 4614–4620, https://doi.org/10.1021/jm990144s

[1560] Merriam-Webster 'Medical Definition of lipofuscin', Accessed Online: https://www.merriam-webster.com/medical/lipofuscin

[1561] Double et al (2008). The comparative biology of neuromelanin and lipofuscin in the human brain. Cellular and Molecular Life Sciences, 65(11), 1669–1682. doi:10.1007/s00018-008-7581-9

[1562] Terman, A., & Brunk, U. T. (1998). Lipofuscin: mechanisms of formation and increase with age. APMIS : acta pathologica, microbiologica, et immunologica Scandinavica, 106(2), 265–276. https://doi.org/10.1111/j.1699-0463.1998.tb01346.x

[1563] Kurz, T., Terman, A., Gustafsson, B., & Brunk, U. T. (2008). Lysosomes in iron metabolism, ageing and apoptosis. Histochemistry and cell biology, 129(4), 389–406. https://doi.org/10.1007/s00418-008-0394-y

[1564] Kakimoto et al (2019). Myocardial lipofuscin accumulation in ageing and sudden cardiac death. Scientific Reports, 9(1). doi:10.1038/s41598-019-40250-0

[1565] Skoczyńska et al (2017). Melanin and lipofuscin as hallmarks of skin aging. Postepy dermatologii i alergologii, 34(2), 97–103. https://doi.org/10.5114/ada.2017.67070

[1566] Moreno-Garcia et al (2018) 'An Overview of the Role of Lipofuscin in Age-Related Neurodegeneration', Front. Neuroscience, https://doi.org/10.3389/fnins.2018.00464

[1567] HARVARD MEDICAL SCHOOL (2006) 'Harvard Medical signs agreement with Merck to develop potential therapy for macular degeneration', PUBLIC RELEASE: 23-MAY-2006, Accessed Online: https://www.eurekalert.org/pub_releases/2006-05/hms-hms052306.php

[1568] Höhn, A., Sittig, A., Jung, T., Grimm, S., & Grune, T. (2012). Lipofuscin is formed independently of macroautophagy and lysosomal activity in stress-induced prematurely senescent human fibroblasts. Free radical biology & medicine, 53(9), 1760–1769. https://doi.org/10.1016/j.freeradbiomed.2012.08.591

[1569] Brunk, U. T., Jones, C. B., & Sohal, R. S. (1992). A novel hypothesis of lipofuscinogenesis and cellular aging based on interactions between oxidative stress and autophagocytosis. Mutation Research/DNAging, 275(3-6), 395–403. doi:10.1016/0921-8734(92)90042-n

[1570] Leon, L. J., & Gustafsson, Å. B. (2016). Staying young at heart: autophagy and adaptation to cardiac aging. Journal of Molecular and Cellular Cardiology, 95, 78–85. doi:10.1016/j.yjmcc.2015.11.006

[1571] Shirakabe et al (2016) 'Aging and Autophagy in the Heart', Circulation Research. 2016;118:1563–1576, https://doi.org/10.1161/CIRCRESAHA.116.307474

[1572] König et al (2017) 'Mitochondrial contribution to lipofuscin formation', Redox Biology, Volume 11, April 2017, Pages 673-681

[1573] Durand, G., & Desnoyers, F. (1980). Acides gras polyinsaturés et vieillissement. Les lipofuscines : structure, origine, évolution [Polyunsaturated fatty acids and aging. Lipofuscins : structure, origin and development]. Annales de la nutrition et de l'alimentation, 34(2), 317–332.

[1574] Sharma et al (2010) 'Mechanisms of 4-Hydroxy-2-nonenal Induced Pro- and Anti-Apoptotic Signaling', Biochemistry 2010, 49, 29, 6263–6275, https://doi.org/10.1021/bi100517x

[1575] Elkin et al (2018) 'Trichloroethylene metabolite S-(1,2-dichlorovinyl)-l-cysteine induces lipid peroxidation-associated apoptosis via the intrinsic and extrinsic apoptosis pathways in a first-trimester placental cell line', Toxicology and Applied Pharmacology, Volume 338, 1 January 2018, Pages 30-42

[1576] Dodson et al (2017) Regulation of autophagy, mitochondrial dynamics, and cellular bioenergetics by 4-hydroxynonenal in primary neurons, Autophagy, 13:11, 1828-1840, DOI: 10.1080/15548627.2017.1356948

[1577] Krohne et al (2010) 'Lipid peroxidation products reduce lysosomal protease activities in human retinal pigment epithelial cells via two different mechanisms of action', Experimental Eye Research, Volume 90, Issue 2, February 2010, Pages 261-266

[1578] Ma et al (2011) 'Aldehyde dehydrogenase 2 (ALDH2) rescues myocardial ischaemia/reperfusion injury: role of autophagy paradox and toxic aldehyde', European Heart Journal, Volume 32, Issue 8, April 2011, Pages 1025–1038, https://doi.org/10.1093/eurheartj/ehq253

[1579] Dolinsky et al (2009) 'Resveratrol Prevents the Prohypertrophic Effects of Oxidative Stress on LKB1', Circulation. 2009;119:1643–1652

[1580] Aschbacher et al (2013). Good stress, bad stress and oxidative stress: insights from anticipatory cortisol reactivity. Psychoneuroendocrinology, 38(9), 1698–1708. https://doi.org/10.1016/j.psyneuen.2013.02.004

[1581] Yin D. (1996). Biochemical basis of lipofuscin, ceroid, and age pigment-like fluorophores. Free radical biology & medicine, 21(6), 871–888. https://doi.org/10.1016/0891-5849(96)00175-x

[1582] Vistoli et al (2013). Advanced glycoxidation and lipoxidation end products (AGEs and ALEs): an overview of their mechanisms of formation. Free radical research, 47 Suppl 1, 3–27. https://doi.org/10.3109/10715762.2013.815348

[1583] Negre-Salvayre, A., Coatrieux, C., Ingueneau, C., & Salvayre, R. (2008). Advanced lipid peroxidation end products in oxidative damage to proteins. Potential role in diseases and therapeutic prospects for the inhibitors. British Journal of Pharmacology, 153(1), 6–20. doi:10.1038/sj.bjp.0707395

[1584] Laberge et al (2015). MTOR regulates the pro-tumorigenic senescence-associated secretory phenotype by promoting IL1A translation. Nature Cell Biology, 17(8), 1049–1061. doi:10.1038/ncb3195

[1585] Narita et al (2011) 'Spatial Coupling of mTOR and Autophagy Augments Secretory Phenotypes', Science 20 May 2011: Vol. 332, Issue 6032, pp. 966-970, DOI: 10.1126/science.1205407

[1586] García-Prat et al (2016). Autophagy maintains stemness by preventing senescence. Nature, 529(7584), 37–42. https://doi.org/10.1038/nature16187

[1587] Höhn, A., Jung, T., Grimm, S., & Grune, T. (2010). Lipofuscin-bound iron is a major intracellular source of oxidants: role in senescent cells. Free radical biology & medicine, 48(8), 1100–1108. https://doi.org/10.1016/j.freeradbiomed.2010.01.030

[1588] Doll, S. and Conrad, M. (2017), Iron and ferroptosis: A still ill-defined liaison. IUBMB Life, 69: 423-434. doi:10.1002/iub.1616

[1589] Julien, S., Biesemeier, A., Kokkinou, D., Eibl, O., & Schraermeyer, U. (2011). Zinc Deficiency Leads to Lipofuscin Accumulation in the Retinal Pigment Epithelium of Pigmented Rats. PLoS ONE, 6(12), e29245. doi:10.1371/journal.pone.0029245

[1590] Cabrera, Á. J. R. (2015). Zinc, aging, and immunosenescence: an overview. Pathobiology of Aging & Age-Related Diseases, 5(1), 25592. doi:10.3402/pba.v5.25592

[1591] Delima et al (2007) 'Potential protective effects of zinc in iron overload', Liver International, Volume27, Issue1, Pages 4-5.

[1592] Yan et al (2017) 'Estrogen deficiency is associated with hippocampal morphological remodeling of early postmenopausal mice', Oncotarget. 2017; 8:21892-21902. https://doi.org/10.18632/oncotarget.15702

[1593] Milner et al (2013). Stress differentially alters mu opioid receptor density and trafficking in parvalbumin-containing interneurons in the female and male rat hippocampus. Synapse, 67(11), 757–772. doi:10.1002/syn.21683

[1594] Aksglaede L, Juul A, Leffers H, Skakkebaek NE, Andersson AM (2006). "The sensitivity of the child to sex steroids: possible impact of exogenous estrogens". *Hum. Reprod. Update*. 12 (4): 341–9.

[1595] Gassman N. R. (2017). Induction of oxidative stress by bisphenol A and its pleiotropic effects. Environmental and molecular mutagenesis, 58(2), 60–71. https://doi.org/10.1002/em.22072

[1596] Lukaszewicz-Hussain, A. (2010). Role of oxidative stress in organophosphate insecticide toxicity – Short review. Pesticide Biochemistry and Physiology, 98(2), 145–150. doi:10.1016/j.pestbp.2010.07.006

[1597] Abdollahi et al (2004) 'Pesticides and oxidative stress: a review.', Med Sci Monit. 2004 Jun;10(6):RA141-7. Epub 2004 Jun 1.

[1598] Ndonwi et al (2019). Gestational Exposure to Pesticides Induces Oxidative Stress and Lipid Peroxidation in Offspring that Persist at Adult Age in an Animal Model. Toxicological Research, 35(3), 241–248. doi:10.5487/tr.2019.35.3.241

[1599] Mascitelli, L., & Goldstein, M. R. (2010). Inhibition of iron absorption by polyphenols as an anti-cancer mechanism. QJM, 104(5), 459–461. doi:10.1093/qjmed/hcq239

[1600] Sharp P. (2004). The molecular basis of copper and iron interactions. The Proceedings of the Nutrition Society, 63(4), 563–569. https://doi.org/10.1079/pns2004386

[1601] Cook, C. I., & Yu, B. P. (1998). Iron accumulation in aging: modulation by dietary restriction. Mechanisms of ageing and development, 102(1), 1–13. https://doi.org/10.1016/s0047-6374(98)00005-0

[1602] Korolchuk et al (2017). Mitochondria in Cell Senescence: Is Mitophagy the Weakest Link?. EBioMedicine, 21, 7–13. https://doi.org/10.1016/j.ebiom.2017.03.020

[1603] Pietrocola et al (2014). Coffee induces autophagy in vivo. Cell cycle (Georgetown, Tex.), 13(12), 1987–1994. https://doi.org/10.4161/cc.28929

[1604] DISCHIA, M., COSTANTINI, C., & PROTA, G. (1996). Lipofuscin-like pigments by autoxidation of polyunsaturated fatty acids in the presence of amine neurotransmitters: the role of malondialdehyde. Biochimica et Biophysica Acta (BBA) - General Subjects, 1290(3), 319–326. doi:10.1016/0304-4165(96)00036-0

[1605] Marcelo et al (2014) 'Beta cyclodextrins bind, stabilize, and remove lipofuscin bisretinoids from retinal pigment epithelium', PNAS April 8, 2014 111 (14) E1402-E1408; first published March 24, 2014 https://doi.org/10.1073/pnas.1400530111

[1606] Song, W., Wang, F., Lotfi, P., Sardiello, M., & Segatori, L. (2014). 2-Hydroxypropyl-β-cyclodextrin promotes transcription factor EB-mediated activation of autophagy: implications for therapy. The Journal of biological chemistry, 289(14), 10211–10222. https://doi.org/10.1074/jbc.M113.506246

[1607] Amar et al (2016). Randomized double blind clinical trial on the effect of oral α-cyclodextrin on serum lipids. Lipids in Health and Disease, 15(1). doi:10.1186/s12944-016-0284-6

[1608] Monji, A., Morimoto, N., Okuyama, I., Yamashita, N., & Tashiro, N. (1994). Effect of dietary vitamin E on lipofuscin accumulation with age in the rat brain. Brain research, 634(1), 62–68. https://doi.org/10.1016/0006-8993(94)90258-5

[1609] Winstanley, E. K., & Pentreath, V. W. (1985). Lipofuscin accumulation and its prevention by vitamin E in nervous tissue: quantitative analysis using snail buccal ganglia as a simple model system. Mechanisms of ageing and development, 29(3), 299–307. https://doi.org/10.1016/0047-6374(85)90069-7

[1610] Katz, M. L., Drea, C. M., & Robison, W. G., Jr (1986). Relationship between dietary retinol and lipofuscin in the retinal pigment epithelium. Mechanisms of ageing and development, 35(3), 291–305. https://doi.org/10.1016/0047-6374(86)90131-4

[1611] Gao, G., Ollinger, K., & Brunk, U. T. (1994). Influence of intracellular glutathione concentration of lipofuscin accumulation in cultured neonatal rat cardiac myocytes. Free radical biology & medicine, 16(2), 187–194. https://doi.org/10.1016/0891-5849(94)90142-2

[1612] Cordero et al (2012). Oxidative stress correlates with headache symptoms in fibromyalgia: coenzyme Q_{10} effect on clinical improvement. PloS one, 7(4), e35677. https://doi.org/10.1371/journal.pone.0035677

[1613] Dai et al (2011). Reversal of mitochondrial dysfunction by coenzyme Q10 supplement improves endothelial function in patients with ischaemic left ventricular systolic dysfunction: a randomized controlled trial. Atherosclerosis, 216(2), 395–401. https://doi.org/10.1016/j.atherosclerosis.2011.02.013

[1614] Klopstock, T., Elstner, M., & Bender, A. (2011). Creatine in mouse models of neurodegeneration and aging. Amino acids, 40(5), 1297–1303. https://doi.org/10.1007/s00726-011-0850-1

[1615] Vistoli et al (2013). Advanced glycoxidation and lipoxidation end products (AGEs and ALEs): an overview of their mechanisms of formation. Free radical research, 47 Suppl 1, 3–27. https://doi.org/10.3109/10715762.2013.815348

[1616] Uribarri et al (2007). Circulating glycotoxins and dietary advanced glycation endproducts: two links to inflammatory response, oxidative stress, and aging. The journals of gerontology. Series A, Biological sciences and medical sciences, 62(4), 427–433. https://doi.org/10.1093/gerona/62.4.427

[1617] Srikanth et al (2011). Advanced glycation endproducts and their receptor RAGE in Alzheimer's disease. Neurobiology of aging, 32(5), 763–777. https://doi.org/10.1016/j.neurobiolaging.2009.04.016

[1618] Simm et al (2007). Advanced glycation endproducts: a biomarker for age as an outcome predictor after cardiac surgery?. Experimental gerontology, 42(7), 668–675. https://doi.org/10.1016/j.exger.2007.03.006

[1619] Gugliucci, A., & Bendayan, M. (1996). Renal fate of circulating advanced glycated end products (AGE): evidence for reabsorption and catabolism of AGE-peptides by renal proximal tubular cells. Diabetologia, 39(2), 149–160. https://doi.org/10.1007/BF00403957

[1620] Zimmerman et al (1995). Neurotoxicity of advanced glycation endproducts during focal stroke and neuroprotective effects of aminoguanidine. Proceedings of the National Academy of Sciences of the United States of America, 92(9), 3744–3748. https://doi.org/10.1073/pnas.92.9.3744

[1621] Schmidt et al (1999) Activation of receptor for advanced glycation end products: a mechanism for chronic vascular dysfunction in diabetic vasculopathy and atherosclerosis. Circ Res. 1999; *84*: 489–497.

[1622] Schmidt et al (1994) Receptor for advanced glycation end products (AGEs) has a central role in vessel wall interactions and gene activation in response to circulating AGE proteins. Proc Natl Acad Sci U S A. 1994; *91*: 8807–8811.

[1623] Brownlee M. (1995) Advanced protein glycosylation in diabetes and aging. Annu Rev Med. 1995; *46*: 223–234.

[1624] Goh, S.-Y., & Cooper, M. E. (2008). The Role of Advanced Glycation End Products in Progression and Complications of Diabetes. The Journal of Clinical Endocrinology & Metabolism, 93(4), 1143–1152. doi:10.1210/jc.2007-1817

[1625] Prasad, A., Bekker, P., & Tsimikas, S. (2012). Advanced glycation end products and diabetic cardiovascular disease. Cardiology in review, 20(4), 177–183. https://doi.org/10.1097/CRD.0b013e318244e57c

[1626] Prasad, A., Bekker, P., & Tsimikas, S. (2012). Advanced glycation end products and diabetic cardiovascular disease. Cardiology in review, 20(4), 177–183. https://doi.org/10.1097/CRD.0b013e318244e57c

[1627] Goldin et al (2006) 'Advanced Glycation End Products: Sparking the Development of Diabetic Vascular Injury', Circulation. 2006;114:597–605

[1628] Uribarri et al (2010). Advanced glycation end products in foods and a practical guide to their reduction in the diet. Journal of the American Dietetic Association, 110(6), 911–16.e12. https://doi.org/10.1016/j.jada.2010.03.018

[1629] Sebeková et al (2001). Plasma levels of advanced glycation end products in healthy, long-term vegetarians and subjects on a western mixed diet. European journal of nutrition, 40(6), 275–281. https://doi.org/10.1007/s394-001-8356-3

[1630] Ahmed, N., & Furth, A. J. (1992). Failure of common glycation assays to detect glycation by fructose. Clinical chemistry, 38(7), 1301–1303.

[1631] Uribarri et al (2010). Advanced glycation end products in foods and a practical guide to their reduction in the diet. Journal of the American Dietetic Association, 110(6), 911–16.e12. https://doi.org/10.1016/j.jada.2010.03.018

[1632] Basciano et al (2005) Fructose, insulin resistance, and metabolic dyslipidemia. Nutr Metab (Lond). 2005;2(1):5. Published 2005 Feb 21. doi:10.1186/1743-7075-2-5

[1633] Stanhope KL, Schwarz JM, Keim NL, et al. (2009) Consuming fructose-sweetened, not glucose-sweetened, beverages increases visceral adiposity and lipids and decreases insulin sensitivity in overweight/obese humans. J Clin Invest. 2009;119(5):1322-1334. doi:10.1172/JCI37385

[1634] DeChristopher, L. R., Uribarri, J., & Tucker, K. L. (2017). Intake of high fructose corn syrup sweetened soft drinks, fruit drinks and apple juice is associated with prevalent coronary heart disease, in U.S. adults, ages 45–59 y. BMC Nutrition, 3(1). doi:10.1186/s40795-017-0168-9

[1635] Levi, B., & Werman, M. J. (1998). Long-Term Fructose Consumption Accelerates Glycation and Several Age-Related Variables in Male Rats. The Journal of Nutrition, 128(9), 1442–1449. doi:10.1093/jn/128.9.1442

[1636] Carvalho et al (2015). High intake of heterocyclic amines from meat is associated with oxidative stress. British Journal of Nutrition, 113(8), 1301–1307. doi:10.1017/s0007114515000628

[1637] Hasegawa et al (1991), Synergistic Enhancement of Glutathione S–Transferase Placental Form–positive Hepatic Foci Development in Diethylnitrosamine–treated Rats by Combined Administration of Five Heterocyclic Amines at Low Doses. Japanese Journal of Cancer Research, 82: 1378-1384. doi:10.1111/j.1349-7006.1991.tb01809.x

[1638] Uribarri J et al (2010) 'Advanced Glycation End Products in Foods and a Practical Guide to Their Reduction in the Diet', Journal of the American Dietetic Association, Volume 110, Issue 6, June 2010, Pages 911-916.e12.

[1639] Boor et al (2009). Regular moderate exercise reduces advanced glycation and ameliorates early diabetic nephropathy in obese Zucker rats. Metabolism: clinical and experimental, 58(11), 1669–1677. https://doi.org/10.1016/j.metabol.2009.05.025

[1640] Abdul, H. M., & Butterfield, D. A. (2007). Involvement of PI3K/PKG/ERK1/2 signaling pathways in cortical neurons to trigger protection by cotreatment of acetyl-L-carnitine and alpha-lipoic acid against HNE-mediated oxidative stress and neurotoxicity: implications for Alzheimer's disease. Free radical biology & medicine, 42(3), 371–384. https://doi.org/10.1016/j.freeradbiomed.2006.11.006

[1641] Nandhini, A. T., Thirunavukkarasu, V., & Anuradha, C. V. (2005). Taurine prevents collagen abnormalities in high fructose-fed rats. The Indian journal of medical research, 122(2), 171–177.

[1642] Urios, P., Grigorova-Borsos, A. M., & Sternberg, M. (2007). Aspirin inhibits the formation of pentosidine, a cross-linking advanced glycation end product, in collagen. Diabetes research and clinical practice, 77(2), 337–340. https://doi.org/10.1016/j.diabres.2006.12.024

[1643] Guiotto, A., Calderan, A., Ruzza, P., & Borin, G. (2005). Carnosine and carnosine-related antioxidants: a review. Current medicinal chemistry, 12(20), 2293–2315. https://doi.org/10.2174/0929867054864796

[1644] Mizutani, K., Ikeda, K., & Yamori, Y. (2000). Resveratrol inhibits AGEs-induced proliferation and collagen synthesis activity in vascular smooth muscle cells from stroke-prone spontaneously hypertensive rats. Biochemical and biophysical research communications, 274(1), 61–67. https://doi.org/10.1006/bbrc.2000.3097

[1645] Tang, Y., & Chen, A. (2014). Curcumin eliminates the effect of advanced glycation end-products (AGEs) on the divergent regulation of gene expression of receptors of AGEs by interrupting leptin signaling. Laboratory investigation; a journal of technical methods and pathology, 94(5), 503–516. https://doi.org/10.1038/labinvest.2014.42

[1646] Sanchis et al (2018). Phytate Decreases Formation of Advanced Glycation End-Products in Patients with Type II Diabetes: Randomized Crossover Trial. Scientific Reports, 8(1). doi:10.1038/s41598-018-27853-9

[1647] Caito, S. W., & Aschner, M. (2016). NAD+Supplementation Attenuates Methylmercury

Dopaminergic and Mitochondrial Toxicity inCaenorhabditis Elegans. Toxicological Sciences, 151(1), 139–149. doi:10.1093/toxsci/kfw030

[1648] Fuhrmeister et al (2016) Fasting-induced liver GADD45 restrains hepatic fatty acid uptake and improves metabolic health. *EMBO Molecular Medicine*, 2016; DOI: 10.15252/emmm.201505801

[1649] Sears, M. E., Kerr, K. J., & Bray, R. I. (2012). Arsenic, Cadmium, Lead, and Mercury in Sweat: A Systematic Review. Journal of Environmental and Public Health, 2012, 1–10. doi:10.1155/2012/184745

[1650] Speroni et al (2003). Efficacy of different Cynara scolymus preparations on liver complaints. Journal of ethnopharmacology, 86(2-3), 203–211. https://doi.org/10.1016/s0378-8741(03)00076-x

[1651] Clifford, T., Howatson, G., West, D., & Stevenson, E. (2015). The Potential Benefits of Red Beetroot Supplementation in Health and Disease. Nutrients, 7(4), 2801–2822. doi:10.3390/nu7042801

[1652] Craig, S. A. (2004). Betaine in human nutrition. The American Journal of Clinical Nutrition, 80(3), 539–549. doi:10.1093/ajcn/80.3.539

[1653] Houghton, C. A., Fassett, R. G., & Coombes, J. S. (2016). Sulforaphane and Other Nutrigenomic Nrf2 Activators: Can the Clinician's Expectation Be Matched by the Reality? Oxidative Medicine and Cellular Longevity, 2016, 1–17. doi:10.1155/2016/7857186

[1654] Egner et al (2014). Rapid and Sustainable Detoxication of Airborne Pollutants by Broccoli Sprout Beverage: Results of a Randomized Clinical Trial in China. Cancer Prevention Research, 7(8), 813–823. doi:10.1158/1940-6207.capr-14-0103

[1655] Chen et al (2016). Dietary broccoli protects against fatty liver development but not against progression of liver cancer in mice pretreated with diethylnitrosamine. Journal of functional foods, 24, 57–62. https://doi.org/10.1016/j.jff.2016.03.028

[1656] Sandau, E., Sandau, P., & Pulz, O. (1996). Heavy metal sorption by microalgae. Acta Biotechnologica, 16(4), 227–235. doi:10.1002/abio.370160402

[1657] Victor et al (2012). Dietary d-limonene alleviates insulin resistance and oxidative stress-induced liver injury in high-fat diet and L-NAME-treated rats. European journal of nutrition, 51(1), 57–68. https://doi.org/10.1007/s00394-011-0182-7

[1658] Yun et al (1995). Non-specific inhibition of cytochrome P450 activities by chlorophyllin in human and rat liver microsomes. Carcinogenesis, 16(6), 1437–1440. https://doi.org/10.1093/carcin/16.6.1437

[1659] LiverTox: Clinical and Research Information on Drug-Induced Liver Injury [Internet]. Bethesda (MD): National Institute of Diabetes and Digestive and Kidney Diseases; 2012-. Available from: https://www.ncbi.nlm.nih.gov/books/NBK547852/

[1660] Lee et al (2016). Turmeric extract and its active compound, curcumin, protect against chronic CCl4-induced liver damage by enhancing antioxidation. BMC complementary and alternative medicine, 16(1), 316. https://doi.org/10.1186/s12906-016-1307-6

[1661] Dulbecco, P., & Savarino, V. (2013). Therapeutic potential of curcumin in digestive diseases. World journal of gastroenterology, 19(48), 9256–9270. https://doi.org/10.3748/wjg.v19.i48.9256

[1662] Lu et al (2016). Effects of Omega-3 Fatty Acid in Nonalcoholic Fatty Liver Disease: A Meta-Analysis. Gastroenterology research and practice, 2016, 1459790. https://doi.org/10.1155/2016/1459790

[1663] Schecter et al (1997). Levels of dioxins, dibenzofurans, PCB and DDE congeners in pooled food samples collected in 1995 at supermarkets across the United States. Chemosphere, 34(5-7), 1437–1447. doi:10.1016/s0045-6535(97)00440-2

[1664] Consumerlab.com 'Is Fish Oil Safe?', Accessed Online: https://www.consumerlab.com/answers/is-fish-oil-safe/fish-oil_contamination/

[1665] Doughman et al (2007) 'Omega-3 Fatty Acids for Nutrition and Medicine: Considering Microalgae Oil as a Vegetarian Source of EPA and DHA', Current Diabetes Reviews, Volume 3 ,

Issue 3 , 2007, DOI : 10.2174/157339907781368968

[1666] Rhee, S.H. et al (2009) 'Principles and clinical implications of the brain–gut–enteric microbiota axis', Nature Reviews. Gastroenterology & Hepatology, Vol 6, p 306-314.

[1667] Wang, Y. and Kasper, L.H. (2014) 'The role of microbiome in central nervous system disorders', Brain Behav Immun, Vol 38, p 1-12.

[1668] Carabotti, M. et al (2015) 'The gut-brain axis: interactions between enteric microbiota, central and enteric nervous systems', Annals & Gastroenterology, Vol 28(2), p 203-209.

[1669] Fasano, A. (2012). "Leaky Gut and Autoimmune Diseases". *Clinical Reviews in Allergy & Immunology* (Review). 42 (1): 71–78.

[1670] Rapin JR, Wiernsperger N (2010). "Possible links between intestinal permeability and food processing: A potential therapeutic niche for glutamine". *Clinics* (Review). 65 (6): 635–43.

[1671] Groschwitz, Katherine R.; Hogan, Simon P. (2009). "Intestinal Barrier Function: Molecular Regulation and Disease Pathogenesis". *Journal of Allergy and Clinical Immunology*. 124 (1): 3–22.

[1672] Takeda S, Sato N, Morishita R (2014) Systemic inflammation, blood-brain barrier vulnerability and cognitive/non-cognitive symptoms in Alzheimer disease: relevance to pathogenesis and therapy, Front Aging Neuroscience, 6:171

[1673] Drago et al (2006). Gliadin, zonulin and gut permeability: Effects on celiac and non-celiac intestinal mucosa and intestinal cell lines. Scandinavian Journal of Gastroenterology, 41(4), 408–419. doi:10.1080/00365520500235334

[1674] Fasano, A. (2011). Zonulin and Its Regulation of Intestinal Barrier Function: The Biological Door to Inflammation, Autoimmunity, and Cancer. Physiological Reviews, 91(1), 151–175. doi:10.1152/physrev.00003.2008

[1675] Fasano et al (2000). Zonulin, a newly discovered modulator of intestinal permeability, and its expression in coeliac disease. The Lancet, 355(9214), 1518–1519. doi:10.1016/s0140-6736(00)02169-3

[1676] Cani et al (2014) 'Glucose metabolism: Focus on gut microbiota, the endocannabinoid system and beyond', DIABETES & METABOLISM, Vol 40 - N° 4, P. 246-257 - septembre 2014.

[1677] Tilg, H., & Kaser, A. (2011). Gut microbiome, obesity, and metabolic dysfunction. Journal of Clinical Investigation, 121(6), 2126–2132. doi:10.1172/jci58109

[1678] Canani, R. et al. (2011). Potential beneficial effects of butyrate in intestinal and extraintestinal diseases. *World Journal of Gastroenterology* 17 (12): 1519–1528.

[1679] Akobeng, AK; Elawad, M; Gordon, M (8 February 2016). "Glutamine for induction of remission in Crohn's disease" (PDF). *Cochrane Database of Systematic Reviews*. 2: CD007348.

[1680] De Filippis et al (2016). High-level adherence to a Mediterranean diet beneficially impacts the gut microbiota and associated metabolome. Gut, 65(11), 1812–1821. https://doi.org/10.1136/gutjnl-2015-309957

[1681] Takechi et al (2013). Nutraceutical agents with anti-inflammatory properties prevent dietary saturated-fat induced disturbances in blood–brain barrier function in wild-type mice. Journal of Neuroinflammation, 10(1). doi:10.1186/1742-2094-10-73

[1682] Zarrinpar, A., Chaix, A., Yooseph, S., & Panda, S. (2014). Diet and feeding pattern affect the diurnal dynamics of the gut microbiome. Cell metabolism, 20(6), 1006–1017. https://doi.org/10.1016/j.cmet.2014.11.008

[1683] Lopez et al (2012) 'Ghrelin Prevents Disruption of the Blood–Brain Barrier after Traumatic Brain Injury', Journal of NeurotraumaVol. 29, No. 2.

[1684] University of Illinois at Urbana-Champaign. (2017, December 4). Exercise changes gut microbial composition independent of diet, team reports. ScienceDaily. Retrieved May 31, 2020 from www.sciencedaily.com/releases/2017/12/171204144757.htm

[1685] Gomez-Gonzalez et al (2013) 'REM Sleep Loss and Recovery Regulates Blood-Brain Barrier

Function', Current Neurovascular Research, Volume 10, Number 3, 2013, pp. 197-207(11).

[1686] He et al (2014). Sleep Restriction Impairs Blood-Brain Barrier Function. Journal of Neuroscience, 34(44), 14697–14706. doi:10.1523/jneurosci.2111-14.2014

[1687] Johansson, B., & Nilsson, B. (1977). The pathophysiology of the blood-brain barrier dysfunction induced by severe hypercapnia and by epileptic brain activity. Acta Neuropathologica, 38(2), 153–158. doi:10.1007/bf00688563

[1688] Haorah, J. (2005). Alcohol-induced oxidative stress in brain endothelial cells causes blood-brain barrier dysfunction. Journal of Leukocyte Biology, 78(6), 1223–1232. doi:10.1189/jlb.0605340

[1689] Chen et al (2008). Caffeine blocks disruption of blood brain barrier in a rabbit model of Alzheimer's disease. Journal of Neuroinflammation, 5(1), 12. doi:10.1186/1742-2094-5-12

[1690] Halberg, F. (1959). "Physiologic 24-hour periodicity: general and procedural considerations with reference to the adrenal cycle". *Zeitschrift für Vitamin- Hormone- und Fermentforschung*. 10: 225–296.

[1691] Halberg et al (1977). "[Glossary of chronobiology (author's transl)]". *Chronobiologia*. 4 Suppl 1: 1–189.

[1692] Hall JC, Rosbash, M, and Young MW. (2017) The Nobel Prize in Physiology or Medicine 2017, Accessed; https://www.nobelprize.org/prizes/medicine/2017/press-release/

[1693]Nagoshi E et al (2004) 'Circadian Gene Expression in Individual Fibroblasts', Cell, VOLUME 119, ISSUE 5, P693-705.

[1694] https://www.intechopen.com/online-first/epigenetic-modulation-of-circadian-rhythms-bmal1-gene-regulation/

[1695] Masri, S., Kinouchi, K., & Sassone-Corsi, P. (2015). Circadian clocks, epigenetics, and cancer. Current opinion in oncology, 27(1), 50–56. https://doi.org/10.1097/CCO.0000000000000153

[1696] Bellet, M. M., & Sassone-Corsi, P. (2010). Mammalian circadian clock and metabolism - the epigenetic link. Journal of cell science, 123(Pt 22), 3837–3848. https://doi.org/10.1242/jcs.051649

[1697] Duffy, J. F., & Wright, K. P. (2005). Entrainment of the Human Circadian System by Light. Journal of Biological Rhythms, 20(4), 326–338.

[1698] Cromie, William (1999). "Human Biological Clock Set Back an Hour". *Harvard Gazette*.

[1699] Gibo, S., & Kurosawa, G. (2020). Theoretical study on the regulation of circadian rhythms by RNA methylation. Journal of theoretical biology, 490, 110140. https://doi.org/10.1016/j.jtbi.2019.110140

[1700] Garcia-Saenz, A. et al (2018) 'Evaluating the Association between Artificial Light-at-Night Exposure and Breast and Prostate Cancer Risk in Spain (MCC-Spain Study)', Environmental Health Perspectives, 126 (04).

[1701] Kondratov R. V. (2007). A role of the circadian system and circadian proteins in aging. Ageing research reviews, 6(1), 12–27. https://doi.org/10.1016/j.arr.2007.02.003

[1702] Kondratova, A. A., & Kondratov, R. V. (2012). The circadian clock and pathology of the ageing brain. Nature reviews. Neuroscience, 13(5), 325–335. https://doi.org/10.1038/nrn3208

[1703] Orozco-Solis, R., & Sassone-Corsi, P. (2014). Circadian clock: linking epigenetics to aging. Current opinion in genetics & development, 26, 66–72. https://doi.org/10.1016/j.gde.2014.06.003

[1704] Libert et al (2012). Deviation of innate circadian period from 24 h reduces longevity in mice. Aging cell, 11(5), 794–800. https://doi.org/10.1111/j.1474-9726.2012.00846.x

[1705] Wyse et al (2010). Association between mammalian lifespan and circadian free-running period: the circadian resonance hypothesis revisited. Biology letters, 6(5), 696–698. https://doi.org/10.1098/rsbl.2010.0152

[1706] Ripperger, J. A., & Merrow, M. (2011). Perfect timing: Epigenetic regulation of the circadian clock. FEBS Letters, 585(10), 1406–1411. doi:10.1016/j.febslet.2011.04.047

[1707] Utiger (1992) 'Melatonin--the hormone of darkness', N Engl J Med. 1992 Nov 5;327(19):1377-9.

[1708] Wurtman (2000) 'Age-Related Decreases in Melatonin Secretion—Clinical Consequences', The Journal of Clinical Endocrinology & Metabolism, Volume 85, Issue 6, 1 June 2000, Pages 2135–2136, https://doi.org/10.1210/jcem.85.6.6660.

[1709] Turner, P. L., & Mainster, M. A. (2008). Circadian photoreception: ageing and the eye's important role in systemic health. British Journal of Ophthalmology, 92(11), 1439–1444. doi:10.1136/bjo.2008.141747

[1710] Esquiva et al (2017) 'Loss of Melanopsin-Expressing Ganglion Cell Subtypes and Dendritic Degeneration in the Aging Human Retina', Front. Aging Neurosci., 04 April 2017, https://doi.org/10.3389/fnagi.2017.00079

[1711] Hofman and Swaab (2006) 'Living by the clock: The circadian pacemaker in older people', Ageing Research Reviews, Volume 5, Issue 1, February 2006, Pages 33-51

[1712] Carrier, J., Monk, T. H., Buysse, D. J., & Kupfer, D. J. (1997). Sleep and morningness-eveningness in the 'middle' years of life (20-59 y). Journal of sleep research, 6(4), 230–237. https://doi.org/10.1111/j.1365-2869.1997.00230.x

[1713] Roenneberg et al (2007). Epidemiology of the human circadian clock. Sleep medicine reviews, 11(6), 429–438. https://doi.org/10.1016/j.smrv.2007.07.005

[1714] Schmidt et al (2012). Adapting test timing to the sleep-wake schedule: effects on diurnal neurobehavioral performance changes in young evening and older morning chronotypes. Chronobiology international, 29(4), 482–490. https://doi.org/10.3109/07420528.2012.658984

[1715] Carrier, J., Monk, T. H., Buysse, D. J., & Kupfer, D. J. (1997). Sleep and morningness-eveningness in the 'middle' years of life (20-59 y). Journal of sleep research, 6(4), 230–237. https://doi.org/10.1111/j.1365-2869.1997.00230.x

[1716] Duffy et al (2002). Peak of circadian melatonin rhythm occurs later within the sleep of older subjects. American journal of physiology. Endocrinology and metabolism, 282(2), E297–E303. https://doi.org/10.1152/ajpendo.00268.2001

[1717] Dijk, D. J., Duffy, J. F., & Czeisler, C. A. (2000). Contribution of circadian physiology and sleep homeostasis to age-related changes in human sleep. Chronobiology international, 17(3), 285–311. https://doi.org/10.1081/cbi-100101049

[1718] Zhdanova et al (2011). Aging of intrinsic circadian rhythms and sleep in a diurnal nonhuman primate, Macaca mulatta. Journal of biological rhythms, 26(2), 149–159. https://doi.org/10.1177/0748730410395849

[1719] Naylor et al (1998). Effects of aging on sleep in the golden hamster. Sleep, 21(7), 687–693. https://doi.org/10.1093/sleep/21.7.687

[1720] Koh, K., Evans, J. M., Hendricks, J. C., & Sehgal, A. (2006). A Drosophila model for age-associated changes in sleep:wake cycles. Proceedings of the National Academy of Sciences of the United States of America, 103(37), 13843–13847. https://doi.org/10.1073/pnas.0605903103

[1721] Ohkubo, R., & Chen, D. (2017). Aging: rewiring the circadian clock. Nature Structural & Molecular Biology, 24(9), 687–688. doi:10.1038/nsmb.3461

[1722] Masri S. (2015). Sirtuin-dependent clock control: new advances in metabolism, aging and cancer. Current opinion in clinical nutrition and metabolic care, 18(6), 521–527. https://doi.org/10.1097/MCO.0000000000000219

[1723] Asher et al (2008). SIRT1 Regulates Circadian Clock Gene Expression through PER2 Deacetylation. Cell, 134(2), 317–328. doi:10.1016/j.cell.2008.06.050

[1724] Satoh et al (2013). Sirt1 extends life span and delays aging in mice through the regulation of Nk2 homeobox 1 in the DMH and LH. Cell metabolism, 18(3), 416–430. https://doi.org/10.1016/j.cmet.2013.07.013

[1725] Chang, H.-C., & Guarente, L. (2013). SIRT1 Mediates Central Circadian Control in the SCN by a Mechanism that Decays with Aging. Cell, 153(7), 1448–1460. doi:10.1016/j.cell.2013.05.027

[1726] Ruckenstuhl et al (2014). Lifespan extension by methionine restriction requires autophagy-dependent vacuolar acidification. PLoS genetics, 10(5), e1004347. https://doi.org/10.1371/journal.pgen.1004347

[1727] Morselli et al (2010). Caloric restriction and resveratrol promote longevity through the Sirtuin-1-dependent induction of autophagy. Cell death & disease, 1(1), e10. https://doi.org/10.1038/cddis.2009.8

[1728] Van Cauter, E., & Plat, L. (1996). Physiology of growth hormone secretion during sleep. The Journal of pediatrics, 128(5 Pt 2), S32–S37. https://doi.org/10.1016/s0022-3476(96)70008-2

[1729] Beaulé et al (2003) 'Melanopsin in the circadian timing system.' J Mol Neurosci. 2003;21(1):73-89.

[1730] Ondrusova et al (2017) 'Subcutaneous white adipocytes express a light sensitive signaling pathway mediated via a melanopsin/TRPC channel axis', Scientific Reportsvolume 7, Article number: 16332 (2017).

[1731] Cheung IN et al (2016) 'Morning and Evening Blue-Enriched Light Exposure Alters Metabolic Function in Normal Weight Adults', PLoS One. 2016 May 18;11(5):e0155601.

[1732] Zhao, Z. C., Zhou, Y., Tan, G., & Li, J. (2018). Research progress about the effect and prevention of blue light on eyes. International journal of ophthalmology, 11(12), 1999–2003. https://doi.org/10.18240/ijo.2018.12.20

[1733] Chen et al (2013). Autophagy protects the retina from light-induced degeneration. The Journal of biological chemistry, 288(11), 7506–7518. https://doi.org/10.1074/jbc.M112.439935

[1734] Mitter et al (2012). Autophagy in the retina: a potential role in age-related macular degeneration. Advances in experimental medicine and biology, 723, 83–90. https://doi.org/10.1007/978-1-4614-0631-0_12

[1735] Feng, J., Chen, Y., Lu, B., Sun, X., Zhu, H., & Sun, X. (2019). Autophagy activated via GRP78 to alleviate endoplasmic reticulum stress for cell survival in blue light-mediated damage of A2E-laden RPEs. BMC ophthalmology, 19(1), 249. https://doi.org/10.1186/s12886-019-1261-4

[1736] Kuse et al (2014). Damage of photoreceptor-derived cells in culture induced by light emitting diode-derived blue light. Scientific reports, 4, 5223. https://doi.org/10.1038/srep05223

[1737] Ogawa et al (2014). Protective effects of bilberry and lingonberry extracts against blue light-emitting diode light-induced retinal photoreceptor cell damage in vitro. BMC complementary and alternative medicine, 14, 120. https://doi.org/10.1186/1472-6882-14-120

[1738] Yang et al (2019). White light emitting diode induces autophagy in hippocampal neuron cells through GSK-3-mediated GR and RORα pathways. Aging, 11(6), 1832–1849. https://doi.org/10.18632/aging.101878

[1739] Lee et al (2014). Blue light-induced oxidative stress in human corneal epithelial cells: protective effects of ethanol extracts of various medicinal plant mixtures. Investigative ophthalmology & visual science, 55(7), 4119–4127. https://doi.org/10.1167/iovs.13-13441

[1740] Choi et al (2016). Amber Light (590 nm) Induces the Breakdown of Lipid Droplets through Autophagy-Related Lysosomal Degradation in Differentiated Adipocytes. Scientific Reports, 6(1). doi:10.1038/srep28476

[1741] Denda, M., & Fuziwara, S. (2008). Visible radiation affects epidermal permeability barrier recovery: selective effects of red and blue light. The Journal of investigative dermatology, 128(5), 1335–1336. https://doi.org/10.1038/sj.jid.5701168

[1742] Zhao, J., Tian, Y., Nie, J., Xu, J., & Liu, D. (2012). Red light and the sleep quality and endurance performance of Chinese female basketball players. Journal of athletic training, 47(6), 673–678. https://doi.org/10.4085/1062-6050-47.6.08

[1743] Zhang et al (2016) 'Antimicrobial blue light inactivation of Candida albicans: In vitro and in vivo studies.' Virulence. 2016 Jul 3;7(5):536-45. doi: 10.1080/21505594.2016.1155015. Epub 2016

Feb 24.

[1744] Skobowiat and Slominski (2015) 'Ultraviolet B (UVB) activates hypothalamic-pituitary-adrenal (HPA) axis in C57BL/6 mice', J Invest Dermatol. 2015 Jun; 135(6): 1638–1648.

[1745] Cui et al (2007) 'Central role of p53 in the suntan response and pathologic hyperpigmentation.' Cell. 2007 Mar 9;128(5):853-64.

[1746] Wu, S., & Sun, J. (2011). Vitamin D, vitamin D receptor, and macroautophagy in inflammation and infection. Discovery medicine, 11(59), 325–335.

[1747] Tavera-Mendoza et al (2017). Vitamin D receptor regulates autophagy in the normal mammary gland and in luminal breast cancer cells. Proceedings of the National Academy of Sciences of the United States of America, 114(11), E2186–E2194. https://doi.org/10.1073/pnas.1615015114

[1748] Jones (2013) 'In U.S., 40% Get Less Than Recommended Amount of Sleep', Gallup, WELL-BEING, DECEMBER 19, 2013, Accessed Online: https://news.gallup.com/poll/166553/less-recommended-amount-sleep.aspx

[1749] Harvard Medical School 'Harvard Heart Letter examines the costs of not getting enough sleep', Accessed Online: https://www.health.harvard.edu/press_releases/sleep_deprivation_problem.htm

[1750] Division of Sleep Medicine at Harvard Medical School 'Sleep and Disease Risk', Accessed Online http://healthysleep.med.harvard.edu/healthy/matters/consequences/sleep-and-disease-risk

[1751] Broussard et al (2012) 'Impaired Insulin Signaling in Human Adipocytes After Experimental Sleep Restriction: A Randomized, Crossover Study. Ann Intern Med. 2012;157:549–557. doi: 10.7326/0003-4819-157-8-201210160-00005

[1752] Mark London (2007) 'The Role of Magnesium in Fibromyalgia', MIT, Accessed Online: http://web.mit.edu/london/www/magnesium.html

[1753] Orzeł-Gryglewska J (2010) 'Consequences of sleep deprivation.' Int J Occup Med Environ Health. 2010;23(1):95-114. doi: 10.2478/v10001-010-0004-9.

[1754] Nedeltcheva et al (2010) 'Insufficient sleep undermines dietary efforts to reduce adiposity', Ann Intern Med. 2010 Oct 5; 153(7): 435–441.

[1755] Ackermann et al (2012). Diurnal Rhythms in Blood Cell Populations and the Effect of Acute Sleep Deprivation in Healthy Young Men. Sleep, 35(7), 933–940. doi:10.5665/sleep.1954

[1756] Mayo Clinic (2018) 'Lack of sleep: Can it make you sick?', Nov. 28, 2018, Accessed 11.07.19 https://www.mayoclinic.org/diseases-conditions/insomnia/expert-answers/lack-of-sleep/faq-20057757

[1757] Leproult and Van Cauter (2015) 'Effect of 1 Week of Sleep Restriction on Testosterone Levels in Young Healthy MenFREE', JAMA. 2011 Jun 1; 305(21): 2173–2174.

[1758] Patrick et al (2017) 'Effects of sleep deprivation on cognitive and physical performance in university students', Sleep Biol Rhythms. 2017; 15(3): 217–225.

[1759] Read (2008) 'Sleep Deprivation', High School Physiology, American Physiological Association, Feb 09, 2018, Accessed Online https://web.archive.org/web/20080209144819/http:/www.apa.org/ed/topss/bryanread.html

[1760] Petrovsky et al (2014) 'Sleep deprivation disrupts prepulse inhibition and induces psychosis-like symptoms in healthy humans', Journal of Neuroscience 2 July 2014, 34 (27) 9134-9140; DOI: https://doi.org/10.1523/JNEUROSCI.0904-14.2014

[1761] Alhola and Polo-Kantola (2007) 'Sleep deprivation: Impact on cognitive performance', Neuropsychiatr Dis Treat. 2007 Oct; 3(5): 553–567.

[1762] Thomas et al (2000) 'Neural basis of alertness and cognitive performance impairments during sleepiness. I. Effects of 24 h of sleep deprivation on waking human regional brain activity', Journal of Sleep Research, Volume9, Issue4, December 2000, Pages 335-352.

[1763] Vartanian et al (2014) 'The effects of a single night of sleep deprivation on fluency and prefrontal cortex function during divergent thinking', Front. Hum. Neurosci., 22 April 2014 |

https://doi.org/10.3389/fnhum.2014.00214

[1764] Jackson (2016) 'Sleep deprivation contributes to false confessions, study confirms', The Christian Science Monitor, February 10, 2016, Accessed 11.07.19. https://www.csmonitor.com/Science/2016/0210/Sleep-deprivation-contributes-to-false-confessions-study-confirms

[1765] Taheri et al (2004) 'Short Sleep Duration Is Associated with Reduced Leptin, Elevated Ghrelin, and Increased Body Mass Index', PLoS Med. 2004 Dec; 1(3): e62.

[1766] Holth et al (2019) 'The sleep-wake cycle regulates brain interstitial fluid tau in mice and CSF tau in humans', Science 22 Feb 2019, Vol. 363, Issue 6429, pp. 880-884, DOI: 10.1126/science.aav2546.

[1767] Collins (2019) 'Sleep loss encourages spread of toxic Alzheimer's protein', NIH Director's Blog, February 12, 2019, Accessed 11.07.19. https://www.nia.nih.gov/news/sleep-loss-encourages-spread-toxic-alzheimers-protein

[1768] Beyer (2019) 'A lack of deep sleep could indicate Alzheimer's development', Medical News Today, Monday 14 January 2019, Accessed Online: https://www.medicalnewstoday.com/articles/324161.php

[1769] Bellesi et al (2017) 'Sleep Loss Promotes Astrocytic Phagocytosis and Microglial Activation in Mouse Cerebral Cortex', Journal of Neuroscience 24 May 2017, 37 (21) 5263-5273; DOI: https://doi.org/10.1523/JNEUROSCI.3981-16.2017.

[1770] Harvard Medical School (2007) 'Sleep and Disease Risk', Division of Sleep Medicine at Harvard Medical School, Accessed 11.07.19 http://healthysleep.med.harvard.edu/healthy/matters/consequences/sleep-and-disease-risk

[1771] Cappuccio et al (2010) 'Sleep Duration and All-Cause Mortality: A Systematic Review and Meta-Analysis of Prospective Studies', Sleep. 2010 May 1; 33(5): 585–592.

[1772] NHLBI 'Sleep Deprivation and Deficiency', NIH, Accessed Online: https://www.nhlbi.nih.gov/health-topics/sleep-deprivation-and-deficiency

[1773] Morbidity and Mortality Weekly Report (2013) 'Drowsy Driving — 19 States and the District of Columbia, 2009–2010, Centers for Disease Control and Prevention, Weekly / Vol. 61 / Nos. 51 & 52, Accessed 11.07.19 https://www.cdc.gov/mmwr/pdf/wk/mm6151.pdf

[1774] Hafner et al (2016) 'Why sleep matters — the economic costs of insufficient sleep: A cross-country comparative analysis', RAND Europe, November 29, 2016.

[1775] Schwartz, Jonathan R.L; Roth, Thomas (24 January 2017). "Neurophysiology of Sleep and Wakefulness: Basic Science and Clinical Implications". *Current Neuropharmacology*. 6 (4): 367–378.

[1776] Fuller Patrick M.; Gooley Joshua J.; Saper Clifford B. (2006). "Neurobiology of the Sleep-Wake Cycle: Sleep Architecture, Circadian Regulation, and Regulatory Feedback". *Journal of Biological Rhythms*. 21 (6): 482–93.

[1777] Derk-Jan Dijk & Dale M. Edgar (1999), "Circadian and Homeostatic Control of Wakefulness and Sleep", in Turek & Zee (eds.), *Regulation of Sleep and Circadian Rhythms*, pp. 111–147'

[1778] Coren, Stanley (1 March 1998). "Sleep Deprivation, Psychosis and Mental Efficiency". *Psychiatric Times*. 15 (3).

[1779] Ross J (1965). "Neurological Findings After Prolonged Sleep Deprivation". *Archives of Neurology*. 12 (4): 399–403.

[1780] Siegel (2008) 'Do all animals sleep?', VOLUME 31, ISSUE 4, P208-213, APRIL 01, 2008. DOI:https://doi.org/10.1016/j.tins.2008.02.001

[1781] Walker et al (2002) 'Practice with sleep makes perfect: sleep-dependent motor skill learning', Neuron. 2002 Jul 3;35(1):205-11.

[1782] Roffwarg et al (1966) 'Ontogenetic development of the human sleep-dream cycle', Science. 1966 Apr 29;152(3722):604-19.

[1783] Tassi, P; Muzet, A (2000). "Sleep inertia". *National Center for Biotechnology Information, U.S. National Library of Medicine*. 4 (4): 341–353.

[1784] Åkerstedt et al (2002). "Awakening from Sleep". Sleep Medicine Reviews. 6 (4): 267–286.

[1785] Peraita-Adrados (2005), "Electroencephalography, Polysomnography, and Other Sleep Recording Systems"; Chapter 5 in Parmeggiani & Velluti.

[1786] Derk-Jan Dijk & Dale M. Edgar (1999), "Circadian and Homeostatic Control of Wakefulness and Sleep", in Turek & Zee (eds.), *Regulation of Sleep and Circadian Rhythms*, pp. 111–147'

[1787] Cappuccio et al (2010) 'Sleep duration and all-cause mortality: a systematic review and meta-analysis of prospective studies.' Sleep. 2010 May;33(5):585-92.

[1788] Dagan Y (2002). "Circadian rhythm sleep disorders (CRSD)" (PDF). Sleep Medicine Reviews. 6 (1): 45–54. doi:10.1053/smrv.2001.0190.

[1789] Falchi et al (2011) 'Limiting the impact of light pollution on human health, environment and stellar visibility', J Environ Manage. 2011 Oct;92(10):2714-22. doi: 10.1016/j.jenvman.2011.06.029. Epub 2011 Jul 13.

[1790] Mizuno (2012) 'Effects of thermal environment on sleep and circadian rhythm', J Physiol Anthropol. 2012; 31(1): 14.

[1791] Lack et al (2008) 'The relationship between insomnia and body temperatures', Sleep Medicine Reviews, Volume 12, Issue 4, August 2008, Pages 307-317.

[1792] National Sleep Foundation 'The Ideal Temperature for Sleep', Accessed Online: https://www.sleep.org/articles/temperature-for-sleep/

[1793] Ngo et al (2013) 'Auditory Closed-Loop Stimulation of the Sleep Slow Oscillation Enhances Memory', Neuron, VOLUME 78, ISSUE 3, P545-553, MAY 08, 2013.

[1794] Zanobetti et al (2010) 'Associations of PM10 with Sleep and Sleep-disordered Breathing in Adults from Seven U.S. Urban Areas', Am J Respir Crit Care Med. 2010 Sep 15; 182(6): 819–825.

[1795] Abou-Khadra (2013) 'Association between PM_{10} exposure and sleep of Egyptian school children.' Sleep Breath. 2013 May;17(2):653-7. doi: 10.1007/s11325-012-0738-7. Epub 2012 Jun 26.

[1796] Wolverton et al (September 1989). Interior landscape plants for indoor air pollution abatement (Report). NASA. NASA-TM-101766.

[1797]Statland, B. E., & Demas, T. J. (1980). Serum caffeine half-lives. Healthy subjects vs. patients having alcoholic hepatic disease. American journal of clinical pathology, 73(3), 390–393. https://doi.org/10.1093/ajcp/73.3.390

[1798] Debono et al (2009). Modified-release hydrocortisone to provide circadian cortisol profiles. The Journal of clinical endocrinology and metabolism, 94(5), 1548–1554. https://doi.org/10.1210/jc.2008-2380

[1799] Cornelis et al (2006) 'Coffee, CYP1A2 genotype, and risk of myocardial infarction.' JAMA. 2006 Mar 8;295(10):1135-41.

[1800] Palatini et al (2009) 'CYP1A2 genotype modifies the association between coffee intake and the risk of hypertension.' J Hypertens. 2009 Aug;27(8):1594-601. doi: 10.1097/HJH.0b013e32832ba850.

[1801] Banno et al (2018) 'Exercise can improve sleep quality: a systematic review and meta-analysis', PeerJ. 2018; 6: e5172.

[1802] Kovacevic et al (2018) 'The effect of resistance exercise on sleep: A systematic review of randomized controlled trials', Sleep Medicine Reviews, Volume 39, June 2018, Pages 52-68.

[1803] Fernstrom and Wurtman (1971) 'Brain serotonin content: physiological dependence on plasma tryptophan levels', Science. 1971 Jul 9;173(3992):149-52.

[1804] Halson (2014) 'Sleep in Elite Athletes and Nutritional Interventions to Enhance Sleep', Sports Med. 2014; 44(Suppl 1): 13–23.

[1805] National Sleep Foundation 'How Alcohol Affects the Quality—And Quantity—Of Sleep', Accessed Online https://www.sleepfoundation.org/articles/how-alcohol-affects-quality-and-quantity-sleep

[1806] Jamshed et al (2019). Early Time-Restricted Feeding Improves 24-Hour Glucose Levels and Affects Markers of the Circadian Clock, Aging, and Autophagy in Humans. Nutrients, 11(6), 1234. https://doi.org/10.3390/nu11061234

[1807] Hutchison et al (2019). Time-Restricted Feeding Improves Glucose Tolerance in Men at Risk for Type 2 Diabetes: A Randomized Crossover Trial. Obesity. doi:10.1002/oby.22449

[1808] Harvard School of Public Health (2007) 'New Study Shows Naps May Reduce Coronary Mortality', Press Releases Archive Home, 2007 Releases, Accessed 15.06.19 https://archive.sph.harvard.edu/press-releases/2007-releases/press02122007.html

[1809] "NASA: Alertness Management: Strategic Naps in Operational Settings". 1995. Archived from the original on 2012-04-19. Retrieved 2019-06-16.

[1810] Tanaka, H; Tamura, N (2015). "Sleep education with self-help treatment and sleep health promotion for mental and physical wellness in Japan". Sleep and Biological Rhythms. 14: 89–99.

[1811] Da Costa, Jacob Medes (January 1871). "On irritable heart; a clinical study of a form of functional cardiac disorder and its consequences". *The American Journal of the Medical Sciences* (61): 18–52.

[1812] Mackenzie, et al (1916-01-18). "Discussions On The Soldier's Heart". *Proceedings of the Royal Society of Medicine, Therapeutical and Pharmacological Section*. 9: 27–60.

[1813] American Psychiatric Association (2013). *Diagnostic and Statistical Manual of Mental Disorders* (5th ed.). Arlington, VA: American Psychiatric Publishing. pp. 271–80.

[1814] Panagioti M, Gooding PA, Triantafyllou K, Tarrier N (April 2015). "Suicidality and posttraumatic stress disorder (PTSD) in adolescents: a systematic review and meta-analysis". *Social Psychiatry and Psychiatric Epidemiology*. 50 (4): 525–37. doi:10.1007/s00127-014-0978-x.

[1815] D'Amico et al (2000). Stress and chronic headache. The Journal of Headache and Pain, 1(S1), S49–S52. doi:10.1007/s101940070026

[1816] Theeler et al (2009). Headache Triggers in the US Military. Headache: The Journal of Head and Face Pain, 50(5), 790–794. doi:10.1111/j.1526-4610.2009.01571.x

[1817] Gross et al (2019). Potential Protective Mechanisms of Ketone Bodies in Migraine Prevention. Nutrients, 11(4), 811. doi:10.3390/nu11040811

[1818] Vachon-Presseau et al (2013). The stress model of chronic pain: evidence from basal cortisol and hippocampal structure and function in humans. Brain, 136(3), 815–827. doi:10.1093/brain/aws371

[1819] Wong et al (2012). Chronic psychosocial stress: does it modulate immunity to the influenza vaccine in Hong Kong Chinese elderly caregivers? AGE, 35(4), 1479–1493. doi:10.1007/s11357-012-9449-z

[1820] GRAHAM, N. M. H., DOUGLAS, R. M., & RYAN, P. (1986). STRESS AND ACUTE RESPIRATORY INFECTION. American Journal of Epidemiology, 124(3), 389–401. doi:10.1093/oxfordjournals.aje.a114409

[1821] Dahlgren, A., Kecklund, G., & Åkerstedt, T. (2005). Different levels of work-related stress and the effects on sleep, fatigue and cortisol. Scandinavian Journal of Work, Environment & Health, 31(4), 277–285. doi:10.5271/sjweh.883

[1822] Drake, C. L., Pillai, V., & Roth, T. (2014). Stress and Sleep Reactivity: A Prospective Investigation of the Stress-Diathesis Model of Insomnia. Sleep, 37(8), 1295–1304. doi:10.5665/sleep.3916

[1823] Carter, J. S., & Garber, J. (2011). Predictors of the first onset of a major depressive episode and changes in depressive symptoms across adolescence: Stress and negative cognitions. Journal of Abnormal Psychology, 120(4), 779–796. doi:10.1037/a0025441

[1824] Hammen, C., Kim, E. Y., Eberhart, N. K., & Brennan, P. A. (2009). Chronic and acute stress and the prediction of major depression in women. Depression and Anxiety, 26(8), 718–723. doi:10.1002/da.20571

[1825] Anderson, G. O. (2010). Loneliness Among Older Adults: A National Survey of Adults 45+. doi:10.26419/res.00064.001

[1826] Song et al (2019). Stress related disorders and risk of cardiovascular disease: population based, sibling controlled cohort study. BMJ, l1255. doi:10.1136/bmj.l1255

[1827] Dobkin, P. L., & Pihl, R. O. (1992). Measurement of Psychological and Heart Rate Reactivity to Stress in the Real World. Psychotherapy and Psychosomatics, 58(3-4), 208–214. doi:10.1159/000288629

[1828] Knight, W. E. J., & Rickard, N. S. (2001). Relaxing Music Prevents Stress-Induced Increases in Subjective Anxiety, Systolic Blood Pressure, and Heart Rate in Healthy Males and Females. Journal of Music Therapy, 38(4), 254–272. doi:10.1093/jmt/38.4.254

[1829] Sanders et al (2006). Role of magnesium in the failure of rhDNase therapy in patients with cystic fibrosis. Thorax, 61(11), 962–966. doi:10.1136/thx.2006.060814

[1830] Devanarayana, N. M., & Rajindrajith, S. (2009). Association between Constipation and Stressful Life Events in a Cohort of Sri Lankan Children and Adolescents. Journal of Tropical Pediatrics, 56(3), 144–148. doi:10.1093/tropej/fmp077

[1831] Hertig et al (2007) 'Daily stress and gastrointestinal symptoms in women with irritable bowel syndrome', Nurs Res. 2007 Nov-Dec;56(6):399-406.

[1832] Kiecolt-Glaser et al (2015) 'Daily Stressors, Past Depression, and Metabolic Responses to High-Fat Meals: A Novel Path to Obesity', Biol Psychiatry. 2015 Apr 1; 77(7): 653–660.

[1833] O'Donnell et al (2004). The health of normally aging men: The Massachusetts Male Aging Study (1987–2004). Experimental Gerontology, 39(7), 975–984. doi:10.1016/j.exger.2004.03.023

[1834] Travison et al (2007) 'A Population-Level Decline in Serum Testosterone Levels in American Men', Journal of Clinical Endocrinology & Metabolism 92(1):196-202, DOI: 10.1210/jc.2006-1375.

[1835] Selby C. (1990). Sex hormone binding globulin: origin, function and clinical significance. Annals of clinical biochemistry, 27 (Pt 6), 532–541. https://doi.org/10.1177/000456329002700603

[1836] Ruokonen, A., Alén, M., Bolton, N., & Vihko, R. (1985). Response of serum testosterone and its precursor steroids, SHBG and CBG to anabolic steroid and testosterone self-administration in man. Journal of Steroid Biochemistry, 23(1), 33–38. doi:10.1016/0022-4731(85)90257-2

[1837] Ding et al (2009). Sex Hormone–Binding Globulin and Risk of Type 2 Diabetes in Women and Men. New England Journal of Medicine, 361(12), 1152–1163. doi:10.1056/nejmoa0804381

[1838] Cangemi, R., Friedmann, A. J., Holloszy, J. O., & Fontana, L. (2010). Long-term effects of calorie restriction on serum sex-hormone concentrations in men. Aging Cell, 9(2), 236–242. doi:10.1111/j.1474-9726.2010.00553.x

[1839] Estour, B., Pugeat, M., Lang, F., Dechaud, H., Pellet, J., & Rousset, H. (1986). Sex hormone binding globulin in women with anorexia nervosa. Clinical endocrinology, 24(5), 571–576. https://doi.org/10.1111/j.1365-2265.1986.tb03287.x

[1840] Hammond, G. L. (2017). Sex Hormone-Binding Globulin and the Metabolic Syndrome. Male Hypogonadism, 305–324. doi:10.1007/978-3-319-53298-1_15

[1841] Panzer et al (2006). Impact of oral contraceptives on sex hormone-binding globulin and androgen levels: a retrospective study in women with sexual dysfunction. The journal of sexual medicine, 3(1), 104–113. https://doi.org/10.1111/j.1743-6109.2005.00198.x

[1842] Zimmerman et al (2014). The effect of combined oral contraception on testosterone levels in healthy women: a systematic review and meta-analysis. Human reproduction update, 20(1), 76–105. https://doi.org/10.1093/humupd/dmt038

[1843] Pincus G, Hoagland H. Effects of administering pregnenolone on fatiguing psychomotor performance. J Aviation Med. 1944;15:98-115.

[1844] Travis J. Rat memory skills boosted by steroid. Science News Nov 1995;148(20):311.

[1845] Flood JF, Morley JE, Roberts E. Memory-enhancing effects in male mice of pregnenolone and steroids metabolically derived from it. Proc Natl Acad Sci. 1992;89(5):1567-1571.

[1846] Brown et al (2014). A randomized, double-blind, placebo-controlled trial of pregnenolone for bipolar depression. Neuropsychopharmacology : official publication of the American College of Neuropsychopharmacology, 39(12), 2867–2873. https://doi.org/10.1038/npp.2014.138

[1847] https://www.questdiagnostics.com/hcp/intguide/EndoMetab/EndoManual_AtoZ_PDFs/Pregnenolone.pdf

[1848] Gordon et al (2015). Ovarian Hormone Fluctuation, Neurosteroids, and HPA Axis Dysregulation in Perimenopausal Depression: A Novel Heuristic Model. American Journal of Psychiatry, 172(3), 227–236. doi:10.1176/appi.ajp.2014.14070918

[1849] Schüssler, P., Kluge, M., Yassouridis, A., Dresler, M., Held, K., Zihl, J., & Steiger, A. (2008). Progesterone reduces wakefulness in sleep EEG and has no effect on cognition in healthy postmenopausal women. Psychoneuroendocrinology, 33(8), 1124–1131. https://doi.org/10.1016/j.psyneuen.2008.05.013

[1850] Melcangi, R. C., Giatti, S., Calabrese, D., Pesaresi, M., Cermenati, G., Mitro, N., Viviani, B., Garcia-Segura, L. M., & Caruso, D. (2014). Levels and actions of progesterone and its metabolites in the nervous system during physiological and pathological conditions. Progress in neurobiology, 113, 56–69. https://doi.org/10.1016/j.pneurobio.2013.07.006

[1851] Hughes, G., Choubey, D. Modulation of autoimmune rheumatic diseases by oestrogen and progesterone. Nat Rev Rheumatol 10, 740–751 (2014). https://doi.org/10.1038/nrrheum.2014.144

[1852] Prior, J. C. (1994). Progesterone and Its Role in Bone Remodelling. Sex Steroids and Bone, 29–56. doi:10.1007/978-3-662-03043-1_3

[1853] Holzer, G., Riegler, E., Hönigsmann, H., Farokhnia, S., & Schmidt, J. B. (2005). Effects and side-effects of 2% progesterone cream on the skin of peri- and postmenopausal women: results from a double-blind, vehicle-controlled, randomized study. The British journal of dermatology, 153(3), 626–634. https://doi.org/10.1111/j.1365-2133.2005.06685.x

[1854] LANDAU, R. L., BERGENSTAL, D. M., LUGIBIHL, K., & KASCHT, M. (1955). THE METABOLIC EFFECTS OF PROGESTERONE IN MAN*†. The Journal of Clinical Endocrinology & Metabolism, 15(10), 1194–1215. doi:10.1210/jcem-15-10-1194

[1855] Lynch, W. J., & Sofuoglu, M. (2010). Role of progesterone in nicotine addiction: evidence from initiation to relapse. Experimental and clinical psychopharmacology, 18(6), 451–461. https://doi.org/10.1037/a0021265

[1856] Habib, F. K., Maddy, S. Q., & Stitch, S. R. (1980). Zinc induced changes in the progesterone binding properties of the human endometrium. Acta endocrinologica, 94(1), 99–106. https://doi.org/10.1530/acta.0.0940099

[1857] Gavaler (1998) 'Alcoholic Beverages as a Source of Estrogens', Alcohol Health & Research World, Vol. 22, No. 3, 1998.

[1858] Abraham G. E. (1983). Nutritional factors in the etiology of the premenstrual tension syndromes. The Journal of reproductive medicine, 28(7), 446–464.

[1859] Zeligs (1998) 'Diet and Estrogen Status: The Cruciferous Connection', Journal of Medicinal Food.Jan 1998.67-82.http://doi.org/10.1089/jmf.1998.1.67

[1860] City of Hope. "Anticancer effect of mushrooms demonstrated." ScienceDaily. ScienceDaily, 6 June 2011. <www.sciencedaily.com/releases/2011/06/110606092736.htm>.

[1861] Cordain L et al (2004) 'Optimal low-density lipoprotein is 50 to 70 mg/dl: lower is better and physiologically normal', J Am Coll Cardiol. 2004 Jun 2;43(11):2142-6.

[1862] Djousse, L. and Gaziano JM (2009) 'Dietary cholesterol and coronary artery disease: a systematic review', Curr Atheroscler Rep. 2009 Nov;11(6):418-22.

[1863] Griffin, JD and Lichenstein, AH (2013) 'Dietary Cholesterol and Plasma Lipoprotein Profiles: Randomized-Controlled Trials', Curr Nutr Rep. 2013 Dec; 2(4): 274–282.

[1864] Price et al (August 2006). "Weight, shape, and mortality risk in older persons: elevated waist-hip ratio, not high body mass index, is associated with a greater risk of death". *Am. J. Clin. Nutr.* 84 (2): 449–60.

[1865] World Health Organization (2008), 'Waist Circumference and Waist–Hip Ratio: Report of a WHO Expert Consultation', Geneva, 8–11 December 2008

[1866] Chen et al (2016). DNA methylation-based measures of biological age: meta-analysis predicting time to death. Aging, 8(9), 1844–1865. https://doi.org/10.18632/aging.101020

[1867] Horvath, S., & Raj, K. (2018). DNA methylation-based biomarkers and the epigenetic clock theory of ageing. Nature Reviews Genetics, 19(6), 371–384. doi:10.1038/s41576-018-0004-3

[1868] Fransquet et al (2019). The epigenetic clock as a predictor of disease and mortality risk: a systematic review and meta-analysis. Clinical Epigenetics, 11(1). doi:10.1186/s13148-019-0656-7

[1869] Tsoukalas, D., Alegakis, A., Fragkiadaki, P., Papakonstantinou, E., Nikitovic, D., Karataraki, A., ... Tsatsakis, A. M. (2017). Application of metabolomics: Focus on the quantification of organic acids in healthy adults. International Journal of Molecular Medicine, 40(1), 112–120. doi:10.3892/ijmm.2017.2983

[1870] Hannibal, L., Lysne, V., Bjørke-Monsen, A.-L., Behringer, S., Grünert, S. C., Spiekerkoetter, U., ... Blom, H. J. (2016). Biomarkers and Algorithms for the Diagnosis of Vitamin B12 Deficiency. Frontiers in Molecular Biosciences, 3. doi:10.3389/fmolb.2016.00027

[1871] van der Heiden, C., Wauters, E. A., Duran, M., Wadman, S. K., & Ketting, D. (1971). Gas chromatographic analysis of urinary tyrosine and phenylalanine metabolites in patients with gastrointestinal disorders. Clinica chimica acta; international journal of clinical chemistry, 34(2), 289–296. https://doi.org/10.1016/0009-8981(71)90182-3

[1872] Vella, A., & Farrugia, G. (1998). D-lactic acidosis: pathologic consequence of saprophytism. Mayo Clinic proceedings, 73(5), 451–456. https://doi.org/10.1016/S0025-6196(11)63729-4

[1873] Finsterer, J., & Zarrouk-Mahjoub, S. (2018). Biomarkers for Detecting Mitochondrial Disorders. Journal of Clinical Medicine, 7(2), 16. doi:10.3390/jcm7020016

[1874] Lord, R. S., & Bralley, J. A. (2008). Clinical applications of urinary organic acids. Part I: Detoxification markers. Alternative medicine review : a journal of clinical therapeutic, 13(3), 205–215.

[1875] Rahe RH, Mahan JL, Arthur RJ (1970). "Prediction of near-future health change from subjects' preceding life changes". *J Psychosom Res.* 14 (4): 401–6. doi:10.1016/0022-3999(70)90008-5.

[1876] Rahe RH, Biersner RJ, Ryman DH, Arthur RJ (1972). "Psychosocial predictors of illness behavior and failure in stressful training". *J Health Soc Behav.* 13 (4): 393–7. doi:10.2307/2136831.

[1877] Rahe RH, Arthur RJ (1978). "Life change and illness studies: past history and future directions". *J Human Stress.* 4 (1): 3–15. doi:10.1080/0097840X.1978.9934972.

[1878] Masuda M, Holmes TH (1967). "The Social Readjustment Rating Scale: a cross-cultural study of Japanese and Americans". *J Psychosom Res.* 11 (2): 227–37. doi:10.1016/0022-3999(67)90012-8.

[1879] Lovibond, P.F.; Lovibond, S.H. (March 1995). "The structure of negative emotional states: Comparison of the Depression Anxiety Stress Scales (DASS) with the Beck Depression and Anxiety Inventories". *Behaviour Research and Therapy.* 33 (3): 335–343. doi:10.1016/0005-7967(94)00075-U.

[1880] Hoenig and Zeidel (2014) 'Homeostasis, the Milieu Intérieur, and the Wisdom of the Nephron', Clin J Am Soc Nephrol. 2014 Jul 7; 9(7): 1272–1281.

[1881] Stetter, Friedhelm; Kupper, Sirko (2002-03-01). "Autogenic Training: A Meta-Analysis of Clinical Outcome Studies". *Applied Psychophysiology and Biofeedback.* 27 (1): 45–98. doi:10.1023/a:1014576505223.

[1882] Hofmann SG, Sawyer AT, Witt AA, Oh D (April 2010). "The effect of mindfulness-based therapy on anxiety and depression: A meta-analytic review". *Journal of Consulting and Clinical Psychology*. 78 (2): 169–83. doi:10.1037/a0018555.

[1883] Lin, et al (2009). Effects of Acupuncture Stimulation on Recovery Ability of Male Elite Basketball Athletes. The American Journal of Chinese Medicine, 37(03), 471–481. doi:10.1142/s0192415x09006989

[1884] Tu, C.-H., MacDonald, I., & Chen, Y.-H. (2019). The Effects of Acupuncture on Glutamatergic Neurotransmission in Depression, Anxiety, Schizophrenia, and Alzheimer's Disease: A Review of the Literature. Frontiers in Psychiatry, 10. doi:10.3389/fpsyt.2019.00014

[1885] Li et al (2016). α-Pinene, linalool, and 1-octanol contribute to the topical anti-inflammatory and analgesic activities of frankincense by inhibiting COX-2. Journal of Ethnopharmacology, 179, 22–26. doi:10.1016/j.jep.2015.12.039

[1886] Li et al (2007) 'Forest Bathing Enhances Human Natural Killer Activity and Expression of Anti-Cancer Proteins', International Journal of Immunopathology and Pharmacology, 3–8. https://doi.org/10.1177/03946320070200S202

[1887] Li et al (2016). Effects of Forest Bathing on Cardiovascular and Metabolic Parameters in Middle-Aged Males. Evidence-Based Complementary and Alternative Medicine, 2016, 1–7. doi:10.1155/2016/2587381

[1888] Ohtsuka et al (1998). Shinrin-yoku (forest-air bathing and walking) effectively decreases blood glucose levels in diabetic patients. International Journal of Biometeorology, 41(3), 125–127. doi:10.1007/s004840050064

[1889] Prisby et al (2008). Effects of whole body vibration on the skeleton and other organ systems in man and animal models: What we know and what we need to know. Ageing Research Reviews, 7(4), 319–329. doi:10.1016/j.arr.2008.07.004

[1890] Sutbeyaz, S., Sezer, N., & Koseoglu, B. (2005).The effect of pulsed electromagnetic fields in the treatment of cervical osteoarthritis: a randomized, double-blind, sham-controlled trial. Rheumatology International, 26(4), 320-324.doi:10.1007/s00296-005-0600-3

[1891] Ganesan et al (2009) 'Low Frequency Pulsed Electromagnetic Field - A Viable Alternative Therapy for Arthritis', Indian Journal of Experimental Biology, Vol 47, December 2009, pp 939-948

[1892] Thomas et al (2007). A randomized, double-blind, placebo-controlled clinical trial using a low-frequency magnetic field in the treatment of musculoskeletal chronic pain. Pain research & management, 12(4), 249–258. https://doi.org/10.1155/2007/626072

[1893] Huang et al (2008) 'Clinical update of pulsed electromagnetic fields on osteoporosis', Chinese Medical Journal, 01 Oct 2008, 121(20):2095-2099

[1894] Cohen et al (2016). Repetitive deep transcranial magnetic stimulation for motor symptoms in Parkinson's disease: A feasibility study. Clinical Neurology and Neurosurgery, 140, 73–78. doi:10.1016/j.clineuro.2015.11.017

[1895] Stranahan et al (2008) 'Diet-induced insulin resistance impairs hippocampal synaptic plasticity and cognition in middle-aged rats.' Hippocampus. 2008;18(11):1085-8. doi: 10.1002/hipo.20470.

[1896] Yang et al (2013) 'High Dose Zinc Supplementation Induces Hippocampal Zinc Deficiency and Memory Impairment with Inhibition of BDNF Signaling', PLoS One. 2013; 8(1): e55384.

[1897] Kleiger et al (1987) 'Decreased heart rate variability and its association with increased mortality after acute myocardial infarction', Am J Cardiol. 1987 Feb 1;59(4):256-62.

[1898] Kloter et al (2018) 'Heart Rate Variability as a Prognostic Factor for Cancer Survival – A Systematic Review', Front Physiol. 2018; 9: 623.

[1899] Mølgaard et al (1991) 'Attenuated 24-h heart rate variability in apparently healthy subjects, subsequently suffering sudden cardiac death', Clin Auton Res. 1991 Sep;1(3):233-7.

[1900] De Souza et al (2014) 'Risk evaluation of diabetes mellitus by relation of chaotic globals to HRV', Complexity, Volume20, Issue3, January/February 2015, Pages 84-92.

[1901] Mani et al (2008) 'Decreased heart rate variability in patients with cirrhosis relates to the presence and degree of hepatic encephalopathy', Am J Physiol Gastrointest Liver Physiol. 2009 Feb; 296(2): G330–G338.

[1902] Brosschot et al (2007) 'Daily worry is related to low heart rate variability during waking and the subsequent nocturnal sleep period', Int J Psychophysiol. 2007 Jan;63(1):39-47. Epub 2006 Oct 3.

[1903] Nickel and Nachreiner (2003) 'Sensitivity and diagnosticity of the 0.1-Hz component of heart rate variability as an indicator of mental workload', Hum Factors. 2003 winter;45(4):575-90.

[1904] Cohen et al (1998) 'Analysis of heart rate variability in posttraumatic stress disorder patients in response to a trauma-related reminder.' Biol Psychiatry. 1998 Nov 15;44(10):1054-9.

[1905] Van De Borne et al (1999). Hyperinsulinemia produces cardiac vagal withdrawal and nonuniform sympathetic activation in normal subjects. American Journal of Physiology-Regulatory, Integrative and Comparative Physiology, 276(1), R178–R183. doi:10.1152/ajpregu.1999.276.1.r178

[1906] Jaiswal et al (2012). Reduced Heart Rate Variability Among Youth With Type 1 Diabetes: The SEARCH CVD study. Diabetes Care, 36(1), 157–162. doi:10.2337/dc12-0463

[1907] Williams et al (2019). Heart rate variability and inflammation: A meta-analysis of human studies. Brain, Behavior, and Immunity, 80, 219–226. doi:10.1016/j.bbi.2019.03.009

[1908] Cooper et al (2015). Heart rate variability predicts levels of inflammatory markers: Evidence for the vagal anti-inflammatory pathway. Brain, Behavior, and Immunity, 49, 94–100. doi:10.1016/j.bbi.2014.12.017

[1909] Vaschillo et al (2008). Heart rate variability response to alcohol, placebo, and emotional picture cue challenges: Effects of 0.1-Hz stimulation. Psychophysiology, 45(5), 847–858. doi:10.1111/j.1469-8986.2008.00673.x

[1910] Jackowska et al (2012) 'Sleep problems and heart rate variability over the working day', Journal of Sleep Research, 21(4). doi:10.1111/jsr.2012.21.issue-4

[1911] Quintana et al (2017). Diurnal Variation and Twenty-Four Hour Sleep Deprivation Do Not Alter Supine Heart Rate Variability in Healthy Male Young Adults. PLOS ONE, 12(2), e0170921. doi:10.1371/journal.pone.0170921

[1912] Chua et al (2012) 'Heart Rate Variability Can Be Used to Estimate Sleepiness-related Decrements in Psychomotor Vigilance during Total Sleep Deprivation', Sleep. 2012 Mar 1; 35(3): 325–334.

[1913] Stein, P. K., & Pu, Y. (2012). Heart rate variability, sleep and sleep disorders. Sleep Medicine Reviews, 16(1), 47–66. doi:10.1016/j.smrv.2011.02.005

[1914] Poehling, Cory P. (2019) "The Effects of Submaximal and Maximal Exercise on Heart Rate Variability," *International Journal of Exercise Science*: Vol. 12 : Iss. 2, Pages 9 - 14. Available at: https://digitalcommons.wku.edu/ijes/vol12/iss2/1

[1915] Cansel et al (2014). The effects of Ramadan fasting on heart rate variability in healthy individuals: A prospective study. Anadolu Kardiyoloji Dergisi/The Anatolian Journal of Cardiology, 14(5), 413–416. doi:10.5152/akd.2014.5108

[1916] Lutfi, M. F., & Elhakeem, R. F. (2016). Effect of Fasting Blood Glucose Level on Heart Rate Variability of Healthy Young Adults. PLOS ONE, 11(7), e0159820. doi:10.1371/journal.pone.0159820

[1917] Mazurak et al (2013). Effects of a 48-h fast on heart rate variability and cortisol levels in healthy female subjects. European Journal of Clinical Nutrition, 67(4), 401–406. doi:10.1038/ejcn.2013.32

[1918] Laukkanen et al (2015) 'Association Between Sauna Bathing and Fatal Cardiovascular and All-Cause Mortality Events', JAMA Intern Med. 2015;175(4):542-548.

[1919] Mäkinen et al (2008). Autonomic Nervous Function During Whole-Body Cold Exposure Before and After Cold Acclimation. Aviation, Space, and Environmental Medicine, 79(9), 875–882. doi:10.3357/asem.2235.2008

[1920] Sasaki et al (2013) 'Cardiac Sympathetic Activity Assessed by Heart Rate Variability Indicates

Myocardial Ischemia on Cold Exposure in Diabetes', Ann Vasc Dis. 2013; 6(3): 583–589.

1921 Sutarto et al (2012) 'Resonant breathing biofeedback training for stress reduction among manufacturing operators', Int J Occup Saf Ergon. 2012;18(4):549-61.

1922 Miller and Goss (2014) 'An Exploration of Physiological Responses to the Native American Flute', ISQRMM 2013, Athens, GA. ArXiv:1401.6004, January 24, 2014. 17 pages, Accessed Online: https://www.flutopedia.com/refs/MillerGoss_2014_PhysioNAF_v2.pdf

1923 Vickhoff et al (2013). Music structure determines heart rate variability of singers. Frontiers in Psychology, 4. doi:10.3389/fpsyg.2013.00334

1924 Van der Zwan et al (2015). Physical Activity, Mindfulness Meditation, or Heart Rate Variability Biofeedback for Stress Reduction: A Randomized Controlled Trial. Applied Psychophysiology and Biofeedback, 40(4), 257–268. doi:10.1007/s10484-015-9293-x

1925 Tsai et al (2014) 'Heart rate variability and meditation with breath suspension', Biomedical Research (2014) Volume 25, Issue 1.

1926 Cohen, M., & Tyagi, A. (2016). Yoga and heart rate variability: A comprehensive review of the literature. International Journal of Yoga, 9(2), 97. doi:10.4103/0973-6131.183712

1927 HORNYAK et al (1991). SYMPATHETIC MUSCLE NERVE ACTIVITY DURING SLEEP IN MAN. Brain, 114(3), 1281–1295. doi:10.1093/brain/114.3.1281

1928 Vanoli et al (1995) 'Heart Rate Variability During Specific Sleep Stages: A Comparison of Healthy Subjects With Patients After Myocardial Infarction', Circulation. 1995;91:1918–1922, https://doi.org/10.1161/01.CIR.91.7.1918

1929 Browning et al (2017) 'The vagus nerve in appetite regulation, mood and intestinal inflammation', Gastroenterology. 2017 Mar; 152(4): 730–744.

1930 Bonaz et al (2018) 'The Vagus Nerve at the Interface of the Microbiota-Gut-Brain Axis', Front Neurosci. 2018; 12: 49.

1931 Browning et al (2017) 'The Vagus Nerve in Appetite Regulation, Mood, and Intestinal Inflammation.' Gastroenterology. 2017 Mar;152(4):730-744. doi: 10.1053/j.gastro.2016.10.046. Epub 2016 Dec 15.

1932 Glover (2014) 'Maternal depression, anxiety and stress during pregnancy and child outcome; what needs to be done', Best Practice & Research Clinical Obstetrics & Gynaecology, Volume 28, Issue 1, January 2014, Pages 25-35.

1933 Panebianco, M; Rigby, A; Weston, J; Marson, AG (3 April 2015). "Vagus nerve stimulation for partial seizures". *The Cochrane Database of Systematic Reviews* (4): CD002896. doi:10.1002/14651858.CD002896.pub2.

1934 Edwards, CA; Kouzani, A; Lee, KH; Ross, EK (September 2017). "Neurostimulation Devices for the Treatment of Neurologic Disorders". *Mayo Clinic Proceedings*. 92 (9): 1427–1444. doi:10.1016/j.mayocp.2017.05.005.

1935 Chang, C.-H., Lane, H.-Y., & Lin, C.-H. (2018). Brain Stimulation in Alzheimer's Disease. Frontiers in Psychiatry, 9. doi:10.3389/fpsyt.2018.00201

1936 Marrosu et al (2007) 'Vagal nerve stimulation improves cerebellar tremor and dysphagia in multiple sclerosis', Mult Scler. 2007 Nov;13(9):1200-2. Epub 2007 Jul 10.

1937 Chakravarthy, K; Chaudhry, H; Williams, K; Christo, PJ (December 2015). "Review of the Uses of Vagal Nerve Stimulation in Chronic Pain Management". *Current Pain and Headache Reports*. 19 (12): 54. doi:10.1007/s11916-015-0528-6.

1938 Sabbah, HN (August 2011). "Electrical vagus nerve stimulation for the treatment of chronic heart failure". *Cleveland Clinic Journal of Medicine*. 78 Suppl 1: S24–9. doi:10.3949/ccjm.78.s1.04.

1939 Groves, Duncan A.; Brown, Verity J. (2005). "Vagal nerve stimulation: A review of its applications and potential mechanisms that mediate its clinical effects". *Neuroscience & Biobehavioral Reviews*. 29 (3): 493–500. doi:10.1016/j.neubiorev.2005.01.004.

[1940] Howland, RH (June 2014). "Vagus Nerve Stimulation". *Current Behavioral Neuroscience Reports*. 1 (2): 64–73. doi:10.1007/s40473-014-0010-5.

[1941] Montano et al (1994) 'Power spectrum analysis of heart rate variability to assess the changes in sympathovagal balance during graded orthostatic tilt', Circulation. 1994 Oct;90(4):1826-31.

[1942] Grossman and Taylor (2007) 'Toward understanding respiratory sinus arrhythmia: relations to cardiac vagal tone, evolution and biobehavioral functions', Biol Psychol. 2007 Feb;74(2):263-85. Epub 2006 Nov 1.

[1943] Lehrer P. M. (2013). How does heart rate variability biofeedback work? resonance, the baroreflex, and other mechanisms. *Biofeedback* 41 26–31. 10.5298/1081-5937-41.1.02

[1944] McCraty and Childre (2010) 'Coherence: bridging personal, social, and global health', Altern Ther Health Med. 2010 Jul-Aug;16(4):10-24.

[1945] Ma et al (2017) 'The Effect of Diaphragmatic Breathing on Attention, Negative Affect and Stress in Healthy Adults', Front Psychol. 2017; 8: 874.

[1946] Breit, S., Kupferberg, A., Rogler, G., & Hasler, G. (2018). Vagus Nerve as Modulator of the Brain–Gut Axis in Psychiatric and Inflammatory Disorders. Frontiers in Psychiatry, 9. doi:10.3389/fpsyt.2018.00044

[1947] Norlén et al (2005). 'The vagus regulates histamine mobilization from rat stomach ECL cells by controlling their sensitivity to gastrin', The Journal of Physiology, 564(3). doi:10.1111/tjp.2005.564.issue-3

[1948] Bravo et al (2011). Ingestion of Lactobacillus strain regulates emotional behavior and central GABA receptor expression in a mouse via the vagus nerve. Proceedings of the National Academy of Sciences, 108(38), 16050–16055. doi:10.1073/pnas.1102999108

[1949] Bonaz et al (2018) 'The Vagus Nerve at the Interface of the Microbiota-Gut-Brain Axis', Front Neurosci. 2018; 12: 49.

[1950] Yuan, P.-Q., Taché, Y., Miampamba, M., & Yang, H. (2001). Acute cold exposure induces vagally mediated Fos expression in gastric myenteric neurons in conscious rats. American Journal of Physiology-Gastrointestinal and Liver Physiology, 281(2), G560–G568. doi:10.1152/ajpgi.2001.281.2.g560

[1951] Mäkinen et al (2008). Autonomic Nervous Function During Whole-Body Cold Exposure Before and After Cold Acclimation. Aviation, Space, and Environmental Medicine, 79(9), 875–882. doi:10.3357/asem.2235.2008

[1952] Wang, Y., Kondo, T., Suzukamo, Y., Oouchida, Y., & Izumi, S.-I. (2010). Vagal Nerve Regulation Is Essential for the Increase in Gastric Motility in Response to Mild Exercise. The Tohoku Journal of Experimental Medicine, 222(2), 155–163. doi:10.1620/tjem.222.155

[1953] Mason et al (2013). Cardiovascular and Respiratory Effect of Yogic Slow Breathing in the Yoga Beginner: What Is the Best Approach? Evidence-Based Complementary and Alternative Medicine, 2013, 1–7. doi:10.1155/2013/743504

[1954] Laine Green, A., & Weaver, D. F. (2014). Vagal stimulation by manual carotid sinus massage to acutely suppress seizures. Journal of Clinical Neuroscience, 21(1), 179–180. doi:10.1016/j.jocn.2013.03.017

[1955] Wa et al (2011) 'Foot reflexology can increase vagal modulation, decrease sympathetic modulation, and lower blood pressure in healthy subjects and patients with coronary artery disease', Altern Ther Health Med. 2011 Jul-Aug;17(4):8-14.

[1956] Khasar et al (2003). Fasting is a physiological stimulus of vagus-mediated enhancement of nociception in the female rat. Neuroscience, 119(1), 215–221. doi:10.1016/s0306-4522(03)00136-2

[1957] Chipponi, JX; Bleier, JC; Santi, MT; Rudman, D (May 1982). "Deficiencies of essential and conditionally essential nutrients". *American Journal of Clinical Nutrition*. 35 (5 Suppl): 1112–1116. doi:10.1093/ajcn/35.5.1112.

[1958] Rodriguez, N. R. (2015). Introduction to Protein Summit 2.0: continued exploration of the impact of high-quality protein on optimal health. The American Journal of Clinical Nutrition, 101(6), 1317S–1319S. doi:10.3945/ajcn.114.083980

[1959] Phillips, S. M., & Van Loon, L. J. (2011). Dietary protein for athletes: from requirements to optimum adaptation. Journal of sports sciences, 29 Suppl 1, S29–S38. https://doi.org/10.1080/02640414.2011.619204

[1960] Sanders, T. A. (2010), The role of fat in the diet – quantity, quality and sustainability. Nutrition Bulletin, 35: 138-146.

[1961] EFSA Panel on Dietetic Products, Nutrition and Allergies (NDA) (2012) 'Scientific Opinion related to the Tolerable Upper Intake Level of eicosapentaenoic acid (EPA), docosahexaenoic acid (DHA) and docosapentaenoic acid (DPA)'. EFSA Journal 2012; 10(7):2815. [48 pp.] doi:10.2903/j.efsa.2012.2815.

[1962] Ottoboni F. et al (2005) 'Ascorbic Acid and the Immune System', Journal of Orthomolecular Medicine, Vol. 20, No. 3, 2005.

[1963] Grandner et al (2014) 'Sleep Symptoms Associated with Intake of Specific Dietary Nutrients', J Sleep Res. Author manuscript; available in PMC 2015 Feb 1.

[1964] Noga et al (2002). An Unexpected Requirement for PhosphatidylethanolamineN-Methyltransferase in the Secretion of Very Low Density Lipoproteins. Journal of Biological Chemistry, 277(44), 42358–42365. doi:10.1074/jbc.m204542200

[1965] Corbin and Zeisel (2012) 'Choline Metabolism Provides Novel Insights into Non-alcoholic Fatty Liver Disease and its Progression', Curr Opin Gastroenterol. 2012 Mar; 28(2): 159–165.

[1966] Hunnicutt et al (2014) 'Dietary Iron Intake and Body Iron Stores Are Associated with Risk of Coronary Heart Disease in a Meta-Analysis of Prospective Cohort Studies, The Journal of Nutrition, Volume 144, Issue 3, March 2014, Pages 359–366, https://doi.org/10.3945/jn.113.185124

[1967] Sun, X., Shan, Z., & Teng, W. (2014). Effects of increased iodine intake on thyroid disorders. Endocrinology and metabolism (Seoul, Korea), 29(3), 240–247. https://doi.org/10.3803/EnM.2014.29.3.240

[1968] Drennan et al (1991) 'Potassium affects actigraph-identified sleep', Sleep. 1991 Aug;14(4):357-60.

[1969] Meltem et al (2009) 'A hypokalemic muscular weakness after licorice ingestion: a case report', Cases J. 2009 Sep 17;2:8053. doi: 10.1186/1757-1626-0002-0000008053.

[1970] Olivero, J. J. (2016). Cardiac Consequences Of Electrolyte Imbalance. Methodist DeBakey Cardiovascular Journal, 12(2), 125–126. doi:10.14797/mdcj-12-2-125

[1971] Stone, M., Martyn, L., & Weaver, C. (2016). Potassium Intake, Bioavailability, Hypertension, and Glucose Control. Nutrients, 8(7), 444. doi:10.3390/nu8070444

[1972] Boyle, N. B., Lawton, C., & Dye, L. (2017). The Effects of Magnesium Supplementation on Subjective Anxiety and Stress-A Systematic Review. Nutrients, 9(5), 429. https://doi.org/10.3390/nu9050429

[1973] Abbasi et al (2012) 'The effect of magnesium supplementation on primary insomnia in elderly: A double-blind placebo-controlled clinical trial', J Res Med Sci. 2012 Dec;17(12):1161-9.

[1974] Dmitrašinović et al (2016). ACTH, Cortisol and IL-6 Levels in Athletes Following Magnesium Supplementation. Journal of Medical Biochemistry, 35(4), 375–384. doi:10.1515/jomb-2016-0021

[1975] Donald R. Davis, Melvin D. Epp & Hugh D. Riordan (2004) Changes in USDA Food Composition Data for 43 Garden Crops, 1950 to 1999, Journal of the American College of Nutrition, 23:6, 669-682, DOI: 10.1080/07315724.2004.10719409

[1976] Thomas D. (2007). The mineral depletion of foods available to us as a nation (1940-2002)--a review of the 6th Edition of McCance and Widdowson. Nutrition and health, 19(1-2), 21–55. https://doi.org/10.1177/026010600701900205

[1977] Mayer, A. (1997), "Historical changes in the mineral content of fruits and vegetables", British Food Journal, Vol. 99 No. 6, pp. 207-211. https://doi.org/10.1108/00070709710181540

[1978] World Health Organization. *Calcium and Magnesium in Drinking Water: Public health significance*. Geneva: World Health Organization Press; 2009.

[1979] *Sedighi et al.* (2014) *'Effect of selenium supplementation on glutathione peroxidase enzyme activity in patients with chronic kidney disease: a randomized clinical trial.'* Nephrourol Mon. 2014 May 4;6(3):e17945. doi: 10.5812/ numonthly. 17945.

[1980] Mattson (2015) ' Toxic Chemicals in Fruits and Vegetables Are What Give', Scientific American, Accessed Online: https://nickrath.weebly.com/uploads/6/5/4/1/6541061/toxic_chemicals_in_fruits_and_vegetables_are_what_give_them_their_health_benefits_-_scientific_american.pdf

[1981] Prasad, A. S. (2008). Zinc in Human Health: Effect of Zinc on Immune Cells. Molecular Medicine, 14(5-6), 353–357. doi:10.2119/2008-00033.prasad

[1982] Rizzo et al (2016). Vitamin B12 among Vegetarians: Status, Assessment and Supplementation. Nutrients, 8(12), 767. doi:10.3390/nu8120767

[1983] Rusher DR, Pawlak R (2013) A Review of 89 Published Case Studies of Vitamin B12 Deficiency. J Hum Nutr Food Sci 1(2): 1008.

[1984] Beydoun et al (2014) 'Serum Nutritional Biomarkers and Their Associations with Sleep among US Adults in Recent National Surveys', PLoS ONE 9(8): e103490. https://doi.org/10.1371/journal.pone.0103490.

[1985] Johnston, C. S., Solomon, R. E., & Corte, C. (1996). Vitamin C depletion is associated with alterations in blood histamine and plasma free carnitine in adults. Journal of the American College of Nutrition, 15(6), 586–591. doi:10.1080/07315724.1996.10718634

[1986] Stanley et al (1993). Renal Handling of Carnitine in Secondary Carnitine Deficiency Disorders. Pediatric Research, 34(1), 89–96. doi:10.1203/00006450-199307000-00021

[1987] Carr et al (2013). Human skeletal muscle ascorbate is highly responsive to changes in vitamin C intake and plasma concentrations. The American Journal of Clinical Nutrition, 97(4), 800–807. doi:10.3945/ajcn.112.053207

[1988] Hemilä, H., & Chalker, E. (2013). Vitamin C for preventing and treating the common cold. Cochrane Database of Systematic Reviews. doi:10.1002/14651858.cd000980.pub4

[1989] Bozdag, H., & Akdeniz, E. (2018). Does severe vitamin D deficiency impact obstetric outcomes in pregnant women with thyroid autoimmunity? The Journal of Maternal-Fetal & Neonatal Medicine, 1–11. doi:10.1080/14767058.2018.1519017

[1990] Han et al (2017) 'Association between Serum Vitamin D Levels and Sleep Disturbance in Hemodialysis Patients', Nutrients. 2017 Feb; 9(2): 139.

[1991] Gominak and Stumpg (2012) 'The world epidemic of sleep disorders is linked to vitamin D deficiency.' Med Hypotheses. 2012 Aug;79(2):132-5. doi: 10.1016/j.mehy.2012.03.031. Epub 2012 May 13.

[1992] Meagher et al (2001) 'Effects of Vitamin E on Lipid Peroxidation in Healthy Persons. JAMA. 2001;285(9):1178–1182. doi:10.1001/jama.285.9.1178

[1993] Brekhman, I. I.; Dardymov, I. V. (1969). "New Substances of Plant Origin which Increase Nonspecific Resistance". *Annual Review of Pharmacology*. 9: 419–430. doi:10.1146/annurev.pa.09.040169.002223.

[1994] European Medicines Agency (2008) "Reflection Paper on the Adaptogenic Concept" (PDF). Committee on Herbal Medicinal Products. 8 May 2008.

[1995] Ayeka, P. (2018). Potential of Mushroom Compounds as Immunomodulators in Cancer Immunotherapy: A Review. *Evidence-based Complementary and Alternative Medicine* 2018: 7271509.

[1996] Lull, C. & Wichers, H. & Savelkoul, H. (2005). Antiinflammatory and immunomodulating properties of fungal metabolites. *Mediators of Inflammation* 2005 (2): 63–80

[1997] Balandaykin, M. & Zmitrovich, I. (2015). Review on Chaga medicinal mushroom, Inonotus obliquus (Higher Basidiomycetes): Realm of medicinal applications and approaches on estimating its resource potential. *International Journal of Medicinal Mushrooms* 17 (2): 95–104.

[1998] Won, D. et al. (2011). Immunostimulating activity by polysaccharides isolated from fruiting body of Inonotus obliquus. *Molecules and Cells* 31 (2): 165-173.

[1999] Kikuchi (2014) “Chaga mushroom-induced oxalate nephropathy”. Clinical Nephrology 81(6):440-444.

[2000] Gao Y et al (2003) 'Effects of ganopoly (a Ganoderma lucidum polysaccharide extract) on the immune functions in advanced-stage cancer patients', Immunol Invest. 2003 Aug;32(3):201-15.

[2001] Bao PP et al (2012) 'Ginseng and Ganoderma lucidum use after breast cancer diagnosis and quality of life: a report from the Shanghai Breast Cancer Survival Study', PLoS One. 2012;7(6):e39343.

[2002] Boh, B. & Berovic, M. & Zhang, J. & Zhi-Bin, L. (2007). Ganoderma lucidum and its pharmaceutically active compounds. *Biotechnology Annual Review* 13: 265–301.

[2003] Lin, Z. (2005). Cellular and molecular mechanisms of immuno-modulation by Ganoderma lucidum. *Journal of Pharmacological Sciences* 99 (2): 144–153. Review.

[2004] Dudhgaonkar, S. & Thyagarajan, A. & Sliva, D. (2009). Suppression of the inflammatory response by triterpenes isolated from the mushroom Ganoderma lucidum. *International immunopharmacology* 9 (11): 1272–1280.

[2005] Finimundy et al (2014) A Review on General Nutritional Compounds and Pharmacological Properties of the Lentinula edodes Mushroom. Food and Nutrition Sciences, 5, 1095-1105.

[2006] Dai, X. et al. (2015). Consuming Lentinula edodes (Shiitake) mushrooms daily improves human immunity: A randomized dietary intervention in healthy young adults. *Journal of the American College of Nutrition* 34 (6): 478-487.

[2007] Hsieh et al (2002). Effects of extracts of Coriolus versicolor (I'm-Yunity™) on cell-cycle progression and expression of interleukins-1β,-6, and-8 in Promyelocytic HL-60 leukemic cells and mitogenically stimulated and nonstimulated human lymphocytes. The Journal of Alternative & Complementary Medicine 8 (5): 591–602.

[2008] Torkelson, C. et al. (2012). Phase 1 clinical trial of Trametes versicolor in women with breast cancer. *ISRN Oncology* 2012: 251632.

[2009] Janjušević, L. et al. (2017). The lignicolous fungus Trametes versicolor (L.) Lloyd (1920): a promising natural source of antiradical and AChE inhibitory agents. Journal of Enzyme Inhibition and Medicinal Chemistry 32 (1): 355–362.

[2010] Blagodatski, A. et al. (2018). Medicinal mushrooms as an attractive new source of natural compounds for future cancer therapy. *Oncotarget* 9 (49): 29259–29274.

[2011] Lu, H. et al. (2011). TLR2 agonist PSK activates human NK cells and enhances the antitumor effect of HER2-targeted monoclonal antibody therapy. *Clinical Cancer Research* 17 (21): 6742–6753.

[2012] Matijašević, D. et al. (2016). The Antibacterial Activity of Coriolus versicolor Methanol Extract and Its Effect on Ultrastructural Changes of Staphylococcus aureus and Salmonella Enteritidis. *Frontiers in Microbiology* 7: 1226.

[2013] Lai PL et al (2015) 'Neurotrophic properties of the Lion's mane medicinal mushroom, Hericium erinaceus (Higher Basidiomycetes) from Malaysia', Int J Med Mushrooms. 2013;15(6):539-54.

[2014] Mori K et al (2009) 'Improving effects of the mushroom Yamabushitake (Hericium erinaceus) on mild cognitive impairment: a double-blind placebo-controlled clinical trial', Phytother Res. 2009 Mar;23(3):367-72.

[2015] Davis, L. & Kuttan, G. (2002). Effect of Withania somnifera on cell mediated immune responses in mice. *Journal of Experimental & Clinical Cancer Research* 21 (4): 585–590.

[2016] Malik, F. et al. (2007). A standardized root extract of Withania somnifera and its major constituent withanolide-A elicit humoral and cell-mediated immune responses by up regulation of

Th1-dominant polarization in BALB/c mice. *Life Sciences* 80 (16): 1525–1538.

[2017] Chandrasekhar, K. & Kapoor, J. & Anishetty, S. (2012). A prospective, randomized double-blind, placebo-controlled study of safety and efficacy of a high-concentration full-spectrum extract of ashwagandha root in reducing stress and anxiety in adults. *Indian Journal of Psychological Medicine* 34 (3): 255–262.

[2018] Andrade et al (2000). A double-blind, placebo-controlled evaluation of the anxiolytic efficacy of an ethanolic extract of withania somnifera. *Indian Journal of Psychiatry* 42 (3): 295–301.

[2019] Mikolai, J. et al. (2009). In vivo effects of ashwagandha (Withania somnifera) extract on the activation of lymphocytes. *The Journal of Alternative and Complementary Medicine* 15 (4): 423-430.

Mikolai, J. et al. (2009). In vivo effects of ashwagandha (Withania somnifera) extract on the activation of lymphocytes. *The Journal of Alternative and Complementary Medicine* 15 (4): 423-430.

[2020] Kang, S. & Min, H. (2012). Ginseng, the 'Immunity Boost': The Effects of Panax ginseng on Immune System. *Journal of Ginseng Research* 36 (4): 354–368.

[2021] Park et al (2012). Potentiation of antioxidative and anti-inflammatory properties of cultured wild ginseng root extract through probiotic fermentation. Journal of Pharmacy and Pharmacology, 65(3), 457–464. doi:10.1111/jphp.12004

[2022] de Andrade et al (2007) 'Study of the efficacy of Korean Red Ginseng in the treatment of erectile dysfunction', Asian J Androl. 2007 Mar;9(2):241-4. Epub 2006 Jul 11.

[2023] Ellis, J. M., & Reddy, P. (2002). Effects of Panax Ginseng on Quality of Life. Annals of Pharmacotherapy, 36(3), 375–379. doi:10.1345/aph.1a245

[2024] Reay, J. L., Kennedy, D. O., & Scholey, A. B. (2005). Single doses of Panax ginseng (G115) reduce blood glucose levels and improve cognitive performance during sustained mental activity. Journal of Psychopharmacology, 19(4), 357–365. doi:10.1177/0269881105053286

[2025] Karuppiah, P., & Rajaram, S. (2012). Antibacterial effect of Allium sativum cloves and Zingiber officinale rhizomes against multiple-drug resistant clinical pathogens. Asian Pacific Journal of Tropical Biomedicine, 2(8), 597–601. doi:10.1016/s2221-1691(12)60104-x

[2026] Mashhadi, N. et al. (2013). Anti-oxidative and anti-inflammatory effects of ginger in health and physical activity: review of current evidence. *International Journal of Preventive Medicine* 4 (Suppl 1): S36–S42.

[2027] Wang et al (2014) 'Biological properties of 6-gingerol: a brief review', Nat Prod Commun. 2014 Jul;9(7):1027-30.

[2028] Black et al (2010) 'Ginger (Zingiber officinale) reduces muscle pain caused by eccentric exercise', J Pain. 2010 Sep;11(9):894-903. doi: 10.1016/j.jpain.2009.12.013. Epub 2010 Apr 24.

[2029] Altman et al (2001) 'Effects of a ginger extract on knee pain in patients with osteoarthritis', Arthritis Rheum. 2001 Nov;44(11):2531-8.

[2030] Ozgoli, G., Goli, M., & Moattar, F. (2009). Comparison of Effects of Ginger, Mefenamic Acid, and Ibuprofen on Pain in Women with Primary Dysmenorrhea. The Journal of Alternative and Complementary Medicine, 15(2), 129–132. doi:10.1089/acm.2008.0311

[2031] Zhu, J., Chen, H., Song, Z., Wang, X., & Sun, Z. (2018). Effects of Ginger (Zingiber officinale Roscoe) on Type 2 Diabetes Mellitus and Components of the Metabolic Syndrome: A Systematic Review and Meta-Analysis of Randomized Controlled Trials. Evidence-Based Complementary and Alternative Medicine, 2018, 1–11. doi:10.1155/2018/5692962

[2032] Alizadeh-Navaei et al (2008) 'Investigation of the effect of ginger on the lipid levels. A double blind controlled clinical trial', Saudi Med J. 2008 Sep;29(9):1280-4.

[2033] Wu et al (2008). Effects of ginger on gastric emptying and motility in healthy humans. European Journal of Gastroenterology & Hepatology, 20(5), 436–440. doi:10.1097/meg.0b013e3282f4b224

[2034] Hewlings, S. & Kalman, D. (2017). Curcumin: A Review of Its' Effects on Human Health.

Foods (Basel, Switzerland) 6 (10): 92.

2035 Moghadamtousi, S. et al. (2014). A review on antibacterial, antiviral, and antifungal activity of curcumin. *Biomed Research International* 2014: 186864.

2036 Auyeung, K. & Han, Q. & Ko, J. (2016). Astragalus membranaceus: a review of its protection against inflammation and gastrointestinal cancers. *The American Journal of Chinese Medicine* 44 (01): 1–22.

2037 Brush, J. et al. (2006). The effect of Echinacea purpurea, Astragalus membranaceus and Glycyrrhiza glabra on CD69 expression and immune cell activation in humans. *Phytotherapy Research* 20 (8): 687–695.

2038 Qin et al (2012). Astragalus membranaceus extract activates immune response in macrophages via heparanase. *Molecules (Basel, Switzerland)* 17 (6): 7232–7240.

2039 Shan, H. & Zheng, X. & Li, M. (2019). The effects of Astragalus Membranaceus Active Extracts on Autophagy-related Diseases. *International Journal of Molecular Sciences* 20 (8): 1904.

2040 Edwards, D., Heufelder, A., & Zimmermann, A. (2012). Therapeutic Effects and Safety of Rhodiola rosea Extract WS® 1375 in Subjects with Life-stress Symptoms - Results of an Open-label Study. Phytotherapy Research, 26(8), 1220–1225. doi:10.1002/ptr.3712

2041 Lekomtseva, Y., Zhukova, I., & Wacker, A. (2017). Rhodiola rosea in Subjects with Prolonged or Chronic Fatigue Symptoms: Results of an Open-Label Clinical Trial. Complementary Medicine Research, 24(1), 46–52. doi:10.1159/000457918

2042 Darbinyan et al (2007). Clinical trial ofRhodiola roseaL. extract SHR-5 in the treatment of mild to moderate depression. Nordic Journal of Psychiatry, 61(5), 343–348. doi:10.1080/08039480701643290

2043 De Bock, K., Eijnde, B. O., Ramaekers, M., & Hespel, P. (2004). Acute Rhodiola Rosea Intake Can Improve Endurance Exercise Performance. International Journal of Sport Nutrition and Exercise Metabolism, 14(3), 298–307. doi:10.1123/ijsnem.14.3.298

2044 Shin, B.-C., Lee, M. S., Yang, E. J., Lim, H.-S., & Ernst, E. (2010). Maca (L. meyenii) for improving sexual function: a systematic review. BMC Complementary and Alternative Medicine, 10(1). doi:10.1186/1472-6882-10-44

2045 Stone, M., Ibarra, A., Roller, M., Zangara, A., & Stevenson, E. (2009). A pilot investigation into the effect of maca supplementation on physical activity and sexual desire in sportsmen. Journal of Ethnopharmacology, 126(3), 574–576. doi:10.1016/j.jep.2009.09.012

2046 Wu YZ et al (2008) 'Ginkgo biloba extract improves coronary artery circulation in patients with coronary artery disease: contribution of plasma nitric oxide and endothelin-1', Phytother Res. 2008 Jun;22(6):734-9.

2047 Hosseinzadeh, H., & Nassiri-Asl, M. (2015). Pharmacological Effects ofGlycyrrhizaspp. and Its Bioactive Constituents: Update and Review. Phytotherapy Research, 29(12), 1868–1886. doi:10.1002/ptr.5487

2048 Hajiaghamohammadi, A. A., Zargar, A., Oveisi, S., Samimi, R., & Reisian, S. (2016). To evaluate of the effect of adding licorice to the standard treatment regimen of Helicobacter pylori. The Brazilian Journal of Infectious Diseases, 20(6), 534–538. doi:10.1016/j.bjid.2016.07.015

2049 Guo et al (2015) 'Promotion of regulatory T cell induction by immunomodulatory herbal medicine licorice and its two constituents', Sci Rep. 2015; 5: 14046.

2050 JY et al (2013) '[Anti-virus research of triterpenoids in licorice].' Bing Du Xue Bao. 2013 Nov;29(6):673-9.

2051 Ozturk et al (2013) 'Polymorphic ventricular tachycardia (Torsades de pointes) due to licorice root tea', Turk Kardiyoloji Dernegi arsivi: Turk Kardiyoloji Derneginin yayin organidir 41(3):241-244, DOI: 10.1016/S0167-5273(12)70487-4

2052 Penn State Hershey Milton S. Hershey Medical Center. 'Licorice', Accessed Online:

http://pennstatehershey.adam.com/content.aspx?productid=107&pid=33&gid=000262

[2053] Cheng et al (2013). Antioxidant and hepatoprotective effects of Schisandra chinensis pollen extract on CCl4-induced acute liver damage in mice. Food and Chemical Toxicology, 55, 234–240. doi:10.1016/j.fct.2012.11.022

[2054] Yan et al (2016). The effect of Schisandra chinensis extracts on depression by noradrenergic, dopaminergic, GABAergic and glutamatergic systems in the forced swim test in mice. Food & Function, 7(6), 2811–2819. doi:10.1039/c6fo00328a

[2055] Park, J. Y., & Kim, K. H. (2016). A randomized, double-blind, placebo-controlled trial of Schisandra chinensis for menopausal symptoms. Climacteric, 19(6), 574–580. doi:10.1080/13697137.2016.1238453

[2056] Zhang et al (2017). The influence of Schisandrin B on a model of Alzheimer's disease using β-amyloid protein Aβ1-42-mediated damage in SH-SY5Y neuronal cell line and underlying mechanisms. Journal of Toxicology and Environmental Health, Part A, 80(22), 1199–1205. doi:10.1080/15287394.2017.1367133

[2057] Coppin et al (2013). Determination of flavonoids by LC/MS and anti-inflammatory activity in Moringa oleifera. Journal of Functional Foods, 5(4), 1892–1899. doi:10.1016/j.jff.2013.09.010

[2058] Mbikay, M. (2012). Therapeutic Potential of Moringa oleifera Leaves in Chronic Hyperglycemia and Dyslipidemia: A Review. Frontiers in Pharmacology, 3. doi:10.3389/fphar.2012.00024

[2059] Chattopadhyay et al (2010). Protective Role of Moringa oleifera (Sajina) Seed on Arsenic-Induced Hepatocellular Degeneration in Female Albino Rats. Biological Trace Element Research, 142(2), 200–212. doi:10.1007/s12011-010-8761-7

[2060] Karadi, R. V., Gadge, N. B., Alagawadi, K. R., & Savadi, R. V. (2006). Effect of Moringa oleifera Lam. root-wood on ethylene glycol induced urolithiasis in rats. Journal of Ethnopharmacology, 105(1-2), 306–311. doi:10.1016/j.jep.2005.11.004

[2061] Ouédraogo et al (2013) 'Protective effect of Moringa oleifera leaves against gentamicin-induced nephrotoxicity in rabbits', Experimental and Toxicologic Pathology, Volume 65, Issue 3, March 2013, Pages 335-339.

[2062] Ingale and Gandhi (2016) 'Effect of aqueous extract of Moringa oleifera leaves on pharmacological models of epilepsy and anxiety in mice', International Journal of Epilepsy 2016; 03(01): 012-019, DOI: 10.1016/j.ijep.2016.02.001

[2063] TAHILIANI, P., & KAR, A. (2000). ROLE OF MORINGA OLEIFERA LEAF EXTRACT IN THE REGULATION OF THYROID HORMONE STATUS IN ADULT MALE AND FEMALE RATS. Pharmacological Research, 41(3), 319–323. doi:10.1006/phrs.1999.0587

[2064] Yamani, H. A., Pang, E. C., Mantri, N., & Deighton, M. A. (2016). Antimicrobial Activity of Tulsi (Ocimum tenuiflorum) Essential Oil and Their Major Constituents against Three Species of Bacteria. Frontiers in Microbiology, 7. doi:10.3389/fmicb.2016.00681

[2065] Cohen and Jamshidi (2017) 'The Clinical Efficacy and Safety of Tulsi in Humans: A Systematic Review of the Literature', Hindawi, Evidence-Based Complementary and Alternative Medicine, Volume 2017 |Article ID 9217567 | 13 pages | https://doi.org/10.1155/2017/9217567

[2066] Chaurasia, A. (2015). Tulsi-A Promising herb in dentistry. Journal of oral medicine, 1, 21-23.

[2067] Rungapamestry V et al (2007) 'Effect of meal composition and cooking duration on the fate of sulforaphane following consumption of broccoli by healthy human subjects', Br J Nutr. 2007 Apr;97(4):644-52.

[2068] Ali SS et al (2004) 'A biologically effective fullerene (C60) derivative with superoxide dismutase mimetic properties', Free Radic Biol Med. 2004 Oct 15;37(8):1191-202.

[2069] Galvan YP et al (2017) 'Fullerenes as Anti-Aging Antioxidants', Curr Aging Sci. 2017;10(1):56-67.

[2070] Kato S et al (2009) 'Biological safety of liposome-fullerene consisting of hydrogenated lecithin, glycine soja sterols, and fullerene-C60 upon photocytotoxicity and bacterial reverse mutagenicity',

Toxicol Ind Health. 2009 Apr;25(3):197-203.

[2071] Singh et al (2012). Acute exposure of apigenin induces hepatotoxicity in Swiss mice. PloS one, 7(2), e31964.

[2072] Stefek M and Karasu C (2011) 'Eye Lens in Aging and Diabetes: Effect of Quercetin', Rejuvenation ResearchVol. 14, No. 5.

[2073] Villareal, D. T., Holloszy, J. O., & Kohrt, W. M. (2000). Effects of DHEA replacement on bone mineral density and body composition in elderly women and men. Clinical Endocrinology, 53(5), 561–568. doi:10.1046/j.1365-2265.2000.01131.x

[2074] Prall, S. P., & Muehlenbein, M. P. (2018). DHEA Modulates Immune Function: A Review of Evidence. Vitamins and Hormones, 125–144. doi:10.1016/bs.vh.2018.01.023

[2075] Villareal, D. T., & Holloszy, J. O. (2006). DHEA enhances effects of weight training on muscle mass and strength in elderly women and men. American Journal of Physiology-Endocrinology and Metabolism, 291(5), E1003–E1008. doi:10.1152/ajpendo.00100.2006

[2076] Clayton et al (2014) 'The Effects of Dehydroepiandrosterone (DHEA) in the Treatment of Depression and Depressive Symptoms in Other Psychiatric and Medical Illnesses: A Systematic Review', Current Drug Targets, Volume 15, Number 9, 2014, pp. 901-914(14)

[2077] Siim Land (2019) 'Benefits of Glycine for Sleep and Anti Aging', Accessed Online: https://siimland.com/benefits-of-glycine-for-sleep-and-anti-aging/

[2078] Schwab, U., Alfthan, G., Aro, A., & Uusitupa, M. (2010). Long-term effect of betaine on risk factors associated with the metabolic syndrome in healthy subjects. European Journal of Clinical Nutrition, 65(1), 70–76. doi:10.1038/ejcn.2010.230

[2079] Kumazawa T et al (1995) 'Levels of pyrroloquinoline quinone in various foods', Biochem J. 1995 Apr 15;307 (Pt 2):331-3.

[2080] Watanabe A et al (1989) 'Nephrotoxicity of pyrroloquinoline quinone in rats', Hiroshima J Med Sci. 1989 Mar;38(1):49-51.

[2081] McCormack D and McFadden D (2013) 'A review of pterostilbene antioxidant activity and disease modification', Oxid Med Cell Longev. 2013;2013:575482.

[2082] Xu YF (2016) 'Effect of Polysaccharide from Cordyceps militaris (Ascomycetes) on Physical Fatigue Induced by Forced Swimming', Int J Med Mushrooms. 2016;18(12):1083-1092.

[2083] Herxheimer and Petrie (2002) 'Melatonin for the prevention and treatment of jet lag', Cochrane Database Syst Rev. 2002;(2):CD001520.

[2084] Ferracioli-Oda et al (2013) 'Meta-analysis: melatonin for the treatment of primary sleep disorders', PLoS One. 2013 May 17;8(5):e63773. doi: 10.1371/journal.pone.0063773. Print 2013.

[2085] Costello et al (2014) 'The effectiveness of melatonin for promoting healthy sleep: a rapid evidence assessment of the literature', Nutr J. 2014; 13: 106.

[2086] Bannai et al (2012) 'The effects of glycine on subjective daytime performance in partially sleep-restricted healthy volunteers', Front Neurol. 2012 Apr 18;3:61. doi: 10.3389/fneur.2012.00061. eCollection 2012.

[2087] Kawai et al (2015) 'The Sleep-Promoting and Hypothermic Effects of Glycine are Mediated by NMDA Receptors in the Suprachiasmatic Nucleus', Neuropsychopharmacology. 2015 May; 40(6): 1405–1416.

[2088] Garlick PJ (2004) 'The nature of human hazards associated with excessive intake of amino acids', J Nutr. 2004 Jun;134(6 Suppl):1633S-1639S; discussion 1664S-1666S, 1667S-1672S. doi: 10.1093/jn/134.6.1633S.

[2089] Starks et al (2008) 'The effects of phosphatidylserine on endocrine response to moderate intensity exercise', J Int Soc Sports Nutr. 2008; 5: 11.

[2090] Levine (1997) 'Controlled trials of inositol in psychiatry', Eur Neuropsychopharmacol. 1997 May;7(2):147-55.

[2091] Shell et al (2010) 'A randomized, placebo-controlled trial of an amino acid preparation on timing and quality of sleep', Am J Ther. 2010 Mar-Apr;17(2):133-9. doi: 10.1097/MJT.0b013e31819e9eab.

[2092] Amsterdam et al (2009) 'A RANDOMIZED, DOUBLE-BLIND, PLACEBO-CONTROLLED TRIAL OF ORAL MATRICARIA RECUTITA (CHAMOMILE) EXTRACT THERAPY OF GENERALIZED ANXIETY DISORDER', J Clin Psychopharmacol. 2009 Aug; 29(4): 378–382.

[2093] Hardy et al (1995) 'Replacement of drug treatment for insomnia by ambient odour', Lancet. 1995 Sep 9;346(8976):701.

[2094] Woelk et al (2010) 'A multi-center, double-blind, randomised study of the Lavender oil preparation Silexan in comparison to Lorazepam for generalized anxiety disorder', Phytomedicine. 2010 Feb;17(2):94-9. doi: 10.1016/j.phymed.2009.10.006. Epub 2009 Dec 3.

[2095] Cases et al (2011) 'Pilot trial of Melissa officinalis L. leaf extract in the treatment of volunteers suffering from mild-to-moderate anxiety disorders and sleep disturbances', Med J Nutrition Metab. 2011 Dec; 4(3): 211–218.

[2096] National Center for Complementary and Integrative Health (NCCIH) 'St. John's Wort and Depression: In Depth', Accessed Online https://nccih.nih.gov/health/stjohnswort/sjw-and-depression.htm

[2097] Murray et al (2001) 'The effect of Li 1370, extract of Ginkgo biloba, on REM sleep in humans', Pharmacopsychiatry. 2001 Jul;34(4):155-7.

[2098] Wheatley (2001) 'Kava and valerian in the treatment of stress-induced insomnia', Phytother Res. 2001 Sep;15(6):549-51.

[2099] Donath et al (2000) 'Critical evaluation of the effect of valerian extract on sleep structure and sleep quality', Pharmacopsychiatry. 2000 Mar;33(2):47-53.

[2100] Bent et al (2006) 'Valerian for Sleep: A Systematic Review and Meta-Analysis', Am J Med. 2006 Dec; 119(12): 1005–1012.

[2101] Wilson JM and Lowery RP et al (2013) 'β-Hydroxy-β-methylbutyrate free acid reduces markers of exercise-induced muscle damage and improves recovery in resistance-trained men', Br J Nutr. 2013 Aug 28;110(3):538-44.

[2102] Wilkinson DJ et al (2013) 'Effects of leucine and its metabolite β-hydroxy-β-methylbutyrate on human skeletal muscle protein metabolism', J Physiol. 2013 Jun 1;591(11):2911-23.

[2103] Błażek et al (2015) 'Sense of Purpose in Life and Escape from Self as the Predictors of Quality of Life in Clinical Samples', J Relig Health. 2015; 54: 517–523.

[2104] Heisel and Flett (2004) 'Purpose in Life, Satisfaction with Life, and Suicide Ideation in a Clinical Sample', Journal of Psychopathology and Behavioral Assessment volume 26, pages127–135(2004)

[2105] Alimujiang et al (2019). Association Between Life Purpose and Mortality Among US Adults Older Than 50 Years. JAMA Network Open, 2(5), e194270. doi:10.1001/jamanetworkopen.2019.4270

[2106] Tomioka, K., Kurumatani, N., & Hosoi, H. (2016). Relationship of Having Hobbies and a Purpose in Life With Mortality, Activities of Daily Living, and Instrumental Activities of Daily Living Among Community-Dwelling Elderly Adults. Journal of Epidemiology, 26(7), 361–370. doi:10.2188/jea.je20150153

[2107] Rhodes et al (2010) 'Recommended summer sunlight exposure levels can produce sufficient (> or =20 ng ml(-1)) but not the proposed optimal (> or =32 ng ml(-1)) 25(OH)D levels at UK latitudes', J Invest Dermatol. 2010 May;130(5):1411-8. doi: 10.1038/jid.2009.417. Epub 2010 Jan 14.

[2108] Parva et al (2018) 'Prevalence of Vitamin D Deficiency and Associated Risk Factors in the US Population (2011-2012)', Cureus. 2018 Jun; 10(6): e2741.

Printed in Great Britain
by Amazon

61262895R00443